VOLUME FIVE HUNDRED AND EIGHTY NINE

METHODS IN ENZYMOLOGY

Enzymes as Sensors

METHODS IN ENZYMOLOGY

Editors-in-Chief

ANNA MARIE PYLE
Departments of Molecular, Cellular and Developmental Biology and Department of Chemistry
Investigator, Howard Hughes Medical Institute
Yale University

DAVID W. CHRISTIANSON
Roy and Diana Vagelos Laboratories
Department of Chemistry
University of Pennsylvania
Philadelphia, PA

Founding Editors

SIDNEY P. COLOWICK and NATHAN O. KAPLAN

VOLUME FIVE HUNDRED AND EIGHTY NINE

METHODS IN ENZYMOLOGY

Enzymes as Sensors

Edited by

RICHARD B. THOMPSON
University of Maryland School of Medicine, Baltimore, MD, United States

CAROL A. FIERKE
University of Michigan, Ann Arbor, MI, United States

ELSEVIER

ACADEMIC PRESS
An imprint of Elsevier

Academic Press is an imprint of Elsevier
50 Hampshire Street, 5th Floor, Cambridge, MA 02139, United States
525 B Street, Suite 1800, San Diego, CA 92101–4495, United States
The Boulevard, Langford Lane, Kidlington, Oxford OX5 1GB, United Kingdom
125 London Wall, London, EC2Y 5AS, United Kingdom

First edition 2017

ISBN: 978-0-12-805406-2
ISSN: 0076-6879

Publisher: Zoe Kruze
Acquisition Editor: Zoe Kruze
Editorial Project Manager: Helene Kabes
Production Project Manager: Magesh Kumar Mahalingam
Cover Designer: Matthew Limbert

Typeset by SPi Global, India

DEDICATION

Dedicated to Dr. Harold J. Bright, Biochemist, Savant, and Peerless Supporter of Young Scientists

CONTENTS

CONTRIBUTORS

Caitlin E. Anderson
University of Washington, Seattle, WA, United States

Silvana Andreescu
Clarkson University, Clarkson, NY, United States

Stijn J.A. Aper
Laboratory of Chemical Biology and Institute for Complex Molecular Systems, Eindhoven University of Technology, Eindhoven, The Netherlands

Remco Arts
Laboratory of Chemical Biology and Institute for Complex Molecular Systems, Eindhoven University of Technology, Eindhoven, The Netherlands

Erfei Bi
University of Pennsylvania Perelman School of Medicine, Philadelphia, PA, United States

Bruce R. Branchini
Connecticut College, New London, CT, United States

Clotilde Capacchione
Institute of Food Science, CNR, Avellino, Italy

Alessandro Capo
Institute of Food Science, CNR, Avellino, Italy

Anthony E.G. Cass
Department of Chemistry, Imperial College London, London, United Kingdom

Sylvia Daunert
Miller School of Medicine, University of Miami, Miami, FL, United States

Sabato D'Auria
Institute of Food Science, CNR, Avellino, Italy

Huai Deng
University of Michigan, Ann Arbor, MI, United States

Sapna Deo
Miller School of Medicine, University of Miami, Miami, FL, United States

Eduard Dumitrescu
Clarkson University, Clarkson, NY, United States

Carol A. Fierke
University of Michigan, Ann Arbor, MI, United States

Rudolf Griss
École Polytechnique Fédérale de Lausanne (EPFL), Institute of Chemical Sciences and Engineering (ISIC); École Polytechnique Fédérale de Lausanne (EPFL), Institute of Bioengineering; National Centre of Competence in Research (NCCR) Chemical Biology, Lausanne, Switzerland

Jiantao Guo
University of Nebraska–Lincoln, Lincoln, NE, United States

Itaru Hamachi
Graduate School of Engineering, Kyoto University, Katsura, Kyoto; Core Research for Evolutional Science and Technology (CREST), Japan Science and Technology Agency, Kawaguchi, Saitama, Japan

Kai Johnsson
École Polytechnique Fédérale de Lausanne (EPFL), Institute of Chemical Sciences and Engineering (ISIC); École Polytechnique Fédérale de Lausanne (EPFL), Institute of Bioengineering; National Centre of Competence in Research (NCCR) Chemical Biology, Lausanne, Switzerland

Tom K. Kerppola
University of Michigan, Ann Arbor, MI, United States

Shigeki Kiyonaka
Graduate School of Engineering, Kyoto University, Katsura, Kyoto, Japan

Rebecca Y. Lai
University of Nebraska-Lincoln, Lincoln, NE, United States

Mid Eum Lee
The Ohio State University, Columbus, OH, United States

Adelia Majoli
Institute of Food Science, CNR, Avellino, Italy

Maarten Merkx
Laboratory of Chemical Biology and Institute for Complex Molecular Systems, Eindhoven University of Technology, Eindhoven, The Netherlands

Robert H. Newman
North Carolina Agricultural and Technical State University, Greensboro, NC, United States

Wei Niu
University of Nebraska–Lincoln, Lincoln, NE, United States

Satoshi Okada
University of Pennsylvania Perelman School of Medicine, Philadelphia, PA, United States; Kyushu University Graduate School of Medical Sciences, Fukuoka, Japan

Amy Palmer
University of Colorado Boulder, Boulder, CO, United States

Hay-Oak Park
The Ohio State University, Columbus, OH, United States

Angela Pennacchio
Institute of Food Science, CNR, Avellino, Italy

Lynn Sanford
University of Colorado Boulder, Boulder, CO, United States

Alberto Schena
Lucentix SA, Lausanne, Switzerland

Kamal G. Shah
University of Washington, Seattle, WA, United States

Sanjiv Sharma
Department of Chemistry, Imperial College London, London, United Kingdom

Tara L. Southworth
Connecticut College, New London, CT, United States

Maria Staiano
Institute of Food Science, CNR, Avellino, Italy

Richard B. Thompson
University of Maryland School of Medicine, Baltimore, MD, United States

Antonio Varriale
Institute of Food Science, CNR, Avellino, Italy

Daniel Wynn
Miller School of Medicine, University of Miami, Miami, FL, United States

Paul Yager
University of Washington, Seattle, WA, United States

Kei Yamaura
Graduate School of Engineering, Kyoto University, Katsura, Kyoto, Japan

Qiuliyang Yu
École Polytechnique Fédérale de Lausanne (EPFL), Institute of Chemical Sciences and Engineering (ISIC); École Polytechnique Fédérale de Lausanne (EPFL), Institute of Bioengineering; National Centre of Competence in Research (NCCR) Chemical Biology, Lausanne, Switzerland

Jin Zhang
University of California, San Diego, San Diego, CA, United States

PREFACE

The Editors are proud to present this volume of *Methods in Enzymology*, focused largely on Enzymes as Sensors. We are fortunate to have gathered a constellation of star investigators to describe their current work in this field. The editors (and many of the authors) have felt for some years that for sensing and many other tasks, biomolecules, such as enzymes, have outstanding attributes. Indeed, biomolecules now represent the method of choice for recognition of many analytes and transducing their presence or level as a signal one can interpret. In this case, necessity has been the mother of invention because answering the essentially analytical questions in biology of how much of an analyte is present, how fast is that amount changing, and how is it distributed in the matrix (often of a living cell), and collecting this information in real time on a submicron scale is a challenge beyond most traditional analytical methods. The proportion of the chapters herein describing methods focused on answering precisely such difficult questions is an index of the inherent power of enzymes and other biological molecules to address such questions.

The volume begins with a chapter by Drs. Sanford and Palmer reviewing one of the central developments of recent years, the use of expressible fluorescent proteins in sensors of many kinds. The second chapter, by Dr. Wynn and colleagues, details a novel development for increasing the ruggedness of biologically derived sensors, namely, encoding them in a very robust form, the bacterial spore. The chapter by Dr. Arts et al. covers a rapidly growing sensing approach with exciting new applications called bioluminescence resonance energy transfer, or BRET. Dr. Staiano and colleagues follow with a chapter on using enzymes as sensors, which focuses on apoenzymes as non-consuming sensors. Drs. Newman and Zhang take a much broader view in the fifth chapter, using diverse approaches to understand signaling networks well above the cellular scale. Conversely, Dr. Okada et al. in their chapter zoom in on a GTPase component of the signaling cascade that controls the polarity of yeast budding. Drs. Niu and Guo describe a powerful approach, incorporating a variety of different, unnatural amino acids into protein sensors for diverse sensing applications. The eighth chapter by Dr. Lai is an outstanding review of electrochemically based nucleic acid sensors and their unique advantages. In the next chapter, Dr. Yamaura et al. describe a very subtle approach for adapting the recognition capabilities of cell surface receptors such as the $GABA_A$ receptor to turn on sensors. The Editors next discuss numerous issues in the sensing of metal ions, with zinc as an archetype. Dr. Dumitrescu and colleagues describe another group of electrochemically transduced sensors, emphasizing determination of key analytes such as neurotransmitters and reactive oxygen species. Drs. Branchini and Southworth in the twelfth chapter also focus on bioluminescent sensing in describing their engineered luciferases for ATP detection. Dr. Johnsson and his colleagues describe a flexible, modular approach to bioluminescent sensing that is well suited for point-of-care applications. In the 14th chapter, Dr. Anderson and her colleagues describe a fascinating approach to sensing in austere environments using paper-based sensing. Dr. Cass and colleagues take us farther afield in describing sensors that incorporate microneedles specifically to gently pierce the skin to provide access to interstitial fluid and its analytes. In the final chapter, Dr. Kerppola and colleagues describe an exciting approach for sensing called bimolecular fluorescence complementation, and its application to studying multiprotein complexes with fly polytene chromosomes.

After more than 60 years, *Methods in Enzymology* as a series, and developing new methods in experimental science in general, is still relevant and worthwhile. Technology development remains a fruitful part of the scientific enterprise, at least as judged by citations: In January, 2017, 13 of the 15 most cited papers in the *Proceedings of the National Academy of Sciences* (http://www.pnas.org/reports/most-cited) were overtly methodological in nature albeit from many fields, and all of the top 10 most cited papers overall are methodological (Van Noorden et al., *Nature* 514, 550 (2014)). Moreover, the highly cited nature and consequent impact of methodological papers has been maintained in the biological and chemical sciences for at least 40 years. In the current "business model" of government/nonprofit funding of academic research, a case could be made that (for instance) Sanger et al.'s invention of dideoxy sequencing (cited 68,000 times and counting) represents a good return on investment.

We hope you enjoy this collection.

RICHARD B. THOMPSON and CAROL A. FIERKE

CHAPTER ONE

Recent Advances in Development of Genetically Encoded Fluorescent Sensors

Lynn Sanford, Amy Palmer[1]
University of Colorado Boulder, Boulder, CO, United States
[1]Corresponding author: e-mail address: amy.palmer@colorado.edu

Contents

Abstract

Genetically encoded fluorescent sensors are essential tools in modern biological research, and recent advances in fluorescent proteins (FPs) have expanded the scope of sensor design and implementation. In this review we compare different sensor platforms, including Förster resonance energy transfer (FRET) sensors, fluorescence-modulated single FP-based sensors, translocation sensors, complementation sensors, and dimerization-based sensors. We discuss elements of sensor design and engineering for each platform, including the incorporation of new types of FPs and sensor screening

Methods in Enzymology, Volume 589
ISSN 0076-6879
http://dx.doi.org/10.1016/bs.mie.2017.01.019

techniques. Finally, we summarize the wide range of sensors in the literature, exploring creative new sensor architectures suitable for different applications.

1. INTRODUCTION

Biological sensors are fundamental to current research, from quantifying metabolites under different perturbations to identifying cell states. Many types of sensors exist with a variety of readouts, with one of the most widely used platforms being fluorescent sensors. Of these, genetically encoded sensors are advantageous for their versatility, modularity, and tunability.

In this review we outline different genetically encoded fluorescent sensor platforms and their protein and peptide components. We particularly focus on how engineering approaches vary among different sensor platforms and how new technologies such as novel fluorescent proteins (FPs) have been used to generate and optimize sensors. We also provide interesting examples of the range of sensors developed over the last 20 years. For the purposes of this review, we define "genetically encoded" to mean that sensors require the addition of no exogenous components, such as unnatural amino acids.

2. FLUORESCENT PROTEINS

The most important components of genetically encoded fluorescent sensors are FPs. A large number of FPs have been discovered and developed over the past two decades, beginning with the cloning, expression, and optimization of green fluorescent protein (GFP) from *Aequorea victoria* in the 1990s (Tsien, 1998). These proteins fit under two broad categories: those that generate an intrinsic chromophore during folding and maturation, and those that bind an exogenous chemical chromophore from the cellular environment. The first category includes by far the most widely used FPs in sensors, although in the past decade innovations in extrinsic chromophore FPs have led to their increased utility, especially in circumstances where oxygen requirements or chromophore maturation kinetics do not favor intrinsic chromophore FPs.

FPs are available in colors spanning the visible range of the spectrum, and each has a number of properties that are relevant to their use in sensors and in cells. Factors such as quantum yield (φ), extinction coefficient (ε), brightness, and photostability affect how effectively FPs can be visualized for single or repeated measurements. Additionally, the p*Ka* of an FP chromophore and redox sensitivity of the FP affect its photophysics in different in vitro and in vivo environments. Finally, whether FPs are monomeric or tend to oligomerize can substantially change how they operate in sensors, as well as whether they perturb the physiology of cells. All of these properties have been heavily engineered and optimized to generate satisfactory FPs for different types of applications.

2.1 Intrinsic Chromophore FPs

The precursors for most intrinsic chromophore FPs were isolated from marine organisms such as jellyfish, coral, and sea anemone (Chudakov, Matz, Lukyanov, & Lukyanov, 2010). The first engineered FPs were derived from GFP, with mutations to improve photostability, brightness, folding, and monomericity, as well as chromophore mutations that altered FP color (Chudakov et al., 2010; Cormack, Valdivia, & Falkow, 1996; Heim, Cubitt, & Tsien, 1995; Zacharias, Violin, Newton, & Tsien, 2002). Since then, several lineages of intrinsic chromophore FPs have emerged and been actively pursued, leading to a wide variety of proteins with different properties (Fig. 1; Table 1). Intrinsic chromophore FPs constitute the bulk of FPs in sensors, due to their extensive characterization, spectral tunability, biophysical stability, and robustness across model systems. There are many excellent reviews of intrinsic chromophore FPs, and we refer the reader to them for more information (Chudakov et al., 2010; Mérola et al., 2014; Stepanenko et al., 2011; Subach & Verkhusha, 2012).

Recent years have seen three major efforts in FP development. The first is the engineering of better FPs for use in Förster resonance energy transfer (FRET) sensors (see Section 3.1.5). The second effort is the generation of brighter and more photostable red and near-infrared (IR) FPs, which have particular use for imaging in vivo (Ng & Lin, 2016). Light of longer wavelengths experiences less scattering and less absorption by endogenous molecules in biological systems, and thus red FPs have a greater signal-to-noise ratio in thick samples such as tissue. The third effort is the development of FP variants that are better suited for different intracellular environments, such as through removal of redox-sensitive surface residues (Costantini et al., 2015; Costantini & Snapp, 2013). FPs developed through these three initiatives have already found use in sensors due to their optimized properties, and further FP development will no doubt lead to further advancement of sensors.

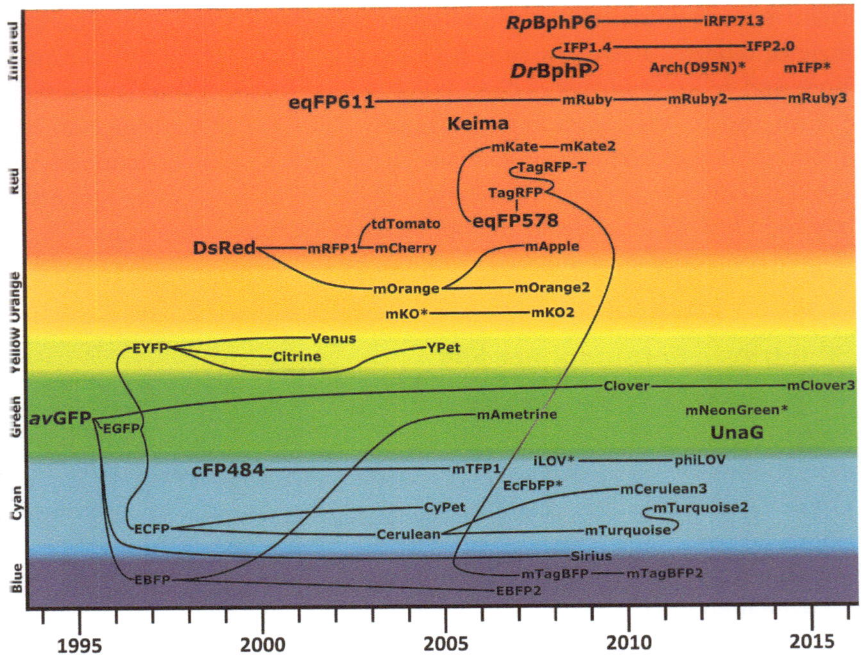

Fig. 1 Timeline of development of major sensor fluorescent proteins. FPs are sorted into general colors but are not ordered by wavelength. *Large text* indicates a naturally isolated parent FP, and * specifies that the parent FP was engineered into the indicated FP in the same publication.

Table 1 lists common intrinsic chromophore FPs, many of which of been incorporated into sensors.

2.2 Extrinsic Chromophore FPs

As with intrinsic chromophore FPs, proteins in this category have been identified as naturally derived FPs and have been engineered to be practically useful for scientific applications (Table 2) (Shcherbakova, Shemetov, Kaberniuk, & Verkhusha, 2015). Extrinsic chromophore FPs can have significant advantages over intrinsic chromophore FPs, including smaller size, fluorescence under anaerobic conditions, or more red-shifted wavelengths. They are, however, less fully characterized and optimized than intrinsic chromophore FPs.

Proteins binding the cofactor flavin mononucleotide (FMN) occur naturally in bacteria, plants, and fungi, and have been heavily developed as optogenetic actuators. A subset of these proteins have specific regions

Table 1 Properties of Common Sensor Intrinsic Chromophore Fluorescent Proteins

FP	$\lambda_{ex}/\lambda_{em}$ (nm)	φ	ε ($M^{-1} \cdot cm^{-1}$)	Brightness[a] $\left(\frac{M^{-1}cm^{-1}}{1000}\right)$	Photostability[b] (s)	p*K*a	Maturation Time (h)	OSER Score[c] (%)	References
EBFP2	383/448	0.56	32,000	17.9	15.31	4.5	0.42	57.0	Ai, Shaner, Cheng, Tsien, and Campbell (2007)
mTagBFP2	399/454	0.64	50,600	32.4	6.21	2.7	0.20	49.8	Subach, Cranfill, Davidson, and Verkhusha (2011)
mTagBFP	399/456	0.63	52,000	32.8	ND	2.7	0.22	ND	Subach et al. (2008)
mTurquoise	434/474	0.84	30,000	25.2	391.51	4.5	ND	93.3	Goedhart et al. (2010)
mTurquoise2	434/474	0.93	30,000	27.9	71.71	3.1	ND	93.8	Goedhart et al. (2012)
Cerulean	433/475	0.62	43,000	26.7	74.63 (mCerulean)	4.7	ND	78.3 (92.6)	Rizzo, Springer, Granada, and Piston (2004)
mCerulean3	433/475	0.87	40,000	34.8	76.83	3.2	ND	91.0	Markwardt et al. (2011)
ECFP	434/477	0.40	32,500	13.0	ND	4.7	ND	ND	Shaner, Steinbach, and Tsien (2005)
CyPet	435/477	0.51	35,000	17.9	ND	5.0	ND	94.0	Nguyen and Daugherty (2005)
mTFP1	462/492	0.85	64,000	54.4	72.34	4.3	ND	92.0	Ai, Henderson, Remington, and Campbell (2006)

Continued

Table 1 Properties of Common Sensor Intrinsic Chromophore Fluorescent Proteins—cont'd

FP	$\lambda_{ex}/\lambda_{em}$ (nm)	φ	ε ($M^{-1} \cdot cm^{-1}$)	Brightness $\left(\frac{M^{-1} cm^{-1}}{1000}\right)$	Photostability (s)	p*K*a	Maturation Time (h)	OSER Score (%)	References
EGFP	489/509	0.60	55,000	33.0	179.21	5.9	0.42	76.5 (98.1)	Heim et al. (1995)
Clover	505/515	0.76	111,000	84.4	61.83	6.2	0.50	72.9 (90.5)	Lam et al. (2012)
mClover3	506/518	0.78	109,000	85.0	ND	6.5	ND	ND	Bajar, Wang, Lam, et al. (2016)
mNeonGreen	506/517	0.80	116,000	92.8	197.22	5.7	<0.16	90.4	Shaner et al. (2013)
mAmetrine	406/526	0.58	45,000	26.1	ND	6.0	0.80	90.0	Ai, Hazelwood, Davidson, and Campbell (2008)
EYFP	514/527	0.61	84,000	51.2	ND	6.5	ND	ND	Shaner et al. (2005)
Venus	515/528	0.57	92,200	52.6	26.46 (mVenus)	6.0	ND	36.5 (83.9)	Nagai et al. (2002)
Citrine	516/529	0.76	77,000	58.5	15.67 (mCitrine)	5.7	ND	36.2 (93.8)	Griesbeck, Baird, Campbell, Zacharias, and Tsien (2001)
Ypet	517/530	0.77	104,000	80.1	30.83	5.6	ND	62.5	Nguyen and Daugherty (2005)
mOrange2	549/565	0.60	58,000	34.8	353.61	6.5	4.50	91.8	Shaner et al. (2008)

mKO2	551/565	0.62	63,800	39.6	531.84	5.5	ND	68.4	Sakaue-Sawano et al. (2008) and Sun et al. (2009)
tdTomato	554/581	0.69	138,000	95.2	31.81	4.7	1.00	57.6	Shaner et al. (2004)
TagRFP-T	555/584	0.41	81,000	33.2	84.74	4.6	1.70	41.2	Shaner et al. (2008)
mApple	568/592	0.49	75,000	36.8	75.92	6.5	0.50	95.3	Shaner et al. (2008)
mRuby	558/605	0.35	112,000	39.2	40.69	4.4	2.80	93.1	Kredel et al. (2009)
mRuby2	559/600	0.38	113,000	42.9	44.19	5.3	2.50	87.4	Lam et al. (2012)
mRuby3	558/592	0.45	128,000	57.6	ND	4.8	<2.50	ND	Bajar, Wang, Lam, et al. (2016)
mRFP1	584/607	0.25	44,000	11.0	26.3	4.5	<1.00	95.8	Campbell et al. (2002)
mCherry	587/610	0.22	72,000	15.8	318.94	<4.5	0.25	95.0	Shaner et al. (2004)
mKeima	440/620	0.24	14,400	3.5	ND	6.5	4.50	ND	Kogure et al. (2006)
mKate2	588/633	0.40	62,500	25.0	51.61	5.4	0.63	81.1	Shcherbo et al. (2009)

[a]Brightness is calculated as $(\varphi \cdot \varepsilon)/1000$.
[b]Photostability represents bleaching $t_{1/2}$ under constant 80 W illumination, systematically performed according to Cranfill et al. (2016). Where indicated, only the monomeric form of the FP was tested.
[c]OSER score is a measure of monomericity, wherein FPs are expressed in the ER and cells are counted to determine the percentage that lack ER morphological abnormalities. Experiments systematically performed according to Cranfill et al. (2016), wherein the authors suggest a score of >90% indicates a monomeric FP. Values in parentheses indicate the OSER score for the engineered monomeric form of the FP.

Table 2 Properties of Extrinsic Chromophore Fluorescent Proteins

FP	$\lambda_{ex}/\lambda_{em}$ (nm)	φ	ε ($M^{-1}cm^{-1}$)	Brightness[a] $\left(\frac{M^{-1}cm^{-1}}{1000}\right)$	Photostability[b] (s)	p*K*a	Maturation Time (h)	Described Quaternary Structure	References
Flavin chromophore									
EcFbFP	448/496	0.44	14,500	6.4	168.6	<4.0	ND	Dimeric	Wingen et al. (2014) and Mukherjee, Walker, Weyant, and Schroeder (2013)
iLOV	447/497	0.44	ND	ND	ND	<4.0	ND	Dimeric	Chapman et al. (2008), Christie et al. (2012), and Mukherjee et al. (2013)
phiLOV2.1	450/497	0.2	ND	ND	778.2	ND	ND	Dimeric	Christie et al. (2012) and Wingen et al. (2014)
Bilirubin chromophore									
UnaG	498/527	0.51	77,300	39.4	ND	<4.0	0.08	ND	Kumagai et al. (2013) and To, Zhang, and Shu (2016)

Biliverdin chromophore									
IFP1.4	684/708	0.077	92,000	7.1	64	4.6	0.28	Weakly dimeric	Yu et al. (2015), Shu et al. (2009), and Filonov et al. (2011)
IFP2.0	690/711	0.07	86,125	6.0	70	ND	1.70	Weakly dimeric	Yu et al. (2015) and Yu, Gustafson, et al. (2014)
mIFP	683/704	0.08	82,000	6.6	400	<4.0	4.60	Monomeric	Yu et al. (2015)
iRFP713	690/713	0.063	98,000	6.2	>1800	4.0	0.57	Dimeric	Yu et al. (2015) and Filonov et al. (2011)
Retinal chromophore									
Arch (D95N)	585/687	0.0004	37,500	0.02	ND	8.9	ND	Monomeric	Kralj, Douglass, Hochbaum, Maclaurin, and Cohen (2012)

[a]Brightness is calculated as $(\varphi \cdot \varepsilon)/1000$.
[b]Photostability represents bleaching $t_{1/2}$ under illumination conditions consistent only within chromophore groups; values are not translatable between groups or between extrinsic and intrinsic chromophore FPs.

known as LOV domains, which are weakly fluorescent and reversibly covalently bind FMN upon UV/blue light illumination to induce conformational shifts (Fan & Lin, 2015). The first generation of FMN-binding FPs (FbFPs), including EcFbFP, Pp1FbFP, and iLOV, were generated with mutations that interrupt this covalent interaction to dramatically increase cyan-green fluorescence (Chapman et al., 2008; Drepper et al., 2007). Further engineering has yielded a second generation of FbFPs, as well as variants such as the fluorescent singlet oxygen generator miniSOG (Christie et al., 2012; Davari et al., 2016; Mukherjee et al., 2015; Shu et al., 2011; Song et al., 2013). While these FPs have substantially lower brightness than either EGFP or ECFP and their derivatives, their distinct advantages are fast turn-on fluorescence, fluorescence in anaerobic conditions, and small size. To date, FbFPs have mostly been used as reporters in hypoxic or anoxic environments, although some have been used as metal sensors or incorporated into oxygen sensors (Chapman et al., 2008; Potzkei et al., 2012; Ravikumar et al., 2016; Ravikumar, Nadarajan, Lee, Rhee, & Yun, 2015; Teng, Wang, Xu, & Xu, 2015).

Another extrinsic chromophore FP was recently characterized by the Miyawaki laboratory. UnaG, isolated from eel, reversibly binds bilirubin as a chromophore and fluoresces in the green (Hayashi & Toda, 2009; Kumagai et al., 2013). This FP is natively brighter and smaller than EGFP, does not require oxygen for fluorescence, and is not pH sensitive in physiological ranges (Kumagai et al., 2013). Furthermore, recent homologs indicate spectral properties can be tuned with mutation, making this a promising new platform for use in vertebrate systems, where bilirubin is endogenous (Gruber et al., 2015). Currently, the incorporation of UnaG has been limited to sensors for bilirubin and hypoxia, but this collection will doubtlessly expand in coming years (Erapaneedi, Belousov, Schäfers, & Kiefer, 2016; Iwatani et al., 2016).

Biliverdin-binding FPs have recently received attention due to their capacity for emitting light in the near-IR region of the spectrum, providing significant advantages for in vivo imaging, as well as offering another spectral window for channel multiplexing. Three lineages of biliverdin-binding FPs have been developed to date (IFP, mIFP, and iRFP), derived from three separate bacterial phytochrome proteins, and all with excitation maxima between 680 and 690 nm and emission maxima around 710 nm (Filonov et al., 2011; Shcherbakova & Verkhusha, 2013; Shu et al., 2009; Yu, Gustafson, et al., 2014). These proteins covalently attach a biliverdin molecule and have been engineered to limit conformational shifts that reduce fluorescence output.

A drawback for the IFP lineage is that the IFP proteins have a weak specificity for biliverdin and thus require exogenous biliverdin or coexpression of biliverdin-generating heme oxygenase for efficient fluorescence (Shu et al., 2009; Yu, Gustafson, et al., 2014). Recently, another biliverdin-binding FP called smURFP was reported that is more blue-shifted than the IFPs and iRFPs, but has high brightness and photostability (Rodriguez et al., 2016). To date, sensors have primarily incorporated IFP1.4, although the rapid development of biliverdin-binding FPs will surely influence sensors in the near future (Nobles, Clark, Green, & Maresso, 2015; To et al., 2015).

Rhodopsin proteins have been extensively studied as light-dependent ion channels but have recently also been explored as FPs in certain conditions. Rhodopsins covalently bind retinal as an extrinsic chromophore and have been observed to be fluorescent, although their fluorescence is generally far too weak for imaging applications. Recently, however, Archaerhodopsin 3 (Arch) from *Halorubrum sodomense* was engineered to emit stronger fluorescence in a voltage-dependent manner and has since successfully been used as a far-red voltage sensor (Flytzanis et al., 2014; Hochbaum et al., 2014; Kralj et al., 2012; Maclaurin, Venkatachalam, Lee, & Cohen, 2013). This is therefore an evolving, specialized class of FPs that may prove important, especially for neuroscience applications (McIsaac, Bedbrook, & Arnold, 2015).

3. SENSOR PLATFORMS

Different properties of experimental systems and targets often necessitate using specific types of sensor components and custom designs. Nevertheless, there are several general classes under which sensors fall, and these often diverge in terms of genetic elements, equipment requirements, and analytical approaches (Fig. 2). Note that we define sensors as dynamic elements that provide a relative or absolute readout of the concentration of a small molecule/ion or protein modification, or that visualize the activity of particular proteins or cellular homeostatic states. Thus, we do not include protein reporter tags or protein–protein interaction tools in the scope of this review.

Across sensor classes, there are major properties that must be considered in designing, validating, or choosing a sensor for a particular purpose (Table 3). Selectivity for the target of interest and sensitivity of a sensor for its substrate are critical. Selectivity describes whether the sensor is avoiding cross talk by only detecting the target analyte or interaction, and

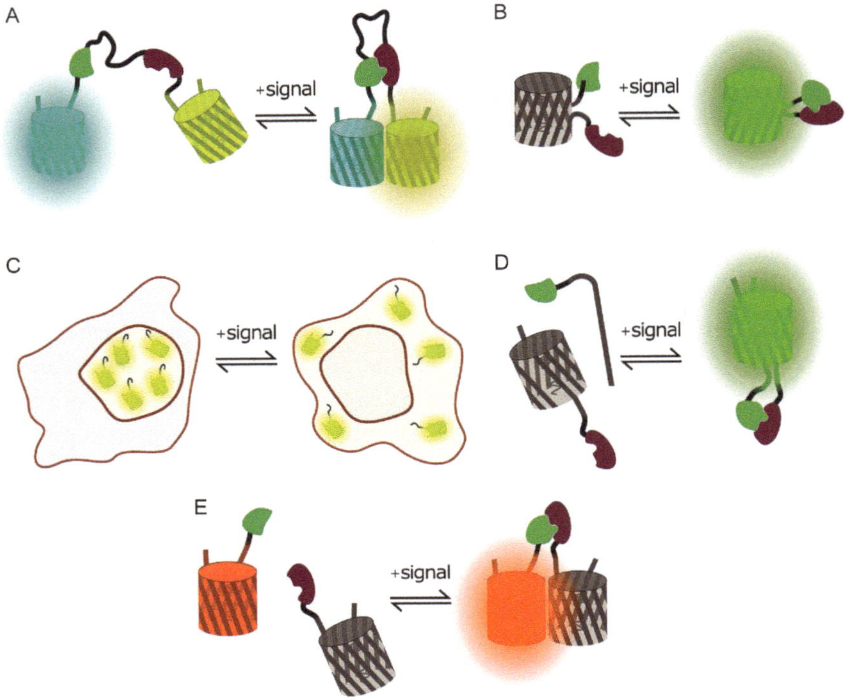

Fig. 2 Different genetically encoded fluorescent sensor platforms, including (A) FRET sensor, (B) fluorescence-modulated single FP sensor, (C) translocation sensor, (D) complementation sensor, and (E) dimerization sensor. "Signal" refers to an analyte, interaction, or process.

sensitivity indicates the lower threshold of analyte or interaction that the sensor can sense. Also important is the extent to which the sensor can perturb the system, either by sequestering the substrate pool itself or by influencing off-target attributes such as trafficking or organelle morphology. From a biophysical standpoint, kinetics, affinity, and reversibility of a sensor factor into what kinds of dynamics can be detected. Quantification of sensor output may be ratiometric, if the "on" and "off" states both have clearly defined readouts (as is the case with FRET sensors), or intensiometric, if a change in analyte or state induces a change in a single state, such as increasing the fluorescence of an FP at a single wavelength. Using a sensor for quantification also relies heavily on the sensor's signal-to-noise and dynamic range. Signal to noise describes the detectable range of signal of a sensor compared to instrumentation noise and autofluorescence, whereas dynamic range in the biological fluorescent sensor community specifies the maximum

Table 3 Advantages and Disadvantages of Genetically Encoded Fluorescent Sensor Classes

Class of Sensor	Kinetics (s)	Reversibility	Ratiometric vs Intensiometric	Signal-to-Noise Ratio	Dynamic Range	Multiplexing
FRET	$<10^1$	Generally reversible, except cleavage FRET sensors	Ratiometric	Variable	Lower	Difficult
Fluorescence-Modulated Single FP	<1	Generally reversible	Either	Variable	Variable	Variable
Translocation	>1	Generally reversible	Ratiometric	Lower	Lower	Easy
Complementation	$>10^2$	Irreversible with intrinsic chromophore FPs	Generally intensiometric	Higher	Higher	Easy
Dimerization	$<10^1$	Generally reversible	Either	Variable	Lower	Difficult

change in signal that can be detected between completely "on" and "off" conditions. Finally, experiments also rely on practicalities such as whether sensors are able to be multiplexed with other sensors or techniques, and what types of equipment setups are required for sensor detection.

3.1 FRET Sensors

FRET is a process that occurs between very close spectrally overlapping fluorophores in which an excited donor fluorophore transfers energy in a nonradiative manner to an acceptor fluorophore. The extent of transfer can be measured either by detecting the fluorescence emitted from the acceptor fluorophore or by measuring the decrease in lifetime of the donor fluorophore (for more details on FRET, please see Hochreiter, Pardo-Garcia, & Schmid, 2015; Shrestha, Jenei, Nagy, Vereb, & Szöllősi, 2015). FRET is a common sensor strategy due to two major advantages: ratiometric measurements and relatively modular composition of FRET sensors. A disadvantage of FRET sensors is that they often suffer from a low dynamic range compared to intensity-based readouts.

The components of genetically encoded FRET sensors are twofold: (1) two FPs wherein the donor FP emission spectrum overlaps substantially with the acceptor FP excitation spectrum, and (2) proteins or peptide segments that alter the proximity or orientation of the two FPs in response to a signal or analyte. There are several different FRET sensor architectures that include these two elements (Fig. 3).

3.1.1 Unimolecular Conformational FRET Sensors

By far the most common architecture of genetically encoded fluorescent sensors, unimolecular conformational FRET sensors most frequently contain two FPs flanking an interaction region. The state of the sensor in the absence of signal or analyte is either "closed," wherein the FPs are in close proximity and the FRET efficiency between them is high, or "open," wherein the FPs have a larger distance between them and therefore a lower FRET efficiency. When the binding or interaction region either binds to an analyte or is modified by an enzyme, the region undergoes a conformational change that transitions the sensor from "open" to "closed," or from "closed" to "open." Quantification for FRET sensors is usually performed by calculating the ratio of acceptor FP fluorescence over donor FP fluorescence, yielding a "FRET ratio." The analyte or interaction can be quantified by comparing the FRET ratio seen in a particular condition to the minimum and maximum observed FRET ratios.

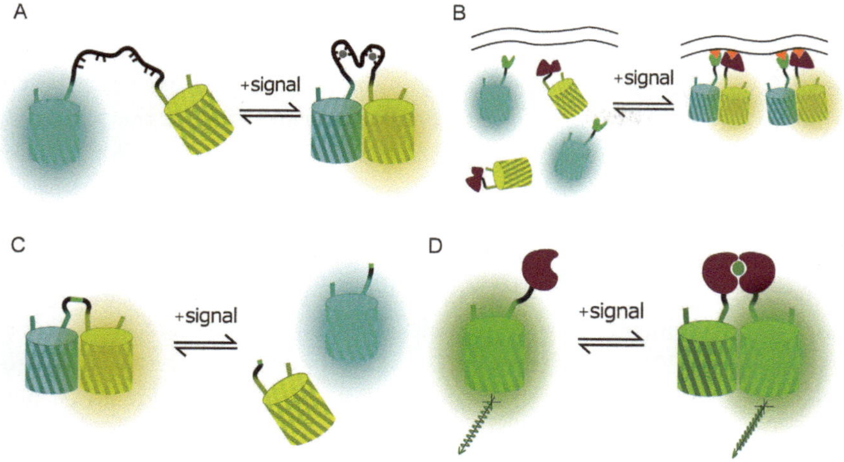

Fig. 3 General strategies for developing FRET sensors. (A) Unimolecular conformational FRET sensors consist of a single protein construct, usually with an FP FRET pair flanking a binding or interaction domain. Conformational changes due to a signal increase or decrease alter proximity of the two FPs and change the resultant FRET efficiency. (B) Bimolecular FRET sensors contain two separate interaction domains attached to an FP FRET pair. A signal alters a direct interaction between the two domains or their interaction with a third set of molecules, altering FP proximity. (C) Cleavage FRET sensors consist of an FP FRET pair connected by a linker containing a proteolytic cleavage site. Cleavage separates the two FPs, causing the sensor to go from a high FRET state to a low FRET state. (D) Homo-FRET sensors dimerize upon signal detection, and FRET between the resultant identical FPs alters the polarization of emitted fluorescent light. "Signal" refers to an analyte, interaction, or process.

Major advantages of unimolecular FRET sensors are that they are ratiometric and usually reversible, they have a stoichiometric ratio of donor to acceptor fluorophores, they have well-established methods for imaging and analysis, and they are easy to design due to a wealth of literature. Drawbacks of unimolecular FRET sensors include relatively low dynamic range due to the high basal level of FRET that results from tethered fluorophores, larger size that may be unsuitable for some cellular environments, and the number of sensor parameters to optimize. There are also a small number of FP pairs that are suitable for live-cell imaging in terms of FRET efficiency, brightness, and photostability, and changing the identities or orientations of FPs in an already established sensor often decreases dynamic range and requires further optimization (Hires, Zhu, & Tsien, 2008; Miranda et al., 2012; Ohta et al., 2016).

While unimolecular FRET sensors have a long history dating back to the first Ca^{2+}-sensing cameleons in 1997, some recent examples illustrate the versatility of this platform (Miyawaki et al., 1997). One sensor for protein crowding, named GimRET, was designed to capitalize on the sensitivity of a YFP variant for crowding by pairing it with insensitive CFP via a short linker (Morikawa et al., 2016). Quenching of the YFP by crowding substantially decreased the FRET ratio of the sensor in cells immediately after cell division. An interesting scaffold for metal signaling was recently developed in the Zn^{2+} sensor eZinCh-2, wherein two out of four Zn^{2+}-binding residues were engineered on the outsides of both Citrine and Cerulean FPs (Hessels et al., 2015). Zn^{2+} binding to all four coordinating residues effectively induces dimerization, and thus a high FRET ratio. Finally, a new approach, called electrochromic FRET (eFRET), was established for the FRET-opsin and QuasAr2 voltage sensors (Gong et al., 2015; Gong, Wagner, Zhong Li, & Schnitzer, 2014; Zou et al., 2014). These sensors use rhodopsin proteins as relatively nonfluorescent FRET acceptors for different intrinsic chromophore FPs and modulate donor fluorescence in an intensiometric manner with voltage-dependent conformational changes that alter absorption of the rhodopsin retinal chromophore.

3.1.2 Bimolecular FRET Sensors

Bimolecular FRET sensors are more specialized than unimolecular FRET sensors, primarily having been developed for sensing aspects of signaling, such as signal-dependent clustering or receptor dissociation. They consist of two interactive regions (domains or full proteins) linked separately to a FRET pair, which then yield a high FRET ratio upon interaction of the two polypeptides. Bimolecular FRET sensors retain the ability to assess signal ratiometrically and reversibly, and furthermore are capable of having a higher dynamic range than unimolecular FRET sensors due to separation of the FRET pair in the inactive or unbound state (Depry, Mehta, Li, & Zhang, 2015). Some bimolecular FRET sensors also are intended to form large complexes or clusters, which results in signal amplification but reduces the quantitative power of the observed signal change, as the ratio of the donor and acceptor fluorophores is not constant. One further drawback of bimolecular FRET sensors is the necessity to use two separate vectors or a split vector to express them in cells, which may result in an imbalance of expression between the two halves of the sensor.

Very recently, a sensor was developed to sense the receptor-mediated dissolution of a G-protein complex (van Unen et al., 2016). The sensor consists of a $G\alpha_{i1}$, $G\alpha_{i2}$, or $G\alpha_{i3}$ subunit attached to mTurquoise2 and $G\gamma_2$ fused to cp173Venus, and was shown to rapidly detect G-protein dispersion induced by the activation of six different G-protein-coupled receptors. Other examples of bimolecular FRET sensors include the NANOMS and NANOPS sensors, which detect plasma membrane localization and clustering of posttranslationally modified proteins (Köhnke et al., 2012; Najumudeen, Köhnke, Solman, Alexandrov, & Abankwa, 2013).

3.1.3 Cleavage FRET Sensors

An even more specialized FRET sensor platform is cleavage-based FRET sensors, which are used to detect the activity of proteases. These have a uniform structure, in which a FRET pair flanks a linker region that contains a substrate cleavage site for a particular protease, yielding a high FRET ratio until protease activity permanently separates the two fluorophores. These sensors benefit from being easy to design, with fewer parameters to optimize than other FRET sensors. Their main drawback is their irreversibility, which limits their ability to sense dynamics in enzyme activity. Cleavage FRET sensors have been developed for caspases, matrix metalloproteases, viral proteases, and more. Most recently, a caspase-3 sensor was developed for in vivo and multiplexed imaging using the far-red FPs mKate2 or eqFP650 and iRFP as a FRET pair (Zlobovskaya et al., 2016).

3.1.4 Homo-FRET Sensors

A rare type of FRET sensor uses the phenomenon of homo-FRET, wherein FRET can occur between two of the same fluorophore. Homo-FRET cannot be detected spectrally but instead is seen through anisotropy measurements, as different orientations of acceptor chromophores introduce variations in fluorescence polarization that do not occur with donor fluorescence only. These sensors are less restricted in fluorophore choice than other FRET sensors, and as single-color sensors are more amenable to multiplexing, but are harder to image and quantify. Homo-FRET sensors have been developed for detecting analytes such as Fe–S clusters, 3′-phosphoinositides, and $NADPH/NADP^+$ (Cameron et al., 2016; Hoff, Goodlitt, Li, Smolke, & Silberg, 2009; Warren, Margineanu, Katan, Dunsby, & French, 2015).

3.1.5 Engineering FRET Sensors

There are three general optimizable elements of FRET sensors: an FP FRET pair, binding or interaction domains for recognition of the analyte or signal, and linker regions that position the other two elements in proper configurations for readout of FRET signals (Fig. 4). These elements can be considered modular for the purposes of designing a sensor but often need to be optimized within the context of the other elements.

As previously mentioned, efficient FP FRET pairs are somewhat limited, although they have been expanded and improved over the past decade. FRET sensors are most commonly composed of cyan and yellow FPs. While early FRET sensors used ECFP and EYFP, the low brightness of ECFP and

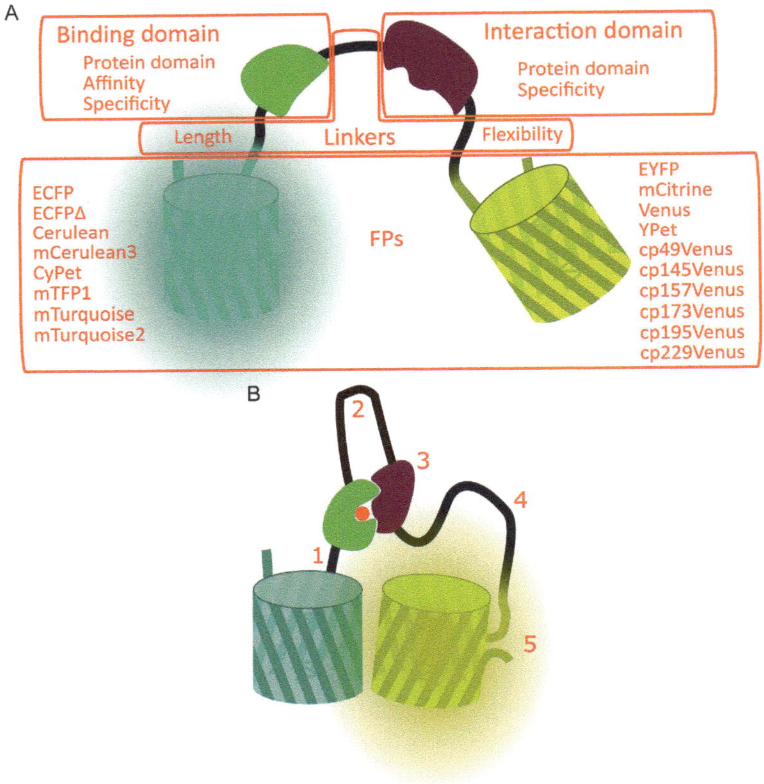

Fig. 4 Engineering FRET sensors. (A) Elements of a cameleon-type unimolecular conformational FRET sensor and possible optimizable characteristics. For FPs, a subset of possible cyan and yellow FPs is suggested. (B) Example of a differently engineered sensor in which the CFP has a C-terminal truncation (1), the linker between binding and interaction domains has been rigidified (2), the interaction domain has been truncated (3), the interaction domain-FP linker has been lengthened (4), and the YFP has been replaced with a circularly permuted Venus (5).

high p*Ka* of EYFP prompted a shift to the incorporation of other cyan and yellow variants (see Fig. 4). Furthermore, circularly permuted FPs (FPs engineered to have the N- and C-termini at different locations along the backbone) can potentially yield higher FRET efficiencies in certain scaffolds (Fritz et al., 2013; Nagai, Yamada, Tominaga, Ichikawa, & Miyawaki, 2004; van der Krogt, Ogink, Ponsioen, & Jalink, 2008). A green and red FP FRET pair has also arisen as a good option for sensors, especially with the advent and further development of Clover and mRuby by the Lin and Chu laboratories, and the engineering of the bright GFP mNeonGreen (Bajar, Wang, Lam, et al., 2016; Lam et al., 2012; Shaner et al., 2013). Other colors of FRET pairs have been developed, including mAmetrine–tdTomato and mOrange2–mCherry, although these and others have not been widely adopted in the literature due to unfavorable properties, including low photostability or quantum yield of donor FPs and overly large spectral overlap between donor and acceptor FPs (Ai et al., 2008; Bajar, Wang, Zhang, Lin, & Chu, 2016; Ding, Ai, Hoi, & Campbell, 2011; Miranda et al., 2012). The biliverdin-binding infrared FPs open up new possible FRET pairs, such as mKate2–iRFP713, which will be useful for multiplexing with existing FRET sensors (Zlobovskaya et al., 2016). For an excellent review of properties of FRET pairs, including those not covered here that have been specifically developed for fluorescence lifetime-based FRET measurements, please see Bajar, Wang, Zhang, et al. (2016).

Binding or interaction domains are very specific to the particular application of the sensor, and thus less generalizable to engineer than the FPs. For FRET sensors that measure a process, such as kinase activity or protease activation, the FRET ratio is indicative of the *extent* of the process, and interaction domains are engineered for efficiency and specificity of interaction. This usually consists of generating a range of protein or peptide truncations and varying their configuration to produce sufficient FRET change (Harvey et al., 2008; Komatsu et al., 2011; Tsou, Zheng, Hsu, Sasaki, & Cantley, 2011). For sensors in which the FRET ratio is intended to represent the *concentration* of an analyte, binding domains may also require engineering to optimize affinity and specificity for the analyte in order to report relevant concentration ranges and avoid analyte cross talk. Altering both specificity and affinity often requires more intimate knowledge of the structure and chemistry of the interaction of the domain with the analyte (Gu et al., 2006; Qin, Dittmer, Park, Jansen, & Palmer, 2011; Tsuyama et al., 2013). Alternatively, an interesting strategy was recently used to successfully develop an arginine sensor by testing reconstructed ancestral-binding domains, although the resultant sensor still required further optimization (Whitfield et al., 2015).

Linker optimization is arguably the most difficult aspect of designing FRET sensors. Linkers are essential for establishing proper distance and orientation between fluorophores, but the sequence and conformational spaces for sensor linkers are enormous. Some linker architectures have been developed for certain sensors and used as starting points for the engineering of other sensors (Deuschle et al., 2005; Harvey et al., 2008; Honda, Sawyer, Cawley, & Dostmann, 2005). There have also been attempts to generate generalizable scaffolds or systematic methods of developing linkers (Deuschle et al., 2005; Jung, Garcia, Kim, Yoon, & Baker, 2015; Komatsu et al., 2011; van Dongen et al., 2007), although these methods can still be labor intensive with relatively low yield (Hires et al., 2008). More recently, the development of different methods for high-throughput screening of FRET changes offer potential for rapid and unbiased optimization of linkers in libraries (Litzlbauer et al., 2015; Ma, Gibson, Dittmer, Jimenez, & Palmer, 2012; Peroza, Boumezbeur, & Zamboni, 2015). High-throughput methods of optimization will likely become a method of choice in the near future for optimizing new and existing FRET sensors.

3.2 Fluorescence-Modulated Single FP Sensors

Genetically encoded sensors using the fluorescence modulation of a single FP as a readout (hereafter referred to as single FP sensors) are more versatile than FRET sensors for many applications and have blossomed in the last decade. Compared to FRET sensors, single FPs are usually smaller, require narrower spectral bandwidth and thus can be more easily multiplexed with other fluorescent channels, and can theoretically be built on a wider range of FPs. They also often have higher dynamic ranges than FRET sensors. The major drawback of single FP sensors is that many of them are intensiometric and therefore can sense only relative changes, although examples of ratiometric single FP sensors do exist.

The components of single FP sensors vary somewhat, in that some require only an FP itself and exploit a natural sensitivity of the FP to analytes, and some attach interaction domains onto the FP to modulate its fluorescence (Fig. 5). In the latter case, the optimizable elements are similar to those of FRET sensors, consisting of an FP, an interaction or binding domain, and linker regions. These sensors usually use a circularly permuted version of an FP, in which the termini are engineered in the middle of the barrel, which

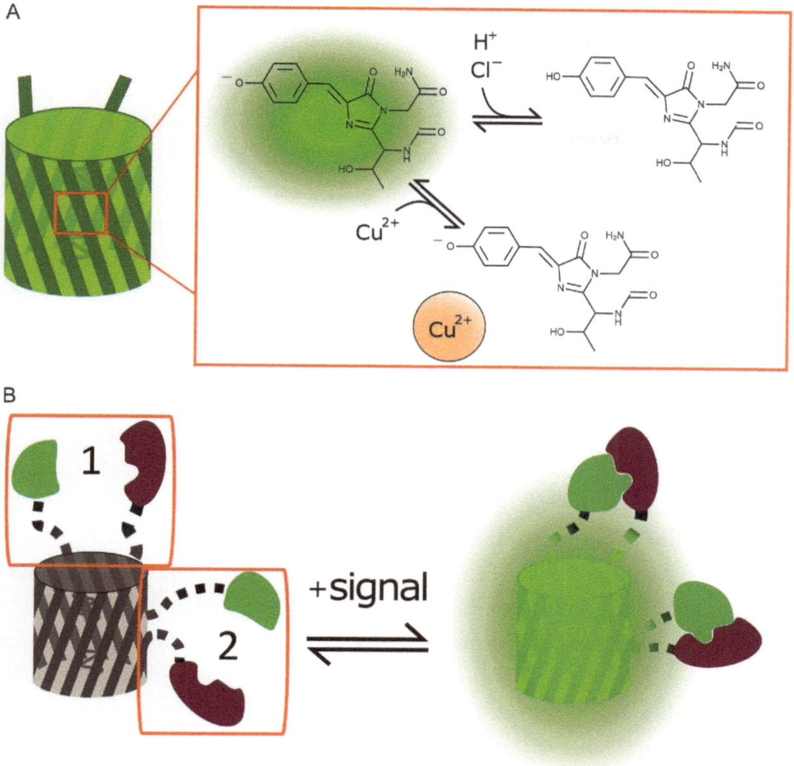

Fig. 5 Fluorescence-modulated single FP sensor strategies. (A) Intrinsic sensitivity single FP sensors show altered fluorescence due to interactions with ions that lead to chromophore protonation or quenching. (B) Conformational single FP sensors consist of domains attached to the termini of the FP. Upon a signal, conformational changes lead to modulation of FP fluorescence, usually due to altered solvent accessibility of the chromophore. Different circularly permuted FPs give rise to different sensor configurations; two possible structures are indicated by the *labeled boxes* 1 and 2.

allows fusions onto the termini to effectively modulate the solvent accessibility of the chromophore. Finally, some sensors attach a reference FP onto another portion of a construct to provide an internal control for ratiometric imaging; as this FP is not involved in sensing, we still consider these sensors to be single FP sensors.

3.2.1 Intrinsic Sensitivity Single FP Sensors

Intrinsic sensitivity single FP sensors usually operate by modulating the p*K*a of the chromophore to be selectively fluorescent. Intrinsic chromophore

FPs have a large blue shift in absorption wavelength upon protonation of the chromophore, and thus are nonfluorescent at standard excitation wavelengths. For this reason, one of the first uses of FPs was as pH sensors (Kneen, Farinas, Li, & Verkman, 1998; Llopis, McCaffery, Miyawaki, Farquhar & Tsien, 1998; Miesenböck, De Angelis, & Rothman, 1998). Mutagenizing residues composing and surrounding the chromophore can, in addition to altering spectral properties, influence chromophore p*Ka* to become more or less sensitive to physiologically relevant pH ranges. Further mutagenesis has proven successful in tuning absorption and emission spectra to produce ratiometric pH sensors (Hanson et al., 2002; Mahon, 2011; Poburko, Santo-Domingo, & Demaurex, 2011). Furthermore, high concentrations of some ions, in particular halides, can induce changes in chromophore p*Ka*, leading to the engineering of FPs as chloride and iodide sensors (Wachter, Yarbrough, Kallio, & Remington, 2000).

Some FPs have been observed to be sensitive to other agents, such as metals. Copper (Cu^{+} and Cu^{2+}) can efficiently quench fluorescence in native iLOV and versions of GFP engineered to have metal-binding ligands on its exterior (Lei et al., 2015; Ravikumar et al., 2015; Yu, Strub, et al., 2014). Hg^{2+}, interestingly, has been shown to induce turn-on fluorescence upon binding to variants of both FbFP and GFPxm, as well as inhibit fluorescence of IFP1.4 by blocking biliverdin binding (Gu, Zhao, Sheng, Bentolila, & Tang, 2011; Jiang et al., 2015; Ravikumar et al., 2016). These examples illustrate that some FP sensitivities may not be predictable, but upon observation can be exploited for generating sensors.

3.2.2 Conformational Single FP Sensors

Conformation-based single FP sensors rely on inducing conformational changes in an FP to modulate solvent accessibility, which can either quench fluorescence or alter the protonation state of the chromophore. Depending on exactly how they modulate chromophore chemistry, these sensors can be either intensiometric (turn-on or turn-off) or ratiometric (due to changes in excitation and emission spectra). The most common scaffold for these sensors uses interaction domains linked to the termini of a circularly permuted FP. The sensor is in an "open" configuration until analyte binding or enzyme activity induces a "closed" configuration that turns on or intensifies fluorescence.

Many examples of conformational single FP sensors exist, with the most famous being different series of Ca^{2+} indicators dating back to G-CaMP and pericam in 2001 (Nagai, Sawano, Park, & Miyawaki, 2001; Nakai,

Ohkura, & Imoto, 2001). A novel example of a single FP sensor is the heme-sensing IFP-HO1, which attaches IFP1.4 to heme oxygenase (Nobles et al., 2015). Heme oxygenase breaks down heme into several components, including biliverdin, which then binds to the IFP1.4 as a chromophore, inducing turn-on fluorescence. Another IFP-based sensor platform recently developed for protease activity utilizes an interesting architecture in which a split version of IFP1.4 that lacks a biliverdin-binding active site cysteine is tethered on one end by a split GFP and on the other by a short linker including a protease sequence flanked by a cysteine (To et al., 2015). Upon cleavage of the linker, the end containing the cysteine is free to associate with the active site of the FP to bind biliverdin and turn on fluorescence. One further example of an intriguing single FP sensor is deAmTrac, in which a modified circularly permuted GFP was inserted into an ammonium channel (Ast, Michele, Kumke, & Frommer, 2015). This sensor displays subtle dual-emission signal when excited at a single wavelength that is modulated by ammonium transport, providing a ratiometric transport sensor.

3.2.3 Engineering Single FP Sensors

Single FPs have some similar components to FRET sensors, and some optimization steps are largely the same, while others are quite different. One major difference between the two platforms lies in which FPs are commonly used. Theoretically, single FP sensors could be engineered from almost any FP. In practice, however, a relatively small number of FPs have been used for almost all single FP sensors. Well-characterized FPs like EGFP, ECFP, YFP, Citrine, and Venus provide the basis for almost all blue/green/yellow single FP sensors, although some have also been developed from intrinsic chromophore FPs mTurquiose2, mTFP1, mNeonGreen, and Sirius and extrinsic chromophore FPs UnaG and iLOV (Chen & Ai, 2016; Gong et al., 2015; Iwatani et al., 2016; Ravikumar et al., 2015; Sugiura et al., 2015). In the red region of the spectrum, mApple has emerged as a major FP for single FP sensors, while some sensors have been based upon mRuby, mCherry, mOrange2, or the long Stokes shift FP mKeima (Carlson & Campbell, 2013; Dana et al., 2016; Tantama, Hung, & Yellen, 2011). IFP1.4 has thus far been the primary far-red FP used in single FP sensors (Nobles et al., 2015; To et al., 2015).

While the list of FPs in single FP sensors is much smaller than the many FPs that exist, a major FP characteristic that must be optimized is circular permutation (Baird, Zacharias, & Tsien, 1999). Most single FP sensors rely on circularly permuted versions of FPs, and while for most of the above FPs a number of permutation possibilities have been established as successful, sensor engineering may require optimization of termini locations.

Most conformational single FP sensors include linkers and binding or interaction domains. The process of optimizing these elements is similar to that used for FRET sensors, in that most methods are low-throughput and often rely on a scaffold developed in previous sensors. Recently, a high-throughput method based on fluorescence-activated cell sorting (FACS) screening of a library of interaction domain variants was successful in generating a sensor for maltose (Nadler, Morgan, Flamholz, Kortright, & Savage, 2016). As is the case with FRET sensors, this and other high-throughput screening methods should facilitate sensor optimization.

3.3 Translocation Sensors

Translocation sensors constitute a class of genetically encoded fluorescent sensors wherein the cellular compartment that contains the FP is the primary sensor readout. These have been developed almost exclusively for signaling processes and usually involve a fluorescent construct translocating between the nucleus and cytosol. The advantages of translocation sensors are that they are relatively easy to construct according to a nuclear/cytosolic template, have comparable or larger dynamic ranges relative to FRET sensors, and can use any FP desired. The major disadvantage to these sensors is that absolute quantification is difficult, and even relative quantification often requires advanced postprocessing of image data. Translocation sensor quantification can also rely on spectrally distinct nuclear stains, which may limit their ability to multiplex with other sensors.

The two major scaffolds for translocation sensors are intended for detecting either kinase activity or protease activity. The kinase scaffold consists of an FP attached to a kinase substrate peptide that contains some combination of a nuclear localization signal (NLS) and nuclear export signal (NES). The strength of these two signal sequences is then modulated by phosphorylation at sites within or adjacent to them and/or the binding of associated factors, which then alters localization of the sensor. The protease scaffold similarly contains NLS and NES sequences connected to an FP, but the signal sequences are located on different sides of a protease cleavage site. When cleavage occurs, the FP translocates to the nucleus. Other types of translocation sensors include sensors that translocate from the cytosol to the plasma membrane upon signal induction, such as those incorporating signaling lipid-binding domains (Gray, Van Der Kaay, & Downes, 1999; Mishina et al., 2012). Additionally, tagged proteins that change localization when activated, such as p53, can also be used as localization sensors for the activation process (Quiñones & Rainov, 2001).

Translocation sensors have been successfully developed for kinases including Akt, JNK, and several human and yeast MAPKs (Anton, Bauer, Keck, Laufer, & Rothbauer, 2014; Durandau, Aymoz, & Pelet, 2015; Gross & Rotwein, 2015; Regot, Hughey, Bajar, Carrasco, & Covert, 2014). Several others detect kinase activity as a proxy for more general cell states, such as T-cell activation and cell cycle quiescence (Lodygin et al., 2013; Spencer et al., 2013). Protease sensors have been developed for caspase-3 and taspase 1 (Bier et al., 2011; Knauer et al., 2005). Given their relative modularity, there is great potential for translocation sensors to be easily engineered and multiplexed, and indeed recent work has shown successful imaging of dynamics of three separate kinases over the course of the cell cycle (Regot et al., 2014).

3.4 Complementation Sensors

FP complementation assays are primarily used for studying protein–protein interactions, but there have been some sensors developed using this technique. Complementation sensors are based on an FP that has been split into two or more fragments and is nonfluorescent in a split state. When a signal is detected, the fragments of the FP are brought into close proximity and complement, maturing or binding a chromophore and becoming fluorescent. These sensors have high dynamic range, as there is low basal fluorescence, and they involve a single FP and are therefore suitable for multiplexing. If based on intrinsic chromophore FPs, however, complementation sensors can have slow turn-on kinetics due to the time required for chromophore maturation, and they are generally irreversible (To, Zhang, & Shu, 2016). Extrinsic chromophore FPs can ameliorate both of these disadvantages, and in recent years split versions of UnaG, IFP1.4, and iRFP have been generated that will likely prove useful in the design of future complementation sensors (Filonov & Verkhusha, 2013; Tchekanda, Sivanesan, & Michnick, 2014; To, Zhang, et al., 2016).

Complementation sensors have been designed for detecting both analytes and signaling processes. One example is a sensor that detects numerous estrogen-based compounds with split-binding domains of ERα attached to split mVenus fragments (McLachlan, Katzenellenbogen, & Zhao, 2011).

Dimerization of the binding domains in the presence of an estrogen-based compound leads to complementation of the FP. An intriguing recent sensor, ZipGFP, was developed for protease activity in which both pieces of split GFP are caged by peptides at each terminus that interact to form a β-sheet and limit accessibility (To, Schepis, et al., 2016). Cleavage of a protease site on one end of the sheet in both fragments leads to uncaging and FP complementation.

3.5 Dimerization Sensors

A new sensor platform was recently developed by the Campbell laboratory that uses dimerization-dependent FPs (Alford, Abdelfattah, Ding, & Campbell, 2012; Alford, Ding, Simmen, & Campbell, 2012). This system utilizes two versions of FPs—one (FPA) that is quenched as a monomer, and one (FPB) that forms no chromophore at all. Heterodimerization between A and B monomers induces FPA fluorescence. The dimerization strategy was successfully developed into sensors for Ca^{2+}, Caspase-3, and mitochondrial-ER contact sites. These sensors were built on similar scaffolds as FRET sensors and had comparable dynamic ranges, although they were intensiometric in nature.

An interesting variation of the dimerization platform is fluorescent protein exchange (FPX), developed by the same laboratory (Ding et al., 2015). They discovered that a single FPB monomer could enhance fluorescence of two different colored FPA monomers, which then led them to develop sensor architectures that allowed for detection of caspase activity, Ca^{2+}, PIP_2 hydrolysis, and PKA activation. In these sensors, analyte binding or enzyme activity preferentially favors one set of dimers (red or green) above the other, allowing for ratiometric imaging. While these sensors can only provide relative quantification, as expression levels and dimerization affinities among the three monomers vary, FPX presents a fascinating new approach to sensor design.

4. TYPES OF SENSORS

The catalog of genetically encoded biosensors is too vast to fully explore in a review, but in this section we aim to cover the range of sensors currently in use and some key sensor applications.

4.1 Ions/Metals

By far the most extensively developed fluorescent biosensors are those for Ca^{2+}, an important signaling ion (for comprehensive reviews, please see Broussard, Liang, & Tian, 2014; Mank & Griesbeck, 2008; Rose, Goltstein, Portugues, & Griesbeck, 2014; Suzuki, Kanemaru, & Iino, 2016; Tang, Reddish, Zhuo, & Yang, 2015). While the dozens of genetically encoded Ca^{2+} sensors vary widely in fluorescence, kinetic, and thermodynamic properties, they share a few common elements. Most Ca^{2+} sensors use the Ca^{2+}-binding domain within the C-terminus of calmodulin as their Ca^{2+}-sensing element and the calmodulin-binding region of myosin light-chain kinase (M13) to induce conformational changes (Miyawaki et al., 1997). These two domains are most commonly flanked by a FRET pair to create a unimolecular Ca^{2+} FRET sensor (cameleon sensor series) or attached to the termini of a circularly permuted FP to generate a single FP Ca^{2+} sensor (GCaMP, RCaMP, GECO sensor series) (Broussard et al., 2014; Palmer, Qin, Park, & McCombs, 2011). Both of these platforms have been heavily optimized over the years by many different labs to improve photophysics, dynamic range, and kinetics, and they have been targeted to a range of cellular locations (Badura, Sun, Giovannucci, Lynch, & Wang, 2014; Choi, Swanson, & Gilroy, 2012; Dana et al., 2016; Suzuki et al., 2016; Waldeck-Weiermair et al., 2015; Wu et al., 2014). Other Ca^{2+}-binding domains have been used in place of calmodulin, such as troponin C (Heim & Griesbeck, 2004; Mank et al., 2006), and several other sensor architectures do exist, including CatchER, in which residues were mutated on the exterior of the EGFP barrel to directly bind Ca^{2+} for turn-on fluorescence (Tang et al., 2011).

The Ca^{2+} sensor field has recently experienced several intriguing innovations. Advances in the development of FPs have led to two new types of Ca^{2+} sensors. The first is a Ca^{2+} signal integration sensor (CAMPARI), which utilizes the traditional calmodulin/M13-binding strategy attached to a circularly permuted version of the photoactivatable FP mEos2 (Fosque et al., 2015). In a bound conformation, irradiation of CAMPARI with blue light induces a permanent shift of fluorescence from green to red, thereby "locking" the Ca^{2+} signal in place. The second new Ca^{2+} sensor platform is derived from FPX technology, wherein dimerization-dependent FPs of different colors are attached to Ca^{2+}-binding domains (Ding et al., 2015). Finally, the push for red-shifted tools for in vivo and multiplexing neuroscience applications has led to improved red single FP Ca^{2+} sensors, the jRCaMP and jRGECO series, which have been successfully used for single-action potential multicolor imaging with GCaMP6f in vitro and in vivo (Dana et al., 2016).

The basic designs of Ca^{2+} sensors have been successfully applied to many other biologically relevant ions and metal-containing molecules. Unimolecular FRET sensors containing a binding domain sandwiched between a FRET pair have been successfully developed for Zn^{2+} (Dittmer, Miranda, Gorski, & Palmer, 2009; Lindenburg, Hessels, Ebberink, Arts, & Merkx, 2013; Miranda et al., 2012; Qin et al., 2011; Vinkenborg et al., 2009), Cu^{+} (Wegner, Arslan, Sunbul, Yin, & He, 2010; Wegner, Sun, Hernandez, & He, 2011), Mg^{2+} (Lindenburg, Vinkenborg, Oortwijn, Aper, & Merkx, 2013), Mo^{2+} (Nakanishi et al., 2013), and heme (Song et al., 2015). The scaffold for conformationally induced fluorescence modulation of a single FP has also been applied to sensors for Cu^{+} (Liang et al., 2012), and very recently, heme (Hanna et al., 2016) and Zn^{2+} (Chen & Ai, 2016; Qin, Sammond, Braselmann, Carpenter, & Palmer, 2016).

Intrinsic chromophore FPs have been found to have innate sensitivity to some ions, most significantly halides. At high enough concentrations, halide anions can permeate the beta barrel and substantially raise the chromophore p*K*a, resulting in a loss of fluorescence (Wachter et al., 2000). EYFP has proven to be very responsive to chloride and iodide and has thus been further engineered as a halide sensor for multiple applications, including imaging inhibitory synaptic activity and studying transmittance of cystic fibrosis transmembrane conductance regulator mutants (Grimley et al., 2013; Vijftigschild, van der Ent, & Beekman, 2013; Zhong, Navaratnam, & Santos-Sacchi, 2014).

4.2 pH

Some early efforts to engineer genetically encoded pH sensors consisted of mutating EGFP in order to modulate its p*K*a, yielding intensiometric responses to pH changes in near-neutral conditions (Kneen et al., 1998; Llopis et al., 1998; Miesenböck et al., 1998). Further development of this idea has generated pH sensors across the visible spectrum (Benčina, 2013; Griesbeck et al., 2001; Llopis et al., 1998; Poëa-Guyon, Pasquier, Mérola, Morel, & Erard, 2013; Shaner et al., 2008; Shcherbo et al., 2009; Shen, Rosendale, Campbell, & Perrais, 2014; Wachter, Elsliger, Kallio, Hanson, & Remington, 1998). Many pH sensors have also been developed in which the change in excitation or emission wavelengths of an FP upon chromophore protonation can be used to ratiometrically quantify pH, and these sensors likewise span a large spectral range (Hanson et al., 2002; Mahon, 2011; Miesenböck et al., 1998; Poburko et al., 2011; Tantama et al., 2011). For detailed reviews of genetically encoded pH sensors, please refer to Benčina (2013) and Germond, Fujita, Ichimura, and Watanabe (2016).

4.3 Metabolites

Sensors have been generated for a wide range of small molecules using the basic FRET sensor template of cyan and yellow FPs flanking a single binding domain. These were pioneered by sensors for the signaling molecules cGMP and cAMP, and in particular the Epac-based cAMP sensors have been well optimized for FP FRET pair and binding domain architecture and affinity (DiPilato, Cheng, & Zhang, 2004; Everett & Cooper, 2013; Honda et al., 2001, 2005; Klarenbeek, Goedhart, Hink, Gadella, & Jalink, 2011; Klarenbeek, Goedhart, van Batenburg, Groenewald, & Jalink, 2015; Nikolaev, Bünemann, Hein, Hannawacker, & Lohse, 2004; Ponsioen et al., 2004; van der Krogt et al., 2008). Similar strategies have been applied to amino acids (Ameen et al., 2016; Hires et al., 2008; Mohsin, Abdin, Nischal, Kardam, & Ahmad, 2013; Mohsin & Ahmad, 2014; Okumoto et al., 2005; Whitfield et al., 2015), sugars (Gam et al., 2015; Kikuta, Hou, Sato, Frommer, & Kikawada, 2015; Lager, Looger, Hilpert, Lalonde, & Frommer, 2006; San Martín et al., 2014; Takanaga, Chaudhuri, & Frommer, 2008), and other important small molecules such as ATP (Imamura et al., 2009; Tsuyama et al., 2013), $NADP^+$ (Zhao, Zhang, Zhang, Tang, & Ye, 2016), phosphate (Gu et al., 2006), inositol triphosphate (Gulyás et al., 2015; Matsu-ura et al., 2006; Nezu, Tanimura, Morita, Shitara, & Tojyo, 2006; Remus et al., 2006; Sato, Ueda, Shibuya, & Umezawa, 2005), citrate (Ewald, Reich, Baumann, Frommer, & Zamboni, 2011), and 2-oxyglutarate (Zhang, Wei, & Ye, 2013).

In order to measure overall metabolic states of a cell, several single FP sensors have been developed that are able to detect ratios of metabolites such as ATP/ADP and NAD^+/NADH. In the case of the ATP/ADP sensor Perceval, phosphonucleotide binding to an attached kinase domain induces a shift in the excitation of cpmVenus, and the ratio is obtained due to differing affinities of the sensor for ATP and ADP (Berg, Hung, & Yellen, 2009).

A different strategy was used for the NAD^+/NADH sensor SoNar, in which NAD^+ or NADH binding leads to homodimerization of a cpYFP construct (Zhao et al., 2015). NAD^+ and NADH binding generate different conformations of the resultant construct, such that they excite the cpYFP at different wavelengths in a ratiometric fashion. SoNar was successfully applied to a high-throughput screen that identified a compound that selectively induced oxidative stress, altered NAD^+/NADH ratio, and prompted apoptosis in lung cancer cells.

4.4 Signaling

Genetically encoded sensors for elucidating activation and dynamics of signaling pathways are extremely diverse, as different signals often require different sensor design strategies. Several of the ions and metabolites covered in previous sections are important for signaling, including Ca^{2+}, cAMP, and inositol triphosphate, and sensors for these molecules have been central to elucidating many signal dynamics. Many sensors have also been developed for enzyme activation, particularly that of kinases and proteases, and other processes, such as signaling complex formation and dissolution. The ability to use one or more of these sensors to monitor the dynamics and effects of cellular signaling is one of the most powerful applications of genetically encoded fluorescent sensors.

Sensors for measuring kinase activity are typically built on the unimolecular FRET sensor model, although translocation-based kinase sensors have also become common in recent years. FRET-based kinase sensors are too numerous to fully describe here, but further information can be found in several excellent reviews (Aoki, Kiyokawa, Nakamura, & Matsuda, 2008; Herbst, Ni, & Zhang, 2009; Nhu Ngoc Van & Morris, 2013; Oldach & Zhang, 2014). In general, FRET sensors for kinase activity are modular, consisting of a FRET FP pair flanking a kinase substrate peptide and phosphopeptide-binding domain, or in some cases two domains of the kinase in question (Braun, Garfield, & Blumberg, 2005; Fujioka et al., 2006; Harvey et al., 2008; Komatsu et al., 2011; Kurokawa et al., 2001; Sasaki, Sato, & Umezawa, 2003; Violin, Zhang, Tsien, & Newton, 2003; Yoshizaki, Mochizuki, Gotoh, & Matsuda, 2007; Zhang, Ma, Taylor, & Tsien, 2001). Activation of the kinase then induces interaction between the sensor domains that changes their conformation, bringing the FPs into closer proximity and increasing FRET. Translocation sensors, in contrast, use kinase substrate peptides in close proximity to NES and NLS signal sequences, such that phosphorylation modulates the ratio of nuclear and cytosolic sensor. Such translocation sensors have been developed for mammalian JNK, p38, ERK1, ERK2, Akt, and CDK2 and yeast MAPKs (Durandau et al., 2015; Gross & Rotwein, 2015; Regot et al., 2014; Spencer et al., 2013). In an ideal experiment, multiple signals can be measured at once, and while some methods have been developed to measure multiple FRET sensors (Grant et al., 2008; Niino, Hotta, & Oka, 2009; Ouyang et al., 2010; Su, Pan, Luo, & Zhang, 2013; Warren et al., 2015), translocation sensors seem most amenable for this purpose (Regot et al., 2014).

Another method of detecting signaling events in cells is by measuring the assembly and disassembly of signaling complexes. Most of the sensors for these events are bimolecular sensors using FRET or complementation methodologies. These sensors in some cases detect specific interactions, such as that of HRas with Raf, or the activation of GPCRs as seen by dissociation of Gα subunits from Gγ subunits (Li, Cheng, & Jin, 2012; Oliveira & Yasuda, 2013; van Unen et al., 2016). In other cases, they detect more general signaling events, such as the presence of PtdIns(4,5)P_2 in membranes or the clustering of myristoylated or prenylated proteins (Ding et al., 2015; Hertel et al., 2011; Köhnke et al, 2012; Najumudeen et al., 2013).

One further major class of signaling sensors is that for protease activity. These sensors generally follow the cleavage FRET sensor design, although different strategies have recently been developed for Caspase and Taspase1 sensors based on single FP, complementation, dimerization, or localization platforms (Ding et al., 2015; Knauer et al., 2011; Luo, Yu, Pu, & Chang, 2001, 2003; To et al., 2015; To, Schepis, et al., 2016; Zlobovskaya et al., 2016). While many proteases operate intracellularly, extracellular proteases such as matrix metalloproteases and tumor necrosis factor-α (TNF-α)-converting enzyme are critical for extracellular signaling processes. This presents a further challenge for sensor design, as many FPs are sensitive to oxidative conditions, such as the extracellular environment. While some sensors have proven successful in detecting extracellular proteolysis events (Chapnick, Bunker, & Liu, 2015; Eichorst, Clegg, & Wang, 2012; Ouyang et al., 2010), further development of extracellular sensors will likely benefit from the advent of oxFPs, which are engineered to be cysteine-less and thus are more resistant to oxidation (Costantini et al., 2015).

4.5 Redox

All redox sensors to date rely on the reversible oxidation of two cysteines to form a disulfide bond. One major design strategy has been to mutate residues proximal to the chromophore of intrinsic chromophore FPs to cysteines, which under oxidative conditions form disulfide bonds with other cysteines on the FP or a connected domain. Conformational changes resulting from these disulfide bonds alter excitation or emission of the FP. A YFP version of this type of redox sensor was the first to be engineered, and a range of colors have since been generated, most of which alter the spectral characteristics of the FP to provide a ratiometric signal (Fan, Chen, & Ai, 2015; Hanson et al., 2004; Ostergaard, Henriksen, Hansen, & Winther, 2001; Sugiura et al., 2015). A recent variant of this strategy was to rely on disulfides to directly link a FRET pair, with FRET signal increasing upon oxidation (Abraham, Santala, Tkachenko, & Karp, 2014). Other engineering strategies for redox sensors follow the canonical conformational single FP and unimolecular FRET sensor designs, wherein oxidation of cysteines on domains or linkers leads to conformational changes that increase fluorescence or FRET signal (Ermakova et al., 2014; Kolossov et al., 2011; Markvicheva et al., 2011; Yano et al., 2010). Most of these sensors are generally responsive to oxidizing and reducing agents, and they have primarily been used to sense H_2O_2 and oxidative stress, as well as the reduced and oxidized glutathione balance in different compartments of the cell (Banach-Latapy et al., 2013; Fan et al., 2015; Huang, Ali, Stein, & Sikes, 2015; Kolossov et al., 2012; Weller, Kizina, Can, Bao, & Müller, 2014). For further information on genetically encoded redox sensors, please refer to Lukyanov and Belousov (2014).

4.6 Force and Crowding

Unimolecular FRET sensors have been useful in investigating biophysical forces acting in cells. Different applications have led to different FRET sensor designs, but often binding domains for proteins such as actin, tubulin, or vinculin are fused onto the outsides of a cassette containing a linked FRET pair, allowing a decrease in FRET to represent a greater distance between the cytoskeletal elements (Grashoff et al., 2010; Iwai & Uyeda, 2008; Knorr, Jackson, Batie, Narmoneva, & Jones, 2016; Meng & Sachs, 2012). A single FP force sensor has also been generated in which different amounts of tension between the termini of circularly permuted YFP alters the excitation spectrum of the FP (Ichimura, Fujita, Yoshizawa, & Watanabe, 2012). Molecular crowding sensors have been successfully engineered by modulating linkers between a FRET pair, and the GimRET sensor relies on sensitivity of a YFP variant to crowding (Boersma, Zuhorn, & Poolman, 2015; Morikawa et al., 2016). Further information on force sensors can be found in two excellent recent reviews (Cost, Ringer, Chrostek-Grashoff, & Grashoff, 2015; Yang et al., 2015).

4.7 Voltage

Genetically encoded voltage sensors represent a prominent field due to their utility in neuroscience research, especially for in vivo applications, and several recent reviews provide excellent coverage of the state of the field (Germond et al., 2016; Gong, 2015; Storace et al., 2016; St-Pierre, Chavarha, & Lin, 2015). Voltage sensors require at least two components: a voltage-sensing domain (VSD) and an FP. In the case of engineered Archaerhodopsin, these two components exist in one membrane protein, and voltage-based modulation of the fluorescence of the retinal chromophore provides the sensor readout (Flytzanis et al., 2014; Hochbaum et al., 2014; Kralj et al., 2012; Maclaurin et al., 2013). Rhodopsins have also been used as FRET acceptors in eFRET sensors, in which voltage modulates the FRET efficiency of the rhodopsin and therefore alters donor fluorescence, which is measured in an intensiometric fashion (Gong et al., 2015, 2014; Zou et al., 2014).

In most other voltage sensors, a VSD from either a voltage-gated potassium channel or a voltage-sensing phosphatase is linked in different permutations to intrinsic chromophore FPs. The ArcLight sensors and their derivatives attach a variant of the GFP-derived super ecliptic pHluorin pH sensor, which in proximity to a VSD is sensitive to voltage (Jin et al., 2012; Jung et al., 2015; Piao, Rajakumar, Kang, Kim, & Baker, 2015). Sensors such as ElectricPk, the ASAP series, and FlicR1 attach the VSD to circularly permuted single FPs, which have an intensiometric response upon a voltage-induced conformational change of the VSD (Abdelfattah et al., 2016; Barnett, Platisa, Popovic, Pieribone, & Hughes, 2012; St-Pierre et al., 2014; Yang et al., 2016). Other series, including the Nabi and Butterfly sensor series, use a similar conformational change in VSD with an intrinsic chromophore FP FRET pair to produce a ratiometric FRET response (Mishina, Mutoh, Song, & Knöpfel, 2014; Sung et al., 2015). The kinetics and dynamic range of all of these types of voltage sensors vary substantially and are rapidly evolving, allowing for consistently better exploration of neuronal systems.

5. CONCLUSION

Genetically encoded fluorescent sensors have become essential tools to understand the homeostasis and dynamics of many different types of cellular molecules and signals. Persistent progress in engineering of sensor component parts, most especially FPs, has in turn expanded the range of possible sensor architectures and improved their properties. The advancement of extrinsic chromophore FPs is motivating the creation of sensors that are usable in previously challenging cellular environments. Furthermore, expansion into the near-infrared region of the spectrum facilitates sensor multiplexing, which can answer a larger range of scientific questions. Finally, the evolving technologies for generating and screening libraries of FPs and sensors will further streamline sensor engineering as the field moves forward.

REFERENCES

Abdelfattah, A. S., Farhi, S. L., Zhao, Y., Brinks, D., Zou, P., Ruangkittisakul, A., et al. (2016). A bright and fast red fluorescent protein voltage indicator that reports neuronal activity in organotypic brain slices. *The Journal of Neuroscience: The Official Journal of the Society for Neuroscience*, *36*(8), 2458–2472.

Abraham, B. G., Santala, V., Tkachenko, N. V., & Karp, M. (2014). Fluorescent protein-based FRET sensor for intracellular monitoring of redox status in bacteria at single cell level. *Analytical and Bioanalytical Chemistry*, *406*, 7195–7204.

Ai, H., Hazelwood, K. L., Davidson, M. W., & Campbell, R. E. (2008). Fluorescent protein FRET pairs for ratiometric imaging of dual biosensors. *Nature Methods*, *5*(5), 401–403.

Ai, H., Henderson, J. N., Remington, S. J., & Campbell, R. E. (2006). Directed evolution of a monomeric, bright and photostable version of *Clavularia* cyan fluorescent protein: Structural characterization and applications in fluorescence imaging. *Biochemical Journal*, *400*(3), 531–540.

Ai, H., Shaner, N. C., Cheng, Z., Tsien, R. Y., & Campbell, R. E. (2007). Exploration of new chromophore structures leads to the identification of improved blue fluorescent proteins. *Biochemistry*, *46*(20), 5904–5910.

Alford, S. C., Abdelfattah, A. S., Ding, Y., & Campbell, R. E. (2012). A fluorogenic red fluorescent protein heterodimer. *Chemistry & Biology*, *19*(3), 353–360.

Alford, S. C., Ding, Y., Simmen, T., & Campbell, R. E. (2012). Dimerization-dependent green and yellow fluorescent proteins. *ACS Synthetic Biology*, *1*(12), 569–575.

Ameen, S., Ahmad, M., Mohsin, M., Qureshi, M. I., Ibrahim, M. M., Abdin, M. Z., et al. (2016). Designing, construction and characterization of genetically encoded FRET-based nanosensor for real time monitoring of lysine flux in living cells. *Journal of Nanobiotechnology*, *14*(1), 49.

Anton, R., Bauer, S. M., Keck, P. R. W. E. F., Laufer, S., & Rothbauer, U. (2014). A p38 substrate-specific MK2-EGFP translocation assay for identification and validation of new p38 inhibitors in living cells: A comprising alternative for acquisition of cellular p38 inhibition data. *PloS One*, *9*(4), e95641.

Aoki, K., Kiyokawa, E., Nakamura, T., & Matsuda, M. (2008). Visualization of growth signal transduction cascades in living cells with genetically encoded probes based on Förster resonance energy transfer. *Philosophical Transactions of the Royal Society of London Series B, Biological Sciences, 363*(1500), 2143–2151.

Ast, C., Michele, R. D., Kumke, M. U., & Frommer, W. B. (2015). Single-fluorophore membrane transport activity sensors with dual-emission read-out. *eLife, 4*, e07113.

Badura, A., Sun, X. R., Giovannucci, A., Lynch, L. A., & Wang, S. S.-H. (2014). Fast calcium sensor proteins for monitoring neural activity. *Neurophotonics, 1*(2), 25008.

Baird, G. S., Zacharias, D. A., & Tsien, R. Y. (1999). Circular permutation and receptor insertion within green fluorescent proteins. *Proceedings of the National Academy of Sciences of the United States of America, 96*(20), 11241–11246.

Bajar, B. T., Wang, E. S., Lam, A. J., Kim, B. B., Jacobs, C. L., Howe, E. S., et al. (2016). Improving brightness and photostability of green and red fluorescent proteins for live cell imaging and FRET reporting. *Scientific Reports, 6*, 20889.

Bajar, B. T., Wang, E. S., Zhang, S., Lin, M. Z., & Chu, J. (2016). A guide to fluorescent protein FRET pairs. *Sensors, 16*(9), 1–24.

Banach-Latapy, A., He, T., Dardalhon, M., Vernis, L., Chanet, R., & Huang, M.-E. (2013). Redox-sensitive YFP sensors for monitoring dynamic compartment-specific glutathione redox state. *Free Radical Biology & Medicine, 65*, 436–445.

Barnett, L., Platisa, J., Popovic, M., Pieribone, V. A., & Hughes, T. (2012). A fluorescent, genetically-encoded voltage probe capable of resolving action potentials. *PloS One*, 7(9), e43454.

Benčina, M. (2013). Illumination of the spatial order of intracellular pH by genetically encoded pH-sensitive sensors. *Sensors, 13*(12), 16736–16758.

Berg, J., Hung, Y. P., & Yellen, G. (2009). A genetically encoded fluorescent reporter of ATP:ADP ratio. *Nature Methods, 6*(2), 161–166.

Bier, C., Knauer, S. K., Klapthor, A., Schweitzer, A., Rekik, A., Krämer, O. H., et al. (2011). Cell-based analysis of structure-function activity of threonine aspartase 1. *The Journal of Biological Chemistry, 286*(4), 3007–3017.

Boersma, A. J., Zuhorn, I. S., & Poolman, B. (2015). A sensor for quantification of macromolecular crowding in living cells. *Nature Methods, 12*(3), 227–229. 1 p following 229.

Braun, D. C., Garfield, S. H., & Blumberg, P. M. (2005). Analysis by fluorescence resonance energy transfer of the interaction between ligands and protein kinase Cδ in the intact cell. *The Journal of Biological Chemistry, 280*(9), 8164–8171.

Broussard, G. J., Liang, R., & Tian, L. (2014). Monitoring activity in neural circuits with genetically encoded indicators. *Frontiers in Molecular Neuroscience*, 7, 97.

Cameron, W. D., Bui, C. V., Hutchinson, A., Loppnau, P., Gräslund, S., & Rocheleau, J. V. (2016). Apollo-NADP(+): A spectrally tunable family of genetically encoded sensors for NADP(+). *Nature Methods, 13*(4), 352–358.

Campbell, R. E., Tour, O., Palmer, A. E., Steinbach, P. A., Baird, G. S., Zacharias, D. A., et al. (2002). A monomeric red fluorescent protein. *Proceedings of the National Academy of Sciences of the United States of America, 99*(12), 7877–7882.

Carlson, H. J., & Campbell, R. E. (2013). Mutational analysis of a red fluorescent protein-based calcium ion indicator. *Sensors (Basel, Switzerland), 13*(9), 11507–11521.

Chapman, S., Faulkner, C., Kaiserli, E., Garcia-Mata, C., Savenkov, E. I., Roberts, A. G., et al. (2008). The photoreversible fluorescent protein iLOV outperforms GFP as a reporter of plant virus infection. *Proceedings of the National Academy of Sciences of the United States of America, 105*(50), 20038–20043.

Chapnick, D. A., Bunker, E., & Liu, X. (2015). A biosensor for the activity of the "sheddase" TACE (ADAM17) reveals novel and cell type-specific mechanisms of TACE activation. *Science Signaling, 8*(365), 1–9.

Chen, Z.-J., & Ai, H.-W. (2016). Single fluorescent protein-based indicators for zinc ion (Zn^{2+}). *Analytical Chemistry, 88*(18), 9029–9036.

Choi, W.-G., Swanson, S. J., & Gilroy, S. (2012). High-resolution imaging of Ca^{2+}, redox status, ROS and pH using GFP biosensors. *The Plant Journal: For Cell and Molecular Biology, 70*(1), 118–128.

Christie, J. M., Hitomi, K., Arvai, A. S., Hartfield, K. A., Mettlen, M., Pratt, A. J., et al. (2012). Structural tuning of the fluorescent protein iLOV for improved photostability. *The Journal of Biological Chemistry, 287*(26), 22295–22304.

Chudakov, D. M., Matz, M. V., Lukyanov, S., & Lukyanov, K. A. (2010). Fluorescent proteins and their applications in imaging living cells and tissues. *Physiological Reviews, 90*(3), 1103–1163.

Cormack, B. P., Valdivia, R. H., & Falkow, S. (1996). FACS-optimized mutants of the green fluorescent protein (GFP). *Gene, 173*(1), 33–38.

Cost, A.-L., Ringer, P., Chrostek-Grashoff, A., & Grashoff, C. (2015). How to measure molecular forces in cells: A guide to evaluating genetically-encoded FRET-based tension sensors. *Cellular and Molecular Bioengineering, 8*(1), 96–105.

Costantini, L. M., Baloban, M., Markwardt, M. L., Rizzo, M., Guo, F., Verkhusha, V. V., et al. (2015). A palette of fluorescent proteins optimized for diverse cellular environments. *Nature Communications, 6*, 7670.

Costantini, L. M., & Snapp, E. L. (2013). Fluorescent proteins in cellular organelles: Serious pitfalls and some solutions. *DNA and Cell Biology, 32*(11), 622–627.

Cranfill, P. J., Sell, B. R., Baird, M. A., Allen, J. R., Lavagnino, Z., de Gruiter, H. M., et al. (2016). Quantitative assessment of fluorescent proteins. *Nature Methods, 13*(7), 557–562.

Dana, H., Mohar, B., Sun, Y., Narayan, S., Gordus, A., Hasseman, J. P., et al. (2016). Sensitive red protein calcium indicators for imaging neural activity. *eLife, 5*, 1–24.

Davari, M. D., Kopka, B., Wingen, M., Bocola, M., Drepper, T., Jaeger, K.-E., et al. (2016). Photophysics of the LOV-based fluorescent protein variant iLOV-Q489K determined by simulation and experiment. *The Journal of Physical Chemistry B, 120*(13), 3344–3352.

Depry, C., Mehta, S., Li, R., & Zhang, J. (2015). Visualization of compartmentalized kinase activity dynamics using adaptable BimKARs. *Chemistry & Biology, 22*(11), 1470–1479.

Deuschle, K., Okumoto, S., Fehr, M., Looger, L. L., Kozhukh, L., & Frommer, W. B. (2005). Construction and optimization of a family of genetically encoded metabolite sensors by semirational protein engineering. *Protein Science: A Publication of the Protein Society, 14*(9), 2304–2314.

Ding, Y., Ai, H., Hoi, H., & Campbell, R. E. (2011). Förster resonance energy transfer-based biosensors for multiparameter ratiometric imaging of Ca^{2+} dynamics and caspase-3 activity in single cells. *Analytical Chemistry, 83*(24), 9687–9693.

Ding, Y., Li, J., Enterina, J. R., Shen, Y., Zhang, I., Tewson, P. H., et al. (2015). Ratiometric biosensors based on dimerization-dependent fluorescent protein exchange. *Nature Methods, 12*(3), 195–198.

DiPilato, L. M., Cheng, X., & Zhang, J. (2004). Fluorescent indicators of cAMP and Epac activation reveal differential dynamics of cAMP signaling within discrete subcellular compartments. *Proceedings of the National Academy of Sciences of the United States of America, 101*(47), 16513–16518.

Dittmer, P. J., Miranda, J. G., Gorski, J. A., & Palmer, A. E. (2009). Genetically encoded sensors to elucidate spatial distribution of cellular zinc. *The Journal of Biological Chemistry, 284*(24), 16289–16297.

Drepper, T., Eggert, T., Circolone, F., Heck, A., Krauß, U., Guterl, J.-K., et al. (2007). Reporter proteins for in vivo fluorescence without oxygen. *Nature Biotechnology, 25*(4), 443–445.

Durandau, E., Aymoz, D., & Pelet, S. (2015). Dynamic single cell measurements of kinase activity by synthetic kinase activity relocation sensors. *BMC Biology, 13*, 55.

Eichorst, J. P., Clegg, R. M., & Wang, Y. (2012). Red-shifted fluorescent proteins monitor enzymatic activity in live HT-1080 cells with fluorescence lifetime imaging microscopy (FLIM). *Journal of Microscopy, 248*(1), 77–89.

Erapaneedi, R., Belousov, V. V., Schäfers, M., & Kiefer, F. (2016). A novel family of fluorescent hypoxia sensors reveal strong heterogeneity in tumor hypoxia at the cellular level. *The EMBO Journal*, *35*(1), 102–113.
Ermakova, Y. G., Bilan, D. S., Matlashov, M. E., Mishina, N. M., Markvicheva, K. N., Subach, O. M., et al. (2014). Red fluorescent genetically encoded indicator for intracellular hydrogen peroxide. *Nature Communications*, *5*, 5222.
Everett, K. L., & Cooper, D. M. F. (2013). An improved targeted cAMP sensor to study the regulation of adenylyl cyclase 8 by Ca^{2+} entry through voltage-gated channels. *PloS One*, *8*(9), e75942.
Ewald, J. C., Reich, S., Baumann, S., Frommer, W. B., & Zamboni, N. (2011). Engineering genetically encoded nanosensors for real-time *in vivo* measurements of citrate concentrations. *PloS One*, *6*(12), e28245.
Fan, Y., Chen, Z., & Ai, H. (2015). Monitoring redox dynamics in living cells with a redox-sensitive red fluorescent protein. *Analytical Chemistry*, *87*(5), 2802–2810.
Fan, L. Z., & Lin, M. Z. (2015). Optical control of biological processes by light-switchable proteins. *WIREs Developmental Biology*, *4*(5), 545–554.
Filonov, G. S., Piatkevich, K. D., Ting, L.-M., Zhang, J., Kim, K., & Verkhusha, V. V. (2011). Bright and stable near-infrared fluorescent protein for *in vivo* imaging. *Nature Biotechnology*, *29*(8), 757–761.
Filonov, G. S., & Verkhusha, V. V. (2013). A near-infrared BiFC reporter for *in vivo* imaging of protein-protein interactions. *Chemistry & Biology*, *20*(8), 1078–1086.
Flytzanis, N. C., Bedbrook, C. N., Chiu, H., Engqvist, M. K. M., Xiao, C., Chan, K. Y., et al. (2014). Archaerhodopsin variants with enhanced voltage-sensitive fluorescence in mammalian and *Caenorhabditis elegans* neurons. *Nature Communications*, *5*, 4894.
Fosque, B. F., Sun, Y., Dana, H., Yang, C.-T., Ohyama, T., Tadross, M. R., et al. (2015). Labeling of active neural circuits *in vivo* with designed calcium integrators. *Science*, *347*(6223), 755–760.
Fritz, R. D., Letzelter, M., Reimann, A., Martin, K., Fusco, L., Ritsma, L., et al. (2013). A versatile toolkit to produce sensitive FRET biosensors to visualize signaling in time and space. *Science Signaling*, *6*(285), 1–13.
Fujioka, A., Terai, K., Itoh, R. E., Aoki, K., Nakamura, T., Kuroda, S., et al. (2006). Dynamics of the Ras/ERK MAPK cascade as monitored by fluorescent probes. *The Journal of Biological Chemistry*, *281*(13), 8917–8926.
Gam, J., Ha, J.-S., Kim, H., Lee, D.-H., Lee, J., & Lee, S.-G. (2015). Ratiometric analyses at critical temperatures can magnify the signal intensity of FRET-based sugar sensors with periplasmic binding proteins. *Biosensors & Bioelectronics*, *72*, 37–43.
Germond, A., Fujita, H., Ichimura, T., & Watanabe, T. M. (2016). Design and development of genetically encoded fluorescent sensors to monitor intracellular chemical and physical parameters. *Biophysical Reviews*, *8*, 121–138.
Goedhart, J., van Weeren, L., Hink, M. A., Vischer, N. O. E., Jalink, K., & Gadella, T. W. J. (2010). Bright cyan fluorescent protein variants identified by fluorescence lifetime screening. *Nature Methods*, 7(2), 137–139.
Goedhart, J., von Stetten, D., Noirclerc-Savoye, M., Lelimousin, M., Joosen, L., Hink, M. A., et al. (2012). Structure-guided evolution of cyan fluorescent proteins towards a quantum yield of 93%. *Nature Communications*, *3*, 751.
Gong, Y. (2015). The evolving capabilities of rhodopsin-based genetically encoded voltage indicators. *Current Opinion in Chemical Biology*, *27*, 84–89.
Gong, Y., Huang, C., Li, J. Z., Grewe, B. F., Zhang, Y., Eismann, S., et al. (2015). High-speed recording of neural spikes in awake mice and flies with a fluorescent voltage sensor. *Science (New York, N.Y.)*, *350*(6266), 1361–1366.
Gong, Y., Wagner, M. J., Zhong Li, J., & Schnitzer, M. J. (2014). Imaging neural spiking in brain tissue using FRET-opsin protein voltage sensors. *Nature Communications*, *5*, 3674.

Grant, D. M., Zhang, W., McGhee, E. J., Bunney, T. D., Talbot, C. B., Kumar, S., et al. (2008). Multiplexed FRET to image multiple signaling events in live cells. *Biophysical Journal, 95*(10), L69–L71.

Grashoff, C., Hoffman, B. D., Brenner, M. D., Zhou, R., Parsons, M., Yang, M. T., et al. (2010). Measuring mechanical tension across vinculin reveals regulation of focal adhesion dynamics. *Nature, 466*(7303), 263–266.

Gray, A., Van Der Kaay, J., & Downes, C. P. (1999). The pleckstrin homology domains of protein kinase B and GRP1 (general receptor for phosphoinositides-1) are sensitive and selective probes for the cellular detection of phosphatidylinositol 3,4-bisphosphate and/or phosphatidylinositol 3,4,5-trisphosphate *in vivo*. *The Biochemical Journal, 344*(Pt. 3), 929–936.

Griesbeck, O., Baird, G. S., Campbell, R. E., Zacharias, D. A., & Tsien, R. Y. (2001). Reducing the environmental sensitivity of yellow fluorescent protein. Mechanism and applications. *The Journal of Biological Chemistry, 276*(31), 29188–29194.

Grimley, J. S., Li, L., Wang, W., Wen, L., Beese, L. S., Hellinga, H. W., et al. (2013). Visualization of synaptic inhibition with an optogenetic sensor developed by cell-free protein engineering automation. *The Journal of Neuroscience: The Official Journal of the Society for Neuroscience, 33*(41), 16297–16309.

Gross, S. M., & Rotwein, P. (2015). Akt signaling dynamics in individual cells. *Journal of Cell Science, 128*(14), 2509–2519.

Gruber, D. F., Gaffney, J. P., Mehr, S., DeSalle, R., Sparks, J. S., Platisa, J., et al. (2015). Adaptive evolution of eel fluorescent proteins from fatty acid binding proteins produces bright fluorescence in the marine environment. *PloS One, 10*(11), e0140972.

Gu, H., Lalonde, S., Okumoto, S., Looger, L. L., Scharff-Poulsen, A. M., Grossman, A. R., et al. (2006). A novel analytical method for *in vivo* phosphate tracking. *FEBS Letters, 580*(25), 5885–5893.

Gu, Z., Zhao, M., Sheng, Y., Bentolila, L. A., & Tang, Y. (2011). Detection of mercury ion by infrared fluorescent protein and its hydrogel-based paper assay. *Analytical Chemistry, 83*(6), 2324–2329.

Gulyás, G., Tóth, J. T., Tóth, D. J., Kurucz, I., Hunyady, L., Balla, T., et al. (2015). Measurement of inositol 1,4,5-trisphosphate in living cells using an improved set of resonance energy transfer-based biosensors. *PloS One, 10*(5), e0125601.

Hanna, D. A., Harvey, R. M., Martinez-Guzman, O., Yuan, X., Chandrasekharan, B., Raju, G., et al. (2016). Heme dynamics and trafficking factors revealed by genetically encoded fluorescent heme sensors. *Proceedings of the National Academy of Sciences of the United States of America, 113*(27), 7539–7544.

Hanson, G. T., Aggeler, R., Oglesbee, D., Cannon, M., Capaldi, R. A., Tsien, R. Y., et al. (2004). Investigating mitochondrial redox potential with redox-sensitive green fluorescent protein indicators. *The Journal of Biological Chemistry, 279*(13), 13044–13053.

Hanson, G. T., McAnaney, T. B., Park, E. S., Rendell, M. E. P., Yarbrough, D. K., Chu, S., et al. (2002). Green fluorescent protein variants as ratiometric dual emission pH sensors. 1. Structural characterization and preliminary application. *Biochemistry, 41*(52), 15477–15488.

Harvey, C. D., Ehrhardt, A. G., Cellurale, C., Zhong, H., Yasuda, R., Davis, R. J., et al. (2008). A genetically encoded fluorescent sensor of ERK activity. *Proceedings of the National Academy of Sciences of the United States of America, 105*(49), 19264–19269.

Hayashi, S., & Toda, Y. (2009). A novel fluorescent protein purified from eel muscle. *Fisheries Science, 75*(6), 1461–1469.

Heim, R., Cubitt, A. B., & Tsien, R. Y. (1995a). Improved green fluorescence. *Nature, 373*(6516), 663–664.

Heim, N., & Griesbeck, O. (2004). Genetically encoded indicators of cellular calcium dynamics based on troponin C and green fluorescent protein. *The Journal of Biological Chemistry*, *279*(14), 14280–14286.
Herbst, K. J., Ni, Q., & Zhang, J. (2009). Dynamic visualization of signal transduction in living cells: From second messengers to kinases. *IUBMB Life*, *61*(9), 902–908.
Hertel, F., Switalski, A., Mintert-Jancke, E., Karavassilidou, K., Bender, K., Pott, L., et al. (2011). A genetically encoded tool kit for manipulating and monitoring membrane phosphatidylinositol 4,5-bisphosphate in intact cells. *PloS One*, *6*(6), e20855.
Hessels, A. M., Chabosseau, P., Bakker, M. H., Engelen, W., Rutter, G. A., Taylor, K., et al. (2015). eZinCh-2: A versatile, genetically-encoded FRET sensor for cytosolic and intra-organelle Zn^{2+} imaging. *ACS Chemical Biology*, *10*(9), 2126–2134.
Hires, S. A., Zhu, Y., & Tsien, R. Y. (2008). Optical measurement of synaptic glutamate spillover and reuptake by linker optimized glutamate-sensitive fluorescent reporters. *Proceedings of the National Academy of Sciences of the United States of America*, *105*(11), 4411–4416.
Hochbaum, D. R., Zhao, Y., Farhi, S. L., Klapoetke, N., Werley, C. A., Kapoor, V., et al. (2014). All-optical electrophysiology in mammalian neurons using engineered microbial rhodopsins. *Nature Methods*, *11*(8), 825–833.
Hochreiter, B., Pardo-Garcia, A., & Schmid, J. A. (2015). Fluorescent proteins as genetically encoded FRET biosensors in life sciences. *Sensors*, *15*(10), 26281–26314.
Hoff, K. G., Goodlitt, R., Li, R., Smolke, C. D., & Silberg, J. J. (2009). Fluorescence detection of a protein-bound 2Fe2S cluster. *ChemBioChem*, *10*(4), 667–670.
Honda, A., Adams, S. R., Sawyer, C. L., Lev-Ram, V., Tsien, R. Y., & Dostmann, W. R. (2001). Spatiotemporal dynamics of guanosine 3',5'-cyclic monophosphate revealed by a genetically encoded, fluorescent indicator. *Proceedings of the National Academy of Sciences of the United States of America*, *98*(5), 2437–2442.
Honda, A., Sawyer, C. L., Cawley, S. M., & Dostmann, W. R. G. (2005). Cygnets: *In vivo* characterization of novel cGMP indicators and *in vivo* imaging of intracellular cGMP. *Methods in Molecular Biology (Clifton, N.J.)*, *307*, 27–43.
Huang, B. K., Ali, S., Stein, K. T., & Sikes, H. D. (2015). Interpreting heterogeneity in response of cells expressing a fluorescent hydrogen peroxide biosensor. *Biophysical Journal*, *109*(10), 2148–2158.
Ichimura, T., Fujita, H., Yoshizawa, K., & Watanabe, T. M. (2012). Engineering strain-sensitive yellow fluorescent protein. *Chemical Communications*, *48*(63), 7871–7873.
Imamura, H., Nhat, K. P. H., Togawa, H., Saito, K., Iino, R., Kato-Yamada, Y., et al. (2009). Visualization of ATP levels inside single living cells with fluorescence resonance energy transfer-based genetically encoded indicators. *Proceedings of the National Academy of Sciences of the United States of America*, *106*(37), 15651–15656.
Iwai, S., & Uyeda, T. Q. P. (2008). Visualizing myosin–actin interaction with a genetically-encoded fluorescent strain sensor. *Proceedings of the National Academy of Sciences of the United States of America*, *105*(44), 16882–16887.
Iwatani, S., Nakamura, H., Kurokawa, D., Yamana, K., Nishida, K., Fukushima, S., et al. (2016). Fluorescent protein-based detection of unconjugated bilirubin in newborn serum. *Scientific Reports*, *6*, 28489.
Jiang, T., Guo, D., Wang, Q., Wu, X., Li, Z., Zheng, Z., et al. (2015). Developing a genetically encoded green fluorescent protein mutant for sensitive light-up fluorescent sensing and cellular imaging of Hg(II). *Analytica Chimica Acta*, *876*, 77–82.
Jin, L., Han, Z., Platisa, J., Wooltorton, J. R. A., Cohen, L. B., & Pieribone, V. A. (2012). Single action potentials and subthreshold electrical events imaged in neurons with a fluorescent protein voltage probe. *Neuron*, *75*(5), 779–785.

Jung, A., Garcia, J. E., Kim, E., Yoon, B.-J., & Baker, B. J. (2015). Linker length and fusion site composition improve the optical signal of genetically encoded fluorescent voltage sensors. *Neurophotonics*, *2*(2), 21012.

Kikuta, S., Hou, B.-H., Sato, R., Frommer, W. B., & Kikawada, T. (2015). FRET sensor-based quantification of intracellular trehalose in mammalian cells. *Bioscience, Biotechnology, and Biochemistry*, *80*(1), 162–165.

Klarenbeek, J. B., Goedhart, J., Hink, M. A., Gadella, T. W. J., & Jalink, K. (2011). A mTurquoise-based cAMP sensor for both FLIM and ratiometric read-out has improved dynamic range. *PloS One*, *6*(4), e19170.

Klarenbeek, J., Goedhart, J., van Batenburg, A., Groenewald, D., & Jalink, K. (2015). Fourth-generation epac-based FRET sensors for cAMP feature exceptional brightness, photostability and dynamic range: Characterization of dedicated sensors for FLIM, for ratiometry and with high affinity. *PloS One*, *10*(4), e0122513.

Knauer, S. K., Fetz, V., Rabenstein, J., Friedl, S., Hofmann, B., Sabiani, S., et al. (2011). Bioassays to monitor Taspase1 function for the identification of pharmacogenetic inhibitors. *PloS One*, *6*(5), e18253.

Knauer, S. K., Moodt, S., Berg, T., Liebel, U., Pepperkok, R., & Stauber, R. H. (2005). Translocation biosensors to study signal-specific nucleo-cytoplasmic transport, protease activity and protein–protein interactions. *Traffic*, *6*(7), 594–606.

Kneen, M., Farinas, J., Li, Y., & Verkman, A. S. (1998). Green fluorescent protein as a noninvasive intracellular pH indicator. *Biophysical Journal*, *74*(3), 1591–1599.

Knorr, J. M., Jackson, J., Batie, M. R., Narmoneva, D. A., & Jones, D. C. (2016). Application of strain and calibration of Förster Resonance Energy Transfer (FRET) emission for *in vitro* live cell response to cytoskeletal deformation. *Journal of Biomechanics*, *49*(14), 3334–3339.

Kogure, T., Karasawa, S., Araki, T., Saito, K., Kinjo, M., & Miyawaki, A. (2006). A fluorescent variant of a protein from the stony coral *Montipora* facilitates dual-color single-laser fluorescence cross-correlation spectroscopy. *Nature Biotechnology*, *24*(5), 577–581.

Köhnke, M., Schmitt, S., Ariotti, N., Piggott, A. M., Parton, R. G., Lacey, E., et al. (2012). Design and application of in vivo FRET biosensors to identify protein prenylation and nanoclustering inhibitors. *Chemistry & Biology*, *19*(7), 866–874.

Kolossov, V. L., Leslie, M. T., Chatterjee, A., Sheehan, B. M., Kenis, P. J. A., & Gaskins, H. R. (2012). Förster resonance energy transfer-based sensor targeting endoplasmic reticulum reveals highly oxidative environment. *Experimental Biology and Medicine (Maywood, NJ)*, *237*(6), 652–662.

Kolossov, V. L., Spring, B. Q., Clegg, R. M., Henry, J. J., Sokolowski, A., Kenis, P. J. A., et al. (2011). Development of a high-dynamic range, GFP-based FRET probe sensitive to oxidative microenvironments. *Experimental Biology and Medicine*, *236*(6), 681–691.

Komatsu, N., Aoki, K., Yamada, M., Yukinaga, H., Fujita, Y., Kamioka, Y., et al. (2011). Development of an optimized backbone of FRET biosensors for kinases and GTPases. *Molecular Biology of the Cell*, *22*(23), 4647–4656.

Kralj, J. M., Douglass, A. D., Hochbaum, D. R., Maclaurin, D., & Cohen, A. E. (2012). Optical recording of action potentials in mammalian neurons using a microbial rhodopsin. *Nature Methods*, *9*(1), 90–95.

Kredel, S., Oswald, F., Nienhaus, K., Deuschle, K., Röcker, C., Wolff, M., et al. (2009). MRuby, a bright monomeric red fluorescent protein for labeling of subcellular structures. *PloS One*, *4*(2), e4391.

Kumagai, A., Ando, R., Miyatake, H., Greimel, P., Kobayashi, T., Hirabayashi, Y., et al. (2013). A bilirubin-inducible fluorescent protein from eel muscle. *Cell*, *153*(7), 1602–1611.

Kurokawa, K., Mochizuki, N., Ohba, Y., Mizuno, H., Miyawaki, A., & Matsuda, M. (2001). A pair of fluorescent resonance energy transfer-based probes for tyrosine phosphorylation of the CrkII adaptor protein *in vivo*. *The Journal of Biological Chemistry*, *276*(33), 31305–31310.

Lager, I., Looger, L. L., Hilpert, M., Lalonde, S., & Frommer, W. B. (2006). Conversion of a putative *Agrobacterium* sugar-binding protein into a FRET sensor with high selectivity for sucrose. *The Journal of Biological Chemistry*, *281*(41), 30875–30883.

Lam, A. J., St-Pierre, F., Gong, Y., Marshall, J. D., Cranfill, P. J., Baird, M. A., et al. (2012). Improving FRET dynamic range with bright green and red fluorescent proteins. *Nature Methods*, *9*(10), 1005–1012.

Lei, C., Wang, Z., Nie, Z., Deng, H., Hu, H., Huang, Y., et al. (2015). Resurfaced fluorescent protein as a sensing platform for label-free detection of copper(II) ion and acetylcholinesterase activity. *Analytical Chemistry*, *87*(3), 1974–1980.

Li, X., Cheng, Z., & Jin, H. (2012). Dynamics of Ras complexes observed in living cells. *Sensors (Basel, Switzerland)*, *12*(7), 9411–9422.

Liang, J., Qin, M., Xu, R., Gao, X., Shen, Y., Xu, Q., et al. (2012). A genetically encoded copper(I) sensor based on engineered structural distortion of EGFP. *Chemical Communications (Cambridge, England)*, *48*(32), 3890–3892.

Lindenburg, L. H., Hessels, A. M., Ebberink, E. H. T. M., Arts, R., & Merkx, M. (2013). Robust red FRET sensors using self-associating fluorescent domains. *ACS Chemical Biology*, *8*(10), 2133–2139.

Lindenburg, L. H., Vinkenborg, J. L., Oortwijn, J., Aper, S. J. A., & Merkx, M. (2013). MagFRET: The first genetically encoded fluorescent Mg^{2+} sensor. *PloS One*, *8*(12), e82009.

Litzlbauer, J., Schifferer, M., Ng, D., Fabritius, A., Thestrup, T., & Griesbeck, O. (2015). Large scale bacterial colony screening of diversified FRET biosensors. *PloS One*, *10*(6), e0119860.

Llopis, J., McCaffery, J. M., Miyawaki, A., Farquhar, M. G., & Tsien, R. Y. (1998). Measurement of cytosolic, mitochondrial, and Golgi pH in single living cells with green fluorescent proteins. *Proceedings of the National Academy of Sciences of the United States of America*, *95*(12), 6803–6808.

Lodygin, D., Odoardi, F., Schläger, C., Körner, H., Kitz, A., Nosov, M., et al. (2013). A combination of fluorescent NFAT and H2B sensors uncovers dynamics of T cell activation in real time during CNS autoimmunity. *Nature Medicine*, *19*(6), 784–790.

Lukyanov, K. A., & Belousov, V. V. (2014). Genetically encoded fluorescent redox sensors. *Biochimica et Biophysica Acta*, *1840*(2), 745–756.

Luo, K. Q., Yu, V. C., Pu, Y., & Chang, D. C. (2001). Application of the fluorescence resonance energy transfer method for studying the dynamics of caspase-3 activation during UV-induced apoptosis in living HeLa cells. *Biochemical and Biophysical Research Communications*, *283*(5), 1054–1060.

Luo, K. Q., Yu, V. C., Pu, Y., & Chang, D. C. (2003). Measuring dynamics of caspase-8 activation in a single living HeLa cell during TNFalpha-induced apoptosis. *Biochemical and Biophysical Research Communications*, *304*(2), 217–222.

Ma, H., Gibson, E. A., Dittmer, P. J., Jimenez, R., & Palmer, A. E. (2012). High-throughput examination of fluorescence resonance energy transfer-detected metal-ion response in mammalian cells. *Journal of the American Chemical Society*, *134*(5), 2488–2491.

Maclaurin, D., Venkatachalam, V., Lee, H., & Cohen, A. E. (2013). Mechanism of voltage-sensitive fluorescence in a microbial rhodopsin. *Proceedings of the National Academy of Sciences of the United States of America*, *110*(15), 5939–5944.

Mahon, M. J. (2011). pHluorin2: An enhanced, ratiometric, pH-sensitive green florescent protein. *Advances in Bioscience and Biotechnology (Print)*, *2*(3), 132–137.

Mank, M., & Griesbeck, O. (2008). Genetically encoded calcium indicators. *Chemical Reviews*, *108*(5), 1550–1564.

Mank, M., Reiff, D. F., Heim, N., Friedrich, M. W., Borst, A., & Griesbeck, O. (2006). A FRET-based calcium biosensor with fast signal kinetics and high fluorescence change. *Biophysical Journal*, *90*(5), 1790–1796.

Markvicheva, K. N., Bilan, D. S., Mishina, N. M., Gorokhovatsky, A. Y., Vinokurov, L. M., Lukyanov, S., et al. (2011). A genetically encoded sensor for H_2O_2 with expanded dynamic range. *Bioorganic & Medicinal Chemistry*, *19*(3), 1079–1084.

Markwardt, M. L., Kremers, G.-J., Kraft, C. A., Ray, K., Cranfill, P. J. C., Wilson, K. A., et al. (2011). An improved cerulean fluorescent protein with enhanced brightness and reduced reversible photoswitching. *PloS One*, *6*(3), e17896.

Matsu-ura, T., Michikawa, T., Inoue, T., Miyawaki, A., Yoshida, M., & Mikoshiba, K. (2006). Cytosolic inositol 1,4,5-trisphosphate dynamics during intracellular calcium oscillations in living cells. *The Journal of Cell Biology*, *173*(5), 755–765.

McIsaac, R. S., Bedbrook, C. N., & Arnold, F. H. (2015). Recent advances in engineering microbial rhodopsins for optogenetics. *Current Opinion in Structural Biology*, *33*, 8–15.

McLachlan, M. J., Katzenellenbogen, J. A., & Zhao, H. (2011). A new fluorescence complementation biosensor for detection of estrogenic compounds. *Biotechnology and Bioengineering*, *108*(12), 2794–2803.

Meng, F., & Sachs, F. (2012). Orientation-based FRET sensor for real-time imaging of cellular forces. *Journal of Cell Science*, *125*(Pt. 3), 743–750.

Mérola, F., Fredj, A., Betolngar, D.-B., Ziegler, C., Erard, M., & Pasquier, H. (2014). Newly engineered cyan fluorescent proteins with enhanced performances for live cell FRET imaging. *Biotechnology Journal*, *9*(2), 180–191.

Miesenböck, G., De Angelis, D. A., & Rothman, J. E. (1998). Visualizing secretion and synaptic transmission with pH-sensitive green fluorescent proteins. *Nature*, *394*(6689), 192–195.

Miranda, J. G., Weaver, A. L., Qin, Y., Park, J. G., Stoddard, C. I., Lin, M. Z., et al. (2012). New alternately colored FRET sensors for simultaneous monitoring of Zn^{2+} in multiple cellular locations. *PloS One*, 7(11), e49371.

Mishina, N. M., Bogeski, I., Bolotin, D. A., Hoth, M., Niemeyer, B. A., Schultz, C., et al. (2012). Can we see PIP(3) and hydrogen peroxide with a single probe? *Antioxidants & Redox Signaling*, *17*(3), 505–512.

Mishina, Y., Mutoh, H., Song, C., & Knöpfel, T. (2014). Exploration of genetically encoded voltage indicators based on a chimeric voltage sensing domain. *Frontiers in Molecular Neuroscience*, 7, 78.

Miyawaki, A., Llopis, J., Heim, R., McCaffery, J. M., Adams, J. A., Ikura, M., et al. (1997). Fluorescent indicators for Ca^{2+} based on green fluorescent proteins and calmodulin. *Nature*, *388*(6645), 882–887.

Mohsin, M., Abdin, M. Z., Nischal, L., Kardam, H., & Ahmad, A. (2013). Genetically encoded FRET-based nanosensor for *in vivo* measurement of leucine. *Biosensors & Bioelectronics*, *50*, 72–77.

Mohsin, M., & Ahmad, A. (2014). Genetically-encoded nanosensor for quantitative monitoring of methionine in bacterial and yeast cells. *Biosensors & Bioelectronics*, *59*, 358–364.

Morikawa, T. J., Fujita, H., Kitamura, A., Horio, T., Yamamoto, J., Kinjo, M., et al. (2016). Dependence of fluorescent protein brightness on protein concentration in solution and enhancement of it. *Scientific Reports*, *6*, 22342.

Mukherjee, A., Walker, J., Weyant, K. B., & Schroeder, C. M. (2013). Characterization of flavin-based fluorescent proteins: An emerging class of fluorescent reporters. *PloS One*, *8*(5), e64753.

Mukherjee, A., Weyant, K. B., Agrawal, U., Walker, J., Cann, I. K. O., & Schroeder, C. M. (2015). Engineering and characterization of new LOV-based fluorescent proteins from *Chlamydomonas reinhardtii* and *Vaucheria frigida*. *ACS Synthetic Biology*, *4*(4), 371–377.
Nadler, D. C., Morgan, S.-A., Flamholz, A., Kortright, K. E., & Savage, D. F. (2016). Rapid construction of metabolite biosensors using domain-insertion profiling. *Nature Communications*, 7, 12266.
Nagai, T., Ibata, K., Park, E. S., Kubota, M., Mikoshiba, K., & Miyawaki, A. (2002). A variant of yellow fluorescent protein with fast and efficient maturation for cell-biological applications. *Nature Biotechnology*, *20*(1), 87–90.
Nagai, T., Sawano, A., Park, E. S., & Miyawaki, A. (2001). Circularly permuted green fluorescent proteins engineered to sense Ca^{2+}. *Proceedings of the National Academy of Sciences of the United States of America*, *98*(6), 3197–3202.
Nagai, T., Yamada, S., Tominaga, T., Ichikawa, M., & Miyawaki, A. (2004). Expanded dynamic range of fluorescent indicators for $Ca^{(2+)}$ by circularly permuted yellow fluorescent proteins. *Proceedings of the National Academy of Sciences of the United States of America*, *101*(29), 10554–10559.
Najumudeen, A. K., Köhnke, M., Solman, M., Alexandrov, K., & Abankwa, D. (2013). Cellular FRET-biosensors to detect membrane targeting inhibitors of N-myristoylated proteins. *PloS One*, *8*(6), e66425.
Nakai, J., Ohkura, M., & Imoto, K. (2001). A high signal-to-noise Ca^{2+} probe composed of a single green fluorescent protein. *Nature Biotechnology*, *19*(2), 137–141.
Nakanishi, Y., Iida, S., Ueoka-Nakanishi, H., Niimi, T., Tomioka, R., & Maeshima, M. (2013). Exploring dynamics of molybdate in living animal cells by a genetically encoded FRET nanosensor. *PloS One*, *8*(3), e58175.
Nezu, A., Tanimura, A., Morita, T., Shitara, A., & Tojyo, Y. (2006). A novel fluorescent method employing the FRET-based biosensor "LIBRA" for the identification of ligands of the inositol 1,4,5-trisphosphate receptors. *Biochimica et Biophysica Acta (BBA)—General Subjects*, *1760*(8), 1274–1280.
Ng, H.-L., & Lin, M. Z. (2016). Structure-guided wavelength tuning in far-red fluorescent proteins. *Current Opinion in Structural Biology*, *39*, 124–133.
Nguyen, A. W., & Daugherty, P. S. (2005). Evolutionary optimization of fluorescent proteins for intracellular FRET. *Nature Biotechnology*, *23*(3), 355–360.
Nhu Ngoc Van, T., & Morris, M. C. (2013). Fluorescent sensors of protein kinases: From basics to biomedical applications. *Progress in Molecular Biology and Translational Science*, *113*, 217–274.
Niino, Y., Hotta, K., & Oka, K. (2009). Simultaneous live cell imaging using dual FRET sensors with a single excitation light. *PloS One*, *4*(6), e6036.
Nikolaev, V. O., Bünemann, M., Hein, L., Hannawacker, A., & Lohse, M. J. (2004). Novel single chain cAMP sensors for receptor-induced signal propagation. *The Journal of Biological Chemistry*, *279*(36), 37215–37218.
Nobles, C. L., Clark, J. R., Green, S. I., & Maresso, A. W. (2015). A dual component heme biosensor that integrates heme transport and synthesis in bacteria. *Journal of Microbiological Methods*, *118*, 7–17.
Ohta, Y., Kamagata, T., Mukai, A., Takada, S., Nagai, T., & Horikawa, K. (2016). Nontrivial effect of the color-exchange of a donor/acceptor pair in the engineering of Förster resonance energy transfer (FRET)-based indicators. *ACS Chemical Biology*, *11*(7), 1816–1822.
Okumoto, S., Looger, L. L., Micheva, K. D., Reimer, R. J., Smith, S. J., & Frommer, W. B. (2005). Detection of glutamate release from neurons by genetically encoded surface-displayed FRET nanosensors. *Proceedings of the National Academy of Sciences of the United States of America*, *102*(24), 8740–8745.

Oldach, L., & Zhang, J. (2014). Genetically encoded fluorescent biosensors for live-cell visualization of protein phosphorylation. *Chemistry & Biology*, *21*(2), 186–197.

Oliveira, A. F., & Yasuda, R. (2013). An improved Ras sensor for highly sensitive and quantitative FRET-FLIM imaging. *PloS One*, *8*(1), e52874.

Ostergaard, H., Henriksen, A., Hansen, F. G., & Winther, J. R. (2001). Shedding light on disulfide bond formation: Engineering a redox switch in green fluorescent protein. *The EMBO Journal*, *20*(21), 5853–5862.

Ouyang, M., Huang, H., Shaner, N. C., Remacle, A. G., Shiryaev, S. A., Strongin, A. Y., et al. (2010). Simultaneous visualization of protumorigenic Src and MT1-MMP activities with fluorescence resonance energy transfer. *Cancer Research*, *70*(6), 2204–2212.

Palmer, A. E., Qin, Y., Park, J. G., & McCombs, J. E. (2011). Design and application of genetically encoded biosensors. *Trends in Biotechnology*, *29*(3), 144–152.

Peroza, E. A., Boumezbeur, A.-H., & Zamboni, N. (2015). Rapid, randomized development of genetically encoded FRET sensors for small molecules. *The Analyst*, *140*(13), 4540–4548.

Piao, H. H., Rajakumar, D., Kang, B. E., Kim, E. H., & Baker, B. J. (2015). Combinatorial mutagenesis of the voltage-sensing domain enables the optical resolution of action potentials firing at 60 Hz by a genetically encoded fluorescent sensor of membrane potential. *The Journal of Neuroscience: The Official Journal of the Society for Neuroscience*, *35*(1), 372–385.

Poburko, D., Santo-Domingo, J., & Demaurex, N. (2011). Dynamic regulation of the mitochondrial proton gradient during cytosolic calcium elevations. *The Journal of Biological Chemistry*, *286*(13), 11672–11684.

Poëa-Guyon, S., Pasquier, H., Mérola, F., Morel, N., & Erard, M. (2013). The enhanced cyan fluorescent protein: A sensitive pH sensor for fluorescence lifetime imaging. *Analytical and Bioanalytical Chemistry*, *405*(12), 3983–3987.

Ponsioen, B., Zhao, J., Riedl, J., Zwartkruis, F., van der Krogt, G., Zaccolo, M., et al. (2004). Detecting cAMP-induced Epac activation by fluorescence resonance energy transfer: Epac as a novel cAMP indicator. *EMBO Reports*, *5*(12), 1176–1180.

Potzkei, J., Kunze, M., Drepper, T., Gensch, T., Jaeger, K.-E., & Büchs, J. (2012). Real-time determination of intracellular oxygen in bacteria using a genetically encoded FRET-based biosensor. *BMC Biology*, *10*, 28.

Qin, Y., Dittmer, P. J., Park, J. G., Jansen, K. B., & Palmer, A. E. (2011). Measuring steady-state and dynamic endoplasmic reticulum and Golgi Zn^{2+} with genetically encoded sensors. *Proceedings of the National Academy of Sciences of the United States of America*, *108*(18), 7351–7356.

Qin, Y., Sammond, D. W., Braselmann, E., Carpenter, M. C., & Palmer, A. E. (2016). Development of an optical Zn^{2+} probe based on a single fluorescent protein. *ACS Chemical Biology*, *11*(10), 2744–2751.

Quiñones, A., & Rainov, N. G. (2001). Identification of genotoxic stress in human cells by fluorescent monitoring of p53 expression. *Mutation Research*, *494*(1–2), 73–85.

Ravikumar, Y., Nadarajan, S. P., Lee, C.-S., Jung, S., Bae, D.-H., & Yun, H. (2016). FMN-based fluorescent proteins as heavy metal sensors against mercury ions. *Journal of Microbiology and Biotechnology*, *26*(3), 530–539.

Ravikumar, Y., Nadarajan, S. P., Lee, C.-S., Rhee, J.-K., & Yun, H.-D. (2015). A new-generation fluorescent-based metal sensor—iLOV protein. *Journal of Microbiology and Biotechnology*, *25*(4), 503–510.

Regot, S., Hughey, J. J., Bajar, B. T., Carrasco, S., & Covert, M. W. (2014). High-sensitivity measurements of multiple kinase activities in live single cells. *Cell*, *157*(7), 1724–1734.

Remus, T. P., Zima, A. V., Bossuyt, J., Bare, D. J., Martin, J. L., Blatter, L. A., et al. (2006). Biosensors to measure inositol 1,4,5-trisphosphate concentration in living cells with spatiotemporal resolution. *The Journal of Biological Chemistry*, *281*(1), 608–616.

Rizzo, M. A., Springer, G. H., Granada, B., & Piston, D. W. (2004). An improved cyan fluorescent protein variant useful for FRET. *Nature Biotechnology*, *22*(4), 445–449.

Rodriguez, E. A., Tran, G. N., Gross, L. A., Crisp, J. L., Shu, X., Lin, J. Y., et al. (2016). A far-red fluorescent protein evolved from a cyanobacterial phycobiliprotein. *Nature Methods*, *13*(9), 763–769.

Rose, T., Goltstein, P. M., Portugues, R., & Griesbeck, O. (2014). Putting a finishing touch on GECIs. *Frontiers in Molecular Neuroscience*, 7, 88.

Sakaue-Sawano, A., Kurokawa, H., Morimura, T., Hanyu, A., Hama, H., Osawa, H., et al. (2008). Visualizing spatiotemporal dynamics of multicellular cell-cycle progression. *Cell*, *132*(3), 487–498.

San Martín, A., Ceballo, S., Baeza-Lehnert, F., Lerchundi, R., Valdebenito, R., Contreras-Baeza, Y., et al. (2014). Imaging mitochondrial flux in single cells with a FRET sensor for pyruvate. *PloS One*, *9*(1), e85780.

Sasaki, K., Sato, M., & Umezawa, Y. (2003). Fluorescent indicators for Akt/protein kinase B and dynamics of Akt activity visualized in living cells. *The Journal of Biological Chemistry*, *278*(33), 30945–30951.

Sato, M., Ueda, Y., Shibuya, M., & Umezawa, Y. (2005). Locating inositol 1,4,5-trisphosphate in the nucleus and neuronal dendrites with genetically encoded fluorescent indicators. *Analytical Chemistry*, 77(15), 4751–4758.

Shaner, N. C., Campbell, R. E., Steinbach, P. A., Giepmans, B. N. G., Palmer, A. E., & Tsien, R. Y. (2004). Improved monomeric red, orange and yellow fluorescent proteins derived from *Discosoma* sp. red fluorescent protein. *Nature Biotechnology*, *22*(12), 1567–1572.

Shaner, N. C., Lambert, G. G., Chammas, A., Ni, Y., Cranfill, P. J., Baird, M. A., et al. (2013). A bright monomeric green fluorescent protein derived from *Branchiostoma lanceolatum*. *Nature Methods*, *10*(5), 407–409.

Shaner, N. C., Lin, M. Z., McKeown, M. R., Steinbach, P. A., Hazelwood, K. L., Davidson, M. W., et al. (2008). Improving the photostability of bright monomeric orange and red fluorescent proteins. *Nature Methods*, *5*(6), 545–551.

Shaner, N. C., Steinbach, P. A., & Tsien, R. Y. (2005). A guide to choosing fluorescent proteins. *Nature Methods*, *2*(12), 905–909.

Shcherbakova, D. M., Shemetov, A. A., Kaberniuk, A. A., & Verkhusha, V. V. (2015). Natural photoreceptors as a source of fluorescent proteins, biosensors, and optogenetic tools. *Annual Review of Biochemistry*, *84*, 519–550.

Shcherbakova, D. M., & Verkhusha, V. V. (2013). Near-infrared fluorescent proteins for multicolor in vivo imaging. *Nature Methods*, *10*(8), 751–754.

Shcherbo, D., Murphy, C. S., Ermakova, G. V., Solovieva, E. A., Chepurnykh, T. V., Shcheglov, A. S., et al. (2009). Far-red fluorescent tags for protein imaging in living tissues. *The Biochemical Journal*, *418*(3), 567–574.

Shen, Y., Rosendale, M., Campbell, R. E., & Perrais, D. (2014). pHuji, a pH-sensitive red fluorescent protein for imaging of exo- and endocytosis. *The Journal of Cell Biology*, *207*(3), 419–432.

Shrestha, D., Jenei, A., Nagy, P., Vereb, G., & Szöllősi, J. (2015). Understanding FRET as a research tool for cellular studies. *International Journal of Molecular Sciences*, *16*(4), 6718–6756.

Shu, X., Lev-Ram, V., Deerinck, T. J., Qi, Y., Ramko, E. B., Davidson, M. W., et al. (2011). A genetically encoded tag for correlated light and electron microscopy of intact cells, tissues, and organisms. *PLoS Biology*, *9*(4), e1001041.

Shu, X., Royant, A., Lin, M. Z., Aguilera, T. A., Lev-Ram, V., Steinbach, P. A., et al. (2009). Mammalian expression of infrared fluorescent proteins engineered from a bacterial phytochrome. *Science (New York, N.Y.)*, *324*(5928), 804–807.

Song, X., Wang, Y., Shu, Z., Hong, J., Li, T., & Yao, L. (2013). Engineering a more thermostable blue light photo receptor *Bacillus subtilis* YtvA LOV domain by a computer aided rational design method. *PLoS Computational Biology*, *9*(7), e1003129.

Song, Y., Yang, M., Wegner, S. V., Zhao, J., Zhu, R., Wu, Y., et al. (2015). A genetically encoded FRET sensor for intracellular heme. *ACS Chemical Biology*, *10*(7), 1610–1615.

Spencer, S. L., Cappell, S. D., Tsai, F.-C., Overton, K. W., Wang, C. L., & Meyer, T. (2013). The proliferation-quiescence decision is controlled by a bifurcation in CDK2 activity at mitotic exit. *Cell*, *155*(2), 369–383.

Stepanenko, O. V., Stepanenko, O. V., Shcherbakova, D. M., Kuznetsova, I. M., Turoverov, K. K., & Verkhusha, V. V. (2011). Modern fluorescent proteins: From chromophore formation to novel intracellular applications. *BioTechniques*, *51*(5), 313–327.

Storace, D., Sepehri Rad, M., Kang, B., Cohen, L. B., Hughes, T., & Baker, B. J. (2016). Toward better genetically encoded sensors of membrane potential. *Trends in Neurosciences*, *39*(5), 277–289.

St-Pierre, F., Chavarha, M., & Lin, M. Z. (2015). Designs and sensing mechanisms of genetically encoded fluorescent voltage indicators. *Current Opinion in Chemical Biology*, *27*, 31–38.

St-Pierre, F., Marshall, J. D., Yang, Y., Gong, Y., Schnitzer, M. J., & Lin, M. Z. (2014). High-fidelity optical reporting of neuronal electrical activity with an ultrafast fluorescent voltage sensor. *Nature Neuroscience*, *17*(6), 884–889.

Su, T., Pan, S., Luo, Q., & Zhang, Z. (2013). Monitoring of dual bio-molecular events using FRET biosensors based on mTagBFP/sfGFP and mVenus/mKOκ fluorescent protein pairs. *Biosensors and Bioelectronics*, *46*, 97–101.

Subach, O. M., Cranfill, P. J., Davidson, M. W., & Verkhusha, V. V. (2011). An enhanced monomeric blue fluorescent protein with the high chemical stability of the chromophore. *PloS One*, *6*(12), e28674.

Subach, O. M., Gundorov, I. S., Yoshimura, M., Subach, F. V., Zhang, J., Grüenwald, D., et al. (2008). Conversion of red fluorescent protein into a bright blue probe. *Chemistry & Biology*, *15*(10), 1116–1124.

Subach, F. V., & Verkhusha, V. V. (2012). Chromophore transformations in red fluorescent proteins. *Chemical Reviews*, *112*(7), 4308–4327.

Sugiura, K., Nagai, T., Nakano, M., Ichinose, H., Nakabayashi, T., Ohta, N., et al. (2015). Redox sensor proteins for highly sensitive direct imaging of intracellular redox state. *Biochemical and Biophysical Research Communications*, *457*(3), 242–248.

Sun, Y., Booker, C. F., Kumari, S., Day, R. N., Davidson, M., & Periasamy, A. (2009). Characterization of an orange acceptor fluorescent protein for sensitized spectral fluorescence resonance energy transfer microscopy using a white-light laser. *Journal of Biomedical Optics*, *14*(5), 54009.

Sung, U., Sepehri-Rad, M., Piao, H. H., Jin, L., Hughes, T., Cohen, L. B., et al. (2015). Developing fast fluorescent protein voltage sensors by optimizing FRET interactions. *PloS One*, *10*(11), e0141585.

Suzuki, J., Kanemaru, K., & Iino, M. (2016). Genetically encoded fluorescent indicators for organellar calcium imaging. *Biophysical Journal*, *111*(6), 1119–1131.

Takanaga, H., Chaudhuri, B., & Frommer, W. B. (2008). GLUT1 and GLUT9 as major contributors to glucose influx in HepG2 cells identified by a high sensitivity intramolecular FRET glucose sensor. *Biochimica et Biophysica Acta (BBA)—Biomembranes*, *1778*(4), 1091–1099.

Tang, S., Reddish, F., Zhuo, Y., & Yang, J. J. (2015). Fast kinetics of calcium signaling and sensor design. *Current Opinion in Chemical Biology*, *27*, 90–97.

Tang, S., Wong, H.-C., Wang, Z.-M., Huang, Y., Zou, J., Zhuo, Y., et al. (2011). Design and application of a class of sensors to monitor Ca^{2+} dynamics in high Ca^{2+}

concentration cellular compartments. *Proceedings of the National Academy of Sciences of the United States of America*, *108*(39), 16265–16270.
Tantama, M., Hung, Y. P., & Yellen, G. (2011). Imaging intracellular pH in live cells with a genetically encoded red fluorescent protein sensor. *Journal of the American Chemical Society*, *133*(26), 10034–10037.
Tchekanda, E., Sivanesan, D., & Michnick, S. W. (2014). An infrared reporter to detect spatiotemporal dynamics of protein-protein interactions. *Nature Methods*, *11*(6), 641–644.
Teng, L., Wang, K., Xu, J., & Xu, C. (2015). Flavin mononucleotide (FMN)-based fluorescent protein (FbFP) as reporter for promoter screening in *Clostridium cellulolyticum*. *Journal of Microbiological Methods*, *119*, 37–43.
To, T.-L., Piggott, B. J., Makhijani, K., Yu, D., Jan, Y. N., & Shu, X. (2015). Rationally designed fluorogenic protease reporter visualizes spatiotemporal dynamics of apoptosis *in vivo*. *Proceedings of the National Academy of Sciences of the United States of America*, *112*(11), 3338–3343.
To, T.-L., Schepis, A., Ruiz-González, R., Zhang, Q., Yu, D., Dong, Z., et al. (2016). Rational design of a GFP-based fluorogenic caspase reporter for imaging apoptosis *in vivo*. *Cell Chemical Biology*, *23*(7), 875–882.
To, T.-L., Zhang, Q., & Shu, X. (2016). Structure-guided design of a reversible fluorogenic reporter of protein-protein interactions. *Protein Science: A Publication of the Protein Society*, *25*(3), 748–753.
Tsien, R. Y. (1998). The green fluorescent protein. *Annual Review of Biochemistry*, *67*, 509–544.
Tsou, P., Zheng, B., Hsu, C.-H., Sasaki, A. T., & Cantley, L. C. (2011). A fluorescent reporter of AMPK activity and cellular energy stress. *Cell Metabolism*, *13*(4), 476–486.
Tsuyama, T., Kishikawa, J., Han, Y.-W., Harada, Y., Tsubouchi, A., Noji, H., et al. (2013). *In vivo* fluorescent adenosine 5′-triphosphate (ATP) imaging of *Drosophila melanogaster* and *Caenorhabditis elegans* by using a genetically encoded fluorescent ATP biosensor optimized for low temperatures. *Analytical Chemistry*, *85*(16), 7889–7896.
van der Krogt, G. N. M., Ogink, J., Ponsioen, B., & Jalink, K. (2008). A comparison of donor-acceptor pairs for genetically encoded FRET sensors: Application to the Epac cAMP sensor as an example. *PloS One*, *3*(4), e1916.
van Dongen, E. M. W. M., Evers, T. H., Dekkers, L. M., Meijer, E. W., Klomp, L. W. J., & Merkx, M. (2007). Variation of linker length in ratiometric fluorescent sensor proteins allows rational tuning of Zn(II) affinity in the picomolar to femtomolar range. *Journal of the American Chemical Society*, *129*(12), 3494–3495.
van Unen, J., Stumpf, A. D., Schmid, B., Reinhard, N. R., Hordijk, P. L., Hoffmann, C., et al. (2016). A new generation of FRET sensors for robust measurement of $G\alpha_{i1}$, $G\alpha_{i2}$ and $G\alpha_{i3}$ activation kinetics in single cells. *PloS One*, *11*(1), e0146789.
Vijftigschild, L. A. W., van der Ent, C. K., & Beekman, J. M. (2013). A novel fluorescent sensor for measurement of CFTR function by flow cytometry. *Cytometry Part A: The Journal of the International Society for Analytical Cytology*, *83*(6), 576–584.
Vinkenborg, J. L., Nicolson, T. J., Bellomo, E. A., Koay, M. S., Rutter, G. A., & Merkx, M. (2009). Genetically encoded FRET sensors to monitor intracellular Zn^{2+} homeostasis. *Nature Methods*, *6*(10), 737–740.
Violin, J. D., Zhang, J., Tsien, R. Y., & Newton, A. C. (2003). A genetically encoded fluorescent reporter reveals oscillatory phosphorylation by protein kinase C. *The Journal of Cell Biology*, *161*(5), 899–909.
Wachter, R. M., Elsliger, M.-A., Kallio, K., Hanson, G. T., & Remington, S. J. (1998). Structural basis of spectral shifts in the yellow-emission variants of green fluorescent protein. *Structure*, *6*(10), 1267–1277.

Wachter, R. M., Yarbrough, D., Kallio, K., & Remington, S. J. (2000). Crystallographic and energetic analysis of binding of selected anions to the yellow variants of green fluorescent protein. *Journal of Molecular Biology*, *301*(1), 157–171.

Waldeck-Weiermair, M., Bischof, H., Blass, S., Deak, A. T., Klec, C., Graier, T., et al. (2015). Generation of red-shifted cameleons for imaging Ca^{2+} dynamics of the endoplasmic reticulum. *Sensors (Basel, Switzerland)*, *15*(6), 13052–13068.

Warren, S. C., Margineanu, A., Katan, M., Dunsby, C., & French, P. M. W. (2015). Homo-FRET based biosensors and their application to multiplexed imaging of signalling events in live cells. *International Journal of Molecular Sciences*, *16*(7), 14695–14716.

Wegner, S. V., Arslan, H., Sunbul, M., Yin, J., & He, C. (2010). Dynamic copper(I) imaging in mammalian cells with a genetically encoded fluorescent copper(I) sensor. *Journal of the American Chemical Society*, *132*(8), 2567–2569.

Wegner, S. V., Sun, F., Hernandez, N., & He, C. (2011). The tightly regulated copper window in yeast. *Chemical Communications*, *47*(9), 2571–2573.

Weller, J., Kizina, K. M., Can, K., Bao, G., & Müller, M. (2014). Response properties of the genetically encoded optical H_2O_2 sensor HyPer. *Free Radical Biology & Medicine*, *76*, 227–241.

Whitfield, J. H., Zhang, W. H., Herde, M. K., Clifton, B. E., Radziejewski, J., Janovjak, H., et al. (2015). Construction of a robust and sensitive arginine biosensor through ancestral protein reconstruction. *Protein Science: A Publication of the Protein Society*, *24*(9), 1412–1422.

Wingen, M., Potzkei, J., Endres, S., Casini, G., Rupprecht, C., Fahlke, C., et al. (2014). The photophysics of LOV-based fluorescent proteins—New tools for cell biology. *Photochemical & Photobiological Sciences*, *13*(6), 875–883.

Wu, J., Prole, D. L., Shen, Y., Lin, Z., Gnanasekaran, A., Liu, Y., et al. (2014). Red fluorescent genetically encoded Ca^{2+} indicators for use in mitochondria and endoplasmic reticulum. *The Biochemical Journal*, *464*(1), 13–22.

Yang, H. H., St-Pierre, F., Sun, X., Ding, X., Lin, M. Z., & Clandinin, T. R. (2016). Subcellular imaging of voltage and calcium signals reveals neural processing in vivo. *Cell*, *166*(1), 245–257.

Yang, C., Zhang, X., Guo, Y., Meng, F., Sachs, F., & Guo, J. (2015). Mechanical dynamics in live cells and fluorescence-based force/tension sensors. *Biochimica et Biophysica Acta*, *1853*(8), 1889–1904.

Yano, T., Oku, M., Akeyama, N., Itoyama, A., Yurimoto, H., Kuge, S., et al. (2010). A novel fluorescent sensor protein for visualization of redox states in the cytoplasm and in peroxisomes. *Molecular and Cellular Biology*, *30*(15), 3758–3766.

Yoshizaki, H., Mochizuki, N., Gotoh, Y., & Matsuda, M. (2007). Akt–PDK1 complex mediates epidermal growth factor-induced membrane protrusion through Ral activation. *Molecular Biology of the Cell*, *18*(1), 119–128.

Yu, D., Baird, M. A., Allen, J. R., Howe, E. S., Klassen, M. P., Reade, A., et al. (2015). A naturally monomeric infrared fluorescent protein for protein labeling *in vivo*. *Nature Methods*, *12*(8), 763–765.

Yu, D., Gustafson, W. C., Han, C., Lafaye, C., Noirclerc-Savoye, M., Ge, W.-P., et al. (2014). An improved monomeric infrared fluorescent protein for neuronal and tumour brain imaging. *Nature Communications*, *5*, 3626.

Yu, X., Strub, M.-P., Barnard, T. J., Noinaj, N., Piszczek, G., Buchanan, S. K., et al. (2014). An engineered palette of metal ion quenchable fluorescent proteins. *PloS One*, *9*(4), e95808.

Zacharias, D. A., Violin, J. D., Newton, A. C., & Tsien, R. Y. (2002). Partitioning of lipid-modified monomeric GFPs into membrane microdomains of live cells. *Science*, *296*(5569), 913–916.

Zhang, J., Ma, Y., Taylor, S. S., & Tsien, R. Y. (2001). Genetically encoded reporters of protein kinase A activity reveal impact of substrate tethering. *Proceedings of the National Academy of Sciences*, *98*(26), 14997–15002.

Zhang, C., Wei, Z.-H., & Ye, B.-C. (2013). Quantitative monitoring of 2-oxoglutarate in *Escherichia coli* cells by a fluorescence resonance energy transfer-based biosensor. *Applied Microbiology and Biotechnology*, *97*(18), 8307–8316.

Zhao, Y., Hu, Q., Cheng, F., Su, N., Wang, A., Zou, Y., et al. (2015). SoNar, a highly responsive NAD^+/NADH sensor, allows high-throughput metabolic screening of anti-tumor agents. *Cell Metabolism*, *21*(5), 777–789.

Zhao, F.-L., Zhang, C., Zhang, C., Tang, Y., & Ye, B.-C. (2016). A genetically encoded biosensor for *in vitro* and *in vivo* detection of $NADP^+$. *Biosensors & Bioelectronics*, *77*, 901–906.

Zhong, S., Navaratnam, D., & Santos-Sacchi, J. (2014). A genetically-encoded YFP sensor with enhanced chloride sensitivity, photostability and reduced pH interference demonstrates augmented transmembrane chloride movement by gerbil prestin (SLC26a5). *PloS One*, *9*(6), e99095.

Zlobovskaya, O. A., Sergeeva, T. F., Shirmanova, M. V., Dudenkova, V. V., Sharonov, G. V., Zagaynova, E. V., et al. (2016). Genetically encoded far-red fluorescent sensors for caspase-3 activity. *BioTechniques*, *60*(2), 62–68.

Zou, P., Zhao, Y., Douglass, A. D., Hochbaum, D. R., Brinks, D., Werley, C. A., et al. (2014). Bright and fast multicoloured voltage reporters via electrochromic FRET. *Nature Communications*, *5*, 4625.

CHAPTER TWO

Engineering Rugged Field Assays to Detect Hazardous Chemicals Using Spore-Based Bacterial Biosensors

Daniel Wynn, Sapna Deo, Sylvia Daunert[1]
Miller School of Medicine, University of Miami, Miami, FL, United States
[1]Corresponding author: e-mail address: sdaunert@med.miami.edu

Contents

Methods in Enzymology, Volume 589
ISSN 0076-6879
http://dx.doi.org/10.1016/bs.mie.2017.02.005

Abstract

Bacterial whole cell-based biosensors have been genetically engineered to achieve selective and reliable detection of a wide range of hazardous chemicals. Although whole-cell biosensors demonstrate many advantages for field-based detection of target analytes, there are still some challenges that need to be addressed. Most notably, their often modest shelf life and need for special handling and storage make them challenging to use in situations where access to reagents, instrumentation, and expertise are limited. These problems can be circumvented by developing biosensors in *Bacillus* spores, which can be engineered to address all of these concerns. In its sporulated state, a whole cell-based biosensor has a remarkably long life span and is exceptionally resistant to environmental insult. When these spores are germinated for use in analytical techniques, they show no loss in performance, even after long periods of storage under harsh conditions. In this chapter, we will discuss the development and use of whole cell-based sensors, their adaptation to spore-based biosensors, their current applications, and future directions in the field.

1. INTRODUCTION

1.1 The Need for Portable, Field-Amenable Technology for Monitoring Environmental Pollutants

The spread and persistence of many different hazardous materials represents a significant threat to both human populations and the environment. These materials range from consumer products and the by-products of industry to naturally occurring elements. Given the toxic and persistent nature of many environmental pollutants there is a need to develop highly selective, inexpensive technologies to quickly, easily, and reliably monitor contamination levels in situ. This is especially important due to the likelihood of biological accumulation, a phenomenon where species high on the food chain become contaminated through consumption of their food (Perkins, 1986). There is also a significant risk of contamination occurring in water sources. Indeed, without quick and sensitive assays, contamination will have an increasingly detrimental effect as the long-term consequences of low-concentration chronic exposure to pollutants are often not as apparent as the symptoms of high-concentration acute exposure, despite remaining quite deadly (Aksoy, 1985; Buchancova et al., 1998; Dallaire et al., 2014; Ihrig, Shalat, & Baynes, 1998; Ludewig & Robertson, 2013; Tsuji, 2015; Wahlang et al., 2014). Additionally, many of the communities affected by pollution are in rural, underdeveloped areas that do not have access to laboratories equipped to perform environmental analysis. These communities will benefit tremendously from the use of portable technologies to monitor their local environment.

1.2 Current Methods of Detection

The range of physical and chemical characteristics found in hazardous materials presents a challenge for researchers since the variance in chemical and physical properties necessitates a number of analytical techniques in order to accurately assess biological and environmental samples for contamination. Historically, one of the most popular techniques was atomic absorption spectrometry, especially for the detection of heavy metals (Soodan, Pakade, Nagpal, & Katnoria, 2014). Despite its initial popularity, the method has largely been replaced by more recent innovations (Soodan et al., 2014). Some of these newer methods involve the use of mass spectrometry (MS). For example, methods designed to assay for organic molecules like polychlorinated biphenyls (PCBs), hydroxylated PCBs (OH-PCBs), benzene, toluene, ethylbenzene, and xylene contamination use gas chromatography-coupled MS (Grosser, 2006). Heavy metal detection is often accomplished using inductively coupled plasma MS (ICP-MS) (Ortega, 2002). In fact, ICP-MS is the "gold standard" for heavy metal analysis since it permits multielemental trace detection (Beauchemin, 2008).

These MS-based techniques are favored because of their excellent sensitivity, selectivity, and their established, reliable methodology. Despite these significant advantages, the nature of environmental pollution demands additional attributes from analytical methods for timely and efficient monitoring of contamination. The instrumentation in combination with the need for highly trained personnel makes the use of traditional methods expensive, often prohibitively so in rural or developing countries. Additionally, MS-based techniques can only be performed in a laboratory setting. Another concern that needs to be considered involves the analysis of complex matrices. This is particularly important in field-based assays designed to analyze biological and environmental samples, since it is difficult to incorporate sample preparation steps without increasing the need for additional materials and instruments. Other techniques, like atomic fluorescence spectrometry, ICP-optical emission spectroscopy, ICP-atomic emission spectrometry, and X-ray fluorescence spectrometry, are widely used and afford very low detection limits, but, unfortunately, also possess some of the same disadvantages as MS (Soodan et al., 2014).

The shortcomings of the current methodologies in field applications have led scientists to explore alternative approaches that employ portable instrument platforms for detection of compounds in the environment. As a result, a host of nanoparticle-based sensors have been developed for a wide range of compounds (Ibrahim, Hayyan, AlSaadi, Hayyan, & Ibrahim, 2016).

Many of these nanoparticles take advantage of surface-enhanced Raman spectroscopy (SERS) (Li, Zhai, Li, & Long, 2014). This technique has been used to develop SERS substrates to detect a wide variety of analytes, including arsenate (Han, Hao, Xu, & Meng, 2011), PCBs (Bantz & Haynes, 2009), pesticides (Pang, Yang, & He, 2016), trinitrotoluene (Mahmoud & Zourob, 2013), and uranium (Ruan, Luo, Wang, & Gu, 2007). Assays using SERS fingerprints as a mode of detection require specialized instrumentation. These tools have been miniaturized and implemented in microfluidic and analytical platforms configured for field studies (Qu et al., 2012), but their use still requires specialized training and access to this technology may be limited by its price. It is worth noting that the nanomaterials themselves may be quite expensive as many of these materials incorporate gold and silver. Although these methods generally have excellent sensitivity and can be performed rapidly, they often have difficulty discriminating between analytes, especially metals (Li et al., 2014). An ideal scenario for field studies would include use of analytical technology that can be adapted to address the need to assay samples containing multiple contaminants. It would also be inexpensive and require minimal training and expertise for its use. Indeed, in some cases, these technologies should be easy to use so that they may be employed by the general public to ensure that contamination in homes and communities is monitored.

2. WHOLE CELL-BASED BIOSENSORS

2.1 Adapting Biology to Serve as Technology

Whole cell-based biosensors (WCBs) have emerged as an alternative to traditional instrumentation in that they are inexpensive to produce, require no special equipment or training beyond basic laboratory facilities and techniques, and exhibit high selectivity and sensitivity of detection. Furthermore, the design and execution of these assays makes them very easy to miniaturize and facilitates their use in a field-based platform. The data produced by these assays can be particularly useful since the living organisms that are employed provide valuable information about the toxicity and bioavailability of the analyte being measured (Belkin, 2003). Many different designs and methods have been developed for WCBs; however, herein we focus on WCBs that have been genetically engineered to exploit the regulatory mechanisms of species of bacteria that have developed resistance to toxic materials.

Many species of bacteria have evolved to become resistant to toxic substances through a variety of mechanisms. For example, in *Escherichia coli*, the plasmid R773 confers resistance to arsenic species by encoding for a number of genes that form an ATP-driven efflux pump that removes arsenic from the cytosol (Carlin, Shi, Dey, & Rosen, 1995). Similar mechanisms have developed in *Synechococcus* to remove cadmium, lead, and zinc from the cell (Erbe, Taylor, & Hall, 1995). Other organisms have developed resistance to toxins through degradation pathways. *Pseudomonas azelaica* has developed a resistance to hydroxybiphenyls by evolving an operon consisting of enzymes that degrade these compounds into carbon and energy sources (Kohler, Kohler-Staub, & Focht, 1988). Similarly, *Pseudomonas putida* has established degradation pathways for volatile organic compounds that ultimately produce Krebs cycle intermediates (Marques & Ramos, 1993).

These resistance mechanisms provide excellent candidates for exploitation in biosensor development, not because of the enzymes that confer resistance but because of the machinery involved in the regulation of the transcription of these enzymes. Two core components form this machinery: a promoter which controls the transcription of the resistance genes, and a constitutively expressed regulatory protein, typically a binding protein. There are two general paradigms in which this occurs as the regulatory protein may be a transcriptional repressor or a transcriptional activator (Snyder & Champness, 1997). If the regulatory gene codes for a transcriptional repressor, the corresponding protein will bind to the promoter and prevent transcription in the absence of the ligand. In the presence of the compound, the repressor will bind its ligand and allow for transcription of the resistance genes; thus, the repressor negatively regulates expression of the resistance genes. Conversely, if the regulatory gene encodes for a transcriptional activator, the corresponding regulatory protein will only bind to the promoter in the presence of ligand, possibly due to a conformational shift upon binding the ligand or oligomerization following binding that allows for DNA binding. This allows for the initiation of transcription of the resistance genes. Gene expression may also be regulated through other mechanisms, but these will not be discussed because they are not pertinent to the development of WCBs. Regulation schemes using transcriptional repressors as sensing elements form the foundation of many WCBs because this particular design is easy to engineer, use, and produces results that are easy to interpret.

WCBs have been developed for a wide range of analytes with diverse chemical and physical properties. Metals like antimony, arsenic, cadmium, chromium, cobalt, copper, lead, mercury, nickel, and zinc have all been detected with WCBs (Collard et al., 1994; Feliciano, Liu, & Daunert, 2006; Klein, Altenbuchner, & Mattes, 1997; Ramanathan, Shi, Rosen, & Daunert, 1997; Shetty, Deo, Liu, & Daunert, 2004; Shetty et al., 2003; Stocker et al., 2003; Tauriainen, Karp, Chang, & Virta, 1998). WCBs have also been engineered to respond to a variety of organic compounds that range from volatile organics to stable, persistent organic pollutants, including benzene, chlorocatechols, ethylbenzene, toluene, middle chain alkanes, naphthalene, organophosphates, phenols, xylene, PCBs, and OH-PCBs (Behzadian, Barjeste, Hosseinkhani, & Zarei, 2011; Feliciano, Xu, et al., 2006; Guan et al., 2000; Kobatake, Niimi, Haruyama, Ikariyama, & Aizawa, 1995; Liu, Germaine, Ryan, & Dowling, 2010; Mulchandani, Mulchandani, Kaneva, & Chen, 1998; Shin, Park, & Lim, 2005; Sticher et al., 1997; Turner et al., 2007). These sensors have found applications in environmental analysis since many of these compounds are dangerous and common environmental contaminants. Indeed, arsenic, lead, mercury, cadmium, benzene, and PCBs are among the most concerning compounds on the Agency for Toxic Substances and Disease Registry (ATSDR) biannual priority list (ATSDR, 2015). Biosensors are also useful for the detection of pesticides since many popular insecticides contain chlorocatechols and organophosphates. Devices have also been developed to monitor water sources by engineering a chip that uses a charge-coupled device to monitor the bioluminescent signal of WCBs (Tsai et al., 2015). In this device, water flows through microfluidic channels and pollutant detection in the water was possible after only 15 min. Although the environmental applications of WCBs have received a great deal of attention, the technology has proven to be useful in many other instances. Biomedical applications have been seen in the detection of amino acids, sugars, and viruses, like tuberculosis and human respiratory syncytial virus (Jacobs et al., 1993; Jaeger, Lindow, Miller, Clark, & Firestone, 1999; Olivo, Collins, Peeples, & Schlesinger, 1998; Shetty, Ramanathan, Badr, Wolford, & Daunert, 1999). The technology has been further exploited to be used as a tool to evaluate clinical samples for bacterial quorum-sensing molecules, and as laboratory tools to demonstrate bacterial proliferation in the gut of humans and animals such as mice (Kumari, Pasini, & Daunert, 2008; Kumari et al., 2006; O'Connor et al., 2015; Raut, Joel, Pasini, & Daunert, 2015; Raut, Pasini, & Daunert, 2013; Struss, Pasini, Ensor, Raut, & Daunert, 2010). WCBs have even been developed to detect undetonated explosive ordnance by engineering sensors responsive to chemicals found in the vapor signature of a landmine (Garmendia, de las Heras, Galvao, & de Lorenzo, 2008; Yagur-Kroll et al., 2014). Kabessa et al. developed a system using WCBs for the remote detection of landmines based on fluorescence (Kabessa et al., 2016). Bacteria are quickly evolving resistance to many different compounds and WCBs take full advantage of this rapidity, as new biosensors can be created as soon as the genes responsible for the resistance to the compounds and the genes that regulate the expression of these enzymes are identified and mapped.

3. DESIGN, CREATION, AND USE OF WCBs

3.1 WCB Design

The core component of these biosensors is a plasmid engineered to contain a sensing element and a reporting element (Figs. 2G and H and 3I). The reporting element is composed of a reporter gene whose transcription is linked to a promoter that allows the reporter gene's expression to be controlled by a regulatory protein that has also been placed in the same plasmid in a manner that allows it to be constitutively expressed (Fig. 1A). This regulatory protein functions as the sensing element (also called a recognition element) in tandem with the promoter. These regulatory genes are specifically chosen for their ability to interact with the intended target analyte of the biosensor and influence the expression of genes linked to a specific promoter. The reporting element, usually a gene whose corresponding protein emits a detectable signal or an enzyme that can be used to produce a signal, is placed under the control of the sensing element's promoter. This means that the addition of the target analyte will induce expression of the reporter (Fig. 1B), which allows for the detection of the target. Given the simplicity of this scheme, it should be no surprise that these sensors can be created using only the most basic molecular cloning techniques.

Engineering a biosensing plasmid requires several steps shared with vectors for insertion and expression of other genes. Selection of the plasmid vector that carries the promoter and genes for the recognition and reporter proteins will primarily depend on the availability of restriction sites and other necessary expression-related genes that the vector contains. Vectors often contain multiple genes in addition to an origin of replication and promoter. Most vectors include a gene conferring antibiotic resistance to allow for selection. Some will already contain a reporter gene. The ideal vector for developing a WCB plasmid would have the following: desirable restriction sites, a resistance gene, the appropriate reporter, and (if required) a promoter that will allow the regulatory gene to be constitutively expressed. Restriction enzymes that produce overhangs are generally preferred for cloning

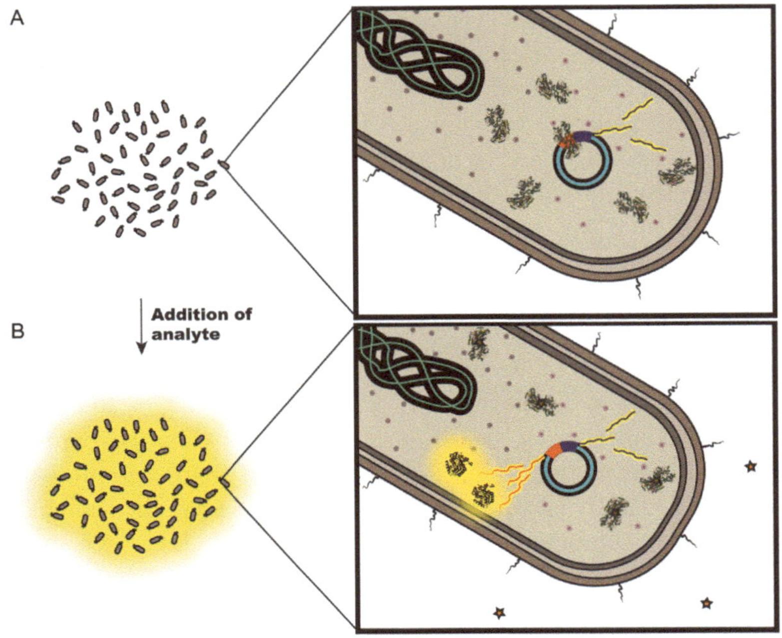

Fig. 1 The whole-cell biosensor and its constitutive parts. (A) In the absence of the target analyte, the transcriptional repressor will bind to the inducible promoter and prevent transcription of the reporter gene. (B) When the analyte (depicted as a *star*) is introduced to the cellular environment, the transcriptional repressor will begin to bind its ligand and allow for the transcription of the reporter from the inducible promoter. The reporter will begin to emit a signal and detection can be accomplished.

applications since they allow for the directional annealing of the insert and the vector (Fig. 2B). Since most inserts produced via restriction digestions will be inserted into the appropriate vector using a DNA ligase (Fig. 2E) to catalyze the formation of the phosphodiester bond, restriction enzymes that produce blunt ends should be avoided if possible. The ligation reaction depends on the association of the insert, vector, and ligase. This is much easier to achieve when the insert and vector have overhangs since the annealing of the two DNA pieces will increase the time of their association and will maintain the reading frame. Blunt-ended ligations are still possible, and even common, but overhangs enhance the chances of a successful ligation and often save the user time in the creation of the desired plasmid. Several considerations need to be taken into account when designing a plasmid for biosensing applications, specifically when there is a need for using restriction endonucleases to isolate the gene of interest. First, enzymes must be chosen

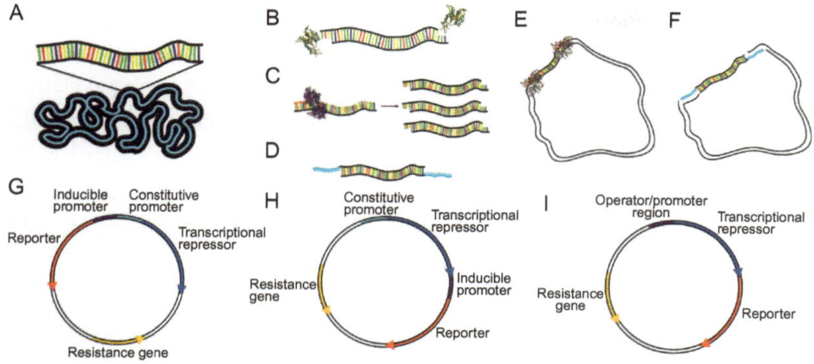

Fig. 2 Creation and design of plasmids engineered for WCBs. (A) Regulatory proteins and inducible promoters are identified and mapped in their native organisms. (B) The target sequence may by isolated from the template using restriction endonucleases. (C) Alternatively, the target may be amplified from the template using PCR. (D) Long 5′ and 3′ overhangs may be incorporated into the insert using PCR to allow for the use of ligation-independent cloning. (E) The insert may be ligated into the desired vector if it contains the appropriate restriction sites. (F) If overhangs have been created for ligation-independent cloning, the insert can be annealed to the vector and transformed into the host organism following treatment with T4 DNA polymerase. (G) In some cases, the transcriptional repressor and inducible promoter may be isolated in a way that incorporates a constitutive promoter to drive expression of the repressor. (H) In other cases, the gene will need to be inserted into a vector that contains a promoter that will allow for constitutive expression of the repressor. (I) Some WCBs have been engineered using a promoter that allows for the autoregulation of the transcriptional repressor.

according to the sequence of the vector. If there is no corresponding restriction site, the enzyme may not be used to cut the vector. Although the vector's sequence may be mutated to allow for the use of a particular enzyme, other methods (such as PCR) may be considerably easier. It is also important to consider the sequence of the insert. If the enzyme also cuts the insert, it is not optimal since it will not be possible to generate overhangs through a restriction digestion. Reaction conditions are also an important consideration, especially when using multiple enzymes. Ensuring that the enzymes function well at the same temperatures and in the same buffers leads to easier protocols, as do other enzyme characteristics like heat inactivation and short reaction times. It should be noted that plasmids can eventually be lost after several generations from a given cell. To overcome this drawback of plasmid-based whole-cell sensors, vectors can be used that recombine into the host's genome to reduce genetic instability and ensure that the genetic material and the properties it endows can be permanently inserted in a given cell.

Generally, the genes of interest are isolated in one of two ways: through the use of restriction endonucleases or a polymerase chain reaction (PCR).

Since the genes are often isolated from the genome of the organism from which they originate, it can be difficult to use restriction enzymes as there may not be restriction sites that allow for the isolation of the target gene in frame and without unnecessary nucleotides on either side of the gene. PCR is often the better choice for isolation (Fig. 2C). It allows genes to be selected regardless of the availability of restriction sites and will increase the flexibility in choosing different vectors. It is also easier to isolate a large amount of the gene since it will be amplified many times over from the original copy. Users can insert restriction sites at regions 5′ and 3′ to the gene to allow for easier insertion into the vector via ligation. It is also possible to use PCR to create inserts that can be inserted into a vector without ligase, perhaps the easiest way to go about cloning the insert into the vector. To do this, PCR primers are designed to create large 5′ and 3′ regions that are complementary to the regions of the vector that are directly up- and downstream of the intended insertion site (Fig. 2D). Both the vector and insert are then treated with T4 DNA polymerase. In the absence of dNTPs, T4 polymerase exhibits 3′ exonucleolytic activity. This can be used to create vary large overhangs (>20 bp) that cause the vector and insert to anneal when mixed (Fig. 2F). The resulting mixture can be transformed into the appropriate host after the addition of dNTPs to halt the nucleolytic activity of T4 polymerase. The machinery of the cell will complete the insertion of the gene with its own ligases. This can be much easier (and less frustrating) than more traditional techniques.

When planning experiments for the isolation of the regulatory gene and the corresponding promoter, it is important to consider the orientation that will be required when inserted into the vector. It is imperative that the regulatory gene is constitutively expressed. As such it must be inserted into the vector in a manner that puts the gene downstream of a promoter that will accommodate this need. Sometimes, the regulatory gene can be isolated with both the promoter it controls (the inducible promoter) and another promoter which will allow it to be constitutively expressed (Fig. 2G) (Turner et al., 2007). Other times, the regulatory gene may autoregulate (Fig. 2I) (Ramanathan et al., 1997). If such a design is not possible, care must be taken to select a vector with a promoter appropriate for this task and insert the gene accordingly (Fig. 2H). Care must also be taken to ensure that the inducible promoter controls the expression of the reporter gene. If using a vector that already contains the reporter gene, the promoter is simply inserted upstream. If using a vector without the desired reporter, it is often easier to perform an overlap extension PCR to splice the regulatory gene, promoter, and reporter gene together. This reduces both the number of restriction sites required and overall effort by requiring only one reaction.

3.2 Selection and Use of Reporter Genes

Selection of the reporter gene depends upon the intended application of the system, its platform, and the equipment available, with each option possessing its own distinct advantages and disadvantages. Many different reporter genes have been employed in WCBs that employ a variety of modes of detection, including electrochemistry, colorimetric, absorbance, fluorescence, and bioluminescence (Close, Ripp, & Sayler, 2009; Daunert et al., 2000). Electrochemical detection of analytes has been accomplished using the *lacZ* gene, which codes for the hydrolase β-galactosidase. Though this method is highly sensitive, it requires both specialized equipment and substrate for β-galactosidase (Biran, Klimentiy, Hengge-Aronis, Ron, & Rishpon, 1999). Colorimetric assays have also been developed using *lacZ* for use in a qualitative assay without any instrumentation or as a quantitative reporter using absorbance. However, this method also requires the addition of a substrate and usually lacks the sensitivity of other methods (Naylor, 1999). Although it may not be entirely necessary, employing host strains where endogenous *lacZ* has been removed will reduce the background signal produced by native expression of the enzyme. Fluorescent reporters, like the well-known green-fluorescent protein, are sensitive and reliable but require instrumentation and are often encumbered by high background signal due to autofluorescence of the cells and other molecules in cell culture media. Therefore, special care must be taken when performing fluorescence-based assays and may require the use of specially prepared buffers and media. Bioluminescence is also a popular mode of detection. Many bioluminescent reporter genes are available for use, including firefly luciferase, *rluc*, *gluc*, *luxAB*, the full lux cassette, and aequorin (Roda et al., 2016). Because of the lack of endogenous reporter expression in most strains of bacteria, reporting elements using bioluminescence are often more sensitive than other methods. In order to produce a viable assay for use in the field, it is important to select a bioluminescent reporter that suits this aim, ideally one with no required exogenous substrate, although stable and easily stored substrates can be used in these assays with tremendous success. Electrochemical, colorimetric, fluorescent, and bioluminescent reporters have been among the most popular methods of detection; however, others have been used. Chloramphenicol acetyltransferase (CAT) has been employed in many cell-based assays and was one of the first reporter genes to be widely used (Lewis, Feltus, Ensor, Ramanathan, & Daunert, 1998). Reporter production is often monitored using radiolabeled chloramphenicol, but the limited linear range and the requirement for radioactive isotopes make CAT an unpopular choice for WCBs (Daunert et al., 2000). Many of the reporters mentioned earlier can be very useful for laboratory-based assays, but their dependence on equipment (fluorimeters, spectrophotometers, luminometers, etc.) and reagents makes them less useful in field-based assays. Consequently, this has traditionally limited the usefulness of fluorescent, bioluminescent, and electrochemical reporters since the assay will neither be as cheap nor as field amenable as is required. Nevertheless, recent advances in adapting these detection approaches using cell phone-based and other technologies show promise for their use in portable field-applicable devices. The use of a colorimetric assay using β-galactosidase is an attractive option as a reporter gene for a truly field-amenable system, despite the potential loss of sensitivity and requirement of substrate addition.

3.3 Reporter Induction Assays With WCBs

The protocol for the assay used to induce reporter expression in WCBs is quite simple, as is illustrated in Fig. 1. Cells harboring the biosensing plasmid are grown to an optimized growth phase, generally measured using optical density, in one of a number of different cell growth media. LB Miller broth, Terrific broth, and minimal media are among the most common media chosen for this task. The analyte of interest is then added at known concentrations and incubated for an optimized amount of time at a specified temperature, with or without shaking. In some cases, the cells may be harvested via centrifugation and suspended in new media or a specially prepared buffer prior to the addition of analyte. The relative volumes of cells, buffer, and analyte will vary from sensor to sensor and are optimized to ensure the greatest analytical performance. Following incubation with the analyte, the reporter expression is evaluated. The method used depends on the assay and the reporter gene. For a luminescent reporter, the mixture of cells and analyte is read on a luminometer following substrate addition, if needed. Fluorescent reporters generally require a centrifugation step to harvest the cells prior to resuspension in a buffer chosen to reduce the matrix effect produced by the autofluorescence of the cells. The signal is then read on a fluorimeter. Colorimetric assays are very similar; however, absorbance is read at a required wavelength on a spectrophotometer or a qualitative assessment of color intensity could be employed depending upon the substrate used. Electrochemical assays are also performed in an analogous manner, but require quantification of reporter production using a specially designed electrode. It is worth noting that some host organisms, especially Gram-positive strains, require lysis before evaluation of reporter expression.

Because of the disparity of hosts and gene constructs, it is difficult to make general statements regarding the analytical characteristics of WCBs. The signal produced can disappear, can appear as new, can shift (for example, if optical, shift to a different wavelength), can increase from a steady-state level, or decrease. For increased sensitivity of detection in the measurement, ideally, it is best to have a biosensor with a reporter system that will produce a dose-dependent response as a result of increasing concentrations of the analyte. The sensitivity of WCBs is generally in the micromolar range, but some WBCs have been produced with limits of detection (LODs) in the nanomolar range (Park, Tsai, & Chen, 2013) and even attomole detection has been reported with bioluminescence detection (Ramanathan et al., 1997). WCBs usually offer exquisite selectivity (He, Yuan, Zhong, Siddikee, & Dai, 2016). WCBs are considerably more rugged than many other methods, especially methods employing enzymes and other proteins due to the robustness of bacterial cells compared with most proteins alone. Changes in buffer composition, temperature, and pH will have little effect, but the addition of some solvents may disrupt analytical performance due to their toxicity.

3.4 Use in Portable Platforms

Assays are most often performed in microtiter plates since they allow for the convenient use of fluorescence-, luminescence-, or absorbance-based detection methods. Although this is acceptable for assays performed in a laboratory, other platforms have been developed to be of more use in the field. Microfluidic platforms constructed of polydimethylsiloxane (PDMS) are among the most popular and have been used with great success for the detection of arsenic and other analytes using WCBs (Date, Pasini, & Daunert, 2010; Rothert et al., 2005). Rothert et al. designed and imparted a pattern onto a PDMS compact disc (CD) that caused sample to flow through channels into wells containing biosensing cells with the application of centrifugal force with rotation. Despite the advantages of PDMS, it can be very expensive and cumbersome to use as the basis for microfluidic devices. Another cheaper, easier method involves the use of filter paper. Biosensing cells were lyophilized onto filter paper and used to assay water and saliva samples for the presence of acyl homoserine lactones (Struss et al., 2010). There are a few different methods to produce these portable analytical devices, but the basic scheme remains the same (shown in Fig. 3): cells containing biosensing plasmids are grown to an optimized optical density (Fig. 3A), dried or immobilized on filter paper (Fig. 3B and C), assays are performed under optimized conditions (Fig. 3D), and a signal is detected as a result of reporter gene production (Fig. 3F).

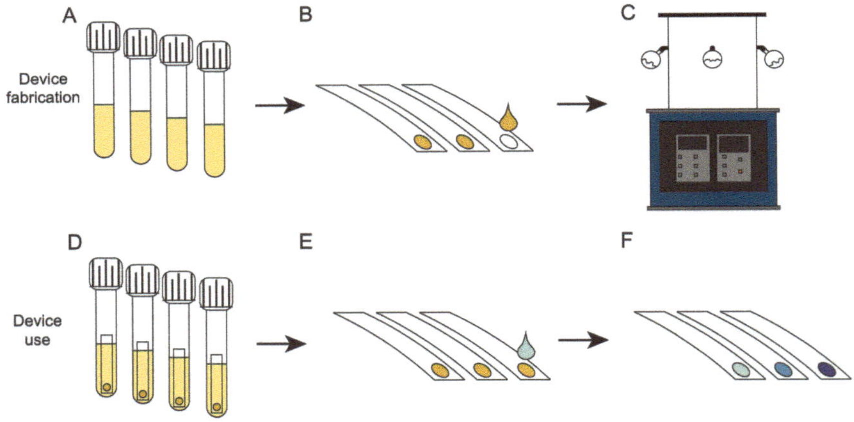

Fig. 3 The development and use of paper strips and microfluidic platforms with WCBs. (A) A WCB strain is grown to an optimized growth phase. (B) After harvesting cells via centrifugation, the cells are resuspended in a drying protectant, cryoprotectant, or immobilizing matrix and applied to the devices. (C) The devices are allowed to dry. If a drying protectant or cryoprotectant was used, they are dried under vacuum or lyophilized. Otherwise, the devices may be left at room temperature to dry or under whatever conditions are required for the polymerization of immobilization matrices. (D) The analytical devices are incubated in media containing either known concentrations of the analyte or samples collected from the field for an optimized period of time. (E) The devices are removed from the media and substrate is applied if required. (F) The signal produced by the reporter is measured. The method used to evaluate reporter expression will depend upon the reporter, with the most popular options being fluorescence, luminescence, or colorimetric signals. An incubation period may be required to allow for full color development when using colorimetric reporters.

In some cases the filter paper is modified with wax printing, where wax designs are melted onto filter paper to create hydrophobic surfaces to allow for laminar flow. This technique can also be used to sequester cells and the applied sample from the rest of the device. Although this increases the cost of device production, wax printing can greatly increase the usefulness of the device because it allows for multiplexing and the application of sample to channels that feed into multiple wells. Since these assays are designed for quick and easy field use, it is advantageous to be able to assay for multiple analytes at once, especially if the device is designed for use by the general public to monitor the safety of the environment and water sources in local communities.

Cell immobilization can be accomplished using many methods but has most commonly been done through lyophilization within a drying protectant, or via an agar immobilization. *lacZ* has been used in WCBs as a reporter gene in microfluidic and paper-based devices in order to create a colorimetric assay that requires no additional instrumentation using the β-galactosidase substrate 5-bromo-4-chloro-3-indolyl-β-D-galactopyranoside, better known as X-Gal (Struss et al., 2010). The use of colorimetric detection may prevent these assays from being employed as quantitative methods in the field; however, software may be used to quantify the intensity of color produced relative to blanks to produce a semi-quantitative method. These software techniques are easy and reliable, but the requirement of specialized software may place some limitations on this technology's use in the field. Strategies to overcome these limitations are discussed later in this chapter.

3.5 Addressing the Concerns of Life Span and Robustness

Although they have many advantages, WCBs have limitations that can make them challenging to use, especially in the context of a field assay. Since these assays employ living cells, all the limitations and weaknesses of the cells are conferred on the assays. The cells require very specific conditions for storage and handling, most likely requiring refrigeration or freezing. Additionally, if the cells are not kept frozen or refrigerated, they may have a limited shelf life, ranging from days to months. They are also vulnerable to a wide variety of factors that can either kill the cells or disrupt their function as biosensors, including excursions in temperature or pH, irradiation, and organic solvents, to name a few. Researchers have developed many strategies to address these limitations with varying degrees of success. Most of these strategies fall into two broad categories: drying methods and immobilization.

Perhaps the most common method uses lyophilization to freeze dry cells in a cryoprotectant solution. Typically, cells are grown to an optimized growth stage, collected, resuspended in the drying protectant, frozen, and dried under vacuum. This method requires careful optimization of both the growth phase chosen to begin lyophilization and the components of the cryoprotectant solution. The growth phase can not only impact the viability of the cells upon rehydration (Costa, Usall, Teixido, Garcia, & Vinas, 2000), it can also interfere with analytical performance. Therefore, it is important for the proper growth phase to be identified as it will vary from species to species and even strain to strain. The composition of the cryoprotectant will vary from strain to strain as well since it has been shown that varying concentrations of sucrose, glucose, fructose, mannose, salts, glycerol, glycol, and DMSO can greatly impact the survivability and analytical capability of the sensing cells (Shin et al., 2005). A cheaper, easier alternative to lyophilization is the use of vacuum drying. Although this method is very similar to lyophilization, it tends to be harsher on the cells and can reduce the survival rate (Manzanera et al., 2002; Manzanera, Vilchez, & Tunnacliffe, 2004; Stocker et al., 2003). It is also worth noting that moisture can reduce the viability and analytical performance of WCBs when drying methods are employed so that, although refrigeration may be unnecessary for short-term storage, care needs to be taken in storing lyophilized cells, especially in humid climates.

Immobilization methods involve encapsulating the sensing cells in solid matrices composed of organic or inorganic polymers in order to sequester the cells and enhance their survivability. Some of the most common matrices are nutrient agar (Kim & Gu, 2003; Lee, Mitchell, Kim, Cullen, & Gu, 2005; Yagur-Kroll et al., 2014), alginate (Corbisier et al., 1999; Elasri & Miller, 1999; Heitzer et al., 1994; Peter, Hutter, Stöllnberger, & Hampel, 1996; Polyak, Geresh, & Marks, 2004), polyvinyl alcohol-based cryogels (Busto, Meza, Ortega, & Perez-Mateos, 2007; López-Fouz et al., 2007; Philp et al., 2003), and inorganic sol-gels (Amoura, Nassif, Roux, Livage, & Coradin, 2007; Lan, Dunn, & Zink, 2005; Nassif et al., 2002; Premkumar et al., 2002; Yu et al., 2005). Generally, this can extend the life span of the cells from weeks to months. Generally, immobilization can extend the shelf life of bacterial sensors to several months when stored at room temperature (Durand, Thouand, Dancheva-Ivanova, Vachon, & DuBow, 2003; Gu, Choi, & Kim, 2001; Shin et al., 2005). Immobilization-based strategies can protect WCBs to some degree, but the cells are still vulnerable to many environmental insults and proper care still needs to be undertaken during storage to ensure full analytical performance, especially with biodegradable polymers and polymers where cross-linking may increase over time. Despite the advances in the methodology used to preserve bacterial biosensors, it is clear that these methods still lack the robustness and life span of ideal field equipment. In order to more completely address these problems, researchers adopted a philosophy similar to those who initially developed WCBs: rather than develop new technologies to solve current problems, they repurposed phenomena common in nature to suit their needs, and turned their attention to natural bacterial spores (Cano & Borucki, 1995). By incorporating the biosensing gene constructs into bacteria that produce endospores, researchers were able to develop rugged biosensors resistant to most environmental factors with a functionally unlimited shelf life (Cano & Borucki, 1995; Date, Pasini, & Daunert, 2007).

4. THE DEVELOPMENT OF SPORE-BASED BACTERIAL BIOSENSORS

4.1 Bacterial Endospore Formation

Some species of bacteria, notably strains of *Bacillus* and *Clostridium*, undergo a stress response called sporulation when they encounter conditions that are unfavorable for growth. This occurs as a multistep process shown in Fig. 4.

Sporulation initiation occurs as a result of nutrient-deficient conditions and a quorum-sensing mechanism indicating high population density. There is no single nutrient that disproportionately influences sporulation, though there is evidence that GTP concentrations play a major role (Molle et al., 2003). Sporulation contains a number of regulatory pathways, many of which are connected and redundant. Sporulation begins with the activation of Spo0A (Burbulys, Trach, & Hoch, 1991), the master regulator of sporulation initiation (Fujita & Losick, 2003) and expression of the sigma factor σ^H (Britton et al., 2002). Following initiation, the cell undergoes asymmetric cell division where a septum forms near the cell pole. Under normal circumstances, cell division is prevented in the polar regions of the cell (Harry, 2001). However, during sporulation, DivIVA recruits the machinery responsible for cell division to the poles (Thomaides, Freeman, Karoui, & Errington, 2001). Two proteins, Spo0J and Soj (Marston & Errington, 1999), are possibly involved in organizing and moving the chromosome toward the poles, where DivIVA will interact with

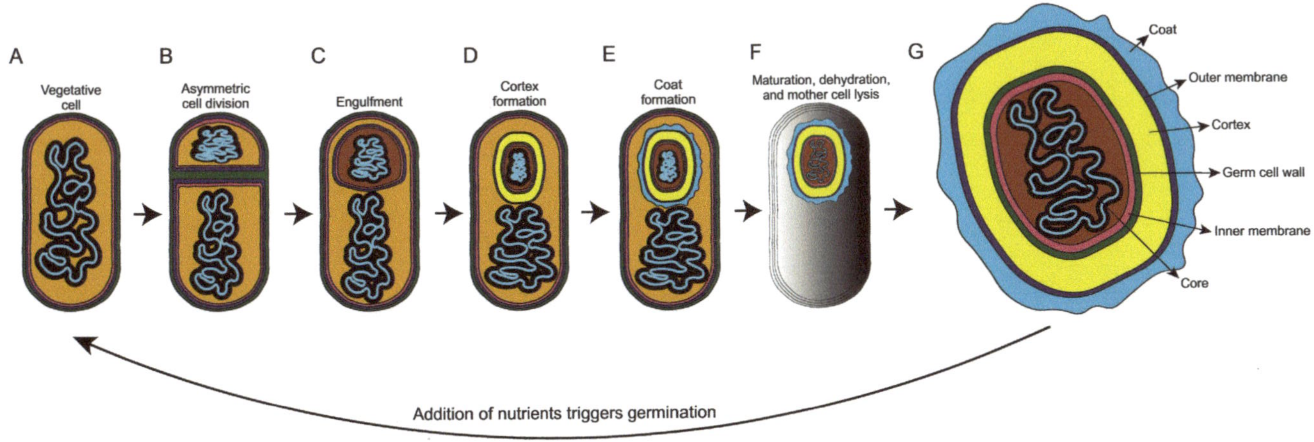

Fig. 4 Depiction of the sporulation cycle in *B. subtilis.* (A) Nutrient deficiency causes the vegetative cell to initiate sporulation. (B) A septum forms toward the pole of the cell during asymmetric cell division. (C) The cell wall material of the septum is degraded and the prespore is engulfed by the plasma membranes of the mother cell. (D) The cortex forms on top of the germ cell wall surrounded the prespore. (E) The coat forms on the exterior of the outer membrane of the cortex. (F) The spore fully matures, dehydrates, and lytic enzymes are released to degrade the mother cell. (G) Enlarged depiction of the fully formed spore.

RacA that has bound the chromosome at positions surrounding *oriC* (Wu & Errington, 2003).

Septum formation between the prespore and the mother cell initiates alterations in gene expression that continue sporulation via the activation of the sigma factors σ^E and σ^F, located in the prespore and mother cell, respectively, ensuring that specific sets of genes are transcribed in each cellular compartment. Following the completion of asymmetric cell division, the cell wall material in the septum is degraded. Enzymes precede the septal membranes as they migrate around the prespore cytosol in a process known as engulfment where the hydrolysis of linkages in the cell wall likely allows the membranes to move around the prespore (Abanes-De Mello, Sun, Aung, & Pogliano, 2002). The membranes then meet and fuse, causing the prespore to be released as a protoplast, which initiates the expression of two more sigma factors, σ^G and σ^K, that facilitate the formation of the spore cortex and the later stages of sporulation (Kunkel, Losick, & Stragier, 1990; Sun, Stragier, & Setlow, 1989). The cortex is comprised of layers of peptidoglycan that make the spores robust and also help to achieve and maintain the dehydration of the spore core. The germ cell wall forms first and has a high degree of cross-linking between peptide side chains. The cortex then forms on top of the cell wall with a much lower level of cross-linking. Once the cortex is formed, the spore coat begins to assemble. The spore coat consists of two layers, composed of between 25 and 35 proteins ranging in size from 8 to 65 kDa. The inner layer has a thin, lamellar-like structure relative to the thick, electron-dense outer coat. Bonds between these coat proteins may have major roles in the hardiness of the spore. SpoVID and CotE will then assemble on top of a SpoIVA layer and form a ring with a small gap between the ring and the prespore (Driks, Roels, Beall, Moran, & Losick, 1994; Price & Losick, 1999). Subsequently, other coat proteins assemble on top of this ring and form the spore coat. Mature spores can then be released from the mother cell following complete dehydration and the release of autolysins to degrade the mother cell.

4.2 The Exceptional Resilience of the Spore

The resistance of spores to radiation, temperature, pH, and chemical insult is conferred by morphological features of the spore. Spore resistance to irradiation is not completely understood, especially resistance to ionizing radiation, but much investigation has been done to elucidate the mechanisms of resistance to UV light at 254 nm. In the late stages of sporulation a number of α/β-type small acid-soluble proteins (SASPs) are expressed in the prespore that saturate the core and bind to DNA (Setlow, 1988). The complexation of SASPs and DNA significantly alters the photochemistry of the DNA in a manner that makes it more resilient. Rather than forming cyclobutane dimers and (6-4)-photoproducts when exposed to 254-nm light, as would typically occur, DNA complexed with SASPs will form "spore photoproducts" (SPs) (Mehl & Begley, 1999), a less lethal form of DNA damage that is quickly repaired at the outset of germination. Consequently, DNA repair mechanisms play a large role in UV resistance. Typical repair mechanisms like nucleotide excision repair and recombination-mediated repair play roles, but SP lyase (Spl), an enzyme unique to spore-forming bacteria, is responsible for the quick repair of SPs. Spl is expressed during sporulation but only becomes active during germination and can reverse thymine dimerization SPs into two thymine residues (Mehl & Begley, 1999). It is important to note that the roles played by α/β-type SASPs and DNA repair mechanisms in resistance to irradiation are also a major feature of spores' resistance to other environmental factors as they prevent accumulated DNA damage from impairing germination and the cellular mechanics of the vegetative cell. This likely explains why spores inactivated by wet heat rarely show significant DNA damage.

Most spores are resistant to temperatures of 95°C, approximately 40°C higher than vegetative cells. This resilience is likely due to the low relative water content of the spore core. Where the cortex and coat contain roughly 80% water (much like a vegetative cell), the core contains less than 50% and even as low as 25% (Gerhardt & Marquis, 1989). The inhibited movement of molecules in the core and reduced protein activity as a result of insufficient hydration are major contributors to the heat resistance of spores. The exact mechanism of core dehydration is unknown, but cortex formation, the addition of pyridine-2,6-dicarboxylic acid (better known as dipicolinic acid or DPA), and the mineralization of the core play significant roles. DPA is produced in the mother cell only during sporulation and is transported into the prespore where it is thought to contribute to core dehydration (Paidhungat, Setlow, Driks, & Setlow, 2000). DPA comprises over 10% of the dry weight of the spore (Setlow, 1994) and also may play a role in radiation resistance as it has been shown to be a photosensitizer (Paidhungat et al., 2000). The high divalent cation content of the spore core, especially the high concentration of Ca^{2+} ions, also plays a role in the dehydration of the core (Gerhardt & Marquis, 1989). This is not fully understood but there may be a relationship between the mineralization of the core and the presence of DPA, which appears to chelate the divalent cations. The spore's high resistance to chemical insult most likely comes from the coat and inner membrane. It is thought that the coat proteins may form a kind of protective barrier. Even more importantly, the low permeability of the inner membrane to both hydrophilic and hydrophobic small molecules prevents many chemicals from penetrating the core (Scherrer, Beaman, & Gerhardt, 1971). The cause of this low permeability is unknown, but it may be related to the compression of the inner membrane relative to normal cell membranes. It is also worth noting that the exceptional DNA repair mechanisms and the dehydration of the core also contribute to chemical resistance as many genotoxic chemicals will have limited effects even if they are able to enter the core. The spores also exhibit no metabolic activity and have an exceptional life span, with reports of spores estimated to be millions of years old (Cano & Borucki, 1995).

4.3 The Return of Endospores to Their Vegetative State

When the spores are introduced to a nutrient-dense environment, they undergo germination and once again become vegetative cells. This occurs via nutrient interaction with Ger receptors on the inner membrane of the cortex (Paidhungat & Setlow, 2000). The signal transduction that occurs during germination is not fully understood, but it is clear that germinant travels through the outer layers of the spore to reach the receptors on the cortex. This may happen via pores in the structure of the spore coat formed by the proteins composing the coat (Moir, Corfe, & Behravan, 2002). This process also involves ion transfers, membrane-associated changes, and ion movements that move water to the core to partially rehydrate it. Signals are then transmitted to outer layers of the spore to activate lytic enzymes to remove the cortex and fully rehydrate the core (Moir et al., 2002). Metabolic activity then resumes as the spores become vegetative cells. The addition of dipicolinic acid and calcium ions induces germination by directly activating cortex lytic enzyme, regardless of Ger receptor activation (Paidhungat & Setlow, 2000). These characteristics make sporulating bacteria an ideal choice of host organism for WCBs as they directly address the major disadvantages of whole-cell sensors in other organisms. The gene construct may be developed using traditional molecular cloning techniques and transformed into the host. Then, sporulation may be initiated using well-developed methodology to store the biosensors. When the WCBs are needed for use, they may then be germinated and used.

4.4 Bacterial Endospores as Analytical Tools

Although the WCB assays largely remain the same, alterations must be made to accommodate the need for germination. This is depicted in Fig. 5. In ideal conditions, germination can occur concurrent with the incubation of the cells and sample but, if this results in inefficient germination or poor reporter induction, a germination step where the spores are incubated with growth media must be undertaken prior to attempting a reporter induction assay. Typically, this takes between 1 and 3 h, but the addition of DPA and calcium chloride may reduce the time required by enhancing germination efficiency.

The ability of *Bacillus* spores to function as biosensors in an analytical assay has been thoroughly demonstrated. Date et al. designed plasmids incorporating operator/promoter regions and the corresponding regulatory proteins from organisms that possess resistance to heavy metals along with reporter proteins that produce a detectable signal (Date et al., 2007). One of these plasmids, pMUTin-23 (Fig. 6A), takes the O/P region and several genes from the *E. coli* resistance plasmid R773 that confers resistance to arsenic and antimony. The plasmid is organized such that the regulatory protein ArsR, a protein capable of binding As(III) and antimonite, regulates the transcription of *lacZ*. The gene encoding for the reductase ArsC was also included in order to reduce As(V) to As(III) and thereby increase the functionality of the assay. The plasmid pSD202 (Fig. 6B) was engineered to contain the gene *smtB*, which encodes for a transcriptional repressor that binds zinc and its corresponding promoter. This promoter was positioned upstream of *egfp*, which functions as the reporter. pMUTin-23 and

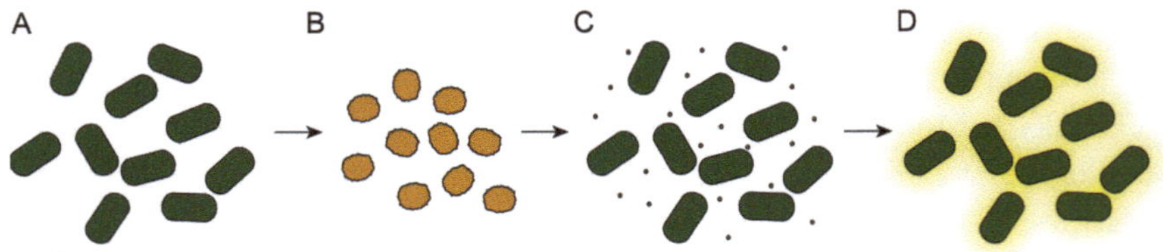

Fig. 5 The preparation and use of sporulated WCBs. (A) Vegetative cells are cultured in specially prepared media for several days to induce and undergo sporulation. (B) The sporulated WCBs may be stored under nearly any conditions prior to use. (C) The spores are incubated in media to induce germination. They are then incubated with either known concentrations of the target analyte or samples collected from the field. In optimal circumstances, these two steps may be performed concurrently. (D) After an optimized incubation period, the signal produced by the reporter may be evaluated. Depending upon the reporter, the cells may need to be lysed to ensure the signal is efficiently read.

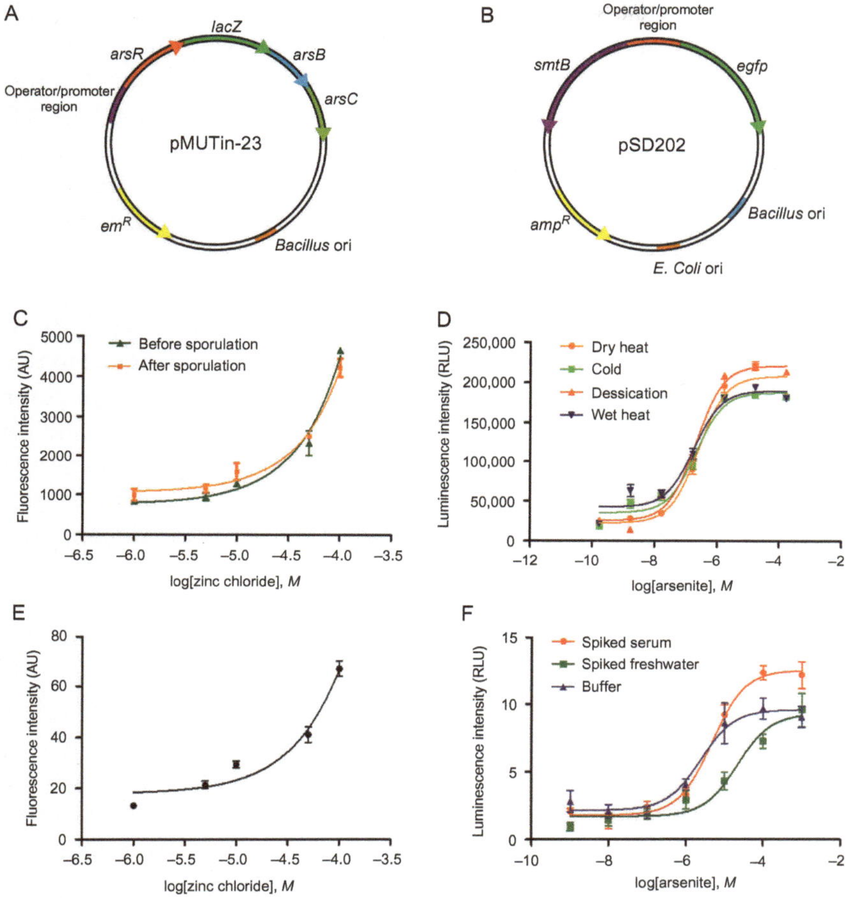

Fig. 6 Detection of arsenite and zinc using sporulated WCBs. (A) Plasmid map of the vector pMUTin-23 used for arsenite detection. (B) Plasmid map of the vector pSD202 used for zinc detection. (C) Calibration curve showing pSD202-harboring cells' response to zinc before and after sporulation. (D) Calibration curves of pMUTin-23 spores stored for 12 months under various conditions. (E) Fluorescent response of pSD202 spores immobilized on a microfluidic CD and measured using an optical fiber. (F) Luminescent response of pMUTin-23 spores immobilized on a microfluidic CD where spiked samples have been prepared in human serum, collected freshwater, and buffer.

pSD202 were transformed into *Bacillus subtilis* and *Bacillus megaterium*, respectively, and reporter induction assays were performed. When compared, the vegetative biosensing *Bacillus* cells performed similarly to whole cell-sensing systems in *E. coli* with LODs ranging from 2.8×10^{-8} to 1×10^{-6} *M*. The authors then compared the analytical characteristics of the sensing cells following sporulation and germination to vegetative cells that had not undergone sporulation, shown in Fig. 6C. As expected, the LODs and dynamic ranges remained consistent for both strains. In order to demonstrate the stability of the spores, the authors then subjected the cells to three sporulation/germination cycles. Again, the sensors remained very consistent, with LODs and dynamic ranges remaining comparable after every cycle. Even after 6 months of storage, the analytical characteristics of the assay remained unchanged. *Bacillus* cells harboring the pMUTin-23-biosensing plasmids have also been shown to be able to survive long-term storage in extreme conditions. In order to demonstrate this, Sangal, Pasini, and Daunert (2011) created a variety of simulated climates designed to replicate some of the more extreme situations under which these biosensors may be employed across the world. The four chosen conditions were dry heat (an oven set to 37°C), wet heat (a water bath set to 28°C with a relative humidity of above 90%), cold (a −20°C freezer), and desiccation (a desiccator at room temperature). Spores were left in these conditions for 12 months. Aliquots were taken at monthly intervals and the analytical performance was evaluated. Unsurprisingly, the LODs and dose–response curves remained consistent across all conditions and time points (Fig. 6D).

Although these experiments prove that sporulated *Bacillus* cells can effectively preserve the biosensing capabilities of these strains, it still needed to be shown that these sensors could function in a context meaningful to researchers and users in the field. Since the matrices seen when analyzing biological and environmental samples are a great deal more complex than the spiked deionized water used in previous experiments, it was very important that the spores be shown to germinate efficiently and retain the same analytical performance when the sample is composed of freshwater or human blood serum. This was demonstrated by using the biosensing strains in simulated freshwater, freshwater collected from various water sources in Indiana and Kentucky, commercially available serum, and serum drawn from healthy volunteers, shown in Fig. 6F. Though some matrix effects were seen when using serum that caused high background signals, LODs remained below $1 \times 10^{-6}\ M$, meaning they were still comparable to the results seen with spiked deionized water and in the *E. coli*-based whole-cell sensors.

These strains have also been used on a centrifugal CD microfluidic device that can easily function as a portable system for sample analysis (Fig. 6E) (Date et al., 2010). Samples are applied to specific reservoirs and the disk is spun at a particular angular frequency to create the flow, which pushes the samples into wells containing immobilized biosensing spores. The signal could then be read using a spectrophotometer with a fiber optic cable. In the case of pSD202, readings were taken at 20-min intervals in order to optimize the incubation time, whereas pMUTin-23-harboring cells were only measured once since the assay requires the addition of substrate. This was done using spiked deionized water, serum, and freshwater. As expected, the cells' analytical performance was comparable to *E. coli*-based systems, vegetative *Bacillus* cells, and assays performed in microtiter plates. Fantino et al. developed what they termed "sposensors" in *B. subtilis* using recognition elements based on the gene *czcD* to detect zinc and *bceA* to detect the antibiotic bacitracin (Fantino, Barras, & Denizot, 2009). Both systems employed *lacZ* as a reporter for colorimetric detection. In these assays, the germination and incubation steps were performed concurrently, making the assay very simple to perform. When spotted on filter paper, zinc and bacitracin detection occurred at 1×10^{-6} and 3.5×10^{-9} *M*, respectively. Given these results, it is clear that the strategy of using *Bacillus* cells as a host for biosensing plasmids and exploiting the robustness and longevity of spores in order to facilitate storage is viable for field-based assays analyzing a variety of matrices.

5. CHALLENGES, INNOVATIONS, AND FUTURE DIRECTIONS

5.1 Achieving Incredible Sensitivity Through Directed Evolution

Since chronic exposure to environmental pollutants at low levels is still a major concern due to possible carcinogenicity and birth defects, among other concerns, it is imperative that WCBs be sensitive enough to detect contamination at very low concentrations. This is also a necessity for the biomedical applications of WCBs, as being able to diagnose and treat diseases at their outset improves the efficacy of many treatments. Although some WCBs may have passed this threshold, many have not. Some researchers have employed directed evolution to address this issue. Researchers were able to increase the sensitivity of a WCB designed to detect dinitrotoluene by performing multiple rounds of random mutagenesis and selecting for the mutations which produced the most sensitive assay with the strongest signal (Yagur-Kroll et al., 2014). The same techniques have been used to push the native selectivity of a binding protein from its native ligand, D-xylose, to dinitrotoluene for use in the detection of landmines (de Las Heras &

de Lorenzo, 2011). This directed evolution strategy using random mutagenesis has promise, since, after several rounds of random mutagenesis have been performed and the characteristics of the regulatory protein and the promoter have been further elucidated, researchers can begin to make rational designs for mutations based upon the data obtained by the directed evolution of the original regulatory protein.

5.2 Employing Cell Phones to Enhance the Utility of Cell-Based Assays

Recent work in the field holds great promise for WCBs, especially spore-based sensors. These advances in methodology and the platforms used make for a bright future for bacterial biosensors because these innovations address the major limitations of WCBs or spore-based sensors and make previously impossible techniques feasible in the field. The chief innovation is the use of cell phone technology in analytical applications. Since smartphones are nearly ubiquitous anywhere in the globe, including remote and difficult to reach locations, they are an outstanding tool for field-based methods because they are relatively inexpensive and, thus, available to almost everyone even in resource-poor countries as well as incorporating powerful light sensing, computational, navigational, and communications capabilities. Recently, researchers have used smartphones to quantify luminescent signals using a smartphone camera (Roda, Guardigli, et al., 2014; Roda, Michelini, et al., 2014). This was initially done using enzyme-based assays, but it has since been shown that the toxicity of a sample may be evaluated using human kidney cells and the bioluminescent signal they produce, quantified using a smartphone (Cevenini, Calabretta, Tarantino, Michelini, & Roda, 2016). It has also been demonstrated that electrochemical and colorimetric signals may be interpreted by smartphones (Aronoff-Spencer et al., 2016; Vashist, Marion Schneider, Zengerle, von Stetten, & Luong, 2015). This is very promising for WCBs as this would provide a cheap and easy to use portable luminometer to be used in conjunction with a paper or microfluidic platform. Previously, the use of these biosensors for field applications has been limited to employing β-galactosidase as a reporter because qualitative colorimetric assays require no instrumentation. This is a practical method and works well in point-of-care, on-site, and field applications, but it generally lacks the sensitivity needed for many important analytes. Also, the fact that it is a semi-quantitative test is a disadvantage when there is a possibility of high background signal due to the "leakiness" of the promoter. This can be a major problem since it is difficult to perform a baseline correction with the naked eye.

With a smartphone luminometer, many different luminescent and electrochemical reporters are now an option outside of the laboratory. Not only would this allow for a more sensitive assay, but it would also make quantitation much more feasible as it would even be possible to estimate the concentration of contaminating materials from samples using a standard curve. Special device cases have been designed and produced using 3D printing to accommodate the analytical devices for use with any type of smartphone (Roda, Guardigli, et al., 2014). Also, software may be developed to interpret results and provide users with information concerning the safety of samples, including correcting the data relative to a blank sample (Cevenini et al., 2016). This creates a platform that is affordable and ideal for use by the general public as it requires materials for its fabrication that are easily attainable and relatively inexpensive. Performance of the assay and interpretation of the results would also require no special training or knowledge while, at the same time, producing a more sensitive assay with greater utility.

6. CONCLUSIONS

Though challenges still remain, WCBs in sporulating host organisms provide an excellent option for use in field-based analytical assays because of their ruggedness, low cost, ease of use, and adaptability to field applications. The ease of creation and use of WCBs together provide many advantages compared to traditional methodologies, including robustness, reliability, easily interpretable results, field amenability, and low cost of manufacture. Moreover, the reagents needed for biosensor detection are not expensive, and thus, the typical cost of the assays is affordable even in developing and resource-poor countries. Many of these advantages are directly related to the simplicity of the system, which allows them to be easily adapted to portable platforms amenable to use in a variety of environmental conditions, including extreme climates and polluted environments. The versatility and ease of use of these analytical technologies make WCBs ideal candidates for further investigations and inspiration in the design of novel technologies that can be used by the general public to ensure the health and safety of people, communities, wildlife, and the environment all around the globe.

ACKNOWLEDGMENTS

The authors would like to thank the National Science Foundation (grant CHE-1506740) and the Government of Israel MOD (P.O. 4440563897) for funding this work. S.D. would also like to thank the Miller School of Medicine of the University of Miami for the Lucille P. Markey Chair in Biochemistry and Molecular Biology.

REFERENCES

Abanes-De Mello, A., Sun, Y. L., Aung, S., & Pogliano, K. (2002). A cytoskeleton-like role for the bacterial cell wall during engulfment of the Bacillus subtilis forespore. *Genes & Development*, *16*(24), 3253–3264. http://dx.doi.org/10.1101/gad.1039902.

Aksoy, M. (1985). Malignancies due to occupational exposure to benzene. *American Journal of Industrial Medicine*, 7(5–6), 395–402.

Amoura, M., Nassif, N., Roux, C., Livage, J., & Coradin, T. (2007). Sol-gel encapsulation of cells is not limited to silica: Long-term viability of bacteria in alumina matrices. *Chemical Communications (Cambridge, England)*, (39), 4015–4017. http://dx.doi.org/10.1039/b711380c.

Aronoff-Spencer, E., Venkatesh, A. G., Sun, A., Brickner, H., Looney, D., & Hall, D. A. (2016). Detection of Hepatitis C core antibody by dual-affinity yeast chimera and smartphone-based electrochemical sensing. *Biosensors and Bioelectronics*, *86*, 690–696. http://dx.doi.org/10.1016/j.bios.2016.07.023.

ATSDR. (2015). *The ATSDR 2015 substance priority list.* Retrieved 8/16/2016, from https://www.atsdr.cdc.gov/spl/.

Bantz, K. C., & Haynes, C. L. (2009). Surface-enhanced Raman scattering detection and discrimination of polychlorinated biphenyls. *Vibrational Spectroscopy*, *50*(1), 29–35. http://dx.doi.org/10.1016/j.vibspec.2008.07.006.

Beauchemin, D. (2008). Inductively coupled plasma mass spectrometry. *Analytical Chemistry*, *80*(12), 4455–4486. http://dx.doi.org/10.1021/ac8006945.

Behzadian, F., Barjeste, H., Hosseinkhani, S., & Zarei, A. R. (2011). Construction and characterization of Escherichia coli whole-cell biosensors for toluene and related compounds. *Current Microbiology*, *62*(2), 690–696. http://dx.doi.org/10.1007/s00284-010-9764-5.

Belkin, S. (2003). Microbial whole-cell sensing systems of environmental pollutants. *Current Opinion in Microbiology*, *6*(3), 206–212. http://dx.doi.org/10.1016/S1369-5274(03)00059-6.

Biran, I., Klimentiy, L., Hengge-Aronis, R., Ron, E. Z., & Rishpon, J. (1999). On-line monitoring of gene expression. *Microbiology*, *145*(Pt. 8), 2129–2133. http://dx.doi.org/10.1099/13500872-145-8-2129.

Britton, R. A., Eichenberger, P., Gonzalez-Pastor, J. E., Fawcett, P., Monson, R., Losick, R., & Grossman, A. D. (2002). Genome-wide analysis of the stationary-phase sigma factor (sigma-H) regulon of Bacillus subtilis. *Journal of Bacteriology*, *184*(17), 4881–4890.

Buchancova, J., Klimentova, G., Knizkova, M., Mesko, D., Galikova, E., Kubik, J., … Jakubis, M. (1998). Health status of workers of a thermal power station exposed for prolonged periods to arsenic and other elements from fuel. *Central European Journal of Public Health*, *6*(1), 29–36.

Burbulys, D., Trach, K. A., & Hoch, J. A. (1991). Initiation of sporulation in B. subtilis is controlled by a multicomponent phosphorelay. *Cell*, *64*(3), 545–552.

Busto, M. D., Meza, V., Ortega, N., & Perez-Mateos, M. (2007). Immobilization of naringinase from Aspergillus niger CECT 2088 in poly(vinyl alcohol) cryogels for the debittering of juices. *Food Chemistry*, *104*(3), 1177–1182. http://dx.doi.org/10.1016/j.foodchem.2007.01.033.

Cano, R. J., & Borucki, M. K. (1995). Revival and identification of bacterial spores in 25- to 40-million-year-old Dominican amber. *Science*, *268*(5213), 1060–1064.

Carlin, A., Shi, W., Dey, S., & Rosen, B. P. (1995). The ars operon of Escherichia coli confers arsenical and antimonial resistance. *Journal of Bacteriology*, *177*(4), 981–986.

Cevenini, L., Calabretta, M. M., Tarantino, G., Michelini, E., & Roda, A. (2016). Smartphone-interfaced 3D printed toxicity biosensor integrating bioluminescent "sentinel cells". *Sensors and Actuators B: Chemical*, *225*, 249–257. http://dx.doi.org/10.1016/j.snb.2015.11.017.

Close, D. M., Ripp, S., & Sayler, G. S. (2009). Reporter proteins in whole-cell optical bioreporter detection systems, biosensor integrations, and biosensing applications. *Sensors (Basel, Switzerland)*, *9*(11), 9147–9174. http://dx.doi.org/10.3390/s91109147.

Collard, J. M., Corbisier, P., Diels, L., Dong, Q., Jeanthon, C., Mergeay, M., ... Wuertz, S. (1994). Plasmids for heavy metal resistance in Alcaligenes eutrophus CH34: Mechanisms and applications. *FEMS Microbiology Reviews*, *14*(4), 405–414.

Corbisier, P., van der Lelie, D., Borremans, B., Provoost, A., de Lorenzo, V., Brown, N. L., ... Mattiasson, B. (1999). Whole cell- and protein-based biosensors for the detection of bioavailable heavy metals in environmental samples. *Analytica Chimica Acta*, *387*(3), 235–244. http://dx.doi.org/10.1016/S0003-2670(98)00725-9.

Costa, E., Usall, J., Teixido, N., Garcia, N., & Vinas, I. (2000). Effect of protective agents, rehydration media and initial cell concentration on viability of Pantoea agglomerans strain CPA-2 subjected to freeze-drying. *Journal of Applied Microbiology*, *89*(5), 793–800.

Dallaire, R., Dewailly, E., Ayotte, P., Forget-Dubois, N., Jacobson, S. W., Jacobson, J. L., & Muckle, G. (2014). Growth in Inuit children exposed to polychlorinated biphenyls and lead during fetal development and childhood. *Environmental Research*, *134*, 17–23. http://dx.doi.org/10.1016/j.envres.2014.06.023.

Date, A., Pasini, P., & Daunert, S. (2007). Construction of spores for portable bacterial whole-cell biosensing systems. *Analytical Chemistry*, *79*(24), 9391–9397. http://dx.doi.org/10.1021/ac701606g.

Date, A., Pasini, P., & Daunert, S. (2010). Integration of spore-based genetically engineered whole-cell sensing systems into portable centrifugal microfluidic platforms. *Analytical and Bioanalytical Chemistry*, *398*(1), 349–356. http://dx.doi.org/10.1007/s00216-010-3930-2.

Daunert, S., Barrett, G., Feliciano, J. S., Shetty, R. S., Shrestha, S., & Smith-Spencer, W. (2000). Genetically engineered whole-cell sensing systems: Coupling biological recognition with reporter genes. *Chemical Reviews*, *100*(7), 2705–2738. http://dx.doi.org/10.1021/cr990115p.

de Las Heras, A., & de Lorenzo, V. (2011). Cooperative amino acid changes shift the response of the sigma(5)(4)-dependent regulator XylR from natural m-xylene towards xenobiotic 2,4-dinitrotoluene. *Molecular Microbiology*, *79*(5), 1248–1259. http://dx.doi.org/10.1111/j.1365-2958.2010.07518.x.

Driks, A., Roels, S., Beall, B., Moran, C. P., Jr., & Losick, R. (1994). Subcellular localization of proteins involved in the assembly of the spore coat of Bacillus subtilis. *Genes & Development*, *8*(2), 234–244.

Durand, M. J., Thouand, G., Dancheva-Ivanova, T., Vachon, P., & DuBow, M. (2003). Specific detection of organotin compounds with a recombinant luminescent bacteria. *Chemosphere*, *52*(1), 103–111. http://dx.doi.org/10.1016/s0045-6535(03)00225-x.

Elasri, M. O., & Miller, R. V. (1999). Study of the response of a biofilm bacterial community to UV radiation. *Applied and Environmental Microbiology*, *65*(5), 2025–2031.

Erbe, J. L., Taylor, K. B., & Hall, L. M. (1995). Metalloregulation of the cyanobacterial smt locus: Identification of SmtB binding sites and direct interaction with metals. *Nucleic Acids Research*, *23*(13), 2472–2478.

Fantino, J. R., Barras, F., & Denizot, F. (2009). Sposensor: A whole-bacterial biosensor that uses immobilized Bacillus subtilis spores and a one-step incubation/detection process.

Journal of Molecular Microbiology and Biotechnology, *17*(2), 90–95. http://dx.doi.org/10.1159/000206634.

Feliciano, J., Liu, Y., & Daunert, S. (2006). Novel reporter gene in a fluorescent-based whole cell sensing system. *Biotechnology and Bioengineering*, *93*(5), 989–997. http://dx.doi.org/10.1002/bit.20808.

Feliciano, J., Xu, S., Guan, X., Lehmler, H. J., Bachas, L. G., & Daunert, S. (2006). ClcR-based biosensing system in the detection of cis-dihydroxylated (chloro-)biphenyls. *Analytical and Bioanalytical Chemistry*, *385*(5), 807–813. http://dx.doi.org/10.1007/s00216-006-0505-3.

Fujita, M., & Losick, R. (2003). The master regulator for entry into sporulation in Bacillus subtilis becomes a cell-specific transcription factor after asymmetric division. *Genes & Development*, *17*(9), 1166–1174. http://dx.doi.org/10.1101/gad.1078303.

Garmendia, J., de las Heras, A., Galvao, T. C., & de Lorenzo, V. (2008). Tracing explosives in soil with transcriptional regulators of Pseudomonas putida evolved for responding to nitrotoluenes. *Microbial Biotechnology*, *1*(3), 236–246. http://dx.doi.org/10.1111/j.1751-7915.2008.00027.x.

Gerhardt, P., & Marquis, R. E. (1989). Spore thermoresistance mechanisms. In I. Smith, R. - Slepecky, & P. Setlow (Eds.), *Regulation of procaryotic development* (pp. 43–63). Washington, DC: American Society for Microbiology.

Grosser, Z. A. (2006). Development of GC/MS in environmental analyses. Increasing computer capability yields improved methods. *Scientific Computing*, *23*(9), 2.

Gu, M. B., Choi, S. H., & Kim, S. W. (2001). Some observations in freeze-drying of recombinant bioluminescent Escherichia coli for toxicity monitoring. *Journal of Biotechnology*, *88*(2), 95–105.

Guan, X., Ramanathan, S., Garris, J. P., Shetty, R. S., Ensor, M., Bachas, L. G., & Daunert, S. (2000). Chlorocatechol detection based on a clc operon/reporter gene system. *Analytical Chemistry*, *72*(11), 2423–2427.

Han, M.-J., Hao, J., Xu, Z., & Meng, X. (2011). Surface-enhanced Raman scattering for arsenate detection on multilayer silver nanofilms. *Analytica Chimica Acta*, *692*(1–2), 96–102. http://dx.doi.org/10.1016/j.aca.2011.02.054.

Harry, E. J. (2001). Bacterial cell division: Regulating Z-ring formation. *Molecular Microbiology*, *40*(4), 795–803.

He, W., Yuan, S., Zhong, W.-H., Siddikee, M. A., & Dai, C.-C. (2016). Application of genetically engineered microbial whole-cell biosensors for combined chemosensing. *Applied Microbiology and Biotechnology*, *100*(3), 1109–1119. http://dx.doi.org/10.1007/s00253-015-7160-6.

Heitzer, A., Malachowsky, K., Thonnard, J. E., Bienkowski, P. R., White, D. C., & Sayler, G. S. (1994). Optical biosensor for environmental on-line monitoring of naphthalene and salicylate bioavailability with an immobilized bioluminescent catabolic reporter bacterium. *Applied and Environmental Microbiology*, *60*(5), 1487–1494.

Ibrahim, R. K., Hayyan, M., AlSaadi, M. A., Hayyan, A., & Ibrahim, S. (2016). Environmental application of nanotechnology: Air, soil, and water. *Environmental Science and Pollution Research*, *23*(14), 13754–13788. http://dx.doi.org/10.1007/s11356-016-6457-z.

Ihrig, M. M., Shalat, S. L., & Baynes, C. (1998). A hospital-based case-control study of stillbirths and environmental exposure to arsenic using an atmospheric dispersion model linked to a geographical information system. *Epidemiology*, *9*(3), 290–294.

Jacobs, W. R., Jr., Barletta, R. G., Udani, R., Chan, J., Kalkut, G., Sosne, G., … Bloom, B. R. (1993). Rapid assessment of drug susceptibilities of Mycobacterium tuberculosis by means of luciferase reporter phages. *Science*, *260*(5109), 819–822.

Jaeger, C. H., 3rd, Lindow, S. E., Miller, W., Clark, E., & Firestone, M. K. (1999). Mapping of sugar and amino acid availability in soil around roots with bacterial sensors of sucrose and tryptophan. *Applied and Environmental Microbiology*, *65*(6), 2685–2690.

Kabessa, Y., Eyal, O., Bar-On, O., Korouma, V., Yagur-Kroll, S., Belkin, S., & Agranat, A. J. (2016). Standoff detection of explosives and buried landmines using fluorescent bacterial sensor cells. *Biosensors & Bioelectronics*, *79*, 784–788. http://dx.doi.org/10.1016/j.bios.2016.01.011.

Kim, B. C., & Gu, M. B. (2003). A bioluminescent sensor for high throughput toxicity classification. *Biosensors & Bioelectronics*, *18*(8), 1015–1021.

Klein, J., Altenbuchner, J., & Mattes, R. (1997). Genetically modified Escherichia coli for colorimetric detection of inorganic and organic Hg compounds. *EXS*, *80*, 133–151.

Kobatake, E., Niimi, T., Haruyama, T., Ikariyama, Y., & Aizawa, M. (1995). Biosensing of benzene derivatives in the environment by luminescent Escherichia coli. *Biosensors & Bioelectronics*, *10*(6-7), 601–605.

Kohler, H. P., Kohler-Staub, D., & Focht, D. D. (1988). Degradation of 2-hydroxybiphenyl and 2,2'-dihydroxybiphenyl by Pseudomonas sp. strain HBP1. *Applied and Environmental Microbiology*, *54*(11), 2683–2688.

Kumari, A., Pasini, P., & Daunert, S. (2008). Detection of bacterial quorum sensing N-acyl homoserine lactones in clinical samples. *Analytical and Bioanalytical Chemistry*, *391*(5), 1619–1627. http://dx.doi.org/10.1007/s00216-008-2002-3.

Kumari, A., Pasini, P., Deo, S. K., Flomenhoft, D., Shashidhar, H., & Daunert, S. (2006). Biosensing systems for the detection of bacterial quorum signaling molecules. *Analytical Chemistry*, *78*(22), 7603–7609. http://dx.doi.org/10.1021/ac061421n.

Kunkel, B., Losick, R., & Stragier, P. (1990). The Bacillus subtilis gene for the development transcription factor sigma K is generated by excision of a dispensable DNA element containing a sporulation recombinase gene. *Genes & Development*, *4*(4), 525–535.

Lan, E. H., Dunn, B., & Zink, J. I. (2005). Nanostructured systems for biological materials. *Methods in Molecular Biology*, *300*, 53–79. http://dx.doi.org/10.1385/1-59259-858-7:053.

Lee, J. H., Mitchell, R. J., Kim, B. C., Cullen, D. C., & Gu, M. B. (2005). A cell array biosensor for environmental toxicity analysis. *Biosensors & Bioelectronics*, *21*(3), 500–507. http://dx.doi.org/10.1016/j.bios.2004.12.015.

Lewis, J. C., Feltus, A., Ensor, C. M., Ramanathan, S., & Daunert, S. (1998). Peer reviewed: Applications of reporter genes. *Analytical Chemistry*, *70*(17), 579A–585A. http://dx.doi.org/10.1021/ac9819638.

Li, D.-W., Zhai, W.-L., Li, Y.-T., & Long, Y.-T. (2014). Recent progress in surface enhanced Raman spectroscopy for the detection of environmental pollutants. *Microchimica Acta*, *181*(1), 23–43. http://dx.doi.org/10.1007/s00604-013-1115-3.

Liu, X., Germaine, K. J., Ryan, D., & Dowling, D. N. (2010). Whole-cell fluorescent biosensors for bioavailability and biodegradation of polychlorinated biphenyls. *Sensors (Basel, Switzerland)*, *10*(2), 1377–1398. http://dx.doi.org/10.3390/s100201377.

López-Fouz, M., Pilar-Izquierdo, M. C., Martínez-Mayo, I., Ortega, N., Pérez-Mateos, M., & Busto, M. D. (2007). Immobilization of Rhodococcus fascians cells in poly(vinyl alcohol) cryogels for the debittering of citrus juices. *Journal of Biotechnology*, *131*(2 Suppl.), S104. http://dx.doi.org/10.1016/j.jbiotec.2007.07.179.

Ludewig, G., & Robertson, L. W. (2013). Polychlorinated biphenyls (PCBs) as initiating agents in hepatocellular carcinoma. *Cancer Letters*, *334*(1), 46–55. http://dx.doi.org/10.1016/j.canlet.2012.11.041.

Mahmoud, K. A., & Zourob, M. (2013). Fe3O4/Au nanoparticles/lignin modified microspheres as effectual surface enhanced Raman scattering (SERS) substrates for highly selective and sensitive detection of 2,4,6-trinitrotoluene (TNT). *Analyst*, *138*(9), 2712–2719. http://dx.doi.org/10.1039/c3an00261f.

Manzanera, M., García de Castro, A., Tøndervik, A., Rayner-Brandes, M., Strøm, A. R., & Tunnacliffe, A. (2002). Hydroxyectoine is superior to trehalose for anhydrobiotic engineering of pseudomonas putida KT2440. *Applied and Environmental Microbiology*, *68*(9), 4328–4333. http://dx.doi.org/10.1128/AEM.68.9.4328-4333.2002.

Manzanera, M., Vilchez, S., & Tunnacliffe, A. (2004). High survival and stability rates of Escherichia coli dried in hydroxyectoine. *FEMS Microbiology Letters*, *233*(2), 347–352. http://dx.doi.org/10.1016/j.femsle.2004.03.005.

Marques, S., & Ramos, J. L. (1993). Transcriptional control of the Pseudomonas putida TOL plasmid catabolic pathways. *Molecular Microbiology*, *9*(5), 923–929.

Marston, A. L., & Errington, J. (1999). Dynamic movement of the ParA-like Soj protein of B. subtilis and its dual role in nucleoid organization and developmental regulation. *Molecular Cell*, *4*(5), 673–682.

Mehl, R. A., & Begley, T. P. (1999). Mechanistic studies on the repair of a novel DNA photolesion: The spore photoproduct. *Organic Letters*, *1*(7), 1065–1066.

Moir, A., Corfe, M. B., & Behravan, J. (2002). Spore germination. *Cellular and Molecular Life Sciences: CMLS*, *59*(3), 403–409. http://dx.doi.org/10.1007/s00018-002-8432-8.

Molle, V., Nakaura, Y., Shivers, R. P., Yamaguchi, H., Losick, R., Fujita, Y., & Sonenshein, A. L. (2003). Additional targets of the Bacillus subtilis global regulator CodY identified by chromatin immunoprecipitation and genome-wide transcript analysis. *Journal of Bacteriology*, *185*(6), 1911–1922.

Mulchandani, A., Mulchandani, P., Kaneva, I., & Chen, W. (1998). Biosensor for direct determination of organophosphate nerve agents using recombinant Escherichia coli with surface-expressed organophosphorus hydrolase. 1. Potentiometric microbial electrode. *Analytical Chemistry*, *70*(19), 4140–4145. http://dx.doi.org/10.1021/ac9805201.

Nassif, N., Bouvet, O., Noelle Rager, M., Roux, C., Coradin, T., & Livage, J. (2002). Living bacteria in silica gels. *Nature Materials*, *1*(1), 42–44.

Naylor, L. H. (1999). Reporter gene technology: The future looks bright. *Biochemical Pharmacology*, *58*(5), 749–757. http://dx.doi.org/10.1016/S0006-2952(99)00096-9.

O'Connor, G., Knecht, L. D., Salgado, N., Strobel, S., Pasini, P., & Daunert, S. (2015). Whole-cell biosensors as tools for the detection of quorum-sensing molecules: Uses in diagnostics and the investigation of the quorum-sensing mechanism. *Advances in Biochemical Engineering/Biotechnology*, *154*, 181–200. http://dx.doi.org/10.1007/10_2015_337.

Olivo, P. D., Collins, P. L., Peeples, M. E., & Schlesinger, S. (1998). Detection and quantitation of human respiratory syncytial virus (RSV) using minigenome cDNA and a Sindbis virus replicon: A prototype assay for negative-strand RNA viruses. *Virology*, *251*(1), 198–205. http://dx.doi.org/10.1006/viro.1998.9419.

Ortega, R. (2002). Analytical methods for heavy metals in the environment. In B. Sarkar (Ed.), *Heavy metals in the environment.* Boca Raton, FL: CRC Press.

Paidhungat, M., & Setlow, P. (2000). Role of ger proteins in nutrient and nonnutrient triggering of spore germination in Bacillus subtilis. *Journal of Bacteriology*, *182*(9), 2513–2519.

Paidhungat, M., Setlow, B., Driks, A., & Setlow, P. (2000). Characterization of spores of Bacillus subtilis which lack dipicolinic acid. *Journal of Bacteriology*, *182*(19), 5505–5512.

Pang, S., Yang, T., & He, L. (2016). Review of surface enhanced Raman spectroscopic (SERS) detection of synthetic chemical pesticides. *TrAC, Trends in Analytical Chemistry*, *85*, 73–82. http://dx.doi.org/10.1016/j.trac.2016.06.017.

Park, M., Tsai, S.-L., & Chen, W. (2013). Microbial biosensors: Engineered microorganisms as the sensing machinery. *Sensors*, *13*(5), 5777.

Perkins, E. J. (1986). The biological accumulation and monitoring of chemical wastes in arctic waters. *Water Science and Technology*, *18*(2), 1–11.

Peter, J., Hutter, W., Stöllnberger, W., & Hampel, W. (1996). Detection of chlorinated and brominated hydrocarbons by an ion sensitive whole cell biosensor. *Biosensors and Bioelectronics*, *11*(12), 1215–1219. http://dx.doi.org/10.1016/0956-5663(96)88086-9.

Philp, J. C., Balmand, S., Hajto, E., Bailey, M. J., Wiles, S., Whiteley, A. S., … Dunbar, S. A. (2003). Whole cell immobilised biosensors for toxicity assessment of a wastewater treatment plant treating phenolics-containing waste. *Analytica Chimica Acta*, *487*(1), 61–74. http://dx.doi.org/10.1016/S0003-2670(03)00358-1.

Polyak, B., Geresh, S., & Marks, R. S. (2004). Synthesis and characterization of a biotin-alginate conjugate and its application in a biosensor construction. *Biomacromolecules*, *5*(2), 389–396. http://dx.doi.org/10.1021/bm034454a.

Premkumar, J. R., Sagi, E., Rozen, R., Belkin, S., Modestov, A. D., & Lev, O. (2002). Fluorescent bacteria encapsulated in sol–gel derived silicate films. *Chemistry of Materials*, *14*(6), 2676–2686. http://dx.doi.org/10.1021/cm020020v.

Price, K. D., & Losick, R. (1999). A four-dimensional view of assembly of a morphogenetic protein during sporulation in Bacillus subtilis. *Journal of Bacteriology*, *181*(3), 781–790.

Qu, L. L., Li, D. W., Xue, J. Q., Zhai, W. L., Fossey, J. S., & Long, Y. T. (2012). Batch fabrication of disposable screen printed SERS arrays. *Lab on a Chip*, *12*(5), 876–881. http://dx.doi.org/10.1039/c2lc20926h.

Ramanathan, S., Shi, W., Rosen, B. P., & Daunert, S. (1997). Sensing antimonite and arsenite at the subattomole level with genetically engineered bioluminescent bacteria. *Analytical Chemistry*, *69*(16), 3380–3384. http://dx.doi.org/10.1021/ac970111p.

Raut, N., Joel, S., Pasini, P., & Daunert, S. (2015). Bacterial autoinducer-2 detection via an engineered quorum sensing protein. *Analytical Chemistry*, *87*(5), 2608–2614. http://dx.doi.org/10.1021/ac504172f.

Raut, N., Pasini, P., & Daunert, S. (2013). Deciphering bacterial universal language by detecting the quorum sensing signal, autoinducer-2, with a whole-cell sensing system. *Analytical Chemistry*, *85*(20), 9604–9609. http://dx.doi.org/10.1021/ac401776k.

Roda, A., Guardigli, M., Calabria, D., Calabretta, M. M., Cevenini, L., & Michelini, E. (2014). A 3D-printed device for a smartphone-based chemiluminescence biosensor for lactate in oral fluid and sweat. *Analyst*, *139*(24), 6494–6501. http://dx.doi.org/10.1039/c4an01612b.

Roda, A., Michelini, E., Cevenini, L., Calabria, D., Calabretta, M. M., & Simoni, P. (2014). Integrating biochemiluminescence detection on smartphones: Mobile chemistry platform for point-of-need analysis. *Analytical Chemistry*, *86*(15), 7299–7304. http://dx.doi.org/10.1021/ac502137s.

Roda, A., Mirasoli, M., Michelini, E., Di Fusco, M., Zangheri, M., Cevenini, L., … Simoni, P. (2016). Progress in chemical luminescence-based biosensors: A critical review. *Biosensors & Bioelectronics*, *76*, 164–179. http://dx.doi.org/10.1016/j.bios.2015.06.017.

Rothert, A., Deo, S. K., Millner, L., Puckett, L. G., Madou, M. J., & Daunert, S. (2005). Whole-cell-reporter-gene-based biosensing systems on a compact disk microfluidics platform. *Analytical Biochemistry*, *342*(1), 11–19. http://dx.doi.org/10.1016/j.ab.2004.10.048.

Ruan, C., Luo, W., Wang, W., & Gu, B. (2007). Surface-enhanced Raman spectroscopy for uranium detection and analysis in environmental samples. *Analytica Chimica Acta*, *605*(1), 80–86. http://dx.doi.org/10.1016/j.aca.2007.10.024.

Sangal, A., Pasini, P., & Daunert, S. (2011). Stability of spore-based biosensing systems under extreme conditions. *Sensors and Actuators B: Chemical*, *158*(1), 377–382. http://dx.doi.org/10.1016/j.snb.2011.06.039.

Scherrer, R., Beaman, T. C., & Gerhardt, P. (1971). Macromolecular sieving by the dormant spore of Bacillus cereus. *Journal of Bacteriology*, *108*(2), 868–873.

Setlow, P. (1988). Small, acid-soluble spore proteins of Bacillus species: Structure, synthesis, genetics, function, and degradation. *Annual Review of Microbiology*, *42*, 319–338. http://dx.doi.org/10.1146/annurev.mi.42.100188.001535.

Setlow, P. (1994). Mechanisms which contribute to the long-term survival of spores of Bacillus species. *Journal of Applied Bacteriology*, *76*, 49S–60S. http://dx.doi.org/10.1111/j.1365-2672.1994.tb04357.x.

Shetty, R. S., Deo, S. K., Liu, Y., & Daunert, S. (2004). Fluorescence-based sensing system for copper using genetically engineered living yeast cells. *Biotechnology and Bioengineering*, *88*(5), 664–670. http://dx.doi.org/10.1002/bit.20331.

Shetty, R. S., Deo, S. K., Shah, P., Sun, Y., Rosen, B. P., & Daunert, S. (2003). Luminescence-based whole-cell-sensing systems for cadmium and lead using genetically engineered bacteria. *Analytical and Bioanalytical Chemistry*, *376*(1), 11–17. http://dx.doi.org/10.1007/s00216-003-1862-9.

Shetty, R. S., Ramanathan, S., Badr, I. H., Wolford, J. L., & Daunert, S. (1999). Green fluorescent protein in the design of a living biosensing system for L-arabinose. *Analytical Chemistry*, *71*(4), 763–768.

Shin, H. J., Park, H. H., & Lim, W. K. (2005). Freeze-dried recombinant bacteria for on-site detection of phenolic compounds by color change. *Journal of Biotechnology*, *119*(1), 36–43. http://dx.doi.org/10.1016/j.jbiotec.2005.06.002.

Snyder, L., & Champness, W. (1997). Regulation of gene expression. In *Molecular genetics of bacteria* (pp. 265–296). Washington, DC: ASM Press.

Soodan, R. K., Pakade, Y. B., Nagpal, A., & Katnoria, J. K. (2014). Analytical techniques for estimation of heavy metals in soil ecosystem: A tabulated review. *Talanta*, *125*, 405–410. http://dx.doi.org/10.1016/j.talanta.2014.02.033.

Sticher, P., Jaspers, M. C., Stemmler, K., Harms, H., Zehnder, A. J., & van der Meer, J. R. (1997). Development and characterization of a whole-cell bioluminescent sensor for bioavailable middle-chain alkanes in contaminated groundwater samples. *Applied and Environmental Microbiology*, *63*(10), 4053–4060.

Stocker, J., Balluch, D., Gsell, M., Harms, H., Feliciano, J., Daunert, S., ... van der Meer, J. R. (2003). Development of a set of simple bacterial biosensors for quantitative and rapid measurements of arsenite and arsenate in potable water. *Environmental Science & Technology*, *37*(20), 4743–4750.

Struss, A., Pasini, P., Ensor, C. M., Raut, N., & Daunert, S. (2010). Paper strip whole cell biosensors: A portable test for the semiquantitative detection of bacterial quorum signaling molecules. *Analytical Chemistry*, *82*(11), 4457–4463. http://dx.doi.org/10.1021/ac100231a.

Sun, D. X., Stragier, P., & Setlow, P. (1989). Identification of a new sigma-factor involved in compartmentalized gene expression during sporulation of Bacillus subtilis. *Genes & Development*, *3*(2), 141–149.

Tauriainen, S., Karp, M., Chang, W., & Virta, M. (1998). Luminescent bacterial sensor for cadmium and lead. *Biosensors and Bioelectronics*, *13*(9), 931–938. http://dx.doi.org/10.1016/S0956-5663(98)00027-X.

Thomaides, H. B., Freeman, M., Karoui, M. E., & Errington, J. (2001). Division site selection protein DivIVA of Bacillus subtilis has a second distinct function in chromosome segregation during sporulation. *Genes & Development*, *15*(13), 1662–1673. http://dx.doi.org/10.1101/gad.197501.

Tsai, H. F., Tsai, Y. C., Yagur-Kroll, S., Palevsky, N., Belkin, S., & Cheng, J. Y. (2015). Water pollutant monitoring by a whole cell array through lens-free detection on CCD. *Lab on a Chip*, *15*(6), 1472–1480. http://dx.doi.org/10.1039/c4lc01189a.

Tsuji, M. (2015). Useful biomarkers for assessing the adverse health effects of PCBs in allergic children: Pediatric molecular epidemiology. *Environmental Health and Preventive Medicine*, *20*(1), 3–11. http://dx.doi.org/10.1007/s12199-014-0419-1.

Turner, K., Xu, S., Pasini, P., Deo, S., Bachas, L., & Daunert, S. (2007). Hydroxylated polychlorinated biphenyl detection based on a genetically engineered bioluminescent whole-cell sensing system. *Analytical Chemistry*, *79*(15), 5740–5745. http://dx.doi.org/10.1021/ac0705162.

Vashist, S. K., Marion Schneider, E., Zengerle, R., von Stetten, F., & Luong, J. H. T. (2015). Graphene-based rapid and highly-sensitive immunoassay for C-reactive protein using a smartphone-based colorimetric reader. *Biosensors and Bioelectronics*, *66*, 169–176. http://dx.doi.org/10.1016/j.bios.2014.11.017.

Wahlang, B., Song, M., Beier, J. I., Cameron Falkner, K., Al-Eryani, L., Clair, H. B., … Cave, M. C. (2014). Evaluation of Aroclor 1260 exposure in a mouse model of diet-induced obesity and non-alcoholic fatty liver disease. *Toxicology and Applied Pharmacology*, *279*(3), 380–390. http://dx.doi.org/10.1016/j.taap.2014.06.019.

Wu, L. J., & Errington, J. (2003). RacA and the Soj-Spo0J system combine to effect polar chromosome segregation in sporulating Bacillus subtilis. *Molecular Microbiology*, *49*(6), 1463–1475.

Yagur-Kroll, S., Lalush, C., Rosen, R., Bachar, N., Moskovitz, Y., & Belkin, S. (2014). Escherichia coli bioreporters for the detection of 2,4-dinitrotoluene and 2,4,6-trinitrotoluene. *Applied Microbiology and Biotechnology*, *98*(2), 885–895. http://dx.doi.org/10.1007/s00253-013-4888-8.

Yu, D., Volponi, J., Chhabra, S., Brinker, C. J., Mulchandani, A., & Singh, A. K. (2005). Aqueous sol-gel encapsulation of genetically engineered Moraxella spp. cells for the detection of organophosphates. *Biosensors & Bioelectronics*, *20*(7), 1433–1437. http://dx.doi.org/10.1016/j.bios.2004.04.022.

CHAPTER THREE

Engineering BRET-Sensor Proteins

Remco Arts[1], Stijn J.A. Aper[1], Maarten Merkx[2]
Laboratory of Chemical Biology and Institute for Complex Molecular Systems, Eindhoven University of Technology, Eindhoven, The Netherlands
[2]Corresponding author: e-mail address: m.merkx@tue.nl

Contents

Abstract

FRET-sensors have become important tools for intracellular imaging, but their dependence on external illumination presents some limitations, such as photobleaching and phototoxicity, which limit measurements over extended periods of time. Fluorescence measurements also suffer from autofluorescence and light scattering, which hampers in vivo imaging and measurements in strongly absorbing and scattering media such as blood. In principle, these issues can be resolved by using sensors based on bioluminescence resonance energy transfer (BRET). The recent development of brighter and more stable luciferases and the concomitant improvement in luciferase substrates have substantially decreased the sensitivity gap between fluorescence and bioluminescence. As a result, the application of BRET-sensors is no longer restricted to measurements on cell populations, but they can also be used for imaging of single living cells, and BRET has started to emerge as an attractive sensor format for point-of-care diagnostics. The aim of this chapter is to first provide a brief overview of the basic design principles for BRET-sensors. Next, important design considerations will be discussed in more detail by describing the development of three different classes of BRET-sensors, both from our own work and that of others. These examples are all based on the NanoLuc luciferase, a bright and very stable blue light-emitting luciferase developed by Promega that has quickly risen to prominence in the bioluminescence field.

[1] Both authors contributed equally to this work.

Methods in Enzymology, Volume 589
ISSN 0076-6879
http://dx.doi.org/10.1016/bs.mie.2017.01.010

1. INTRODUCTION

Technological advances in both probe development and imaging technology have revolutionized the possibilities of fluorescence and other optical detection methods for interrogating the molecular details of life with increasing sensitivity and resolution (Ozawa, Yoshimura, & Kim, 2013). An important advantage of optical detection methods is their noninvasiveness allowing, e.g., intracellular imaging in living cells with high spatial and temporal resolution. Protein engineering has made essential contributions to these fields, not only by contributing fluorescent and bioluminescent proteins with an ever increasing range of properties (color, brightness, and optical switching/activation) but also by contributing sensor proteins whose optical properties are modulated by specific molecular interactions. A successful and well-known class of genetically encoded fluorescent sensor proteins are those based on the modulation of Förster resonance energy transfer (FRET) between a donor and an acceptor fluorescent domain (Campbell, 2009; Newman, Fosbrink, & Zhang, 2011). The strong dependence of the energy transfer efficiency on both the distance and the orientation of the fluorophores provides a generic mechanism to translate conformational changes in a receptor domain into a change in fluorescent spectral properties (Lindenburg & Merkx, 2014; Tamura & Hamachi, 2014). An important advantage of FRET-based sensors is that their response is ratiometric, meaning that the ratio between the donor and acceptor emission provides a measure of the ligand concentration that is independent of (changes in) the absolute sensor concentration and less sensitive to background fluorescence compared to intensity-based sensors.

While FRET-sensors have become important tools for intracellular imaging, the requirement for external illumination also presents some limitations. Prolonged exposure to excitation light can result in photobleaching and the generation of toxic radicals, both of which limit extended measurements (Boute, Jockers, & Issad, 2002; Welsh & Noguchi, 2012). External illumination can also give rise to autofluorescence and light scattering, hampering in vivo imaging and other applications in strongly absorbing or scattering media such as cell suspensions and blood. Finally, fluorescence is ill-suited for use in optogenetics or cells that are endogenously light sensitive. In principle, all of these issues can be resolved by using sensors based on bioluminescence resonance energy transfer (BRET) (Bacart, Corbel, Jockers, Bach, & Couturier, 2008; De, Arora, & Jasani, 2014; Machleidt et al., 2015; Paley & Prescher, 2014). In these sensors, the donor fluorescent domain is replaced by a luciferase enzyme, which generates light by catalyzing the oxidation of specific organic substrate molecules. The absence of autofluorescence and scattering provides for a very low background, which by itself makes bioluminescent assays very sensitive. However, a key difference between fluorescence and bioluminescence is the number of photons that are generated. Excitation of a single fluorophore generates thousands of photons per second. In contrast, the light output of bioluminescence is much lower since generation of each photon requires the oxidation of a new substrate molecule.

Thus far, bioluminescence has, therefore, been used predominantly as a reporter gene, both in cell-based assays and in small animal in vivo imaging studies. The recent development of brighter and more stable luciferases and the concomitant improvement in luciferase substrates have substantially decreased the sensitivity gap between fluorescence and bioluminescence, however (Saito & Nagai, 2015). As a result, the use of BRET-sensors is no longer restricted to intracellular population-based assays, but they can also be used for imaging of single living cells, and have started to emerge as an attractive sensor format for point-of-care diagnostics. The aim of this chapter is to first provide a brief overview of the basic design principles for BRET-sensors. Next, important design considerations will be discussed in more detail by describing the development of different classes of BRET-sensors, both from our own work and that of others. These examples are all based on the NanoLuc luciferase, a bright and very stable blue light-emitting luciferase developed by Promega that has quickly risen to prominence in the bioluminescence field (Dixon et al., 2016; England, Ehlerding, & Cai, 2016; Hall et al., 2012; Machleidt et al., 2015; Stoddart et al., 2015).

1.1 BRET (and FRET)

FRET and BRET are both based on resonance energy transfer (RET), a photophysical phenomenon, where excitation energy is transferred via nonradiative dipole–dipole coupling from a donor to an acceptor chromophore. The efficiency of RET is strongly dependent on the distance between the chromophores, as described by the Förster equation:

$$E = \frac{R_0^6}{R_0^6 + r^6}, \tag{1}$$

where the Förster distance, R_0, is given by:

$$R_0 = 0.21\left[\kappa^2 Q_D n^{-4} J(\lambda)\right]^{1/6}. \quad (2)$$

Energy transfer efficiency, thus, not only depends on the distance between the chromophores, but it is also dependent on the quantum yield of the donor (Q_D), the spectral overlap between donor emission and acceptor absorption ($J(\lambda)$), the refractive index of the medium (n), and an orientation factor κ^2, which is related to the relative orientation of the donor emission and acceptor absorption dipole moments. The Förster distance for commonly used FRET- and BRET-pairs is typically between 40 and 60 Å (Dacres, Wang, Dumancic, & Trowell, 2010; Patterson, Piston, & Barisas, 2000). The orientation factor κ^2 is one of the most important contributors to changes in energy transfer efficiency in FRET- and BRET-sensors, but unlike distance, it is not easily controlled. Therefore, optimization of the ratiometric response of these sensors often involves varying the relative order of domains (N- and C-terminal fusions), the variation of linker lengths, and the use of circularly permuted protein domains. This is particularly important for those systems where the change in distance between the chromophores in the on- and off-state is not very large.

1.2 Luciferases for Integration in BRET-Sensors

An important consideration when designing a BRET-sensor is the choice of the donor luciferase. Until recently, the most popular luciferases were firefly luciferase and Renilla luciferase(De et al., 2014; Saito & Nagai, 2015). Firefly luciferase is a 61-kDa protein that emits orange light (λ_{max} = 562 nm) by oxidation of D-luciferin, which requires ATP as a cosubstrate. While the orange emission is attractive for in vivo imaging and BRET-sensors have been developed with red-shifted fluorescent protein acceptors, the dim emission and its dependence on ATP and Mg^{2+} have hampered its broad application as a donor in BRET-sensors. Renilla luciferase is a 36-kDa enzyme originally isolated from the sea pansy *Renilla reniformis*. An important advantage of Renilla luciferase and other luciferases from marine sources is that they do not require ATP and catalyze the direct oxidation of coelenterazine and other imidazopyrazinone-like substrates to form a dioxetane intermediate. By release of CO_2, this unstable intermediate decomposes to an excited state coelenteramide anion, which emits blue light of approximately 480 nm. Protein engineering efforts have yielded a number of Renilla luciferase variants with improved stability, brightness, and emission wavelength (Loening, Fenn, Wu, & Gambhir, 2006; Loening,

Wu, & Gambhir, 2007). Another way to enhance the performance of bioluminescent reporter systems is by modification of the luciferase substrates, which has yielded substrates showing increased brightness, a reduced sensitivity for autooxidation, different spectral properties, or enhanced bioavailability (Jiang, Du, & Li, 2016). Also, a range of "caged" substrates have been developed that can be used to report on the presence of specific enzymes such as β-galactosidase, glycosidase, and caspases (Adams & Miller, 2014; Jiang et al., 2016). Most BRET-sensors developed to date are based on Renilla luciferase or its improved variants such as the brighter RLuc8. The choice of the acceptor domain is often a compromise between good spectral overlap of the donor emission and acceptor excitation on the one hand, and maximal spectral separation between donor and acceptor emission on the other hand. While GFP has a better spectral overlap with the Renilla luciferase emission, YFP, or one of its optimized variants such as Venus or Citrine, are typically preferred as the acceptor domain, because of better spectral separation between the donor and acceptor emission. For example, the group of Nagai developed a bioluminescent Ca^{2+} indicator consisting of the calmodulin–M13 domain sandwiched in between Venus and RLuc8 (Saito et al., 2010). The BRET-based response of this sensor was optimized by testing circularly permuted variants of Venus and by changing the order of the domains, yielding a variant with a ~60% dynamic range that allowed single-cell Ca^{2+} imaging. Other groups developed BRET-sensors for cGMP and cAMP imaging by inserting specific ligand-binding domains in between GFP or circularly permuted Citrine and Renilla luciferase (Biswas, Sopory, & Visweswariah, 2008; Jiang et al., 2007).

In addition to their modest photon output, BRET-sensors using Renilla luciferase and other classical luciferases have some other drawbacks. They typically show a burst of luminescence followed by a rapid decrease in signal, which hampers time-dependent measurements. Both Renilla and firefly luciferases also have limited thermodynamic stability, reducing their applicability in intracellular measurements. Many of these drawbacks have been successfully addressed by the development of NanoLuc (Hall et al., 2012). NanoLuc is a small (19 kDa) engineered luciferase based on the catalytic domain of a luciferase isolated from the deep sea shrimp *Oplophorus gracilirostris*. Following several rounds of directed evolution and screening for improved brightness, thermodynamic stability and substrate preference, a thermodynamically very stable luciferase was obtained that in combination with the substrate furimazine was shown to be 100-fold brighter than Renilla and firefly luciferases (Figs. 1 and 2A, C, and D). NanoLuc produces

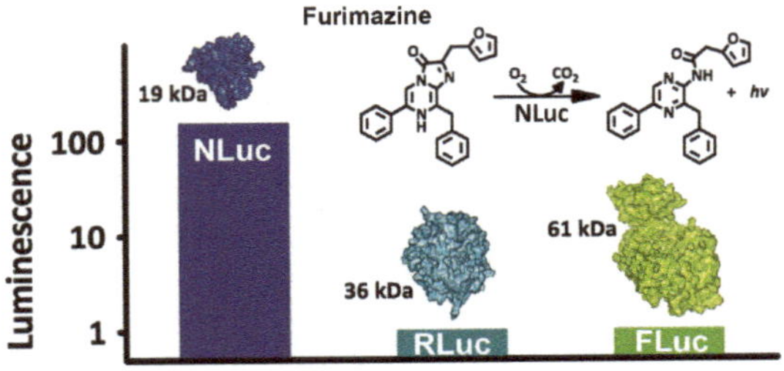

Fig. 1 Relative luminescence brightness of NanoLuc (NLuc) compared to Renilla luciferase (RLuc) and firefly luciferase (FLuc), when converting the optimized imidazopyrazinone substrate furimazine. In addition, the sizes and molecular weights of the luciferases are indicated. *Reprinted with permission from Hall, M. P., Unch, J., Binkowski, B. F., Valley, M. P., Butler, B. L., Wood, M. G., et al. (2012). Engineered luciferase reporter from a deep sea shrimp utilizing a novel imidazopyrazinone substrate.* ACS Chemical Biology, 7, *1848–1857.*

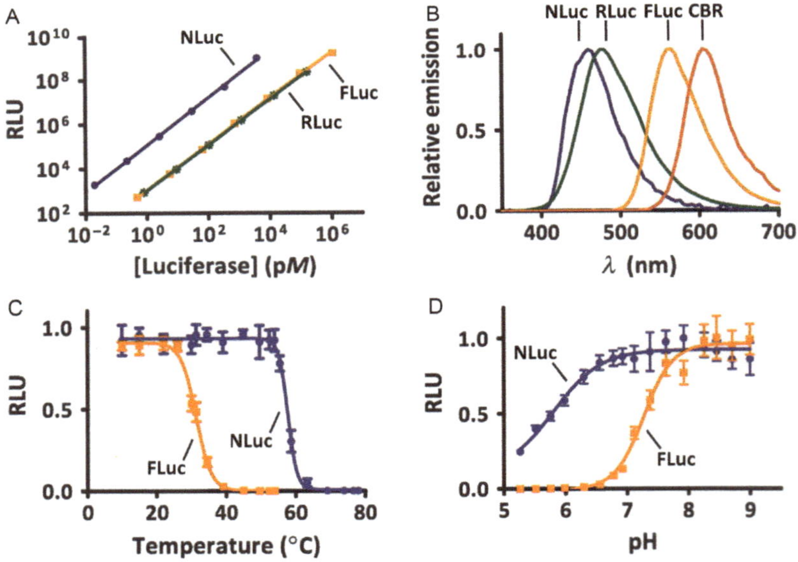

Fig. 2 (A) Bioluminescence emission intensities of NanoLuc (NLuc), Renilla luciferase (RLuc), and firefly luciferase (FLuc), for a range of luciferase concentrations. *RLU*, relative luminescence units. (B) Bioluminescence emission spectra of NLuc, RLuc, FLuc, and click beetle red luciferase (CBR), when converting substrates furimazine, coelenterazine, and D-luciferin (both FLuc and CBR), respectively. (C and D) Sensitivity of NLuc and FLuc to increasing temperature (C) or pH (D). *Reprinted with permission from Hall, M. P., Unch, J., Binkowski, B. F., Valley, M. P., Butler, B. L., Wood, M. G., et al. (2012). Engineered luciferase reporter from a deep sea shrimp utilizing a novel imidazopyrazinone substrate.* ACS Chemical Biology, 7, *1848–1857.*

a glow-type luminescence that is relatively stable over time. Its emission peak at 460 nm is similar to that of Renilla luciferase (Fig. 2B), making it an ideal donor to cyan and green fluorescent proteins and synthetic dyes such as Cy3. However, the relatively long tail in the emission peak above 500 nm also allows it to be used as a BRET donor for orange/red fluorescent proteins and red-shifted dyes. NanoLuc, thus, represents an attractive donor luciferase for BRET-based assays, either in screening for protein–protein interactions or when used in BRET-sensors (Griss et al., 2014; Machleidt et al., 2015; Robers et al., 2015; Schena, Griss, & Johnsson, 2015; Stoddart et al., 2015). In a recent study, Perroy and coworkers directly compared the performance of RLuc8 and NanoLuc as BRET donors in a BRET-sensor for ERK activity. The latter was found to be superior in terms of spatial and temporal resolution, signal stability, and dynamic range (Goyet, Bouquier, Ollendorff, & Perroy, 2016).

2. LUCIDs AND LUMABS: BRET-SENSORS FOR DIAGNOSTIC APPLICATIONS

Until recently, luciferases were not considered for incorporation in point-of-care diagnostic technology, mainly because of their modest brightness and stability. The 100-fold increase in brightness and glow-type luminescence offered by NanoLuc, combined with its remarkable thermal stability and pH insensitivity, have changed this situation, making NanoLuc-based BRET-sensors an attractive new choice class of sensor proteins for use in point-of-care applications. Here, we discuss two recently developed BRET-sensor platforms, LUCiferase-based Indicators of Drugs (LUCIDs) and LUMABS that allow the detection of small-molecule drugs and antibodies, respectively.

2.1 LUCIDs

Point-of-care diagnostic tests for the detection of small-molecule drugs allow monitoring of drug concentrations in individual patients, enabling the administration of a patient-optimized dose. In 2014, Johnsson and coworkers reported on the development of LUCIDs, a series of NanoLuc-based BRET-sensor proteins that can detect a variety of small-molecule drugs (Griss et al., 2014). The LUCID sensor architecture was adopted from Snifit sensors, a series of semisynthetic FRET-based sensor proteins developed by the same group (Figs. 3A and 4A). Both sensors

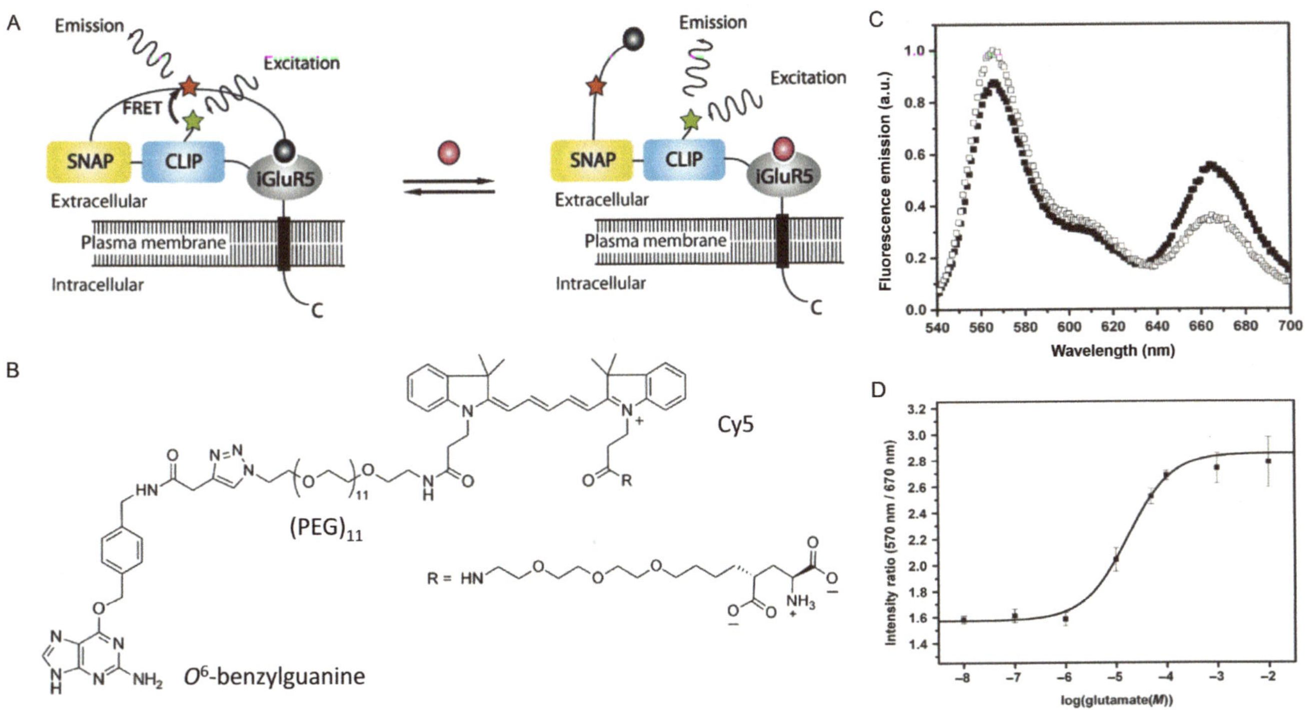

Fig. 3 (A) Sensing mechanism of Snifit-iGluR5 sensor for glutamate detection. The Snifit sensor consists of the glutamate binding protein iGluR5-S1S2 and a CLIP-tag, which is fused to a SNAP-tag via a 15-proline linker. The synthetic part that is linked to the SNAP-tag consists of fluorophore Cy5 (*red star*) and a glutamate analog (*gray ball*). The CLIP-tag is labeled with a DY-547 fluorophore (*green star*). Glutamate (*pink ball*) binding to iGluR5-S1S2 displaces the glutamate analog, decreasing FRET between the two fluorophores. (B) Chemical structure of the synthetic molecule that is attached to the SNAP-tag of the Snifit sensor. (C) Fluorescence emission spectra of Snifit-iGluR5 at low (10 n*M*; *closed circle*) and high (10 m*M*; *open circle*) glutamate levels. (D) Fluorescence emission ratio of Snifit-iGluR5 in the presence of a range of glutamate concentrations. *Reprinted with permission from Brun, M. A., Tan, K. T., Griss, R., Kielkowska, A., Reymond, L., & Johnsson, K. (2012). A semisynthetic fluorescent sensor protein for glutamate.* Journal of the American Chemical Society, 134, *7676–7678.*

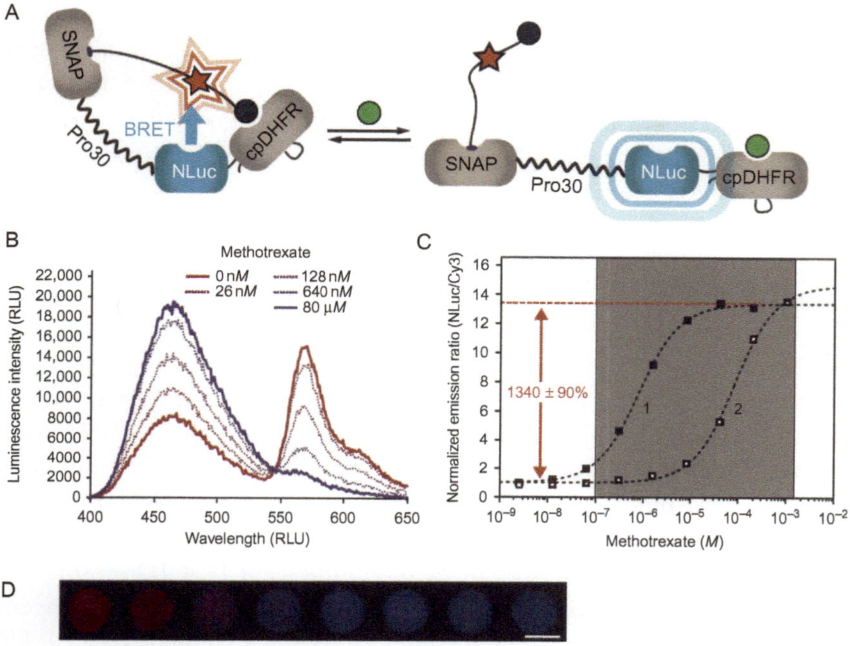

Fig. 4 (A) Sensing mechanism of the LUCID sensor for methotrexate detection. The LUCID sensor consists of a fusion of circularly permuted dihydrofolate reductase (cpDHFR) and NanoLuc (NLuc), which is connected to a SNAP-tag via a rigid 30-proline linker. A synthetic molecule containing fluorophore Cy3 (*red star*) and a DHFR inhibitor (either trimethoprim (TMP) or methotrexate; *gray ball*) is fused to the SNAP-tag. Methotrexate (*green ball*) binding to cpDHFR displaces the inhibitor, decreasing BRET between NLuc and Cy3. (B) Bioluminescence emission spectra of the methotrexate LUCID sensor in human serum that was spiked with a range of methotrexate concentrations. *RLU*, relative luminescence units. (C) Affinity variation of methotrexate LUCID sensors by changing the intramolecular inhibitor (1: trimethoprim; 2: methotrexate) allows analyte quantification over a broad concentration regime. (D) Bioluminescence emission of the methotrexate LUCID sensor was recorded using a digital camera, showing a clear color change as a consequence of increasing methotrexate concentrations in human serum. *Reprinted with permission from Griss, R., Schena, A., Reymond, L., Patiny, L., Werner, D., Tinberg, C. E., et al. (2014). Bioluminescent sensor proteins for point-of-care therapeutic drug monitoring.* Nature Chemical Biology, 10, *598–603.*

are based on competition between intramolecular and intermolecular binding to a ligand-binding domain, and switch from a high-FRET/BRET closed conformation (intramolecular ligand binding) to a low-FRET/BRET state upon intermolecular ligand binding. In the Snifit sensors, the ligand-binding domain is genetically fused to a fluorescent donor protein or a CLIP-tag domain, followed by a linker and SNAP-tag domain

(Fig. 3A) (Brun et al., 2012; Brun, Tan, Nakata, Hinner, & Johnsson, 2009). The SNAP- and CLIP-tags are self-labeling protein domains that can be specifically functionalized with a molecule of interest. In Snifit, the CLIP-tag is labeled with a donor fluorophore, while the SNAP-tag is used to introduce both an acceptor fluorophore and a ligand that can form an intramolecular complex with the receptor domain (Fig. 3B). In the absence of the analyte of interest, the Snifit sensor adopts a closed conformation, keeping the fluorophores in close proximity, which leads to efficient FRET. Binding of analyte displaces the intramolecular ligand from the receptor domain, increasing the distance between the two fluorophores and thereby decreasing FRET (Fig. 3C and D). In recent years, the ratiometric response of the Snifit sensors has been further optimized by introducing a relatively rigid 30 amino-acid polyproline linker between the CLIP- and the SNAP-tag domains (Brun et al., 2011), and by introducing donor fluorophores directly into the ligand-binding domain by the site-specific incorporation of non-natural amino acids (Xue, Prifti, & Johnsson, 2016).

The main difference between the bioluminescent LUCID sensors and the fluorescence-based Snifit sensors is the replacement of the fluorescent donor domain by NanoLuc (Fig. 4A). In LUCIDs, Cy3 is used as the BRET acceptor because its excitation spectrum shows sufficient overlap with the NanoLuc emission, whereas its emission peak at 565 nm is well separated from that of NanoLuc. The first LUCID sensor was designed to detect the anticancer drug methotrexate. Bacterial dihydrofolate reductase (DHFR) was used as the receptor domain, whereas trimethoprim (TMP), a weaker binding analog of methotrexate, was selected as a suitable intramolecular ligand. The initial design showed a 218% change in emission ratio upon addition of methotrexate with a C50 (concentration at which the signal change is 50%) of 0.75 ± 0.04 μ*M*. Since the BRET efficiency in the open, analyte-bound state was already very low, the dynamic range was further improved by optimizing the BRET efficiency in the closed state. This was achieved by using a circularly permuted dihydrofolate reductase (cpDHFR) variant, which relocated the receptor domain's N-terminus to a loop right next to the active site. Since the acceptor fluorophore is located next to the intramolecular ligand this permutation decreased the distance between NanoLuc and the acceptor fluorophore in the closed state, resulting in a sixfold increase of the dynamic range to $1340 \pm 90\%$ (Fig. 4B and C). Conveniently, the concentration regime in which LUCIDs are most responsive can be adjusted by changing the affinity of the intramolecular ligand. In the sensor described earlier, the sensor affinity could be effectively attenuated by replacing TMP with methotrexate, increasing the C50 to $85 \pm 6\ \mu M$ (Fig. 4C). This ability to rationally adjust the C50s by tuning the affinity of the competing, intramolecular ligand binding is important, because it allows one to broaden the responsive concentration regime, in this case between 100 n*M* and 500 μ*M*.

An important challenge associated with using optical sensors in blood is the problem of patient-to-patient variation in autofluorescence and absorbance. Even though the lack of autofluorescence and diminished light scattering makes bioluminescence more compatible with detection in blood plasma, Johnsson and coworkers still observed that the emission spectrum of LUCID was sensitive to physiologically relevant concentrations of the light-absorbing bilirubin when using the LUCID sensors in blood serum. An effective strategy to suppress the sensitivity to bilirubin concentration, which is known to vary between individuals, was to perform the assay on filter paper to minimize the distance that the bioluminescent light has to travel through blood plasma. Using a simple digital camera installed in a Styrofoam box to exclude ambient light, the transition from red to blue induced by increasing methotrexate concentrations was easily observed and the emission ratios could be determined using ImageJ software (Fig. 4D). A comparison of this paper-based LUCID assay with a standard polarization immunoassay on 30 patient samples yielded very comparable results.

To demonstrate the modularity of the LUCID system, the methotrexate-specific DHFR domain was replaced by ligand-binding domains targeting a range of other small molecules, including tacrolimus, sirolimus, cyclosporin A, topiramate, and digoxin. Although the modular design of the sensor proteins allowed straightforward exchange of receptor domains and intramolecular ligands, in all cases optimization of BRET in the absence of analyte was required. In some cases, such as in the cyclosporine A-binding LUCID, this was achieved through circular permutation of the receptor domain. In the digoxin-binding LUCID, the order of the domains in the sensor was inverted because in this case, the active site of a receptor domain was closer to the C- than to the N-terminus.

To extend the range of analytes to, e.g., protein targets, the Johnsson group also reported a second class of BRET-sensors called CLASH (Chemical Ligand-Associated Steric Hindrance). CLASH proteins have the same overall structure as LUCIDs, following the SNAP-Pro30-NanoLuc-receptor domain design. Whereas LUCIDs and Snifits rely on competition between intra- and intermolecular ligand binding to the same

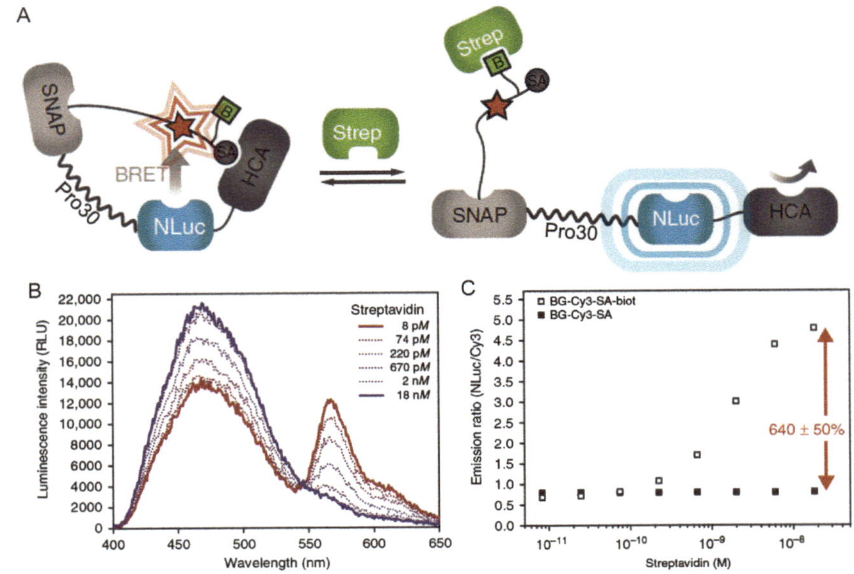

Fig. 5 (A) Sensing mechanism of a CLASH sensor for the detection of streptavidin. This CLASH sensor (CLASH-Strep/HCA) consists of human carbonic anhydrase II (HCA) and NanoLuc (NLuc), which is fused to a SNAP-tag via a 30-proline linker. The synthetic part that is linked to the SNAP-tag consists of fluorophore Cy3 (*red star*) and two ligands, biotin (B) and benzene sulfonamide (SA). The binding of streptavidin (Strep) to biotin prevents the mutually exclusive interaction between benzene sulfonamide and HCA, which leads to a decrease of BRET between NLuc and Cy3. (B) Bioluminescence emission spectra of CLASH-Strep/HCA for increasing streptavidin concentrations. *RLU*, relative luminescence units. (C) Bioluminescence emission ratio of CLASH-Strep/HCA (*open square*) or a control that lacks the tethered biotin ligand in the synthetic part (*closed square*), obtained for a titration of streptavidin. *Reprinted with permission from Schena, A., Griss, R., & Johnsson, K. (2015). Modulating protein activity using tethered ligands with mutually exclusive binding sites.* Nature Communications, 6, *7830.*

ligand-binding domain, the CLASH sensors contain two ligands, one of which can bind intramolecularly to a receptor domain, while the other binds to the target protein. Because of steric hindrance, binding of the synthetic ligands to their respective receptor domains is mutually exclusive (Schena et al., 2015). Fig. 5A shows an example of a proof-of-principle design for the detection of streptavidin. This sensor used human carbonic anhydrase as the receptor domain, which binds intramolecularly to the synthetic ligand benzenesulfonamide. The other ligand that was conjugated to the SNAP-tag was biotin. Binding of streptavidin to the biotin ligand disrupts the intramolecular interaction between the carbonic anhydrase and the sulfonamide ligand, switching the sensor from a high to a low-BRET state (Fig. 5B and C). Using a similar design, a CLASH sensor sensitive to DHFR was obtained by using methotrexate instead of biotin as the competing ligand.

2.2 LUMABS—Smartphone-Based Antibody Detection

Therapeutic antibodies represent an important class of newly introduced drugs and have been particularly successful in cancer therapy and the treatment of inflammatory diseases. The clearance rates of therapeutic antibodies are known to differ between individuals and correlate with survival rates. Being able to define a personalized dose per patient based on continuous monitoring of therapeutic antibody levels would, therefore, be beneficial both for the individual patient and for the hospital's budget. Our group recently developed a new platform of BRET-sensors for antibody detection named LUMABS (Arts et al., 2016). These NanoLuc-based sensor proteins allow measurement of antibodies directly in blood plasma, eliminating the need for any washing steps. LUMABS are not only attractive for therapeutic antibody monitoring but could also be used for the point-of-care diagnosis of infectious diseases or monitoring the success of vaccination campaigns. The sensor principle employed in LUMABS is based on an approach first introduced by our group in the development of FRET-based antibody sensor proteins called AbSense (Fig. 6). This fluorescent sensor consists of self-associating variants of Cerulean and Citrine connected by a semiflexible linker, with antibody-specific epitope sequences incorporated at both ends of the linker (Golynskiy, Rurup, & Merkx, 2010). In the absence of antibody, Cerulean and Citrine form an intramolecular complex through specifically introduced hydrophobic interactions, giving rise to efficient FRET (Fig. 6A). Bivalent binding of an antibody to the epitope sequences in the linker disrupts the intramolecular interaction of the two fluorescent domains, resulting in a large decrease in FRET efficiency. Although the AbSense sensor showed a robust 5.5-fold change in emission ratio that allowed detection of antibodies directly in solution without washing steps, the sensor suffered from the general drawbacks of fluorescence-based sensors discussed earlier, notably the occurrence of autofluorescence and scattering. In addition, the requirement for external illumination of the fluorescent proteins made the concept less attractive for point-of-care applications. In the LUMABS platform, we replaced the donor fluorescent domain Cerulean by NanoLuc, whereas mNeonGreen was used as the fluorescent acceptor domain (Fig. 6B). mNeonGreen is one of the brightest fluorescent proteins

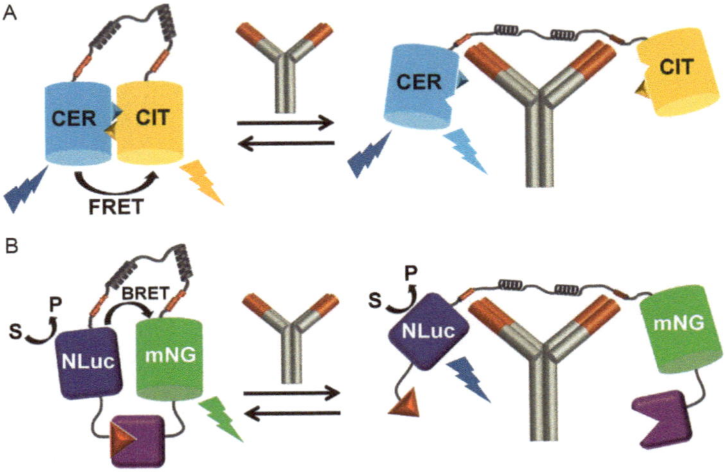

Fig. 6 Sensing mechanisms of AbSense (A) and LUMABS (B) sensors. Antibody binding to the two epitopes disrupts the interaction between self-associating fluorescent domains Cerulean (CER) and Citrine (CIT) (A), or increases the distance between NanoLuc (NLuc) and mNeonGreen (mNG), which are kept in close proximity by helper domains (B), allowing FRET- and BRET-based readout, respectively. *Adapted with permission from Arts, R., den Hartog, I., Zijlema, S. E., Thijssen, V., van der Beelen, S. H. E., & Merkx, M. (2016). Detection of antibodies in blood plasma using bioluminescent sensor proteins and a smartphone.* Analytical Chemistry, 88, *4525-4532; Golynskiy, M. V., Rurup, W. F., & Merkx, M. (2010). Antibody detection by using a FRET-based protein conformational switch.* ChemBioChem, 11, *2264–2267.*

and its excitation spectrum strongly overlaps with the emission of NanoLuc, ensuring efficient energy transfer. To ensure efficient BRET in the absence of antibody, helper domains that keep the BRET partners in close proximity were introduced on both protein termini. The introduction of antiparallel leucine zippers as helper domains increased the BRET efficiency in the absence of antibody, while still allowing the formation of the low-BRET antibody complex. Further, optimization of the BRET efficiency in the absence of antibody was achieved by replacing the leucine zipper domains by an SH3 domain from *Saccharomyces cerevisiae* on the N-terminus and a proline-rich peptide on the C-terminus. The first LUMABS sensor developed using this design targeting the anti-HIV1-p17 antibody (HIV-LUMABS) showed a robust change in bioluminescence emission ratio of ~200%. In the absence of antibody, an intense mNeonGreen emission peak was observed, which almost completely disappeared upon addition of the target antibody. A titration experiment in which the emission ratio was measured as a function of antibody concentration revealed a dissociation constant for the antibody-sensor complex of 80 p*M*. This high affinity is due to the bivalent interaction between the antibody and the sensor, which substantially enhances the affinity and specificity of antibody binding compared to the monovalent interaction. As in the LUCID sensors, it is important to be able to generate a series of sensors with different affinities that, when used together, allow reliable measurements over a wide analyte concentration range. In LUMABS, the most straightforward way to adapt the sensor affinity is by tuning the interaction strength of the antibody–antigen interaction. Mutating the aspartic acid at position three of the epitope (ELDRWEKIRLR) in HIV-LUMABS to alanine attenuated the sensor's affinity tenfold (0.6 n*M*), expanding the range of antibody sensor concentrations from the low p*M* to the low n*M* regime. Because of the bivalent interaction between the LUMABS sensor and antibody, an x-fold change in monovalent antigen–antibody interaction will translate into an x^2-fold change in the sensor's affinity. However, the monovalent antibody–antigen interaction cannot be weakened indefinitely, as a K_d of ~1 μ*M* is required for the sensor to switch from the closed to the open state. For example, deletion of the first four amino acids of the epitope sequences in HIV-LUMABS, resulted in a variant that was no longer responsive and remained in the high BRET closed state, even though the shortened epitope still bound the antibody with a K_d of 3 μ*M*. In fact, this property may render the LUMABS sensors inherently less sensitive to nonspecific background binding. As in the LUCID system, the modular design of the LUMABS sensors allows one to change the sensor's target specificity simply by replacing the epitope sequences in the linker domain. This was demonstrated by creating HA-LUMABS and DEN1-LUMABS sensors that target antibodies raised against the hemagglutinin (HA) tag and the NS1 protein of dengue virus 1, respectively. HA-LUMABS had an antibody affinity comparable with that of HIV-LUMABS. DEN1-LUMABS showed a tenfold lower affinity, which is due to the lower monovalent affinity of the DEN1 epitope.

An important aspect of the applicability of LUMABS is its performance in blood plasma. The emission spectra of HIV-LUMABS obtained in undiluted blood plasma in the absence and presence of antibody were shown to be very similar to those obtained in buffer (Fig. 7A). The emission ratio could be monitored continuously for at least 60 min. While the absolute luminescence signal will decrease with time because of substrate turnover, the emission ratio remains remarkably stable (Fig. 7B). Importantly, HIV-LUMABS still responds to the addition of antibody after being incubated in blood plasma for 30 min, showing that the sensor is stable over this

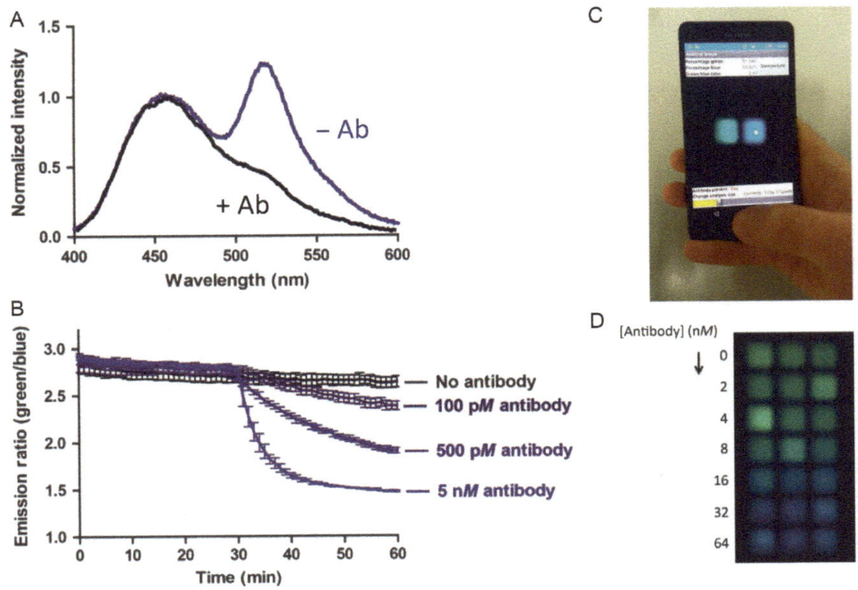

Fig. 7 (A) Bioluminescence emission spectrum of the LUMABS sensor for detection of anti-HIV1-p17 antibody (HIV-LUMABS) in blood plasma in the absence (– Ab) or presence (+ Ab) of 1 n*M* of antibody. (B) Bioluminescence emission ratio of HIV-LUMABS in blood plasma, which was spiked with the indicated levels of anti-HIV1-p17 at $t=30$ min. (C) Antibody detection in blood plasma can be recorded using a conventional mobile phone camera. The emission ratio can be determined using an Android application that determines the RGB values in the picture. Here, the bioluminescent emission of HIV-LUMABS (1 n*M* sensor) in blood plasma in the absence and presence of 2 n*M* of antibody is shown. (D) Mobile phone photograph showing the color changes of DEN1-LUMABS (LUMABS sensor for detection of antidengue-1; 1 n*M* sensor) in blood plasma in response to increasing antibody concentrations. *Reprinted with permission from Arts, R., den Hartog, I., Zijlema, S. E., Thijssen, V., van der Beelen, S. H. E., & Merkx, M. (2016). Detection of antibodies in blood plasma using bioluminescent sensor proteins and a smartphone.* Analytical Chemistry, 88, *4525-4532.*

period of time. To test the sensitivity of the LUMABS assay, the sensor's luminescence was recorded using the photo camera of a conventional mobile phone (Fig. 7C). Even using a relatively low sensor concentration of 1 n*M*, the bioluminescent signal was readily detected, showing a clear transition from green–blue to blue (Fig. 7D). A smartphone application that determines the RGB values in the selected wells was developed that allowed the determination of the green/blue ratio from these images. Mobile phone-based antibody detection proved comparably reliable to plate reader-based antibody detection, when detecting nanomolar amounts of antidengue virus 1 antibodies using the DEN1-LUMABS sensor.

3. BRET/FRET-SENSORS FOR INTRACELLULAR MEASUREMENTS

BRET-sensors are often developed starting from an existing FRET-sensor by replacing the fluorescent domains by a luciferase and a matching fluorescent acceptor domain. The examples discussed earlier demonstrate that this strategy can be successful, but in both cases the original FRET-sensors underwent a well-defined and robust conformational change between two mutually exclusive conformations. Many FRET-sensors rely on more subtle allosteric mechanisms; however, the development of these FRET-sensors typically involved extensive optimization of linker lengths, testing various domain orientations, and circular permutations. For these FRET-sensors, simple replacement of the donor fluorescent protein by a luciferase domain typically does not yield a useful BRET-sensor. Instead of replacing the donor fluorescent domain by a luciferase, we recently introduced an alternative design strategy in which a luciferase domain is simply fused close to the donor fluorescent domain. This approach allows the introduction of bioluminescent detection, while retaining the previously optimized FRET-sensor properties. NanoLuc is an attractive luciferase for this approach because its emission spectrum overlaps with that of cyan fluorescent proteins such as Cerulean, the donor fluorescent domain used in most FRET-sensors developed to date.

The feasibility of the BRET/FRET approach was explored by fusing NanoLuc to the donor Cerulean domain of two previously developed FRET-sensors for intracellular Zn^{2+} imaging, the eCALWY and eZinCh-2 sensors (Aper, Dierickx, & Merkx, 2016). In these sensors, bioluminescent excitation of Cerulean takes place via BRET, and this energy is subsequently transferred to the yellow acceptor Citrine via FRET (Figs. 8A and 9A). As the emission spectrum of NanoLuc also shows some overlap with the excitation peak of Citrine, direct BRET between NanoLuc and Citrine additionally affects the sensor's response to ligand binding. NanoLuc was attached to the N-terminus of Cerulean in eCALWY-1 using a short, flexible GSGSG-linker to allow independent folding of NanoLuc and Cerulean. In the eCALWY sensor, the two ligand-binding domains Atox1 and WD4, which together can coordinate one Zn^{2+} ion using four cysteines, are fused via a long, flexible linker and flanked by self-associating Cerulean and Citrine domains (Vinkenborg et al., 2009). In the absence of Zn^{2+}, the interaction of the fluorescent

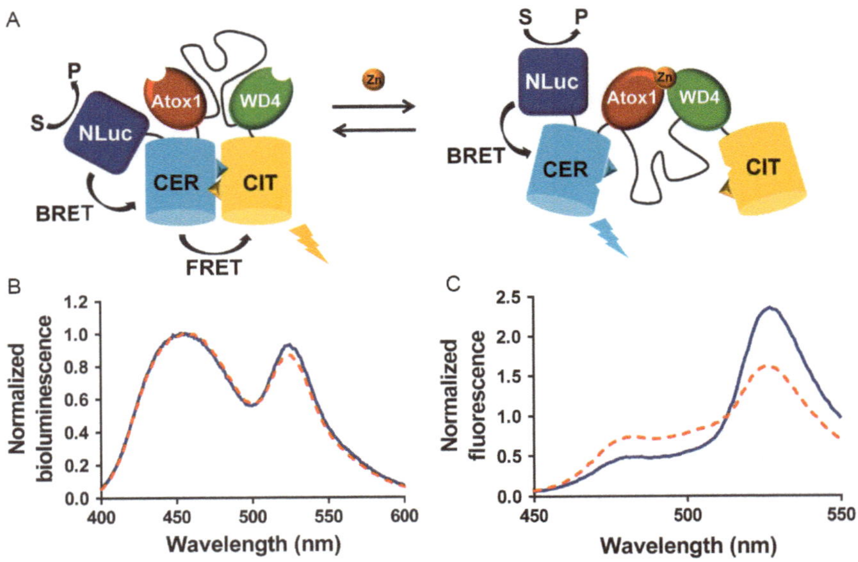

Fig. 8 (A) Sensing mechanism of the BLCALWY-1 sensor. In BLCALWY-1, Cerulean (CER) and Citrine (CIT) each displays two self-associating residues on their exterior, inducing a high-FRET state in the absence of Zn^{2+}. Coordination of one Zn^{2+} ion by two cysteines in both Atox1 and WD4 increases the distance between the fluorescent domains, resulting in a decrease in FRET. The N-terminally fused NanoLuc (NLuc) domain ensures bioluminescent excitation of Cerulean (and Citrine) via BRET. (B and C) Bioluminescence (B: normalized to emission at 455 nm) and fluorescence (C: normalized to emission at 513 nm; excitation at 430 nm) emission spectra of BLCALWY-1 in Zn^{2+}-depleted (*blue, straight*) and Zn^{2+}-saturated states (*red, dashed*). *Adapted with permission from Aper, S. J., Dierickx, P., & Merkx, M. (2016). Dual readout BRET/FRET sensors for measuring intracellular zinc.* ACS Chemical Biology, 11, *2854–2864.*

domains results in a high-FRET state, which is disrupted by Zn^{2+} complexation, inducing a decrease of FRET (Fig. 8A). The bioluminescence spectrum of the new sensor called BLCALWY-1 showed a combined emission peak of NanoLuc and Cerulean, and a peak originating from Citrine (Fig. 8B). Zn^{2+} binding led to a small, but significant drop (~7%) of the bioluminescent emission ratio of both peaks. Both FRET- and BRET-based readout of the sensor resulted in a similar K_d of ~4 pM for Zn^{2+} binding, which corresponded well to previous reports on eCALWY-1 (Vinkenborg et al., 2009). The introduction of NanoLuc, thus, had a minimal effect on the binding and switching properties of the sensor. However, the change of the bioluminescent emission ratio was much smaller than the ~2-fold decrease observed for the fluorescent

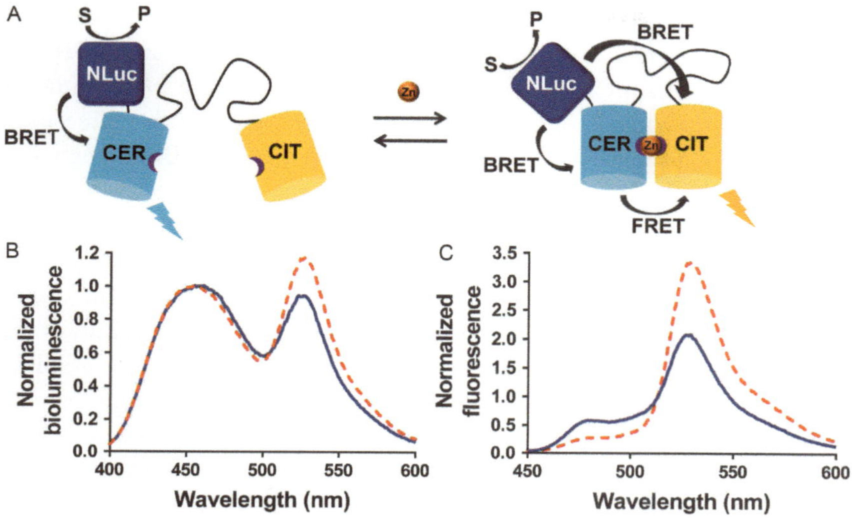

Fig. 9 (A) Sensing mechanism of the BLZinCh-1 sensor. In BLZinCh-1, Cerulean (CER) and Citrine (CIT) each displays two Zn^{2+}-coordinating amino acids (His-206 and Cys-208) on their surface, which together form a high-affinity tetrahedral Zn^{2+} coordination site. The binding of Zn^{2+} brings the two fluorescent domains in close proximity, increasing FRET. The N-terminally fused NanoLuc (NLuc) domain ensures bioluminescent excitation of Cerulean and Citrine via BRET. (B and C) Bioluminescence (B: normalized to emission at 455 nm) and fluorescence emission spectra (C: normalized to emission at 513 nm; excitation at 430 nm) of BLZinCh-1 in Zn^{2+}-depleted (*blue, straight*) and Zn^{2+}-saturated states (*red, dashed*). *Adapted with permission from Aper, S. J., Dierickx, P., & Merkx, M. (2016). Dual readout BRET/FRET sensors for measuring intracellular zinc.* ACS Chemical Biology, 11, *2854–2864.*

emission ratio (Fig. 8C). This may be explained by direct energy transfer from NanoLuc to Citrine, which could have been affected differently by Zn^{2+} binding, and therefore, could have counteracted the spectral changes related to the decrease in FRET. In addition, because of incomplete BRET between NanoLuc and Cerulean, the change of the Citrine/Cerulean emission ratio had to be detected on top of a large, Zn^{2+}-insensitive background emission from NanoLuc.

The second Zn^{2+} FRET-sensor tested, eZinCh-2, is based on a different sensor mechanism. In contrast to the eCALWY sensors, eZinCh-2 does not contain distinct Zn^{2+}-binding domains, but has a Zn^{2+}-binding histidine and cysteine residue located on the surface of both Cerulean and Citrine, which are fused via a long, flexible linker (Fig. 9A) (Hessels et al., 2015). The coordination of one Zn^{2+} ion in between the two histidines and two cysteines brings the two fluorescent domains close together, inducing a ~5-fold increase of FRET. eZinCh-2 was converted into BRET/FRET-sensor BLZinCh-1 by connecting NanoLuc to the N-terminus of its Cerulean domain. A tight fusion was made leaving out the three most C-terminal NanoLuc residues to maximize the efficiency of BRET between NanoLuc and Cerulean (BLZinCh-1). The crystal structure of NanoLuc showed that these residues were not essential for the structure and function of the luciferase (PDB: 5IBO). The BLZinCh-1 sensor demonstrated a robust decrease of the bioluminescent emission ratio (25%) upon Zn^{2+} binding (Fig. 9B). Again, the attachment of NanoLuc to the N-terminus of the sensor did not affect the Zn^{2+} binding properties, because the affinities observed using BRET- and FRET-based detection were similar (Fig. 9C). The larger dynamic range of the BLZinCh-1 sensor compared to the BLCALWY-1 sensor can be explained by the different sensor mechanisms. Zn^{2+} binding to BLZinCh-1 brings both Cerulean and NanoLuc in closer proximity of Citrine, enhancing FRET between the fluorescent domains, and BRET between NanoLuc and Citrine. Therefore, both energy transfer pathways contribute to an increase of the bioluminescent emission ratio. To investigate the importance of each of the energy transfer pathways, a variant called BLZinCh-3 was constructed, in which the Cerulean chromophore was mutated to eliminate its fluorescence, while the Zn^{2+} binding properties of the fluorescent domain were preserved. BLZinCh-3 showed an increased dynamic range of ~50%, which demonstrates that direct energy transfer from NanoLuc to Citrine significantly contributed to the bioluminescent response of the sensor. The enhanced dynamic range of BLZinCh-3 can be attributed to the improved spectral separation of the NanoLuc and Citrine emission peaks, which led to a significantly lower bioluminescent emission ratio in the absence of Zn^{2+}.

Because bioluminescent detection is not affected by autofluorescence and light scattering, BRET-sensors are expected to be superior compared to FRET-sensors in high-throughput intracellular measurements on cell suspensions. Providing both bioluminescent and fluorescent readout, the BRET/FRET-sensors allowed a head-to-head comparison of both detection strategies in a plate reader. HeLa cells transfected with BLZinCh-1 displayed very similar bioluminescence spectra for a range of cell dilutions, allowing sensitive determination of intracellular Zn^{2+} levels with high signal-to-noise ratio and reproducibility (Fig. 10A and C). In contrast, the fluorescence spectra were dominated by a large scattering peak, preventing readout of the Zn^{2+}-induced FRET response (Fig. 10B).

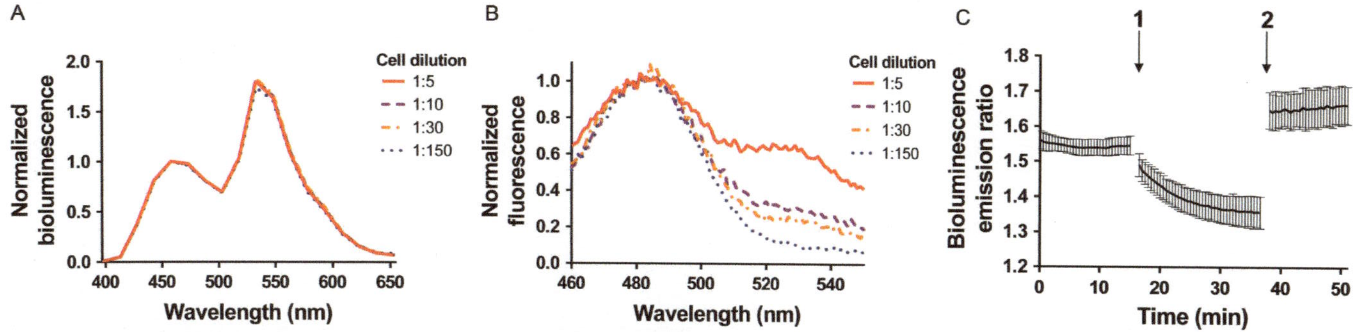

Fig. 10 Bioluminescence (A) and fluorescence (B) emission spectra of suspensions of HeLa cells expressing the BLZinCh-1 sensor. Two days after transfection, cells were trypsinized and resuspended in 1 mL live-cell imaging buffer. From this stock, various dilutions were prepared as indicated. (C) Bioluminescence emission ratio (emission 500–545 nm/emission 400–455 nm) of HeLa cells expressing BLZinCh-1, in resting state and after subsequent additions of 50 μ*M* of membrane-permeable Zn^{2+} chelator TPEN (1), and 100 μ*M* Zn^{2+} together with 5 μ*M* of Zn^{2+} transporter pyrithione (2). *Adapted with permission from Aper, S. J., Dierickx, P., & Merkx, M. (2016). Dual readout BRET/FRET sensors for measuring intracellular zinc.* ACS Chemical Biology, 11, *2854–2864.*

BRET-based detection is, thus, preferred over FRET for intracellular measurements in suspension. The absolute bioluminescent emission ratios obtained in HeLa cells were also very similar to those measured in vitro for the purified sensor proteins using the same instrument.

Due to their limited photon output, to date only a few studies have been reported in which BRET-sensors were used for live-cell imaging on single cells, often requiring the use of highly sensitive detectors and long integration times (Goyet et al., 2016; Saito et al., 2012, 2010; Takai et al., 2015). To test whether the improved brightness of NanoLuc-based BRET-sensors is also beneficial for BRET-based intracellular imaging in living cells, we performed imaging experiments on HeLa cells transfected with BLZinCh-1. Using a recently developed microscope that is optimal for bioluminescence detection, integration times of less than 1 s were found to be sufficient to obtain reliable ratiometric images and observe changes in intracellular Zn^{2+} following exposure to Zn^{2+} chelators or excess Zn^{2+}(Fig. 11) (Aper et al., 2016). The Zn^{2+} levels obtained using BRET detection were similar

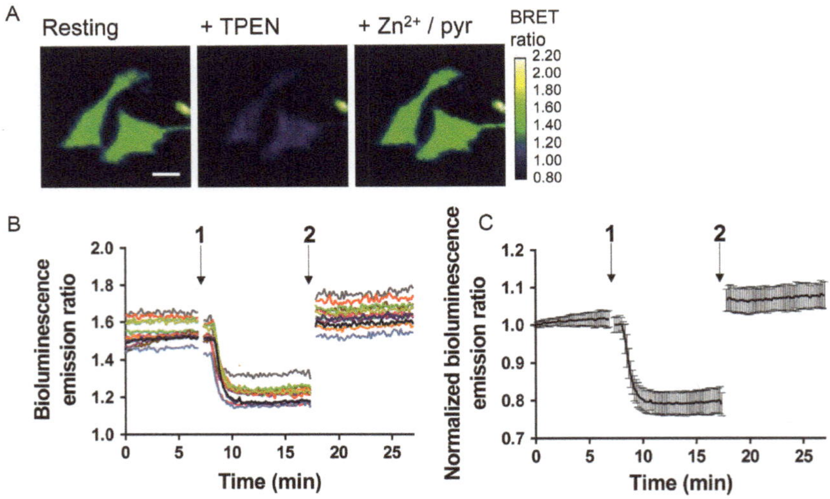

Fig. 11 (A) False-colored ratiometric images of two HeLa cells expressing BLZinCh-1, in resting state and after additions of 50 μ*M* TPEN and 100 μ*M* Zn^{2+}/5 μ*M* pyrithione. Scale bar = 20 μm. (B) Bioluminescence emission ratio traces of 12 cells expressing BLZinCh-1, in resting state and after additions of 50 μ*M* TPEN (1) and 100 μ*M* Zn^{2+}/5 μ*M* pyrithione (2). (C) Average bioluminescence emission ratio of the data shown in (B), after normalization of the emission ratio at t = 0 min. *Reprinted with permission from Aper, S. J., Dierickx, P., & Merkx, M. (2016). Dual readout BRET/FRET sensors for measuring intracellular zinc.* ACS Chemical Biology, 11, *2854–2864.*

to those previously determined using the eZinCh-2 FRET-sensor (Hessels et al., 2015; Hessels, Taylor, & Merkx, 2016), but BRET-based detection resulted in lower cell-to-cell variation in emission ratio because of the absence of background fluorescence. The BLZinCh sensors, thus, not only allow straightforward monitoring of intracellular Zn^{2+} levels using a plate reader, but should also be very useful for all those imaging applications where laser excitation is undesirable, such as prolonged imaging, imaging in optogenetics, imaging in light-sensitive cells, and in vivo imaging.

4. A BRIGHT FUTURE FOR BRET-SENSORS

The development of bright and stable luciferases such as NanoLuc and the continuing improvements in luciferase substrates have significantly expanded the possible applications of BRET-based sensors. Two important new areas for BRET-based detection that have recently become feasible were discussed in this chapter, the use of BRET-sensors for point-of-care diagnostics and the development of BRET/FRET-sensors for intracellular measurements.

While bioluminescence is still underrepresented in point-of-care diagnostics, it has several inherent properties that are very attractive for performing optical measurements in complex matrices such as blood. An important advantage of bioluminescence detection over fluorescence-based assay formats is that the signal can be measured directly in solution without the requirement for a sample processing step and/or the need for complex illumination setups. In fact, bioluminescence detection can be done using a simple digital camera such as the one present on smartphones (McCracken & Yoon, 2016). BRET-based sensors carry the additional advantage of ratiometric detection, providing internal calibration that largely obviates the need for precise calibrations necessary for intensity-based luminescent assays. The LUCID and LUMABS sensors discussed in this chapter provide generic platforms for the detection of either small molecules or antibodies, respectively. Their modular design allows straightforward adaptation to different analytes and systematic tuning of sensor affinity, while also facilitating the incorporation of improved luciferases, acceptor, and helper domains when they become available in the future. The results obtained for the LUCID sensors suggest that the ratiometric response of LUMABS could be further improved by bringing the donor and acceptor closer together via circular permutation of either NanoLuc or the acceptor fluorescent domain. While the use of more red-shifted fluorophores will most likely decrease the BRET efficiency, better spectral separation of the emission peaks could still prove beneficial for distinguishing signal changes upon antibody binding. New color variants are also interesting because they would facilitate multiplexed assays in which multiple analytes can be quantified simultaneously. An important step toward the application of LUCID, LUMABS, and other BRET-based diagnostic assays is the translation from a laboratory-based assay to a fully implemented point-of-care assay. This will require investigating the long-term stability of both sensor and substrate, and the development of cheap and robust test strips that allow multiplexed detection and can be easily read using, e.g., a smartphone camera.

The increased brightness of NanoLuc-based BRET-sensors has also enabled bioluminescent ratiometric imaging on single living cells. BRET-based intracellular imaging is particularly attractive for monitoring cellular processes over prolonged periods of time, studying cellular processes that are light sensitive such as circadian rhythms, or for use in combination with optogenetic tools. At present, an advantage of fluorescence imaging over bioluminescence imaging is the higher spatial resolution that can be obtained using confocal fluorescence microscopy. However, Ozawa and coworkers very recently reported a "confocal" version of bioluminescence imaging based on light-activatable caged luciferase substrates (Hattori et al., 2016). These new technologies and the further development of dedicated bioluminescence microscopes will further expand the possibilities offered by BRET-based sensor proteins.

An attractive approach to develop BRET-sensors is to append NanoLuc to previously developed CFP/YFP-based FRET-sensors, yielding dual output BRET/FRET-sensors that allow both fluorescent and bioluminescent detection. A further increase in BRET between NanoLuc and the donor fluorescent domain would be expected to improve the dynamic range of the BRET signal, which currently is still suppressed by the presence of a substantial background of NanoLuc emission. With the recently released crystal structure of NanoLuc as a guide, it should be possible to further increase BRET efficiency by changing the position of the luciferase, or by using different linkers and circularly permuted variants of either NanoLuc or the fluorescent domain. An important application area for BRET-based sensors is in vivo imaging, but the blue light produced by NanoLuc is suboptimal for this purpose, despite its inherent brightness. For these applications, BRET-sensors in which both donor and acceptor emit in the orange–red part of the optical spectrum would be highly desirable. An important recent advance that has improved the brightness of Renilla luciferase and at the same time allowed the tuning of its emission wavelength is to tightly fuse this luciferase domain to a bright fluorescent acceptor domain. Nagai and coworkers showed that these so-called Nanolanterns consisting of RLuc8 (S257G) and a fluorescent acceptor protein produce up to 20-fold more photons than wild-type Renilla luciferase because of a higher overall quantum efficiency, allowing their application in single-cell bioluminescence imaging and the development of bioluminescent sensors for Ca^{2+}, cAMP, and ATP with a dynamic range of 130%–300% (Saito et al., 2012; Takai et al., 2015). Very recently, the same strategy was used to develop bright NanoLuc color variants that emit cyan, green, yellow, orange, and red light. This also allowed the development of a green luminescent Ca^{2+} sensor with increased brightness (Suzuki et al., 2016). The Nanolantern-based sensor proteins are essentially intensiometric, however, since they rely on modulation of the activity of the luciferase domain, and not on modulation of the BRET efficiency between the luciferase and the acceptor fluorescent domain. In analogy to the BRET/FRET-sensor approach described in this chapter, the Nanolanterns could be used as luminescent donors to construct BRET/FRET-sensors that allow both bioluminescent and fluorescent ratiometric detection. Another elegant approach was recently reported by Lin and coworkers, who fused NanoLuc to CyOFP1, a bright orange–red fluorescent protein that is excitable by cyan light (Chu et al., 2016). The high spectral overlap between NanoLuc emission and CyOFP excitation ensures efficient BRET when NanoLuc is tightly coupled to a CyOFP1 domain. Due to the large Stokes shift of CyOFP1, this fusion protein produces bright emission at 589 nm. Using NanoLuc-CyOFP as a luminescent donor in combination with an efficient red fluorescent acceptor domain would yield a bright red-shifted BRET-sensor that, like a Nanolantern-based system, would allow both BRET- and FRET-based detection.

ACKNOWLEDGMENTS

The work on BRET-sensor development in our group was supported by an ERC starting Grant (ERC-2011-StG 280255).

REFERENCES

Adams, S. T., Jr., & Miller, S. C. (2014). Beyond D-luciferin: Expanding the scope of bioluminescence imaging in vivo. *Current Opinion in Chemical Biology*, *21*, 112–120.

Aper, S. J., Dierickx, P., & Merkx, M. (2016). Dual readout BRET/FRET sensors for measuring intracellular zinc. *ACS Chemical Biology*, *11*, 2854–2864.

Arts, R., den Hartog, I., Zijlema, S. E., Thijssen, V., van der Beelen, S. H. E., & Merkx, M. (2016). Detection of antibodies in blood plasma using bioluminescent sensor proteins and a smartphone. *Analytical Chemistry*, *88*, 4525–4532.

Bacart, J., Corbel, C., Jockers, R., Bach, S., & Couturier, C. (2008). The BRET technology and its application to screening assays. *Biotechnology Journal*, *3*, 311–324.

Biswas, K. H., Sopory, S., & Visweswariah, S. S. (2008). The GAF domain of the cGMP-binding, cGMP-specific phosphodiesterase (PDE5) is a sensor and a sink for cGMP. *Biochemistry*, *47*, 3534–3543.

Boute, N., Jockers, R., & Issad, T. (2002). The use of resonance energy transfer in high-throughput screening: BRET versus FRET. *Trends in Pharmacological Sciences*, *23*, 351–354.

Brun, M. A., Griss, R., Reymond, L., Tan, K. T., Piguet, J., Peters, R. J., et al. (2011). Semisynthesis of fluorescent metabolite sensors on cell surfaces. *Journal of the American Chemical Society*, *133*, 16235–16242.

Brun, M. A., Tan, K. T., Griss, R., Kielkowska, A., Reymond, L., & Johnsson, K. (2012). A semisynthetic fluorescent sensor protein for glutamate. *Journal of the American Chemical Society*, *134*, 7676–7678.

Brun, M. A., Tan, K. T., Nakata, E., Hinner, M. J., & Johnsson, K. (2009). Semisynthetic fluorescent sensor proteins based on self-labeling protein tags. *Journal of the American Chemical Society*, *131*, 5873–5884.

Campbell, R. E. (2009). Fluorescent-protein-based biosensors: Modulation of energy transfer as a design principle. *Analytical Chemistry*, *81*, 5972–5979.

Chu, J., Oh, Y., Sens, A., Ataie, N., Dana, H., Macklin, J. J., et al. (2016). A bright cyan-excitable orange fluorescent protein facilitates dual-emission microscopy and enhances bioluminescence imaging in vivo. *Nature Biotechnology*, *34*, 760–767.

Dacres, H., Wang, J., Dumancic, M. M., & Trowell, S. C. (2010). Experimental determination of the Förster distance for two commonly used bioluminescent resonance energy transfer pairs. *Analytical Chemistry*, *82*, 432–435.

De, A., Arora, R., & Jasani, A. (2014). Engineering aspects of bioluminescence resonance energy transfer systems. In W. Cai (Ed.), *Engineering in translational medicine* (pp. 257–300). London: Springer London.

Dixon, A. S., Schwinn, M. K., Hall, M. P., Zimmerman, K., Otto, P., Lubben, T. H., et al. (2016). NanoLuc complementation reporter optimized for accurate measurement of protein interactions in cells. *ACS Chemical Biology*, *11*, 400–408.

England, C. G., Ehlerding, E. B., & Cai, W. B. (2016). NanoLuc: A small luciferase is brightening up the field of bioluminescence. *Bioconjugate Chemistry*, *27*, 1175–1187.

Golynskiy, M. V., Rurup, W. F., & Merkx, M. (2010). Antibody detection by using a FRET-based protein conformational switch. *ChemBioChem*, *11*, 2264–2267.

Goyet, E., Bouquier, N., Ollendorff, V., & Perroy, J. (2016). Fast and high resolution single-cell BRET imaging. *Scientific Reports*, *6*, 28231.

Griss, R., Schena, A., Reymond, L., Patiny, L., Werner, D., Tinberg, C. E., et al. (2014). Bioluminescent sensor proteins for point-of-care therapeutic drug monitoring. *Nature Chemical Biology*, *10*, 598–603.

Hall, M. P., Unch, J., Binkowski, B. F., Valley, M. P., Butler, B. L., Wood, M. G., et al. (2012). Engineered luciferase reporter from a deep sea shrimp utilizing a novel imidazopyrazinone substrate. *ACS Chemical Biology*, *7*, 1848–1857.

Hattori, M., Kawamura, G., Kojima, R., Kamiya, M., Urano, Y., & Ozawa, T. (2016). Confocal bioluminescence imaging for living tissues with a caged substrate of luciferin. *Analytical Chemistry*, *88*, 6231–6238.

Hessels, A. M., Chabosseau, P., Bakker, M. H., Engelen, W., Rutter, G. A., Taylor, K. M., et al. (2015). EZinCh-2: A versatile, genetically encoded FRET sensor for cytosolic and intraorganelle Zn^{2+} imaging. *ACS Chemical Biology*, *10*, 2126–2134.

Hessels, A. M., Taylor, K. M., & Merkx, M. (2016). Monitoring cytosolic and ER Zn^{2+} in stimulated breast cancer cells using genetically encoded FRET sensors. *Metallomics*, *8*, 211–217.

Jiang, L. I., Collins, J., Davis, R., Lin, K. M., DeCamp, D., Roach, T., et al. (2007). Use of a cAMP BRET sensor to characterize a novel regulation of cAMP by the sphingosine 1-phosphate/G13 pathway. *The Journal of Biological Chemistry*, *282*, 10576–10584.

Jiang, T., Du, L., & Li, M. (2016). Lighting up bioluminescence with coelenterazine: Strategies and applications. *Photochemical & Photobiological Sciences*, *15*, 466–480.

Lindenburg, L., & Merkx, M. (2014). Engineering genetically encoded FRET sensors. *Sensors (Basel)*, *14*, 11691–11713.

Loening, A. M., Fenn, T. D., Wu, A. M., & Gambhir, S. S. (2006). Consensus guided mutagenesis of Renilla luciferase yields enhanced stability and light output. *Protein Engineering, Design & Selection*, *19*, 391–400.

Loening, A. M., Wu, A. M., & Gambhir, S. S. (2007). Red-shifted Renilla reniformis luciferase variants for imaging in living subjects. *Nature Methods*, *4*, 641–643.

Machleidt, T., Woodroofe, C. C., Schwinn, M. K., Mendez, J., Robers, M. B., Zimmerman, K., et al. (2015). NanoBRET—A novel BRET platform for the analysis of protein-protein interactions. *ACS Chemical Biology*, *10*, 1797–1804.

McCracken, K. E., & Yoon, J.-Y. (2016). Recent approaches for optical smartphone sensing in resource-limited settings: A brief review. *Analytical Methods*, *8*, 6591–6601.

Newman, R. H., Fosbrink, M. D., & Zhang, J. (2011). Genetically encodable fluorescent biosensors for tracking signaling dynamics in living cells. *Chemical Reviews*, *111*, 3614–3666.

Ozawa, T., Yoshimura, H., & Kim, S. B. (2013). Advances in fluorescence and bioluminescence imaging. *Analytical Chemistry*, *85*, 590–609.

Paley, M. A., & Prescher, J. A. (2014). Bioluminescence: A versatile technique for imaging cellular and molecular features. *Medicinal Chemistry Communications*, *5*, 255–267.

Patterson, G. H., Piston, D. W., & Barisas, B. G. (2000). Förster distances between green fluorescent protein pairs. *Analytical Biochemistry*, *284*, 438–440.

Robers, M. B., Dart, M. L., Woodroofe, C. C., Zimprich, C. A., Kirkland, T. A., Machleidt, T., et al. (2015). Target engagement and drug residence time can be observed in living cells with BRET. *Nature Communications*, *6*, 10091.

Saito, K., Chang, Y. F., Horikawa, K., Hatsugai, N., Higuchi, Y., Hashida, M., et al. (2012). Luminescent proteins for high-speed single-cell and whole-body imaging. *Nature Communications*, *3*, 1262.

Saito, K., Hatsugai, N., Horikawa, K., Kobayashi, K., Matsu-Ura, T., Mikoshiba, K., et al. (2010). Auto-luminescent genetically-encoded ratiometric indicator for real-time Ca^{2+} imaging at the single cell level. *PloS One*, *5*, e9935.

Saito, K., & Nagai, T. (2015). Recent progress in luminescent proteins development. *Current Opinion in Chemical Biology*, *27*, 46–51.

Schena, A., Griss, R., & Johnsson, K. (2015). Modulating protein activity using tethered ligands with mutually exclusive binding sites. *Nature Communications*, *6*, 7830.

Stoddart, L. A., Johnstone, E. K., Wheal, A. J., Goulding, J., Robers, M. B., Machleidt, T., et al. (2015). Application of BRET to monitor ligand binding to GPCRs. *Nature Methods*, *12*, 661–663.

Suzuki, K., Kimura, T., Shinoda, H., Bai, G., Daniels, M. J., Arai, Y., et al. (2016). Five colour variants of bright luminescent protein for real-time multicolour bioimaging. *Nature Communications*, 7, 13718.

Takai, A., Nakano, M., Saito, K., Haruno, R., Watanabe, T. M., Ohyanagi, T., et al. (2015). Expanded palette of nano-lanterns for real-time multicolor luminescence imaging. *Proceedings of the National Academy of Sciences of the United States of America*, *112*, 4352–4356.

Tamura, T., & Hamachi, I. (2014). Recent progress in design of protein-based fluorescent biosensors and their cellular applications. *ACS Chemical Biology*, *9*, 2708–2717.

Vinkenborg, J. L., Nicolson, T. J., Bellomo, E. A., Koay, M. S., Rutter, G. A., & Merkx, M. (2009). Genetically encoded FRET sensors to monitor intracellular Zn^{2+} homeostasis. *Nature Methods*, *6*, 737–740.

Welsh, D. K., & Noguchi, T. (2012). Cellular bioluminescence imaging. *Cold Spring Harbor Protocols*, *2012*, 852–866.

Xue, L., Prifti, E., & Johnsson, K. (2016). A general strategy for the semisynthesis of ratiometric fluorescent sensor proteins with increased dynamic range. *Journal of the American Chemical Society*, *138*, 5258–5261.

CHAPTER FOUR

Enzymes as Sensors

Maria Staiano, Angela Pennacchio, Antonio Varriale, Alessandro Capo, Adelia Majoli, Clotilde Capacchione, Sabato D'Auria[1]
Institute of Food Science, CNR, Avellino, Italy
[1]Corresponding author: e-mail address: sabato.dauria@cnr.it

Contents

Abstract

Over the last few decades the development of new technologies, the fabrication of new materials, and the introduction of nanotechnologies created new trends in a series of advances that produced innovations in biological sensing devices with a wide range of application from health, security, defense, food, and medicine, to the environment. Specificity, low cost, rapidity, sensitivity, and multiplicity are some of the reasons for their growth, and their commercial success is expected to increase in the next future.

Biosensors are devices in which the recognition part of the target molecule is accomplished by biological macromolecules such as proteins, enzymes, antibodies, aptamers, etc. These biomolecules are able to bind to the target molecules with high selectivity and specificity. The interaction between the target molecule and the specific biomolecule is reflected as a change of the biomolecule structural features. The extent of this change is strictly related to the biosensor response. Fluorescence spectroscopy,

Methods in Enzymology, Volume 589
ISSN 0076-6879
http://dx.doi.org/10.1016/bs.mie.2017.01.015

due to its sensitivity, is often used as the principal technique to monitor biological interactions, and thus the biosensor response as well. Both the intrinsic ultraviolet fluorescence of protein, arising from aromatic amino acids (tryptophan, tyrosine, and phenylalanine), and extrinsic fluorescent labels emitting in the visible region of the spectrum together allow for very flexible transduction of the analyte recognition, suitable for many different applications.

This chapter focuses special attention on enzymes as practically unmatched recognition elements for biosensors and emphasizes the potential advantages of customized biosensor devices using *apo-* or *holo* forms of enzymes also isolated from thermophile sources.

1. INTRODUCTION

Over the last few decades the scientific communities synergistically combined technologies from various fields including physics, chemistry, molecular biology, engineering, and bioelectronics, in order to develop innovative biosensor devices. This area is very attractive, both from a commercial and an academic standpoint, because biosensors may have a wide range of applications in health, pharmaceuticals, homeland security, food safety and quality, defense, security application, and environmental monitoring. Both the research and practical applications are expected to increase, in the near future. Actually the global market in the field of biosensors is around US$13 billion for 2015 (Turner, 2013).

The reasons for the growth and success of biosensors include their high specificity, low cost, rapidity, sensitivity, multiplicity, and portability, to name just a few. In fact, these features are of fundamental importance for reliable detection of the target molecules.

The history of biosensors starts in 1962 with scientist L.C. Clark and the invention of the enzyme electrode to measure glucose. However, the scientist K. Cammann used the term "biosensor" for the first time in 1977, and later in 1999 it was included in a technical report by IUPAC. According to the IUPAC paper: "A chemical sensor is a device that transforms chemical information, ranging from the concentration of a specific sample component to total composition analysis, into an analytical signal. Chemical sensors usually contain two basic components connected in series: a chemical recognition system and a physicochemical transducer" (Hulanicki, Geab, & Ingman, 1991). When the recognition system utilizes a biological sensing element, the sensor is termed a biosensor; the term is synonymous with "biological sensor."

It can be divided into the following essential elements:

(a) an analyte, a substance of interest that needs to be detected;

(b) a biological system, bioreceptor, or molecular recognition element (MRE) that acts as a sensing element, that is, a molecule that specifically recognizes the target analyte. The MRE can be an enzyme, a cell, an aptamer, a sequence of deoxyribonucleic acid (DNA), or an antibody;

(c) a transducer system is the element that detects the signal, converting one form of energy into another. Its role is to convert the biorecognition event into a measurable signal. The types of transducer include at least piezoelectric, electrochemical, thermometric, magnetic, and optical;

(d) an electric system that amplifies the signal, quantifies it, and transfers it to an operator.

When the recognition biomolecule interacts with the target analyte, a biomolecule conformational change occurs, and an electric or optical signal is produced and detected by a transducer (Fig. 1).

It is possible to identify at least five different types of biosensors based on transduction principle:

(1) calorimetric biosensors (heat is released or absorbed upon interaction);

(2) potentiometric biosensors (production of an electrical potential due to changed distribution of electrons);

(3) acoustic wave biosensors (change in mass of the biological component as a result of the interaction with the analyte);

(4) amperometric biosensors (measure of electron flux due to redox reaction);

(5) optical biosensors (light produced or absorbed during reaction).

A particular case is represented by the immunochromatographic biosensors. This biosensor type does not exactly use any of transducer strategies listed before, but it is based on the analyte flowing in liquid over stationary lateral flow strips, where the speed of analyte movement is determined by the interaction with (usually) antibody immobilized on the strips, and the (optical) readout is actually the position of the analyte band after a fixed time period. Among the most popular lateral flow strip products are the patient-operated (home) pregnancy and glucose tests (Bahadır & Sezgintürk, 2015).

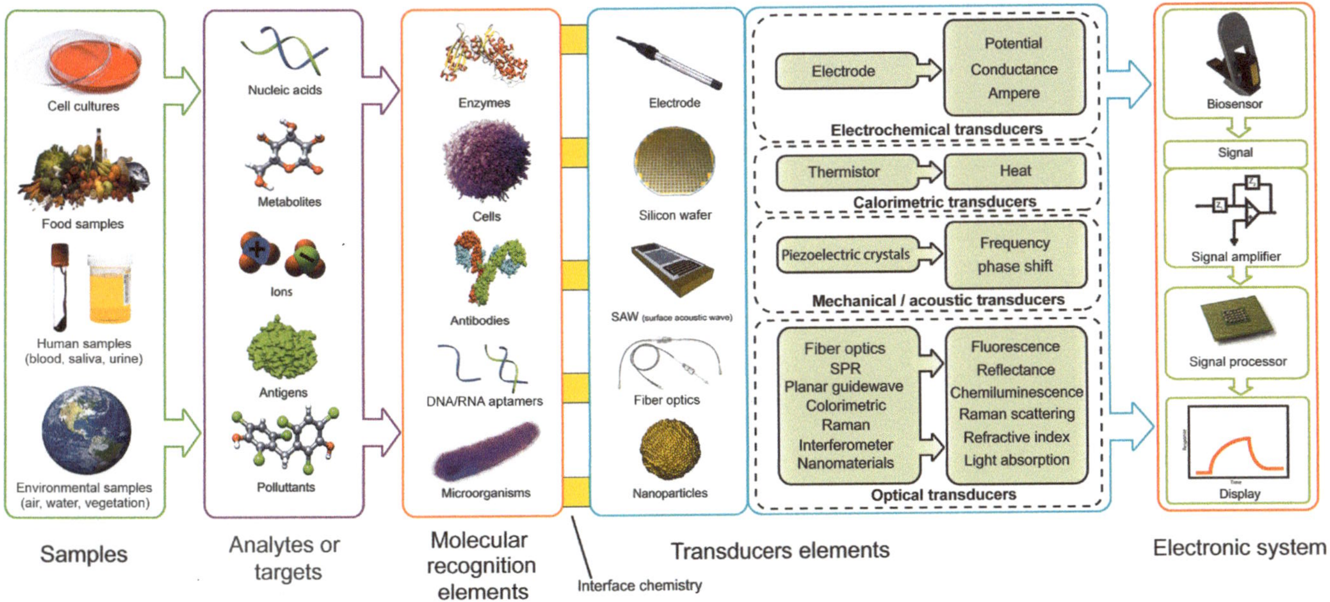

Fig. 1 Schematic representation of a biosensor platform. From *left* to *right*, in each panel, are shown only some samples, analytes or targets, molecular recognition elements (MREs), transducers element, and electronic systems. It is possible to couple the different elements in numerous combinations and obtain several biosensors.

2. OPTICAL BIOSENSORS

Here, we will focus on the class of optical biosensors, and in particular on fluorescence biosensors used for qualitative and quantitative measurements of biomolecular interactions. Fluorescence activity is more suitable for ultrasensing applications to detect and quantify chemical or biological compounds than other, more conventional methods (Martins et al., 2013). The detection of molecular changes upon biological interactions occurs through the measurement of photons involved in the process of absorption and emission in the UV, visible, or near-infrared spectral regions of light.

Biosensors and biosensor test formats, respectively, can be classified into labeled and label-free types depending on whether the output signal caused by analyte binding is obtained using labeled compounds or not.

3. LABEL-FREE DETECTION AND FLUORESCENCE-BASED DETECTION

3.1 Label-Free Detection

This type of detection uses unlabeled target molecules. This choice may be due to the difficulty of performing the labeling procedure. In fact, labeling methods are not always favorable because probes may occupy the binding sites of molecule with a steric hindrance and, consequently, to interfere with the binding process. To avoid this limitation, alternative detection methods based on label-free detection can be utilized. Surface plasmon resonance (SPR), nuclear magnetic resonance analysis, microsurface-enhanced Raman scattering, microsphere cavities, liquid crystal sensors, and calorimetric arrays all represent examples of label-free detection strategies (Gopinath, 2010). An SPR array is based on the evanescent field detection principle and measures the variation of refractive index (RI) upon the occurrence of molecular interaction, in femtoliter to nanoliter volumes of sample, providing information about the specificity, affinity, binding kinetics, and the concentration levels of specific analyte within the 1–10 μm range (Fan et al., 2008). Another label-free biosensor type is based upon variations in RI that occur upon binding; the biorecognition molecules, such as antibodies, aptamers, etc., are immobilized on the surface of chip, and molecular interactions are measured following the RI unit variation (Hoa, Kirk, & Tabrizian, 2007).

3.2 Detection Using Optical Labels

The recent advancements in bioinformatics and probe chemistry, such as the ability to study protein dynamics in a timescale of microseconds, and the commercial availability of new fluorescent organic labels, open promising opportunities to study the role of proteins in cell function, and to deduce the existence of new regulatory mechanisms of proteins in cell life.

The biosensing research area and in particular the design of novel, highly performing biosensors are one of the most promising applications of the advances in chemical biology.

In actual applications, novel proteins usable as biomarkers, and/or the synthesis of new chemical probes for labeling proteins and antibodies, are essential for moving toward rapid diagnoses directly on real matrices such as whole blood (D'Auria, 2013).

Today it is now possible to choose the best method and the best label strategy in order to satisfy the required application. Many parameters are important to consider for selecting suitable organic fluorophore or quantum dot labels with the appropriate properties such as brightness, lifetime, size, and photostability (Toseland, 2013).

In designing a biosensor it is also possible to separate the optical properties of the biosensor (due to the fluorescence properties of the selected fluorophore) from binding specificity and affinity characteristics that are defined by the biomolecule used (D'Auria, Gryczynski, Gryczynski, Rossi, & Lakowicz, 2000).

Multiple biological recognitions integrated in a single device allow the detection of multiple analytes in the same cartridge. This miniaturized device is called "Lab On A Chip" (LOC) platform. The LOC approach began to spread widely at the end of the 1980s and beginning of the 1990s, and it has since found wide application in molecular biology, proteomics, cell biology, and chemistry.

The emerging technologies of LOC devices and nanosensors offer new possibilities such as fully automatic operation, which permits performing several hundred analyses simultaneously on the same chip to detect specific targets in real-time. This reduces human error, increases throughput, and permits performing analyses on site in remote areas without infrastructure. For instance, a current approach is based on electron beam irradiation, to create protein nanopatterns on a porous silicon wafer; other approaches have also been demonstrated for the production of micro- and nanopatterns of biomolecules on solid substrates (Borini et al., 2007).

3.3 Fluorescence-Based Detection

Fluorescent biosensors are usually based on the use of fluorescent organic molecules, nanoparticles, proteins, or combinations of these. They are designed and engineered to change their fluorescent properties in response to external influences or chemical species including pH fluctuations, metal ion homeostasis, cell signaling, membrane potential differences, phosphorylation, ubiquitination, redox reactions, apoptosis, and molecular interactions.

Usually proteins exhibit intrinsic fluorescence due to the presence of the aromatic amino acids tryptophan, tyrosine, and phenylalanine. It is also possible to label proteins, selecting the biomolecule site where to attach the fluorescence probe. Conformational change, protein–protein interaction, single-molecule studies, protein counting, and cell localization are potential applications. There are different strategies to detect and quantify these changes.

Fluorescence correlation spectroscopy (FCS) measures fluctuations of a small number of molecules in a focused laser beam. FCS is based on the analysis of time-dependent intensity fluctuations that are the result of some dynamic process, typically translational diffusion into and out of a small volume (femtoliter) defined by a focused laser beam and a confocal aperture (Varriale et al., 2007). Förster resonance energy transfer (RET) is used to understand the structure and function of biological macromolecules. It is based on the transfer of energy occurring between excited molecule (donor) and ground-state molecule (acceptor) via long-range dipole–dipole interaction up to a maximum distance of 150 Å or so. This large value has been recently reported in studies that used a donor–multiacceptor pair (antenna effect) to enhance the efficiency of transfer of energy so that it is measurable even at distances exceeding 150 Å. This method permits analysis of a variety of larger biological systems and can be used for planning specific FRET experiments above 100 Å (Walczewska-Szewc, Bojarski & D'Auria, 2013).

Measurement of steady-state and time-resolved fluorescence anisotropy offers many opportunities to study molecular orientation, mobility, and diffusion in labeled biological processes and systems. In fact these phenomena are due to transition moments for absorption and emission along specific directions fixed with respect to the fluorophore structure. The rate of rotational diffusion depends on the viscosity of the solvent, and the size and shape of the molecule (Lakowicz, 2006).

4. WHY CHOOSE BIOMOLECULES FOR SENSING?

As reported in scientific literature, it is difficult to design and synthesize in a laboratory organic molecules able to bind specifically to analytes that are structurally more complex than simple ions (Na^+ and K^+) (D'Auria & Lakowicz, 2001). Even if we assume that this kind of synthesis is possible, there is no guarantee that structure changes useful for sensing will occur upon their binding with target analyte. To circumvent this and other limitations related to the use of small molecule fluorescent probes, modern biotechnology has instead resorted to the idea of using proteins and enzymes as components of sensors for biochemical analytes. The idea is to exploit the extremely wide range of selective affinities sculpted into the various proteins by biological evolution. The use of biomolecules in biosensors presents many advantages:

(1) high specificity binding (for instance, enzymes are easily able to discriminate the enantiomer L from enantiomer R of sugar molecules);
(2) solubility in aqueous buffers;
(3) the possibility to use recombinant proteins for high yields;
(4) the possibility to manipulate the properties of genes using site-directed mutagenesis for new molecules, and/or using genetically encoded tags;
(5) the possibility to label proteins such as antibodies with fluorescent probes;
(6) the opportunity to use highly stable biomolecules isolated from thermophilic organisms.

Among biological molecules, enzyme molecules deserve special mention. In fact, enzymes can be easily manipulated to abolish their substrate consumption (apoenzymes) or to increase their stability (thermostable enzymes). This makes them interesting molecules to use for developing customized biosensors.

5. ENZYMES AS MREs

The function of a robust and operationally stable biosensor depends on the biochemical specificity and stability of the MRE. MREs described to date include biologically derived molecular species (e.g., an antibody, a receptor, an enzyme, a protein, or a nucleic acid) or larger portions of living biological systems (e.g., cells, tissue, microorganisms, organelles, or whole organisms).

The biochemical specificity of MRE for a biosensor is crucial since it is responsible for binding the analyte of interest. Enzymes are indispensable biomolecules, involved in all aspects of the cellular metabolism of living organisms. They are selective catalysts necessary to accelerate biological processes, in some cases utilizing other substances as cofactors. They are not consumed by the chemical reaction, and they remain unaffected and available for new reactions. Enzymes are found in all organisms, where they are involved in a broad spectrum of metabolic functions. They catalyze a large number of reactions displaying distinct substrate specificity, good efficiency, and outstanding selectivity, a key requirement for a biosensor. The catalytic activity, sensitivity, affinity, and high chemical specificity provided by enzymes allow for much lower limits of detection of different analytes than would be obtained with different binding techniques (Vo-Dinh & Cullum, 2000).

The use of an enzyme as MRE in a biosensor often ensures reliable function under varying environmental conditions over an extended time period. Since the sensor's operational lifetime is typically limited by the stability of the enzyme, then the use of stable enzymes is an important criterion in the biosensor development (D'souza, 2001; Gouda, Singh, Rao, Thakur, & Karanth, 2003). The main factors affecting enzyme stability and activity are pH, ionic strength, chemical inhibitors, solvent polarity (if not aqueous), and temperature. These limitations can be overcome by using thermostable enzymes isolated from extremophile microorganisms: organisms that grow naturally under extreme conditions.

Extremophiles provide a unique source of novel, robust enzymes, which are more rugged than their mesophilic counterparts under the conditions often encountered where the biosensor is to be employed.

6. THERMOPHILIC ENZYMES

Natural evolution has selected some enzymes to perform their physiological functions under extreme conditions. They operate optimally at temperatures, salinities, pHs, and in solvents that may be quite far from ordinary physiological environments (Cowan et al., 2010; Egorova & Antranikian, 2005).

In particular, the ability to retain native protein structure and function at high temperature is due to higher stabilization of these proteins, resulting from a balance between flexibility (important for activity) and rigidity due their amino acid composition: lower levels of asparagine, glutamine, cysteine, methionine, and tryptophan, which are more susceptible to deamidation or oxidation at high temperatures, and higher levels of isoleucine, alanine, and proline, which should provide more densely packed hydrophobic cores, and extra stability of the loops (D'Auria et al., 1998).

Furthermore, the use of enzymes from thermophilic microorganisms has been frequently proposed and demonstrated as a solution to the problem of enzymatic stability, most readily assessed in enzymes by loss of catalytic activity.

Enzymes isolated from thermophilic organisms are more robust, being stable at high temperatures for long time periods and in the presence of organic solvents. The ability of the enzyme to catalyze reactions in the presence of organic solvents is important because different analytes may require a cosolvent due to low solubility in aqueous media. In fact many of them used in biosensors are more soluble at higher temperatures and in the presence of organic solvents (Pennacchio et al., 2013).

Initially, thermophilic enzymes were isolated directly from microorganisms, including eubacteria and archaea. Nowadays, they are often produced in soluble form in *Escherichia coli* by biological engineering.

The use of thermophilic enzymes has revolutionized the world of biotechnology. Perhaps the best known example is the use of thermostable DNA polymerase for in vitro DNA amplification, the polymerase chain reaction. Replacing (for example) *E. coli* DNA polymerase with the thermostable DNA polymerase purified from the thermophilic bacterium, *Thermus aquaticus* (*Taq*), that can survive extended incubation at 95°C (Saiki et al., 1988), enables the newly polymerized strands to be dissociated by melting without destroying the polymerase. This permits many cycles of amplification without adding more polymerase, simplifying the procedure, and making it amenable to automation.

7. BIOLOGICAL ENGINEERING

Natural enzymes seldom fulfill all the requirements of an MRE (Meyer, 2006; Polizzi, Bommarius, Broering, & Chaparro-Riggers, 2007; Schoemaker, Mink, & WubboLts, 2003) and require improvement of one or more functional characteristics. Using different tools, it is possible to design enzymes able to work in a variety of environmental conditions of temperature, pH, and salinity, or to act on molecules slightly different in structure from their natural substrates. Protein engineering has made it possible to manipulate virtually any enzyme property, either through site-directed mutagenesis or through chemical modification (Jäckel, Kast, & Hilvert, 2008; Kurtovic & Mannervik, 2009; Reetz, 2007; Schmidt, Böttcher, & Bornscheuer, 2009). Genetic engineering using recombinant DNA technology is now a routine and rapid process, permitting the rapid cloning and overexpression of target enzyme genes in suitable hosts (Meiring et al., 2011; Mutondo, Huddy, Bauer, Tuffin, & Cowan, 2010). The development of metagenomic gene discovery has made enzymes from organisms in thermophilic environments readily accessible, even when the parent microorganisms are not cultivatable (Samuelson, 2011; Smith, 1996).

However, the use of enzymes as MREs presents the disadvantage of substrate consumption, a feature that it is not always useful in a detection process. Biosensors based on the use of coenzyme-depleted enzymes can overcome this problem, since they do not complete the catalytic cycle. Apoenzymes do not consume substrate, and they can exhibit conformational changes upon the binding of the analyte that permit its detection.

7.1 Apoenzymes as Nonconsuming Substrate Sensors

Apoenzymes, inactive form of enzymes, are still able to bind substrate with an affinity comparable to holoenzyme, but they are not able to transform substrate into product (Fig. 2). The binding of substrate to apoenzyme may result in a conformational change that can be easily detected by fluorescence measurements.

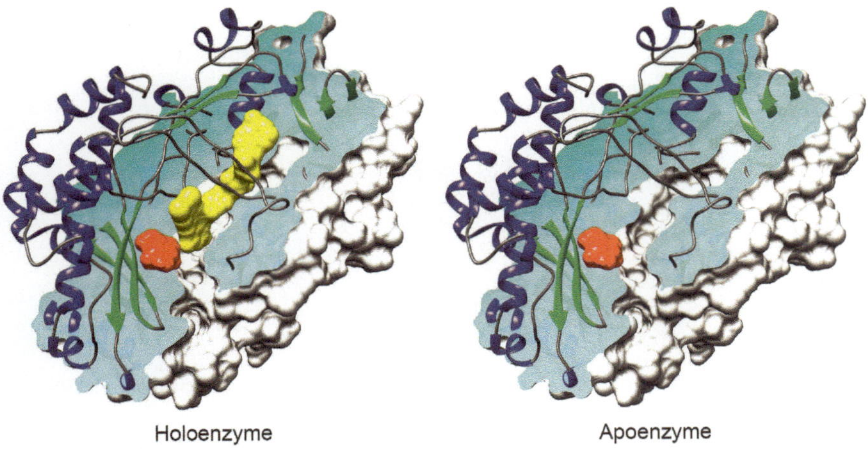

Fig. 2 Glucose oxidase from *Aspergillus niger* in apo and holo forms. The FAD is shown in *yellow* and the glucose in *red*.

Glucose sensors, for continuous monitoring of blood glucose, can use coenzyme-depleted enzymes (apoenzymes) as nonconsuming substrate sensors. Examples of apoenzymes used for glucose sensing are the apoglucose oxidase from *Aspergillus niger*, the apoglucose dehydrogenase from the thermophilic microorganism *Thermoplasma acidophilum*, and the apoglucokinase from the thermophilic eubacterium *Bacillus stearothermophilus* (de Champdore′, Staiano, Rossi, & D'Auria, 2007).

7.2 Apoglucose Oxidase From *A. Niger*

Glucose oxidase (GO) from *A. niger* is a flavoprotein that oxidizes β-D-glucose with high specificity. In holo form this enzyme cannot be used as a reversible sensor to estimate glucose concentration in the blood or urine samples, since glucose is consumed. To prevent glucose oxidation, the FAD cofactor was removed from the GO structure. The apo-GO was still able to bind glucose but not to transform it.

In Fig. 3 are shown the apo-GO absorbance and emission spectra. The absorbance spectrum presents a maximum at 278 nm due to the contribution of the aromatic amino acid residues. The emission spectrum displays a maximum at 340 nm, characteristic of partially shielded tryptophan residues. The absence of absorption at wavelengths above 300 nm in the absorbance spectrum demonstrates that the FAD molecule has been completely removed from the GO structure. The decrease of intrinsic tryptophan fluorescence intensity indicates the binding of glucose in the active site of

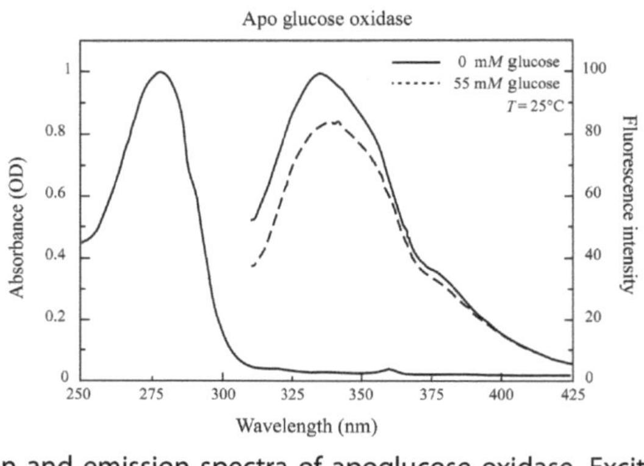

Fig. 3 Absorption and emission spectra of apoglucose oxidase. Excitation at 298 nm. The protein concentration was 0.05 mg/mL. *Adapted from D'Auria, S., Herman, P., Rossi, M., & Lakowicz, J. R. (1999). The fluorescence emission of the apo-glucose oxidase from* Aspergillus niger *as probe to estimate glucose concentrations.* Biochemical and Biophysical Research Communications, 263, *550–553.*

apo-GO. Indeed, the apo-GO is still able to bind glucose, and the Trp 426 involved in this binding event exhibits the observed fluorescence quenching (D'Auria, Herman, Rossi, & Lakowicz, 1999; Hecht, Kalisz, Hendle, Schmid, & Schomburg, 1993; Meyer, Wohlfahrt, Knablein, & Schomburg, 1998; Wohlfahrt et al., 1999). These results suggest that apo-GO can be used in the apoenzyme form as a nonconsuming glucose biosensor (D'Auria et al., 1999).

7.3 Apoglucose Dehydrogenase From the Thermophilic Microorganism *T. Acidophilum*

Glucose dehydrogenase (GD) from the thermoacidophilic archaeon *T. acidophilum* is an enzyme resistant to high temperatures and organic solvents and catalyzes glucose oxidation with high specificity. The apoform of the enzyme still binds glucose with an affinity comparable to the holoenzyme and in the presence of nonpolar solvents displays increased enzyme activity (D'Auria, Di Cesare, et al., 2000). The presence of acetone causes a change in the local protein environment leading to an increase of fluorescence emission intensity. Addition of glucose to ANS-GD in the presence of the optimal acetone concentration resulted in a decrease in ANS emission intensity (ANS is 1-anilino-8-sulfonic acid). This decrease can be also used to measure glucose concentration (D'Auria, Di Cesare, et al., 2000).

7.4 Glucokinase From the Thermophilic Eubacterium *B. Stearothermophilus*

Glucokinase (GK) from the thermophilic organism *B. stearothermophilus* (recently renamed *Geobacillus stearothermophilus*) is known to display long-term stability (D'Auria et al., 2002). The enzyme is stable at room temperature for several days, unlike the hexokinase from *Saccharomyces cerevisiae*, which has also been used as an optical glucose sensor.

Fig. 4 shows the enzyme activity and intrinsic fluorescence as a function of time at room temperature of both the thermophilic and mesophilic enzymes. The yeast hexokinase is quite unstable, it loses enzyme activity over several days, and its fluorescence intensity decreases. In contrast, the thermophilic GK is still enzymatically active over 1 month's time, and the fluorescence intensity remains almost unchanged.

A competitive glucose assay based on RET was developed using the unmodified protein and its intrinsic tryptophan emission as the donor. *O*-nitrophenyl-gluco-pyranoside (ONPG) was used as RET acceptor, a molecule containing the absorbing nitrophenyl group.

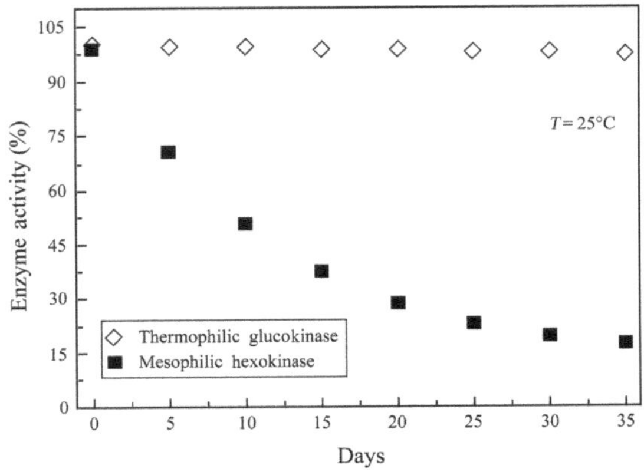

Fig. 4 Stability of the thermophilic glucokinase. The incubation was at room temperature. *Adapted from Scognamiglio, V., Staiano, M., Rossi, M., & D'Auria, S. (2004). Protein-based biosensors for diabetic patients.* Journal of Fluorescence, 14, *491–498.*

The interaction between GK and ONPG produced a decrease of tryptophan emission intensity. Addition of glucose displaced ONPG from the GK-binding site and allowed for the recovery of fluorescence emission intensity, indicating that the intensity changes were due to the binding of glucose to GK-binding site (D'Auria et al., 2002).

The above described examples indicate that enzyme which use glucose as substrate, if deprived of their cofactors, can be used as reversible, non-consuming substrate biosensors.

8. CONCLUSIONS

In conclusion, biosensors are highly valuable devices for measuring a wide array of analytes. They can provide cost-effective, easy-to-use, sensitive, and highly accurate detection devices. Biosensors can be used in a variety of research and commercial applications, ranging from clinical to environmental and agricultural. Some examples of these applications include: general healthcare monitoring, screening for disease, clinical analysis and diagnosis of disease, veterinary and agricultural applications, as well as industrial processing and monitoring and environmental pollution control.

In the field of bio/sensing there has been rapid growth in the development and application of naturally occurring MREs. Clearly, biomolecules are the most widely used recognition element; in particular enzymes are the main favorite in the fabrication of various biosensors due to their highly specific binding affinities and catalytic activities. In the literature it is reported that some receptors offer both advantages but also disadvantages. Sophisticated chemical modification and molecular biological techniques have been applied in the refinement of their binding, chemical and thermal stability, and in their immobilization properties.

Considering the need for reliable and robust sensing in many applications, much effort has been expended to design innovative and novel enzymatic sensor designs. However, other properties of enzyme sensors such as poor stability and critical operational conditions, as well as sensitivity to pH and temperature variation, restrict their wider utility for real-time applications.

With the goals of improved stability, reliability, selectivity, and sensitivity, the scientific community exploited advancements in genetic engineering to develop recombinant enzymes or to modify the catalytic/allosteric enzymatic sites to adapt enzymes to sensing. A successful result of this approach was the development of apoenzymes as MRE for obtaining non-consuming substrate biosensors.

ACKNOWLEDGMENTS

This project was partially funded by CNR project "INTEROMICS" and CTN01-00230-248064-Safe&Smart project.

REFERENCES

Bahadır, E. B., & Sezgintürk, M. K. (2015). Applications of commercial biosensors in clinical, food, environmental, and biothreat/biowarfare analyses. *Analytical Biochemistry*, *478*, 107–120.

Borini, S., Staiano, M., Rocchia, M., Rossi, A. M., Rossi, M., & D'Auria, S. (2007). Advanced nanotechnological approaches for designing protein-based "lab-on-chips" sensors on porous silicon wafer. *Recent Patents on DNA & Gene Sequences*, *1*, 1–7.

Cowan, D. A., Kirby, B. M., Meiring, T. L., Ferrer, M., Guazzaroni, M. E., Golyshina, O. V., et al. (2010). Enzymes from extreme environments. In R. H. Baltz, J. E. Davies, & A. L. Demain (Eds.), *Manual of industrial microbiology and biotechnology* (3rd ed., pp. 512–517). USA: American Society for Microbiology.

D'Auria, S. (2013). Interview with Sabato D'Auria. Section editor for the chemical biology section in BMC biochemistry. *BMC Biochemistry*, *14*, 14.

D'Auria, S., Gryczynski, Z., Gryczynski, I., Rossi, M., & Lakowicz, J. R. (2000). A protein biosensor for lactate. *Analytical Biochemistry*, *283*, 83–88.

D'Auria, S., Herman, P., Rossi, M., & Lakowicz, J. R. (1999). The fluorescence emission of the apo-glucose oxidase from *Aspergillus niger* as probe to estimate glucose concentrations. *Biochemical and Biophysical Research Communications*, *263*, 550–553.

D'Auria, S., & Lakowicz, J. R. (2001). Enzyme fluorescence as sensing tool: New perspectives in biotechnology. *Current Opinion in Biotechnology*, *12*, 99–104.

D'souza, S. F. (2001). Microbial biosensors. *Biosensors and Bioelectronics*, *16*, 337–353.

D'Auria, S., Di Cesare, N., Gryczynski, Z., Gryczynski, I., Rossi, M., & Lakowicz, J. R. (2000). A thermophilic apoglucose dehydrogenase as nonconsuming glucose sensor. *Biochemical and Biophysical Research Communications, 274*, 727–731.

D'Auria, S., Di Cesare, N., Staiano, M., Gryczynski, Z., Rossi, M., & Lakowicz, J. R. (2002). A novel fluorescence competitive assay for glucose determinations by using a thermostable glucokinase from the thermophilic microorganism *Bacillus stearothermophilus*. *Analytical Biochemistry, 303*, 138–144.

D'Auria, S., Moracci, M., Febbraio, F., Tanfani, F., Nucci, R., & Rossi, M. (1998). Structure-function studies on beta-glycosidase from *Sulfolobus solfataricus*. Molecular bases of thermostability. *Biochimie, 80*, 949–957.

de Champdore', M., Staiano, M., Rossi, M., & D'Auria, S. (2007). Proteins from extremophiles as stable tools for advanced biotechnological applications of high social interest. *Journal of the Royal Society, Interface, 4*, 183–191.

Egorova, K., & Antranikian, G. (2005). Industrial relevance of thermophilic Archaea. *Current Opinion in Microbiology, 8*, 649–655.

Fan, X., White, I. M., Shopova, S. I., Zhu, H., Suter, J. D., & Sun, Y. (2008). Sensitive optical biosensors for unlabeled targets: A review. *Analytica Chimica Acta, 620*, 8–26.

Gopinath, S. C. B. (2010). Biosensing applications of surface plasmon resonance-based Biacore technology. *Sensors and Actuators B: Chemical, 150*, 722–733.

Gouda, M. D., Singh, S. A., Rao, A. G., Thakur, M. S., & Karanth, N. G. (2003). Thermal inactivation of glucose oxidase. Mechanism and stabilization using additives. *The Journal of Biological Chemistry, 278*, 24324–24333.

Hecht, H. J., Kalisz, H. M., Hendle, J., Schmid, R. D., & Schomburg, D. (1993). Crystal structure of glucose oxidase from *Aspergillus niger* refined at 2.3 A resolution. *Journal of Molecular Biology, 229*, 153–172.

Hoa, X. D., Kirk, A. G., & Tabrizian, M. (2007). Towards integrated and sensitive surface plasmon resonance biosensors: A review of recent progress. *Biosensors and Bioelectronics, 23*, 151–160.

Hulanicki, A., Geab, S., & Ingman, F. (1991). Chemical sensors definitions and classification. *Pure and Applied Chemistry, 63*(9), 1247–1250.

Jäckel, C., Kast, P., & Hilvert, D. (2008). Protein design by directed evolution. *Annual Review of Biophysics, 37*, 153–173.

Kurtovic, S., & Mannervik, B. (2009). Identification of emerging quasi-species in directed enzyme evolution. *Biochemistry, 48*, 9330–9339.

Lakowicz, J. R. (2006). *Principles of fluorescence spectroscopy* (3rd ed., pp. 353–382). USA: Springer. ISBN-10: 0-387-31278-1.

Martins, T. D., Chaves Ribeiro, A. C., de Camargo, H. S., da Costa Filho, P. A., Mesquita Cavalcante, H. P., & Dias, D. L. (2013). New insights on optical biosensors: Techniques, construction and application. In T. Rinken (Ed.), *State of the art in biosensors—General aspects*. InTech. chapter 5, ISBN: 978-953-51-1004-0.

Meiring, T. L., Bauer, R., Scheepers, I., Ohloff, C., Tuffin, I. M., & Cowan, D. A. (2011). Metagenomics and beyond: Current approaches and integration with complementary technologies. In D. Marco (Ed.), *Metagenomics: Current innovations and future trends* (pp. 1–19). Granada, Spain: Horizon Press.

Meyer, H. P. (2006). Chemocatalysis biocatalysis (biotransformation): Some thoughts of a chemist and of a biotechnologist. *Organic Process Research and Development, 10*, 572–580.

Meyer, M., Wohlfahrt, G., Knablein, J., & Schomburg, D. (1998). Aspects of the mechanism of catalysis of glucose oxidase: A docking, molecular mechanics and quantum chemical study. *Journal of Computer-Aided Molecular Design, 12*, 425–440.

Mutondo, M., Huddy, R. J., Bauer, R., Tuffin, I. M., & Cowan, D. A. (2010). Metagenomic gene discovery. In R. W. Li (Ed.), *Metagenomics and its applications in agriculture, biomedicine and environmental studies* (pp. 287–320). Hauppauge, NY: Ed. Nova Science Publ. chapter 15.

Pennacchio, A., Sannino, V., Sorrentino, G., Rossi, M., Raia, C. A., & Esposito, L. (2013). Biochemical and structural characterization of recombinant short-chain NAD(H)-dependent dehydrogenase/reductase from Sulfolobus acidocaldarius highly enantioselective on diaryl diketone benzil. *Applied Microbiology and Biotechnology*, *97*, 3949–3964.
Polizzi, K. M., Bommarius, A. S., Broering, J. M., & Chaparro-Riggers, J. F. (2007). Stability of biocatalysts. *Current Opinion in Chemical Biology*, *11*, 220–225.
Reetz, M. T. (2007). Controlling the selectivity and stability of proteins by new strategies in directed evolution: The case of organocatalytic enzymes. *Ernst Schering Foundation Symposium Proceedings*, *2*, 321–340.
Saiki, R. K., Gelfand, D. H., Stoffel, S., Scharf, S. J., Higuchi, R., Horn, G. T., et al. (1988). Primer-directed enzymatic amplification of DNA with a thermostable DNA polymerase. *Science*, *239*(4839), 487–491.
Samuelson, J. C. (2011). Recent developments in difficult protein expression: A guide to *E. coli* strains, promoters, and relevant host mutations. *Methods in Molecular Biology*, *705*, 195–209.
Schmidt, M., Böttcher, D., & Bornscheuer, U. T. (2009). Protein engineering of carboxyl esterases by rational design and directed evolution. *Protein and Peptide Letters*, *16*, 1162–1171.
Schoemaker, H. E., Mink, D., & WubboLts, M. G. (2003). Dispelling the myths—Biocatalysis in industrial synthesis. *Science*, *299*, 1694–1697.
Smith, D. W. (1996). Problems of translating heterologous genes in expression systems: The role of tRNA. *Biotechnology Progress*, *12*, 417–422.
Toseland, C. P. (2013). Fluorescent labeling and modification of proteins. *Journal of Chemical Biology*, *6*, 85–95.
Turner, A. P. (2013). Biosensors: Sense and sensibility. *Chemical Society Reviews*, *42*, 3184–3196.
Varriale, A., Rossi, M., Staiano, M., Terpetschnig, E., Barbieri, B., Rossi, M., et al. (2007). Fluorescence correlation spectroscopy assay for gliadin in food. *Analytical Chemistry*, *79*, 4687–4689.
Vo-Dinh, T., & Cullum, B. (2000). Biosensors and biochips: Advances in biological and medical diagnostics. *Fresenius' Journal of Analytical Chemistry*, *366*, 540–551.
Walczewska-Szewc, K., Bojarski, P., & D'Auria, S. (2013). Extending the range of FRET—The Monte Carlo study of the antenna effect. *Journal of Molecular Modeling*, *19*, 4195–4201.
Wohlfahrt, G., Witt, S., Hendle, J., Schomburg, D., Kalisz, H. M., & Hecht, H. U. (1999). 1.8 and 1.9 Å resolution structures of the *Penicillium amagasakiense* and *Aspergillus niger* glucose oxidases as a basis for modelling substrate complexes. *Acta Crystallographica. Section D, Biological Crystallography*, *55*, 967–977.

CHAPTER FIVE

Integrated Strategies to Gain a Systems-Level View of Dynamic Signaling Networks

Robert H. Newman*[,1], Jin Zhang[†,1]
*North Carolina Agricultural and Technical State University, Greensboro, NC, United States
[†]University of California, San Diego, San Diego, CA, United States
[1]Corresponding authors: e-mail address: rhnewman@ncat.edu; jzhang32@ucsd.edu

Contents

Abstract

In order to survive and function properly in the face of an ever changing environment, cells must be able to sense changes in their surroundings and respond accordingly. Cells process information about their environment through complex signaling networks composed of many discrete signaling molecules. Individual pathways within these networks are often tightly integrated and highly dynamic, allowing cells to respond to a given stimulus (or, as is typically the case under physiological conditions, a combination of stimuli) in a specific and appropriate manner. However, due to the size and complexity of many cellular signaling networks, it is often difficult to predict how cellular signaling networks will respond under a particular set of conditions. Indeed, crosstalk between individual signaling pathways may lead to responses that

Methods in Enzymology, Volume 589
ISSN 0076-6879
http://dx.doi.org/10.1016/bs.mie.2017.01.016

are nonintuitive (or even counterintuitive) based on examination of the individual pathways in isolation. Therefore, to gain a more comprehensive view of cell signaling processes, it is important to understand how signaling networks behave at the systems level. This requires integrated strategies that combine quantitative experimental data with computational models. In this chapter, we first examine some of the progress that has recently been made toward understanding the systems-level regulation of cellular signaling networks, with a particular emphasis on phosphorylation-dependent signaling networks. We then discuss how genetically targetable fluorescent biosensors are being used together with computational models to gain unique insights into the spatiotemporal regulation of signaling networks within single, living cells.

1. INTRODUCTION

The ability of cells to receive and rapidly process information about their environment is critical, both for their survival and for their proper function. Indeed, the disruption of cellular signaling mechanisms is believed to underlie a wide range of pervasive diseases, including cancer, diabetes, heart disease, and various autoimmune disorders (Guo, 2014; Kooij et al., 2014; Ortsater, Grankvist, Honkanen, & Sjoholm, 2014; Roskoski, 2016; Ruprecht & Lemeer, 2014; Salzer, Santos-Valente, Keller, Warnatz, & Boztug, 2016; Yang & Yang, 2016). Cellular information processing is achieved through the coordinated action of a diverse collection of signaling enzymes and small molecule second messengers that form extensive, highly integrated signaling networks. At the heart of most eukaryotic signaling networks are protein kinases and phosphatases, whose opposing activities are believed to govern the phosphorylation status of ~30%–40% of cellular proteins (Cohen, 2000; Hu, Neiswinger, Zhang, Zhu, & Qian, 2015).

Inside the cell, protein kinases, phosphatases, and their respective substrates are organized into complex phosphorylation networks that must be precisely regulated in cellular space and time. In recent years, tremendous progress has been made in identifying the structure and organization of cellular phosphorylation networks. As will be discussed in Section 2, this has been accomplished using both "top-down" approaches, such as tandem mass spectrometry (MS/MS) and mass cytometry, that start at the cellular level and "work their way down" to the protein level, and "bottom-up" approaches, such as functional protein microarrays, that start at the protein level and "work their way up" to the cellular level (Newman, Zhang, & Zhu, 2014). Together, these approaches provide complementary information about specific sites of phosphorylation and the corresponding kinase(s) mediating the phosphorylation event, respectively (Newman et al., 2014). This information can be integrated into detailed maps of kinase–substrate relationships (KSRs) that describe which sites can be phosphorylated by a given kinase (or dephosphorylated by a given phosphatase). However, while such maps provide valuable insights about the global organization of cellular phosphorylation networks, in order to better understand how modulation of these networks correlates with a particular physiological response, it is also important to know when and where these phosphorylation events are likely to occur inside the cell. Indeed, the tight spatial and temporal regulation of cellular phosphorylation networks is believed to underlie the specificity of cellular signal transduction. Therefore, in order to gain a more comprehensive, systems-level understanding of dynamic phosphorylation networks, the static maps described earlier must be supplemented with information about the spatial and temporal regulation of the signaling enzymes within these networks. This requires an integrated approach that combines information from large-scale phosphoproteomics experiments with computational models and kinetic data from live-cell imaging experiments designed to measure the activity profiles of specific pools of kinases and phosphatases within intact networks (Fig. 1).

In this chapter, we first examine several emerging high-throughput technologies that are being used to gain insights into the organization and regulation of cellular phosphorylation networks at the systems level. We then describe how this information is being integrated into computational models of signaling networks to provide a conceptual framework for studying complex interactions between components within the networks. Finally, we turn our attention to the application of genetically targetable fluorescent biosensors to the study of endogenous signaling networks, highlighting how these molecular tools are being used in conjunction with computational models to provide unique insights into the spatiotemporal regulation of dynamic signaling networks at the single-cell level. Throughout the chapter, we will highlight both the advantages and the limitations of the various approaches and, importantly, describe how they are being used together to gain a more comprehensive understanding of dynamic cellular signaling networks.

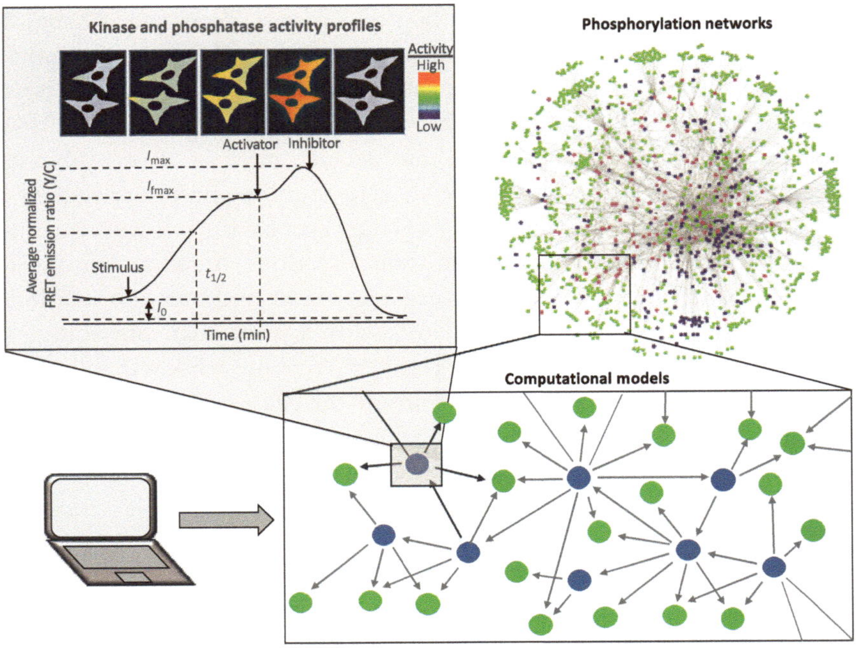

Fig. 1 Integrated strategies to study dynamic phosphorylation networks. Information about the topology of intracellular phosphorylation networks obtained from large-scale phosphoproteomics studies (*upper right*) can be combined with live-cell imaging data describing the spatiotemporal regulation of key signaling enzymes in the network, such as protein kinases and phosphatases (*upper left*), to generate computational models (*bottom*) that can be used to predict the systems-level properties of dynamic phosphorylation networks. Importantly, the resulting computational models can be used to generate testable hypotheses that serve to further refine model parameters.

2. STRATEGIES TO CHARACTERIZE GLOBAL CHANGES IN THE PHOSPHORYLATION STATUS OF CELLULAR PROTEINS

To date, over 270,000 nonredundant phosphorylation sites (also called phosphosites) have been identified, primarily from humans and mice (PhosphositePlus database http://www.phosphosite.org as of 9/2/16) (Hornbeck et al., 2015). However, information about how the majority of these sites are modulated in response to various cellular stimuli is not currently available. Likewise, in many cellular contexts, it is unclear to what extent crosstalk and other systems-level regulatory mechanisms impact phosphorylation dynamics at these sites. Finally, for the vast majority of sites identified in vivo, the identity of the cognate kinase(s) mediating a given phosphorylation event is not currently known. Answers to these questions require systems-level analyses of cellular phosphorylation networks.

Traditionally, changes in the phosphorylation status of cellular proteins have been measured using standard immunological methods, such as Western blotting, enzyme-linked immunosorbent assays, and immunofluorescence (IF). Indeed, due to their specificity and sensitivity, these "tried and true" techniques have been, and will continue to be, staples within the cell signaling community. However, these techniques are only able to provide detailed information about one (or, at most, several) molecular species of interest during a given experiment. As a consequence, even with the recent development of enabling technologies that dramatically reduce the time needed to conduct an experiment using these techniques, their throughput still remains too low to realistically analyze the hundreds, or even thousands, of phosphoproteins required to gain a truly systems-level view of cellular phosphorylation networks. Therefore, to complement and extend these traditional approaches, several high-throughput methods have recently been developed that are able to examine global changes in the phosphorylation status of many cellular proteins simultaneously. These techniques, which can be used to simultaneously monitor time- and/or context-dependent changes in the phosphorylation status of hundreds to thousands of different molecular species, include several ensemble-based approaches, such as quantitative MS/MS (Olsen & Mann, 2013; von Stechow, Francavilla, & Olsen, 2015), protein microarrays (Sutandy, Qian, Chen, & Zhu, 2013; Zhang & Pelech, 2012), and microwestern arrays (Ciaccio, Wagner, Chuu, Lauffenburger, & Jones, 2010), as well as a number of emerging technologies, such as mass cytometry (Bodenmiller et al., 2012), single-cell western arrays (Hughes et al., 2014), and multiplexed cyclic IF (Lin, Fallahi-Sichani, & Sorger, 2015) that provide insights into the regulation of phosphorylation dynamics at single-cell resolution. As described later, together these approaches are providing important information about both the organization and the global regulation of cellular phosphorylation networks. We and others have recently reviewed this topic in detail (Day, Sosale, & Lazzara, 2016; Di Palma & Bodenmiller, 2015; Gingras & Wong, 2016; Grecco, Imtiaz, & Zamir, 2016; Landry, Clarke, & Lee, 2015; Newman et al., 2014; Sutandy et al., 2013; Tape, 2016; Terfve & Saez-Rodriguez, 2012); therefore, here, we will only provide a brief overview of select quantitative phosphoproteomics approaches and highlight how they are being used to gain insights into the global changes in the phosphorylation status of cellular proteins under various cellular conditions. We begin by examining several ensemble methods designed to analyze cellular material (e.g., cell lysates) that has been pooled from a population of cells. We then discuss several recently developed technologies that are providing new information about phosphorylation networks within individual cells.

2.1 Ensemble Approaches Based on Quantitative Mass Spectrometry

As alluded to earlier, the past decade has witnessed an explosion in the number of phosphosites that have been identified in vivo and in situ. The vast majority of these newly identified phosphosites have been characterized using MS/MS-based approaches. This revolution in the field of phosphoproteomics has been fueled by several factors, including the development of upfront enrichment methods designed to isolate relatively rare phosphoproteins; the introduction of novel fragmentation protocols, such as higher-energy collision dissociation (HCD) (Olsen et al., 2007) that preserve phosphate monoester bonds during the ionization/fragmentation step, improvements in the sensitivity and accuracy of mass analyzers, and the development of sophisticated software packages for the analysis of mass spectra (Newman et al., 2014). However, though standard MS/MS protocols have become well suited for identifying and annotating phosphopeptides from complex cellular mixtures, it is often difficult to monitor *changes* in the phosphorylation status of cellular proteins using traditional MS/MS workflows. This can be attributed to many factors, including the complexity of biological samples, low fractional stoichiometries of many phosphosites inside the cell, and run-to-run variations that can occur at several steps during phosphopeptide enrichment protocols. To overcome these challenges, researchers have developed several quantitative MS methods, such as stable isotope labeling of amino acids in cell culture (SILAC) (Ong, 2012) and isobaric tags for relative and absolute quantitation (iTRAQ) (Evans et al., 2012) that make it possible to directly compare phosphorylation profiles of multiple samples in a single experiment (Fig. 2A). These approaches, which rely on isotopic labeling of protein and peptide fragments, respectively, have quickly become cornerstones in the field of phosphoproteomics.

During a SILAC experiment, cellular proteins are first metabolically labeled by growing the cells in culture media containing either a stable isotope of a particular "heavy" amino acid(s) (usually ^{13}C-Arg and/or ^{13}C-Lys)

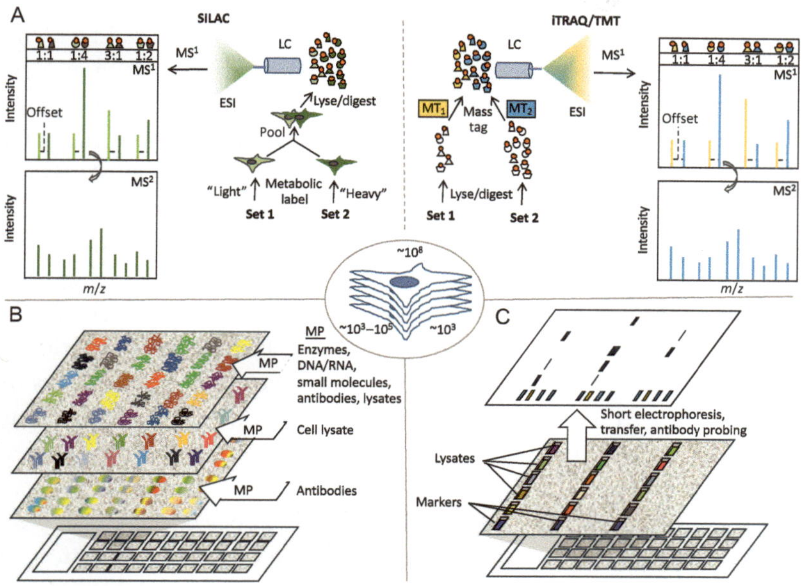

Fig. 2 Ensemble methods to study global changes in the phosphorylation status of cellular proteins.(A) Quantitative mass spectrometry (MS) approaches, such as SILAC (*left*) and iTRAQ/TMT (*right*), allow changes in the relative levels of thousands of phosphoproteins to be measured in a single experiment. In a SILAC experiment, cellular proteins are differentially labeled by growing cells in the presence of either a "heavy" isotope of a particular amino acid (*dark green*) or its naturally occurring "light" counterpart (*light green*). Cells are then pooled, lysed, and digested before being subjected to phosphoenrichment and liquid chromatography (LC). Following chromatographic separation, fragments are ionized via electron spray ionization (ESI) and analyzed by tandem MS. During the first stage of mass analysis (MS^1), the relative abundance of each phosphoprotein is determined based on peak intensities. Peaks containing heavy and light isomers of a given fragment are offset by a known amount, depending on the mass difference between the amino acid isotopes used for metabolic labeling. Finally, the identity of each fragment is determined during the second stage mass analysis (MS^2). The workflow for iTRAQ/TMT (*right*) is similar to that of SILAC, except proteolytic fragments are not labeled with isobaric tags (MT_1 and MT_2) until after cells have been lysed and subject to proteolysis. Once labeled, the proteolytic fragments are pooled, enriched, and analyzed by LC-MS/MS, as described for SILAC. (B) Approaches based on protein microarrays. Functional protein microarrays (*top*) are composed of purified proteins or protein domains immobilized on a functionalized glass surface in a spatially defined manner. Typically, individual proteins are printed in duplicate or triplicate on the arrays. Functional protein microarrays can be used to study interactions between the proteins immobilized on their surfaces and a variety of biomolecules in the mobile phase (MP), including active enzymes (to study global enzyme–substrate relationships), DNA/RNA (to assess the DNA/RNA binding properties of proteins), small molecules

Continued

or its more common, "light" counterpart (Fig. 2A, left). Following lysis, the lysates from each set of cells are mixed together and subjected to identical downstream processing steps, including proteolytic digestion, liquid chromatography (LC)/gel-based separation and phosphopeptide enrichment, followed by MS/MS-based detection. Similarly, iTRAQ and related techniques, such as tandem mass tagging (TMT) (Dayon & Sanchez, 2012; Jia, Andaya, & Leary, 2012), use isotopic labeling to distinguish the cellular contents derived from different samples during analysis (Fig. 2A, right). However, as opposed to SILAC, the isotopic labeling step during iTRAQ and TMT experiments occurs after the cells have been lysed and their lysates subjected to protease digestion. Though this increases the potential for side reactions that can complicate analysis, it also makes these approaches amenable to analysis of primary cells (in addition to cultured cells). Once labeled, the fragments are pooled, processed, and analyzed in a manner similar to SILAC.

The key to isotopic labeling approaches is that, because the isotopic isomers derived from differentially labeled cells are chemically identical, they behave identically during upfront enrichment steps and LC-MS/MS analysis. However, importantly, the mass difference between the isotopic isomers leads to a measurable difference in their respective mass-to-charge (m/z) ratios, allowing them to be distinguished from one another during mass analysis. As a consequence, relative changes in the phosphorylation status at a given site can be determined based on the ratio of the intensities between the heavy and light isotopes of fragments containing that site. Thus, quantitative MS approaches represent powerful tools for tracking system-wide changes in phosphorylation networks in a variety of cellular contexts.

Fig. 2—cont'd (protein–small molecule interactions), antibodies (antibody recognition), and whole-cell lysates. Meanwhile, analytical protein microarrays (*middle*) contain a series of antibodies immobilized on their surface. These arrays are treated with cell lysates in the MP in order to measure the relative abundance of various proteins under a given condition. Finally, reverse phase protein arrays (RPPAs; *bottom*) are composed of a small amount of cell lysate obtained from cells under different conditions and/or from different patients. Each RPPA is treated with a specific antibody in the MP. (C) Microwestern arrays are similar to RPPAs except the proteins in each lysate can be resolved from one another during a short electrophoresis step. Following electrophoresis, the proteins are transferred to a nitrocellulose membrane and probed with various antibodies. *Center*. The approximate number of cells required per assay is shown (e.g., MS/MS-based analysis typically requires $\sim 10^8$ cells/experiment while protein microarrays typically use between 10^5 cells (analytical microarrays) and 10^3 cells (RPPA) per assay).

For instance, Humphrey et al. recently used SILAC to generate dynamic maps of the insulin-sensitive phosphoproteome in 3T3-L1 adipocytes (Humphrey et al., 2013). To this end, the authors employed a multiplexed SILAC strategy to identify those phosphosites whose levels were altered at various times after insulin stimulation. In total, these studies identified over 37,200 phosphosites on 5705 distinct proteins, 15% of which were regulated in an insulin-dependent manner. Moreover, by combining the observed phosphorylation profiles at each time point with a network of well-characterized phosphorylation sites, these studies uncovered functionally linked temporal clusters within the network. Interestingly, all of the well-characterized substrates of some kinases (e.g., Akt) were found in one temporal cluster while all those of other kinases (e.g., p70S6K) were found in another cluster. This suggests that, in some contexts, kinase activation alone may be the rate-limiting step in substrate phosphorylation. As we will see in Section 3, this has potentially important implications for systems-level modeling of dynamic phosphorylation networks using kinetic data obtained using FRET-based kinase activity reporters.

Though temporally resolved phosphoproteomes have the potential to offer important insights into the global regulation of phosphorylation networks, until recently, dynamic profiling of phosphoproteomes has been restricted to relatively few time points and/or biological replicates. This is due, in large part, to the requirement for time-consuming fractionation steps prior to mass analysis. Therefore, to improve the temporal resolution afforded by MS/MS studies, the Mann laboratory recently developed EasyPhos, a streamlined MS/MS-based phosphoproteomics workflow that is optimized for phosphoprotein digestion and enrichment while eliminating labor-intensive desalting steps (Humphrey, Azimifar, & Mann, 2015). Consequently, EasyPhos, which is designed to be compatible with parallel 96-well plate formats, dramatically increases the throughput of MS/MS-based phosphoproteomic analyses. For instance, using this approach, the authors were able to analyze 91 biologically distinct liver tissue phosphoproteomes in parallel. These analyses, which consisted of between 6 and 10 replicates at each of 10 time points after insulin stimulation (including 4 early time points at 5, 10, 15, and 30 s), revealed functionally linked temporal nodes similar to those observed in 3T3-L1 adipocytes. They also uncovered points of signal integration with the fibroblast growth factor receptor (FGFR) signaling pathway and a novel phosphorylation site on Akt2 S478. The latter, which is likely phosphorylated by mammalian target of rapamycin complex 2 (mTORC2) based on consensus site information, was proposed to play a role in a feed-forward mechanism that could lead to Akt2 ultrasensitivity during insulin signaling.

By providing a wealth of information about the specific phosphosites present under different cellular conditions, large-scale MS-based studies such as those described earlier are helping to define the dynamic phosphoproteome. These data promise to provide important insights into the cellular targets of protein kinases and phosphatases during both physiological and pathological signaling processes. However, due to the high cost of instrumentation and the high level of technical expertise required to conduct MS/MS experiments, advanced MS/MS approaches often require substantial upfront investment and/or dedicated core facilities that are not available to many researchers. Moreover, compared to many of the other phosphoproteomics techniques outlined later, MS/MS-based strategies require a relatively large number of cells for analysis (e.g., a typical experiment requires $\sim 10^8$ cells per lysate). Finally, and perhaps most importantly, like most other "top-down" approaches, MS/MS-based experiments are unable to definitively match a given phosphorylation event with the upstream kinase catalyzing the phosphorylation. This KSR information is critical for the construction of functional phosphorylation networks. Therefore, to complement and extend the information provided by MS/MS-based studies, other high-throughput platforms have recently been developed that are amenable to the study of phosphorylation networks.

2.2 Ensemble Approaches Based on Protein Microarrays

Recently, several approaches based on protein microarray technology have emerged as powerful tools for probing the phosphorylation states of a large number of cellular proteins simultaneously. Briefly, protein microarrays are composed of a collection of proteins immobilized on a functionalized glass surface in a spatially defined manner. Currently, three main classes of protein microarrays are broadly available—functional protein microarrays, analytical microarrays, and reverse phase protein microarrays (RPPAs)—and all three have been useful for studying cellular phosphorylation networks (Fig.2B) (Sutandy et al., 2013; Zhang & Pelech, 2012).

2.2.1 Functional Protein Microarrays

We will begin by examining functional protein microarrays, which are composed of purified full-length proteins (and/or protein domains) immobilized on their surfaces (Fig. 2B, top). These arrays have proven to be highly versatile tools for studying signaling processes on a global scale. For instance, functional protein microarrays have been used to study a variety of biochemical processes related to signal transduction, including protein–protein interactions (Fasolo, Im, & Snyder, 2015; Popescu et al., 2007; Zhu et al., 2001), protein–nucleic acid interactions (Hu, Xie, Blackshaw, Qian, & Zhu, 2011; Hu et al., 2009; Sutandy, Hsiao, & Chen, 2016), and protein small-molecule binding (Huang et al., 2004; Kung et al., 2009). Importantly, because they are also amenable to activity assays designed to determine enzyme–substrate relationships, functional protein microarrays have been particularly useful for determining the global substrate specificity of key signaling enzymes (Cox et al., 2015; Jeong, Rho, & Zhu, 2011; Lu, Lin, Boeke, & Zhu, 2013; Sun et al., 2016). For example, to gain a better understanding of the organization of human phosphorylation networks, we recently examined the global substrate profiles of 289 unique human kinases using functional protein microarrays composed of ~4200 full-length human proteins (Newman et al., 2013). These studies, which used in-depth bioinformatics analysis to identify 3656 physiologically likely KSRs from among the over 24,000 raw KSRs identified based solely on signal intensity, more than doubled the number of KSRs that had been identified to that point (Hu et al., 2014; Newman et al., 2013). Moreover, by combining our KSR dataset with information both about in vivo phosphorylation sites determined by MS/MS and consensus phosphorylation motifs for each of the kinases in the network, we were able to construct a high-resolution map of phosphorylation networks that connects 230 kinases to 2591 in vivo phosphorylation sites on 652 substrates (Newman et al., 2013). This information will be particularly useful for understanding the functional consequences of phosphorylation and for developing new tools, such as phospho-specific antibodies and novel FRET-based kinase activity reporters, used to study dynamic phosphorylation networks. Moreover, the generalizability of this approach suggests that it will be useful for examining the global substrate profiles of enzymes involved in SUMOylation (Cox et al., 2015), ubiquitylation (Jeong et al., 2011), acetylation (Lu et al., 2013), and *O*-glycosylation (Sun et al., 2016), which together will offer clues about potential points of intersection between different signaling modalities.

Though functional protein microarrays represent an attractive, high-throughput platform for the rapid identification of a large number of direct enzyme–substrate relationships, there are several caveats that must be considered when using this approach. For instance, when using functional protein microarrays, a stringent cutoff is typically used to reduce the rate of false positives. As a consequence, substrate identification may be biased toward those proteins that contain multiple sites of modification. This is because these substrates will exhibit higher signal-to-noise ratios than other bona fide substrates that may only be modified at a single site and, therefore, may exhibit a signal intensity that falls below the cutoff. Likewise, proteins that are immobilized in greater quantities on the microarrays may exhibit higher signal-to-noise ratios or may be preferentially modified by the enzyme of interest. On the other hand, recognition of some substrates may be impaired due to misfolding of the substrates during purification or by immobilization of substrates in orientations that are inaccessible to the modifying enzyme.

Importantly, because direct enzyme–substrate relationships are generally determined using purified protein components in the mobile phase, interactions that are dependent on auxiliary factors, such as scaffold proteins, cellular cofactors, or other posttranslational modifications, may be missed using functional protein microarrays and other bottom-up approaches that use purified components. This has led to the development of several complementary approaches wherein more complex solutions, such as preformed protein complexes (Hu et al., 2015) or even whole-cell lysates (Goodwin et al., 2016; Woodard et al., 2013), are used in place of purified proteins in the mobile phase. This can be combined with computational approaches to identify novel KSRs that are dependent on auxiliary interactions. For instance, Hu et al. recently employed a bioinformatics strategy to identify novel scaffold proteins involved in the coordination of kinase signaling pathways (Hu et al., 2015). To this end, the authors mined protein–protein interaction databases to identify those proteins that interacted with each of the components within a given linear KSR pathway. These studies, which involved over 1100 known kinase-substrate pairs, identified 212 putative scaffold proteins, of which 194 had not been previously reported. They then used functional protein microarrays composed of >17,000 unique human proteins to validate their predictions. To this end, two putative scaffolds, activating transcription factor 2 (ATF2) and peptidylprolyl *cis/trans* isomerase, NIMA-interacting-1 (PIN1), that are predicted to be involved in casein kinase II (CKII) and c-Jun N-terminal kinase (Jnk) signaling, respectively, were tested. Consistent with their hypothesis, select substrates were only phosphorylated by the kinase of interest in the presence of the corresponding scaffold protein, suggesting that the scaffold protein was indeed facilitating phosphorylation of the substrate by the kinase of interest. These studies will be particularly valuable for understanding the spatial organization of phosphorylation networks and, as will be discussed in Section 3, can be used in conjunction with genetically targetable FRET-based biosensors to study the regulation of specific pools of a given cellular kinase in situ.

2.2.2 Analytical Protein Microarrays

The next class of protein microarrays, known as analytical protein microarrays, is composed of a collection of purified "capture" proteins that bind to distinct analytes (e.g., a phosphorylated residue on a cellular protein) that are present in the mobile phase (Fig. 2B, middle). Though antibodies are by far the most common type of capture proteins utilized on analytical protein microarrays, other biomolecules that bind their targets through specific, high affinity interactions, such as nanobodies and even RNA aptamers, have also been used successfully (Chen, Nakamoto, Niwa, & Corn, 2012; Even-Desrumeaux, Baty, & Chames, 2010). Meanwhile, the mobile phase typically consists of a complex protein solution, such as cell or tissue lysates, under nondenaturing conditions.

In a typical analytical protein microarray experiment, a collection of proteins derived from a population of $\sim 10^5$ cells are either prelabeled with a fluorescent dye before being incubated with the microarray or added directly to the arrays without prior labeling (Hughes et al., 2014; Zhang & Pelech, 2012). Following incubation, the array is washed and, if the lysate was not labeled beforehand, a second fluorescently labeled antibody that recognizes a protein of interest is added to the arrays. Multiplexing of this secondary detection step can provide information about the phosphorylation status of many different proteins using a single array. Importantly, while such a sandwich assay format potentially increases the time required to complete the experiment, the use of a second antibody for detection can substantially increase the sensitivity and specificity of the assay while reducing artifacts stemming from cross-reactivity of immobilized capture antibodies, poor labeling of low abundance proteins in the lysate, and/or disruption of the epitope through conjugation of the small-molecule fluorescent dye (Sutandy et al., 2013). By examining several lysates in parallel (e.g., at different times after treatment or from different patient samples), information about changes in the phosphorylation levels of many proteins in a given system can be obtained in a reasonable timeframe.

As for all antibody-based techniques used for the analysis of phosphorylation networks, one of the primary obstacles associated with the development of analytical protein microarrays is the availability of high-quality, phospho-specific capture antibodies that do not cross-react with other protein species. To help overcome this limitation, Koerber et al. recently described an innovative, structure-guided approach for the development of phospho-specific antibodies (Koerber, Thomsen, Hannigan, Degrado, & Wells, 2013). Using this approach, which begins with a library of single-chain variable fragments (scFv's) containing an engineered anion-binding pocket (or "nest") and then uses phage display to enrich for scFv's that recognize a specific immobilized phosphopeptide, the authors were able to identify >50 antibodies that exhibit high specificity for the phosphorylated forms of their respective targets. This generalizable approach has the potential to dramatically expand the toolkit of reliable phospho-specific antibodies, which will not only facilitate the development of analytical protein microarrays but also promises to increase the utility of other approaches based on immunological detection of phosphorylated residues (Stoevesandt & Taussig, 2013).

In addition to antibody specificity, other factors can also impact the interpretation of data obtained from experiments designed to study global phosphorylation profiles using analytical protein microarrays. For instance, the relatively low fractional stoichiometries of many phosphorylated proteins, coupled with differences in avidity among capture antibodies, can contribute to false negatives using this approach—particularly when a given protein is expressed at relatively low levels in the cells under study. On the other hand, because lysates must be applied to the array under nondenaturing conditions, protein–protein interactions may lead to the formation of protein complexes that could cause false positives. Therefore, to guard against such artifacts, it is important that any hits identified during microarray experiments be validated using traditional immunological methods, such as Western blotting. Nonetheless, due to their ability to probe the phosphorylation status of a large number of cellular proteins simultaneously, analytical protein microarrays can provide important information about global changes in phosphorylation networks under a variety of conditions.

2.2.3 Reverse Phase Protein Microarrays

Like analytical protein microarrays, RPPAs represent powerful tools for tracking changes in cellular phosphorylation profiles in cell lysates. However, instead of antibodies, the solid phase of RPPAs contains lysates derived from cell or tissue extracts (Fig. 2B, bottom). Accordingly, each microarray contains a collection of lysates obtained from various sources (e.g., from different clinical samples or from cells treated with a drug for various times and/or at different doses). Once immobilized, changes in the expression levels and/or phosphorylation status of select targets in the lysates are assessed by probing the arrays with an antibody that is specific for the epitope of interest. Since many microarrays can be probed in parallel, a large number of targets can be tested simultaneously. Such a multiplexing strategy has the benefit of reducing the effects of interassay variability, thereby increasing confidence in the relationships observed between different species in a network.

Since each spot on the array contains only a few nanograms of lysate, many arrays can be prepared using only a small amount of starting material (Pierobon, Wulfkuhle, Liotta, & Petricoin, 2015). This feature is particularly attractive when studying phosphorylation profiles in lysates derived from precious samples, such as patient samples from clinical studies. For instance, Shull et al. recently used RPPA's composed of B-cell lysates from 24 different donors to gain a better understanding of the molecular profiles associated with chronic lymphocytic leukemia (CLL) (Shull et al., 2015). To accomplish this, the authors used 167 different antibodies to compare the levels of various cellular proteins and phosphoproteins in lysates obtained from either CLL patients or healthy controls. Interestingly, nonsupervised hierarchical clustering revealed a clear molecular signature among all CLL samples, regardless of the genetic subtype of CLL being tested. This may suggest that, at the network level, a common molecular profile exists across all CLL subtypes. In support of this notion, supervised hierarchical analysis, which focused on differences in probe intensities between CLL patients and healthy controls, revealed increased expression of several kinases involved in cancer progression in CLL lysates, including B-Raf, Akt, mTOR, Lck, Btk and Syk, as well as several other proteins, such as SF2/ASF and eIF4G, which play important roles in RNA processing and translation initiation, respectively. Moreover, increased phosphorylation of several cellular proteins, including PDK1, 4E-BP1, p70S6K, BAD, and PRAS40, was also observed. Interestingly, interrogation of the molecular functions associated with those proteins that were differentially expressed/phosphorylated between the two groups revealed an overrepresentation of proteins involved in signaling events controlling translation initiation. Together, these observations suggest that increased protein translation may be a key event involved in the initiation/progression of CLL.

2.3 Ensemble Approaches Based on Microwestern Arrays

As alluded to earlier, one of the most important considerations for most antibody-based approaches is antibody specificity. In the case of RPPAs, antibody specificity is particularly important because each spot on the array contains a complex mixture of thousands of different cellular proteins, each of which could lead to a false-positive response if the antibody being used for detection has not been carefully validated. In this respect, RPPAs are akin to dot blots which, unlike Western blots, do not permit a second level of discrimination based on the electrophoretic mobility of the protein(s) recognized by the antibody. Therefore, to increase the accuracy of the RPPA approach, Ciaccio and colleagues recently developed so-called microwestern arrays that combine many of the attractive features of RPPA's with the ability to separate proteins in the lysates based on their apparent molecular weights (Fig. 2C) (Ciaccio et al., 2010). Not only can the increased resolution afforded by microwestern arrays reduce the incidence of false positives caused by antibody cross-reactivity, but they can also potentially expand the repertoire of antibodies that are available for analysis. For instance, previous studies demonstrated that only 4 of the 34 (<12%) antibodies evaluated behaved identically using both RPPA and traditional Western blotting protocols (Sevecka & MacBeath, 2006). In contrast, when a set of eight antibodies—four of which had produced consistent results using both RPPA and Western blotting and another four that had exhibited substantial reduction in their dynamic ranges in the RPPA due to cross-reactivity—were compared side by side using microwestern arrays and traditional "macrowestern" blotting, all of the antibodies tested produced similar quantitative results using both methods (Ciaccio et al., 2010). Though this approach is relatively new, it has already been used to gain important insights into a number of phosphorylation-dependent signaling processes, including those downstream of the epidermal growth factor receptor (EGFR) following EGF stimulation (Ciaccio et al., 2010) and the cytokine receptor during the host–pathogen response in macrophages (Huang et al., 2015).

2.4 Approaches That Provide Single-Cell Resolution

Together, the ensemble approaches described earlier provide important information about global changes in cellular phosphorylation profiles. However, because they report responses that are averaged across many cells within the population, ensemble methods lack the ability to distinguish cell-to-cell heterogeneities that can provide important clues about the regulatory mechanisms underlying a particular cellular response. For instance, cell-to-cell differences in the phase and/or the amplitude of a given response are likely to mask oscillatory and "switch-like" behaviors that encode critical information related to cellular function. In fact, emerging evidence suggests that several signaling systems are characterized by bistable (or multistable) responses (Albeck et al., 2008; Barr, Heldt, Zhang, Bakal, & Novak, 2016; Fosbrink, Aye-Han, Cheong, Levchenko, & Zhang, 2010; Montero et al., 2015; Spencer, Gaudet, Albeck, Burke, & Sorger, 2009). As a consequence, responses that appear to be graded when measured using ensemble approaches may, in fact, reflect differences in the *percentage* of cells that are in the "on" state rather than the degree to which a particular pathway is activated (Bagowski & Ferrell, 2001; Landry et al., 2015; Ni et al., 2011). Therefore, to better understand, how information is processed within individual phosphorylation networks, researchers have developed a series of analytical tools designed to study phosphorylation dynamics at single-cell resolution.

2.4.1 Single-Cell Westerns

As outlined in Section 2.3, microwesterns provide increased resolution of cellular species compared to RPPA's. However, because $\sim 10^3$ cells are required per lysate, critical information about single cells is obscured using this approach. Therefore, to extend the microwestern array platform to single-cell applications, Hughes et al. recently developed single-cell western arrays (Hughes et al., 2014). These arrays, which consist of thousands of 20 μm microwells patterned on a photoactivatable polyacrylamide gel attached to a solid glass support, have the potential to offer exciting new insights into systems-level changes in global phosphorylation profiles with single-cell resolution (Fig. 3A). For instance, as a proof of concept, the authors examined the phosphorylation status of extracellular-regulated kinase 1/2 (Erk1/2) and its upstream activator, MEK1/2, following fibroblast growth factor-2 (FGF-2)-mediated stimulation of neuronal stem cells. Interestingly, although these studies demonstrated an overall increase in the ratios of pMEK1/2 to total MEK1/2, and pErk1/2 to total Erk1/2, a large degree of heterogeneity was observed between the responses from cell to cell. This is consistent with the notion that each cell has a unique cellular profile based on its individual signaling history that may be masked by ensemble methods that pool material from thousands of cells.

2.4.2 Mass Cytometry

Flow cytometry, which is perhaps the most widely used method to study signaling dynamics within single cells, has provided important information

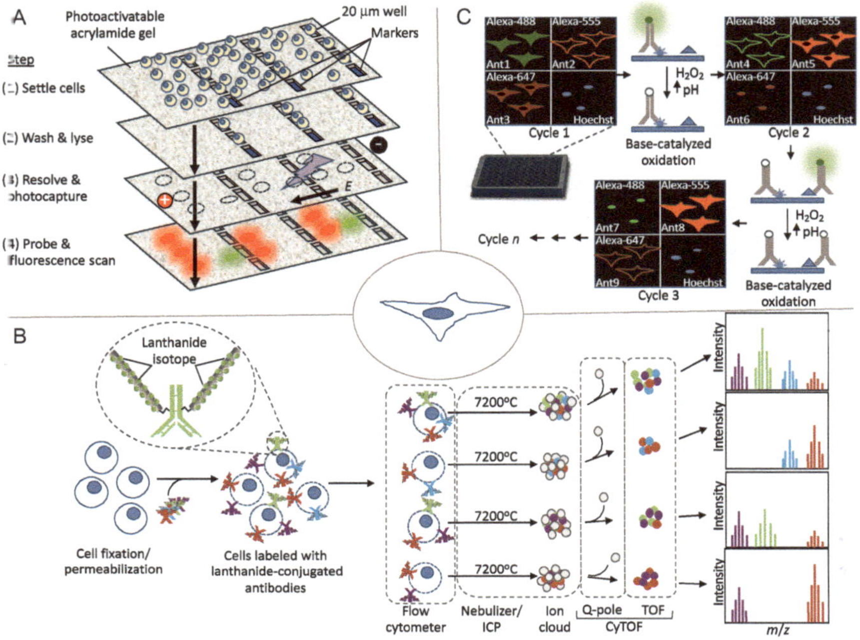

Fig. 3 Single-cell methods to study global changes in phosphorylation status of cellular proteins.(A) Single-cell Western blotting. Individual cells are first applied to a photoactivatable acrylamide gel immobilized on a glass support. Once individual cells have settled into 20 μm wells patterned on the surface, excess cells are removed and the remaining cells are lysed in a RIPA buffer. An electric field (*E*) is then applied to the gel in order to resolve proteins present in the lysates. After electrophoresis, the proteins are fixed in the gel by irradiation (photocapture) before being probed with fluorescently labeled antibodies. (B) Mass cytometry. Following fixation and permeabilization, cells are incubated with up to 40 distinct antibodies, each labeled with a different isotope of a lanthanide metal ion (*red*, *green*, *blue*, or *purple circles*). The cells are then sorted using a flow cytometer before being nebulized and atomized/ionized using an inductively coupled plasma (ICP) torch containing superheated argon plasma. Atomization generates ion clouds that are introduced into a CyTOF mass cytometer, composed of a quadrupole (Q-pole) and a time-of-flight (TOF) mass analyzer, for elemental analysis. Signal intensities are proportional to the concentration of individual target molecules in each cell. (C) Cyclic immunofluorescence. During the first cycle, cells are fixed/permeabilized and treated with a set of antibodies labeled with spectrally distinct antibodies (e.g., Alexa-488 (*green*), Alexa-555 (*red*), and Alexa-647 (*far-red*)), as well as Hoechst stain (*blue*). Following imaging, the fluorophores are inactivated via base-catalyzed oxidation. The cycle is repeated using a second set of fluorescently labeled antibodies along with Hoechst stain as an internal control. This process is repeated through multiple rounds to build a multichannel image.

about signaling dynamics at the single-cell level. However, spectral overlap between fluorescently labeled antibodies effectively limits the number of cellular components that can be monitored simultaneously using this approach to less than 20 species (with upper limits of ~12 species for most systems) (Tape, 2016). This, coupled with time-consuming sample preparation and measurement steps, substantially reduces the throughput of flow cytometric methods. Together, these considerations limit the utility of flow cytometry for systems-level analysis of cellular signaling networks. Therefore, to expand the number of distinct species that can be measured per cell, researchers have recently developed mass cytometry approaches that combine the single-cell capabilities of flow cytometry with the sensitivity of mass spectrometry (Fig. 3B) (Bandura et al., 2009; Baranov, Quinn, Bandura, & Tanner, 2002). For instance, to overcome the issue of spectral overlap, antibodies used in mass cytometry applications are conjugated to lanthanide metal ions rather than fluorescent dyes. To this end, each antibody is conjugated to a distinct lanthanide isotope via an acrylic acid polymer functionalized with multiple copies of a trivalent metal chelator, such as diethylenetriaminepentaacetic acid (Bjornson, Nolan, & Fantl, 2013; Di Palma & Bodenmiller, 2015). Cells are then permeabilized and labeled with the lanthanide-tagged antibodies before being sorted using a flow cytometer and nebulized into single-cell droplets. Once nebulized, each cell droplet is atomized/ionized in a stream of superheated argon plasma to generate single-cell ion clouds that are shuttled into a tandem quadrupole-time-of-flight mass cytometer (CyTOF) for elemental analysis. Since the lanthanide metal isotopes used for analysis do not exist at appreciable levels in most cell types, this approach benefits from very low background signal. This, coupled with the fact that between 100 and 150 metal ions can be conjugated to each antibody, results in very high signal-to-noise ratios, allowing a limit of detection of only a few hundred copies of a given phosphosite per cell (Di Palma & Bodenmiller, 2015). This feature is particularly important for measuring substoichiometric changes in the phosphorylation state of low abundance cellular proteins. Currently, ~40 distinct lanthanide isotopes are available for mass cytometry applications, which permit the simultaneous analysis of subnetworks or key nodes within signaling networks (Bjornson et al., 2013; Di Palma & Bodenmiller, 2015; Gingras & Wong, 2016). Importantly, because a subset of the antibodies can be used to recognize distinct cell types within a complex heterocellular environment, mass cytometry can provide unique insights into the regulation of phosphorylation networks in complex cellular microenvironments more similar to those found in vivo (Tape, 2016). For instance, Fernandez and Maecker recently developed a panel of ~30 antibodies designed to track changes in the phosphorylation status of eight major nodes involved in cytokine, antigen, and toll-like receptor signaling across all major white blood cell lineages, including monocytes, and several B and T cell lineages (Fernandez & Maecker, 2015). However, because cellular phosphorylation networks can be perturbed during dissociation of cells from complex tissue matrices, mass cytometry and other flow-based single-cell strategies are better suited for cells that are naturally nonadherent, such as the hematopoietic lineages alluded to earlier (Tape, 2016). Moreover, like all of the methods discussed to this point, since mass cytometry requires cell lysis prior to detection, it offers little information about the spatial distribution of phosphorylation-dependent signaling events inside cells.

2.4.3 High-Content Imaging Approaches

Traditionally, antibody-based imaging approaches, such as IF, have been used to examine the spatial distribution of cellular signaling enzymes, such as protein kinases and phosphatases, and their downstream substrates. These studies have provided evidence of spatially defined signaling modules that are thought to be important for coordinated information flow inside the cell.[a] However, classical IF suffers from several limitations that reduce its utility for the systems-level analysis of dynamic cellular signaling networks. For instance, the throughput of classical IF is somewhat limited compared to many high-throughput technologies. Moreover, like flow cytometry, the number of distinct molecular species that can be reliably imaged in a single cell using standard IF-based protocols is limited by spectral overlap between fluorophores. To overcome these limitations, traditional IF protocols have been adapted in several innovative ways. For instance, microfluidic technologies, such as that used by the ImStain system developed by the Levchenko lab, can improve the throughput of IF experiments substantially (Cheong, Wang, & Levchenko, 2009). Likewise, to increase the number of distinct molecular species that can be monitored in a single experiment, Giesen et al. recently developed imaging mass cytometery (Giesen et al., 2014). This approach, which combines the multiplicity and precision of mass cytometry with a sophisticated laser ablation system that aerosolizes antibody-stained

[a] In this context, a signaling module represents a collection of signaling molecules (e.g., receptors, signaling enzymes, small molecule second messengers, and downstream effectors) that form an integrated, functional circuit within the larger network.

cellular regions in a spatially defined manner and feeds them into the CyTOF system, permits the analysis of up to ~40 distinct species in formalin-fixed tissue sections at a spatial resolution of ~1 μm (Di Palma & Bodenmiller, 2015). However, because aerosolization and subsequent atomization effectively destroys the cell present at each coordinate, analysis of additional species beyond the number of currently available lanthanide isotopes is not feasible.

As an alternative single-cell imaging strategy, Lin et al. recently developed cyclic immunofluorescence (CycIF) (Lin et al., 2015). CycIF is a generally applicable IF imaging method that involves repeated rounds of IF imaging using 3–5 fluorescence channels (plus an additional channel to visualize DNA) per round (Fig. 3C). Each round of imaging is followed by fluorophore inactivation using a combination of hydrogen peroxide, high pH, and irradiation with white light. Importantly, because the inactivation step is relatively straightforward, CycIF is compatible with 96- and 384-well formats, substantially increasing the throughput of CycIF experiments. Using this approach, the authors were able to track changes in the abundance of six cellular species, including different phosphorylation states of the S6 ribosomal protein (p-S6$^{S240/244}$ and p-S6$^{S235/S236}$), following treatment of BRAF^{V600E} melanoma cells with intermediate doses of the RAF inhibitor, vemurafenib. The relative levels of each species led to the identification of two distinct subpopulations in vemurafenib-treated BRAF^{V600E} cells, one corresponding to proliferating cells in which the Erk/Rsk pathway is highly active and the other representing a quiescent, apoptosis-resistant population characterized by high TORC1 activity. In this study, while no loss of immunogenicity was observed after five cycles of staining and fluorophore inactivation, it remains to be seen whether visualization of the tens to hundreds of species required for systems-level analysis of signaling networks will be hampered by destruction of epitopes and/or modification of the cellular architecture during subsequent rounds of peroxide treatment.

3. STRATEGIES TO TRACK KINASE AND PHOSPHATASE ACTIVITY PROFILES IN LIVING CELLS

Together, the strategies outlined earlier provide a wealth of information about which cellular proteins are modified in response to various environmental changes. However, in order to gain a mechanistic understanding of how cellular phosphorylation networks are regulated under physiological conditions, this information must be supplemented with information about the spatial and temporal regulation of the kinases and phosphatases governing the phosphorylation state of these proteins. However, the methods described earlier generally do not provide direct information about the regulation of the signaling enzymes, themselves. Moreover, in most cases, these approaches lack the spatial and/or temporal resolution necessary to capture key aspects of phosphorylation-dependent signaling at the single-cell level. For example, due to the high cost and/or technical demands associated with many of the aforementioned methods, it is often not practical to take a large number of measurements over relatively short time intervals (e.g., every 30 s or 1 min). However, many important signaling processes, such as the oscillatory behaviors exhibited by protein kinase C (PKC), cAMP-dependent protein kinase (PKA), Ca^{2+}/calmodulin-dependent kinase II (CaMKII), and the Ca^{2+}/calmodulin-dependent protein phosphatase, calcineurin (CaN), occur on timescales of seconds to minutes (Markoulaki, Matson, & Ducibella, 2004; Mehta et al., 2014; Ni et al., 2011; Violin, Zhang, Tsien, & Newton, 2003). As a consequence, these behaviors may appear skewed or could be missed altogether during experiments that lack sufficient temporal resolution. Likewise, when using ensemble-based approaches like MS/MS and microwestern arrays, cell-to-cell heterogeneities are likely to mask dynamic behaviors that provide key information pertaining to a given cellular response. Finally, and perhaps most importantly, because each of the methods described earlier requires either cell lysis or cell fixation/permeabilization prior to analysis, they are not able to monitor real-time changes in single, living cells. Thus, they offer only a "snapshot" of the dynamic changes in kinase and phosphatase activities that characterize most phosphorylation-dependent signal networks.

3.1 Genetically Targetable Fluorescent Biosensors

The considerations outlined earlier are nontrivial, since critical information is often encoded in the dynamic properties (e.g., frequency and duration) and spatial patterning (e.g., compartmentalization) of signaling activities within the cell (Cheong & Levchenko, 2010; Ganesan & Zhang, 2012; Mehta et al., 2014). This, coupled with the fact that stochastic differences in expression patterns and signaling "history" exist between individual cells, suggests that significant cell-to-cell heterogeneities may exist within phosphorylation networks that can only be captured by tracking kinase and phosphatase activities at multiple times within individual, living cells. This requires molecular tools that are able to monitor dynamic changes in the activity profiles of these enzymes with high spatiotemporal resolution. For instance, we and others have developed a series of genetically targetable fluorescent biosensors to track the activities of select signaling molecules within the endogenous cellular environment of single, living cells (Allen et al., 2008; Mehta & Zhang, 2011; Newman, Fosbrink, & Zhang, 2011). These biosensors, which can be directed to specific subcellular regions through the incorporation of a targeting motif (e.g., a nuclear localization sequence) or a component of a signaling complex (e.g., a scaffold protein) (Kunkel & Newton, 2014; Zhang, Ma, Taylor, & Tsien, 2001), are able to monitor real-time changes in the activity profiles of specific pools of a given signaling enzyme or second messenger under a variety of cellular conditions (Gao & Zhang, 2010; Kunkel & Newton, 2009). Importantly, as discussed in more detail later, the high spatiotemporal resolution afforded by these sensors provides kinetic data that can be integrated into computational models that provide insights into the behaviors of entire signaling networks (Greenwald, Polanowska-Grabowska, & Saucerman, 2014; Ni et al., 2011; Saucerman et al., 2006; Violin et al., 2008).

All fluorescent biosensors developed to date contain two basic elements: (1) a "sensor unit," which undergoes a conformational change in response to a given cellular stimulus (e.g., phosphorylation or enzyme activation) and (2) a "reporter unit," which converts the resulting conformational change into a change in the spectral properties of an attached fluorescent protein (FP) color variant(s) (Frommer, Davidson, & Campbell, 2009; Newman & Zhang, 2014) (Fig. 4A). While the reporter unit typically consists of either a circularly permuted variant of an FP (in the case of single fluorophore reporters) or a pair of FP color variants that are able to undergo FRET (in the case of FRET-based reporters), the sensor unit can take several forms depending on the cellular parameter being measured. For example, FRET-based biosensors that exploit an intrinsic conformational change in a cellular protein have been developed to track changes in the phosphatase activity of CaN (Mehta et al., 2014; Newman & Zhang, 2008), as well as the levels of several small molecule second messengers, such as Ca^{2+} (Geiger et al., 2012; Liu et al., 2011), cyclic AMP (cAMP) (DiPilato & Zhang, 2009; Klarenbeek, Goedhart, van Batenburg, Groenewald, & Jalink, 2015), cGMP (Honda et al., 2001; Niino, Hotta, & Oka, 2010; Nikolaev, Bunemann, Hein, Hannawacker, & Lohse, 2004; Russwurm et al., 2007), and nitric oxide (Pearce et al., 2000), involved in the regulation of cellular signaling enzymes. In other cases, the conformational change is generated via an engineered molecular switch. For instance, the sensor unit used by most FRET-based

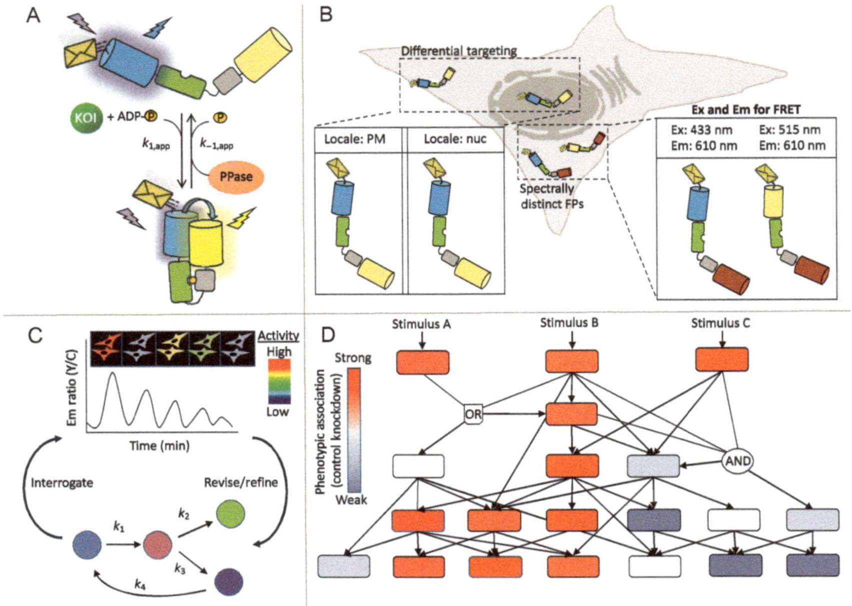

Fig. 4 Molecular and computational tools to gain a systems-level view of dynamic phosphorylation networks. (A) General design of a genetically targetable FRET-based kinase activity reporter. The biosensor consists of a "sensor unit" composed of a substrate domain (*gray rectangle*) linked to a phosphoamino acid binding domain (*green*) via a flexible linker. The sensor domain is sandwiched between a "reporter unit" consisting of two fluorescent protein (FP) color variants (*cyan* and *yellow cylinders*) that can undergo FRET. The sensor can be targeted to different subcellular regions via a targeting motif (envelope). Upon phosphorylation by the kinase of interest (KOI; *green circle*), the reporter undergoes a conformational change that brings the FP's into close proximity to one another, increasing FRET between them (*blue arrow*). Importantly, dephosphorylation by a phosphatase (PPase; *orange oval*) returns the biosensor to its open conformation. Apparent rate constants ($k_{1,app}$ and $k_{-1,app}$) can be determined based on the observed activity profiles. Lightning bolts represent photons of light at the excitation and emission maxima for the respective FP's (*purple*: CFP Ex: 433 nm; *cyan*: CFP Em: 477; *yellow*: YFP Em: 528 nm). (B) Coimaging strategies. FRET-based biosensors can be used to track the activities of two or more signaling enzymes simultaneously. For instance, two reporters that utilize the same FP FRET pair can be targeted to distinct subcellular regions (e.g., the plasma membrane (PM) and nucleus (nuc)). Alternatively, the activity profiles of two enzymes in the same compartment can be monitored simultaneously using reporters that utilize spectrally distinct FPs (e.g., CFP/RFP and YFP/RFP). Donor excitation and acceptor emission maxima for commonly used FP FRET pairs (e.g., Cerulean/mCherry and Venus/mCherry) are shown. Other coimaging strategies, such as FRET-fluorescence lifetime imaging (FRET-FLIM), are not shown. (C) Kinetic data obtained from live-cell imaging experiments can be used to construct computational models of individual signaling circuits based on ODEs/PDEs. The resulting models

kinase activity reporters is based on a modular design consisting of a "substrate domain" containing a sequence that is specifically phosphorylated by the kinase of interest (e.g., a consensus phosphorylation motif) and a "switching domain" encoding a phosphoamino acid-binding domain (PAABD) that binds the phosphorylated form of the substrate domain (Fig.4A). The substrate domain is attached to the PAABD by a flexible linker such that, upon phosphorylation by the kinase of interest, the biosensor undergoes an induced conformational change that alters the distance and/or orientation of flanking FP color variants that are able to undergo FRET (i.e., the "reporter unit"). While the reporter units of most FRET-based kinase and phosphatase reporters utilize a FP FRET pair consisting of variants of cyan FP (CFP) and yellow FP (YFP), other FP combinations, such as green/red FP (GFP/RFP), yellow/orange FP (YFP/OFP), YFP/RFP, and CFP/RFP, have also been used successfully (Ai, Baird, Shen, Davidson, & Campbell, 2014; Day & Davidson, 2009; Lam et al., 2012; Newman et al., 2011; Sample, Newman, & Zhang, 2009). The availability of spectrally distinct reporters facilitates coimaging experiments designed to measure changes in the activity profiles of multiple cellular signaling enzymes (e.g., two kinases or a kinase and a phosphatase) simultaneously in the same cell (Carlson & Campbell, 2009; Woehler, 2013) (Fig.4B). Such information is particularly valuable for understanding crosstalk between signaling pathways within phosphorylation networks. However, due to spectral overlap, currently it is only possible to track the activities of three to four distinct signaling enzymes simultaneously, depending on their subcellular localization (Woehler, 2013).

Using genetically targetable FRET-based activity reporters, researchers have uncovered important details about both the kinetics and the spatial distribution of endogenous kinase and phosphatase activities in a variety of cellular contexts (Allen et al., 2008; Cazabat et al., 2014; Fosbrink et al., 2010;

can be used to generate hypotheses that can be interrogated using live-cell imaging approaches, the results of which can be used to further refine the model. Such an iterative approach can provide insights into complex signaling behaviors, such as oscillations. (D) Logic-based ordinary differential equation models are useful for studying larger phosphorylation networks. Analysis of these models (e.g., via sensitivity analysis) can reveal functional relationships within the network that may not be apparent from network topology alone (*warmer colored boxes*). These relationships can be evaluated using live-cell imaging experiments. Importantly, because each connection is represented by a differential equation, generalized parameters used for initial model development can be refined as experimental data become available, thereby improving the predictive power of the model.

Gao & Zhang, 2009; Hukasova, Silva Cascales, Kumar, & Lindqvist, 2012; Mehta et al., 2014; Ni et al., 2011; Tobias & Newton, 2016). For instance, recently it was proposed that, aside from simply bringing the atypical PKC family member, PKCζ, into close proximity with its substrates, some scaffold proteins, such as partitioning-defective protein 6 (Par6) and p62, may actually regulate the activity of PKCζ directly (Graybill, Wee, Atwood, & Prehoda, 2012; Tsai et al., 2015). This is believed to occur as a consequence of binding interactions between PKCζ and a Phox and Bem1 (PB1) domain on the scaffold protein. According to this model, binding to the scaffold protein induces conformational changes in PKCζ that displaces its autoinhibitory domain, leading to its activation. To investigate this hypothesis further and to gain insights into the impact that Par6 and p62 have on the localized activity of PKCζ during insulin signaling in living cells, Tobias and Newton generated fusion constructs composed of the FRET-based PKC activity reporter, CKAR, linked to the PB1 domains of Par6 and p62 (Tobias & Newton, 2016). Interestingly, these studies revealed that binding to the PB1 domain of Par6 ($PB1^{Par6}$) activated PKCζ approximately four times more efficiently than did binding to the corresponding PB1 domain of p62 ($PB1^{p62}$). This, coupled with live-cell imaging experiments that demonstrated that insulin promotes interactions between PKCζ and p62 (and, subsequently, with the insulin receptor substrate-1 (IRS-1)), suggests that the binding of PKCζ to specific scaffolds can both relocalize and differentially tune its activity during the insulin response.

3.2 Integration of Live-Cell Imaging Data and Mechanistic Computational Models to Study Individual Signaling Pathways and Circuits

To better understand how spatiotemporal parameters impact endogenous signaling processes, information from live-cell imaging experiments such as those described earlier can be integrated into quantitative mathematical models of signaling pathways. These models, which are typically based on ordinary and partial differential equations (ODEs and PDEs, respectively), can reveal unexpected relationships and regulatory modules within the system that play important roles in controlling many aspects of cell physiology (Fig. 4C). Importantly, predictions made by computational models can lead to testable hypotheses that, in turn, can be evaluated experimentally using FRET-based reporters (Ganesan & Zhang, 2012; Landry et al., 2015; Ni et al., 2011; Sample et al., 2009; Tobias & Newton, 2016; Yaniv et al., 2015). For instance, to better understand crosstalk between Ca^{2+}- and cAMP/PKA-dependent signaling pathways during insulin secretion in pancreatic β cells, Ni et al. first used the FRET-based PKA reporter, A-kinase activity reporter 4 (AKAR4), to identify oscillatory changes in PKA activity in individual MIN6 β cells treated with the K^+ channel inhibitor, tetraethylammonium chloride (TEA) (Mehta et al., 2014; Ni et al., 2011). This oscillatory behavior, which was likely masked at the population level due to averaging artifacts, were shown to be synchronized with both Ca^{2+} and cAMP oscillations based on coimaging experiments with the Ca^{2+}-sensitive fluorescent indicator, Fura-2, and the FRET-based cAMP biosensor, indicator of cAMP using Epac (ICUE), respectively. To gain further insights into the molecular mechanisms underlying this Ca^{2+}–cAMP–PKA oscillatory circuit, the authors then constructed a mathematical model based on a detailed description of biochemical events involving these three molecular components. Not only was the model able to capture qualitative features of the experimental data but, importantly, it also made predictions about how the oscillatory circuit would respond to various pharmacological perturbations. These predictions were confirmed by subsequent live-cell imaging experiments, which together suggest that PKA may function as a frequency modulator during insulin secretion in β cells.

Moreover, based on model predictions, the authors were also able to identify other cellular factors that contributed to the observed oscillatory behaviors. The activity of one of these factors, CaN, was recently shown to be synchronized with Ca^{2+} oscillations in a manner similar to PKA in TEA-stimulated MIN6 β cells (Mehta et al., 2014). Interestingly, these studies, which utilized targeted versions of the CaN activity reporter 2 (CaNAR2) to study CaN dynamics in different subcellular regions, identified distinct zones of CaN activity in response to TEA treatment. For example, while cytoplasmic and plasma membrane-targeted versions of CaNAR2 revealed integrative, step-like increases in CaN activity coincident with each Ca^{2+} spike, ER- and mitochondria-targeted versions of the reporter oscillated in register with the Ca^{2+} oscillations. Together, these studies highlight the unique perspectives that can be gained by integrating live-cell imaging data and computational models of cellular signaling networks. Indeed, such an integrated approach has provided key insights into the molecular mechanisms underlying (1) hysteresis and bistability of Aurora B kinase activity during mitosis in HeLa cells (Zaytsev et al., 2016); (2) acceleration of Ras-mediated activation of Raf by the scaffold protein, suppressor of clear homolog (Shoc2), in EGF-stimulated HeLa cells (Matsunaga-Udagawa et al., 2010); (3) cAMP/PKA-mediated regulation of spontaneous action potential generation in sinoatrial node cardiac pacemaker cells (Yaniv et al., 2015); and (4) the existence of spatially restricted phosphorylation gradients downstream of the β-adrenergic receptor in cardiomyocytes (Saucerman et al., 2006), to name a few.

3.3 Perspectives: Strategies to Model Dynamics of Large, Highly Integrated Signaling Networks

To date, the construction and evaluation of computational models using FRET-based biosensors has been applied primarily to the study of individual signaling pathways within a given network, rather than to the broader network itself. This is due, in large part, to the fact that the detailed biochemical parameters necessary to build mechanistic models based on ODEs/PDEs are not currently available for many of the signaling enzymes in the network. Though FRET-based biosensors are well suited to provide this information, live-cell imaging protocols are not particularly amenable to high-throughput analysis (though some progress is being made in this area; e.g., see Maglica, Ozdemir, & McKinney, 2015). Moreover, despite their unique ability to track kinase and phosphatase activity in real-time and at single-cell resolution, FRET-based biosensors may not currently exist for key signaling enzymes within some networks. Finally, very little information about crosstalk between signaling pathways exists and, when it does, it is often cell-type specific. Together, these considerations make it difficult to build mechanistic models of integrated phosphorylation networks "from the ground up" using mechanistic data alone.

How, then, can we gain mechanistic insights into larger, more complex networks? One popular approach involves the development of logic-based computational models. Logic-based models, which use a series of logical operators (e.g., "AND", "OR", and "NOR") to describe the relationships between the activities of protein intermediates, have been used successfully to model large, tightly integrated systems like phosphorylation networks. For instance, logic-based models based on Boolean logic, fuzzy logic, and extreme pathways analysis have been useful frameworks for modeling feasible solution spaces, feedback loops, and global network topologies in a variety of cellular systems (Landry et al., 2015; Terfve & Saez-Rodriguez, 2012). The versatility of these models stems from the fact they do not require rate parameters to be measured or estimated. However, though this improves their scalability, it can be a double-edged sword when studying highly dynamic systems such as phosphorylation networks. For example, because logic-based models assume that the influences of all protein states in the system are binary and equal in magnitude, they do not generally provide information about dynamic spatiotemporal relationships that underlie the regulation of cellular phosphorylation networks (Landry et al., 2015). Moreover, aside from nuclear vs cytoplasmic localization, these models rarely incorporate spatial information that may be critical to achieving signaling specificity within the system.

To overcome these limitations, methods have recently been developed that combine elements of logic-based models with those of kinetic models to create so-called logic-based differential equation models (Henriques, Rocha, Saez-Rodriguez, & Banga, 2015; Kraeutler, Soltis, & Saucerman, 2010; Ryall et al., 2012; Zeigler, Richardson, Holmes, & Saucerman, 2016). These hybrid models retain the scalability of logic models while providing semimechanistic information about dynamic, systems-level behaviors within the network. For instance, interrogation of logic-based differential equation models (e.g., via sensitivity analysis) can provide a global view of quantitative functional relationships between each species in the network, often predicting key signaling enzymes whose overall influence on the system is greater than would be expected based on the network topology alone (Kraeutler et al., 2010; Ryall et al., 2012; Zeigler et al., 2016). The basic approach used by these methods is to construct a logic model and then estimate kinetic parameters within the system by converting binary relationships into continuous logic-based ODEs (Yaniv et al., 2015). The variables within these ODEs, which can take on any value between 0 and 1, are generally estimated based on limited experimental data or assigned default values determined empirically for the system under study. For instance, the Saucerman lab recently described a logic-based differential equation approach based on a normalized Hill modeling framework (Kraeutler et al., 2010). This approach, which has been used to model signaling networks in both cardiomyocytes and cardiac fibroblasts (Kraeutler et al., 2010; Ryall et al., 2012; Zeigler et al., 2016), uses normalized Hill functions (i.e., logic-based ODEs) to describe individual reactions in the network and "AND" and "OR" logical operators to describe crosstalk between pathways. Accordingly, each interaction in the network is assigned a reaction weight (W), a half-maximal effective concentration (EC_{50}), and a Hill coefficient (n). Meanwhile, the activity of each species is controlled by a reaction time constant, τ, and the maximal fractional activation of the species, I_{fmax} (Fig. 1). In most cases, values for these parameters are generalized due to the limited availability of quantitative biochemical data for many of the species in the network. For instance, in the cardiomyocyte model, all species were assigned default parameters, where W, τ, and I_{fmax} were each set to 1, and EC_{50} and n were set to 0.5 and 1.4, respectively. Likewise, all initial activities (Y_0) were arbitrarily set at 25% of the maximum activity. Despite these simplifying global assumptions, the resulting model was able to approximate dynamic behaviors within the system and correctly predict qualitative input–output relationships for many of the reactions reported in the literature. Notably, experimental data either in support of or counter to model predictions could not be found for over 50% of the predicted relationships, suggesting that substantial gaps still exist in the literature (Ryall et al., 2012). This is consistent with the notion that detailed biochemical and kinetic data are not currently available for many of the components in the system.

Importantly, because the parameters used to construct logic-based differential equation models are described by ODEs, they can be updated and refined as experimental data become available. Thus, logic-based differential equation models represent an attractive general modeling framework to construct semimechanistic models of cellular phosphorylation networks that can be tested in live cells using genetically targetable FRET-based kinase and phosphatase activity reporters, and subsequently refined using experimentally derived kinetic parameters (Fig. 4D). Moreover, the targetability of FRET-based reporters allows distinct pools of the signaling enzymes under study to be monitored, adding a crucial spatial component to the network models. Together, this information can be used to make predictions and generate testable hypotheses about complex network behaviors, such as the dynamic control of crosstalk and signal integration within the system. For instance, using sensitivity analysis, it is possible to predict signaling pathways that are likely to interact in response to a given cellular stimulus or set of stimuli. These predictions can be evaluated directly in living cells using FRET-based activity reporters, either by employing coimaging strategies to monitor the activity profiles of individual components in each of the pathways or by activating one pathway and measuring changes in the activity of a representative kinase/phosphatase in the other pathway. Incongruities between model predictions and experimental results can help inform further model refinement (e.g., revisions to the value of n for a given reaction) and the generation of new experimental hypotheses. Such an integrated, iterative approach represents a generalizable method to gain a systems-level view of dynamic phosphorylation networks.

4. CONCLUSIONS

To better understand how cells convert information about their internal and external environments into an appropriate response, it will be necessary to understand how cellular phosphorylation networks function at the systems level. This requires both information about the global phosphoproteome under various cellular conditions and details about how the activities of signaling enzymes that control their phosphorylation state are regulated in cellular space and time. To this end, researchers have developed a series of complementary strategies that, together, are beginning to reveal exciting new details about the organization and regulation of cellular phosphorylation networks, both at the population level and within single cells. By integrating this information into computational models of cellular signaling networks, researchers are gaining new insights into the molecular mechanisms underlying the complex, systems-level behaviors that lead to a given functional cellular response. Such an integrated approach promises not only to provide a clearer picture of how highly dynamic, tightly integrated phosphorylation networks function under normal physiological conditions, but it also has the potential to deepen our understanding of disease progression and improve the efficacy of pharmacological interventions in accord with systems pharmacology paradigms.

ACKNOWLEDGMENTS

This work is supported by National Institutes of Health Grants R35 CA197622, R01 DK073368, R01 MH111516, and R01 GM111665 (to J.Z.); 1SC2GM113784-01 (to R.H.N.); and National Science Foundation Grant DBI-1038160 (to R.H.N.).

REFERENCES

Ai, H. W., Baird, M. A., Shen, Y., Davidson, M. W., & Campbell, R. E. (2014). Engineering and characterizing monomeric fluorescent proteins for live-cell imaging applications. *Nature Protocols*, *9*(4), 910–928. http://dx.doi.org/10.1038/nprot.2014.054.

Albeck, J. G., Burke, J. M., Aldridge, B. B., Zhang, M., Lauffenburger, D. A., & Sorger, P. K. (2008). Quantitative analysis of pathways controlling extrinsic apoptosis in single cells. *Molecular Cell*, *30*(1), 11–25. http://dx.doi.org/10.1016/j.molcel.2008.02.012.

Allen, M. D., DiPilato, L. M., Ananthanarayanan, B., Newman, R. H., Ni, Q., & Zhang, J. (2008). Dynamic visualization of signaling activities in living cells. *Science Signaling*, *1*(37), pt6. http://dx.doi.org/10.1126/scisignal.137pt6.

Bagowski, C. P., & Ferrell, J. E., Jr. (2001). Bistability in the JNK cascade. *Current Biology*, *11*(15), 1176–1182.

Bandura, D. R., Baranov, V. I., Ornatsky, O. I., Antonov, A., Kinach, R., Lou, X., … Tanner, S. D. (2009). Mass cytometry: Technique for real time single cell multitarget immunoassay based on inductively coupled plasma time-of-flight mass spectrometry. *Analytical Chemistry*, *81*(16), 6813–6822. http://dx.doi.org/10.1021/ac901049w.

Baranov, V. I., Quinn, Z., Bandura, D. R., & Tanner, S. D. (2002). A sensitive and quantitative element-tagged immunoassay with ICPMS detection. *Analytical Chemistry*, *74*(7), 1629–1636.

Barr, A. R., Heldt, F. S., Zhang, T., Bakal, C., & Novak, B. (2016). A dynamical framework for the all-or-none G1/S transition. *Cell Systems*, *2*(1), 27–37. http://dx.doi.org/10.1016/j.cels.2016.01.001.

Bjornson, Z. B., Nolan, G. P., & Fantl, W. J. (2013). Single-cell mass cytometry for analysis of immune system functional states. *Current Opinion in Immunology*, *25*(4), 484–494. http://dx.doi.org/10.1016/j.coi.2013.07.004.

Bodenmiller, B., Zunder, E. R., Finck, R., Chen, T. J., Savig, E. S., Bruggner, R. V., … Nolan, G. P. (2012). Multiplexed mass cytometry profiling of cellular states perturbed by small-molecule regulators. *Nature Biotechnology*, *30*(9), 858–867. http://dx.doi.org/10.1038/nbt.2317.

Carlson, H. J., & Campbell, R. E. (2009). Genetically encoded FRET-based biosensors for multiparameter fluorescence imaging. *Current Opinion in Biotechnology*, *20*(1), 19–27. http://dx.doi.org/10.1016/j.copbio.2009.01.003.

Cazabat, L., Ragazzon, B., Varin, A., Potier-Cartereau, M., Vandier, C., Vezzosi, D., … Vandecasteele, G. (2014). Inactivation of the Carney complex gene 1 (PRKAR1A) alters spatiotemporal regulation of cAMP and cAMP-dependent protein kinase: A study using genetically encoded FRET-based reporters. *Human Molecular Genetics*, *23*(5), 1163–1174. http://dx.doi.org/10.1093/hmg/ddt510.

Chen, Y., Nakamoto, K., Niwa, O., & Corn, R. M. (2012). On-chip synthesis of RNA aptamer microarrays for multiplexed protein biosensing with SPR imaging measurements. *Langmuir*, *28*(22), 8281–8285. http://dx.doi.org/10.1021/la300656c.

Cheong, R., & Levchenko, A. (2010). Oscillatory signaling processes: The how, the why and the where. *Current Opinion in Genetics & Development*, *20*(6), 665–669. http://dx.doi.org/10.1016/j.gde.2010.08.007.

Cheong, R., Wang, C. J., & Levchenko, A. (2009). High content cell screening in a microfluidic device. *Molecular & Cellular Proteomics*, *8*(3), 433–442. http://dx.doi.org/10.1074/mcp.M800291-MCP200.

Ciaccio, M. F., Wagner, J. P., Chuu, C. P., Lauffenburger, D. A., & Jones, R. B. (2010). Systems analysis of EGF receptor signaling dynamics with microwestern arrays. *Nature Methods*, 7(2), 148–155. http://dx.doi.org/10.1038/nmeth.1418.

Cohen, P. (2000). The regulation of protein function by multisite phosphorylation—A 25 year update. *Trends in Biochemical Sciences*, *25*(12), 596–601.

Cox, E., Uzoma, I., Guzzo, C., Jeong, J. S., Matunis, M., Blackshaw, S., & Zhu, H. (2015). Identification of SUMO E3 ligase-specific substrates using the HuProt human proteome microarray. *Methods in Molecular Biology*, *1295*, 455–463. http://dx.doi.org/10.1007/978-1-4939-2550-6_32.

Day, R. N., & Davidson, M. W. (2009). The fluorescent protein palette: Tools for cellular imaging. *Chemical Society Reviews*, *38*(10), 2887–2921. http://dx.doi.org/10.1039/b901966a.

Day, E. K., Sosale, N. G., & Lazzara, M. J. (2016). Cell signaling regulation by protein phosphorylation: A multivariate, heterogeneous, and context-dependent process. *Current Opinion in Biotechnology*, *40*, 185–192. http://dx.doi.org/10.1016/j.copbio.2016.06.005.

Dayon, L., & Sanchez, J. C. (2012). Relative protein quantification by MS/MS using the tandem mass tag technology. *Methods in Molecular Biology*, *893*, 115–127. http://dx.doi.org/10.1007/978-1-61779-885-6_9.

Di Palma, S., & Bodenmiller, B. (2015). Unraveling cell populations in tumors by single-cell mass cytometry. *Current Opinion in Biotechnology*, *31*, 122–129. http://dx.doi.org/10.1016/j.copbio.2014.07.004.

DiPilato, L. M., & Zhang, J. (2009). The role of membrane microdomains in shaping beta2-adrenergic receptor-mediated cAMP dynamics. *Molecular BioSystems*, *5*(8), 832–837. http://dx.doi.org/10.1039/b823243a.

Evans, C., Noirel, J., Ow, S. Y., Salim, M., Pereira-Medrano, A. G., Couto, N., … Wright, P. C. (2012). An insight into iTRAQ: Where do we stand now? *Analytical and Bioanalytical Chemistry*, *404*(4), 1011–1027. http://dx.doi.org/10.1007/s00216-012-5918-6.

Even-Desrumeaux, K., Baty, D., & Chames, P. (2010). Strong and oriented immobilization of single domain antibodies from crude bacterial lysates for high-throughput compatible cost-effective antibody array generation. *Molecular BioSystems*, *6*(11), 2241–2248. http://dx.doi.org/10.1039/c005279e.

Fasolo, J., Im, H., & Snyder, M. P. (2015). Probing high-density functional protein microarrays to detect protein-protein interactions. *Journal of Visualized Experiments, 102*, e51872. http://dx.doi.org/10.3791/51872.

Fernandez, R., & Maecker, H. (2015). Cytokine-stimulated phosphoflow of whole blood using CyTOF mass cytometry. *Bio-Protocol*, *5*(11), e1495.

Fosbrink, M., Aye-Han, N. N., Cheong, R., Levchenko, A., & Zhang, J. (2010). Visualization of JNK activity dynamics with a genetically encoded fluorescent biosensor. *Proceedings of the National Academy of Sciences of the United States of America*, *107*(12), 5459–5464. http://dx.doi.org/10.1073/pnas.0909671107.

Frommer, W. B., Davidson, M. W., & Campbell, R. E. (2009). Genetically encoded biosensors based on engineered fluorescent proteins. *Chemical Society Reviews*, *38*(10), 2833–2841. http://dx.doi.org/10.1039/b907749a.

Ganesan, A., & Zhang, J. (2012). How cells process information: Quantification of spatiotemporal signaling dynamics. *Protein Science*, *21*(7), 918–928. http://dx.doi.org/10.1002/pro.2089.

Gao, X., & Zhang, J. (2009). Akt signaling dynamics in plasma membrane microdomains visualized by FRET-based reporters. *Communicative & Integrative Biology*, *2*(1), 32–34.

Gao, X., & Zhang, J. (2010). FRET-based activity biosensors to probe compartmentalized signaling. *Chembiochem*, *11*(2), 147–151. http://dx.doi.org/10.1002/cbic.200900594.

Geiger, A., Russo, L., Gensch, T., Thestrup, T., Becker, S., Hopfner, K. P., … Griesbeck, O. (2012). Correlating calcium binding, Forster resonance energy transfer, and conformational change in the biosensor TN-XXL. *Biophysical Journal*, *102*(10), 2401–2410. http://dx.doi.org/10.1016/j.bpj.2012.03.065.

Giesen, C., Wang, H. A., Schapiro, D., Zivanovic, N., Jacobs, A., Hattendorf, B., … Bodenmiller, B. (2014). Highly multiplexed imaging of tumor tissues with subcellular resolution by mass cytometry. *Nature Methods*, *11*(4), 417–422. http://dx.doi.org/10.1038/nmeth.2869.

Gingras, A. C., & Wong, C. J. (2016). Proteomics approaches to decipher new signaling pathways. *Current Opinion in Structural Biology*, *41*, 128–134. http://dx.doi.org/10.1016/j.sbi.2016.07.008.

Goodwin, C. R., Woodard, C. L., Zhou, X., Pan, J., Olivi, A., Xia, S., … Laterra, J. (2016). Microarray-based phospho-proteomic profiling of complex biological systems. *Translational Oncology*, *9*(2), 124–129. http://dx.doi.org/10.1016/j.tranon.2016.02.001.

Graybill, C., Wee, B., Atwood, S. X., & Prehoda, K. E. (2012). Partitioning-defective protein 6 (Par-6) activates atypical protein kinase C (aPKC) by pseudosubstrate displacement. *The Journal of Biological Chemistry*, *287*(25), 21003–21011. http://dx.doi.org/10.1074/jbc.M112.360495.

Grecco, H. E., Imtiaz, S., & Zamir, E. (2016). Multiplexed imaging of intracellular protein networks. *Cytometry. Part A*, *89*(8), 761–775. http://dx.doi.org/10.1002/cyto.a.22876.

Greenwald, E. C., Polanowska-Grabowska, R. K., & Saucerman, J. J. (2014). Integrating fluorescent biosensor data using computational models. *Methods in Molecular Biology*, *1071*, 227–248. http://dx.doi.org/10.1007/978-1-62703-622-1_18.

Guo, S. (2014). Insulin signaling, resistance, and the metabolic syndrome: Insights from mouse models into disease mechanisms. *The Journal of Endocrinology*, *220*(2), T1–T23. http://dx.doi.org/10.1530/JOE-13-0327.

Henriques, D., Rocha, M., Saez-Rodriguez, J., & Banga, J. R. (2015). Reverse engineering of logic-based differential equation models using a mixed-integer dynamic optimization approach. *Bioinformatics*, *31*(18), 2999–3007. http://dx.doi.org/10.1093/bioinformatics/btv314.

Honda, A., Adams, S. R., Sawyer, C. L., Lev-Ram, V., Tsien, R. Y., & Dostmann, W. R. (2001). Spatiotemporal dynamics of guanosine 3',5'-cyclic monophosphate revealed by a genetically encoded, fluorescent indicator. *Proceedings of the National Academy of Sciences of the United States of America*, *98*(5), 2437–2442. http://dx.doi.org/10.1073/pnas.051631298.

Hornbeck, P. V., Zhang, B., Murray, B., Kornhauser, J. M., Latham, V., & Skrzypek, E. (2015). PhosphoSitePlus, 2014: Mutations, PTMs and recalibrations. *Nucleic Acids Research*, *43*(Database issue), D512–D520. http://dx.doi.org/10.1093/nar/gku1267.

Hu, J., Neiswinger, J., Zhang, J., Zhu, H., & Qian, J. (2015). Systematic prediction of scaffold proteins reveals New design principles in scaffold-mediated signal transduction. *PLoS Computational Biology*, *11*(9), e1004508. http://dx.doi.org/10.1371/journal.pcbi.1004508.

Hu, J., Rho, H. S., Newman, R. H., Zhang, J., Zhu, H., & Qian, J. (2014). PhosphoNetworks: A database for human phosphorylation networks. *Bioinformatics*, *30*(1), 141–142. http://dx.doi.org/10.1093/bioinformatics/btt627.

Hu, S., Xie, Z., Blackshaw, S., Qian, J., & Zhu, H. (2011). Characterization of protein-DNA interactions using protein microarrays. *Cold Spring Harbor Protocols*, *2011*(5), 499–507. http://dx.doi.org/10.1101/pdb.prot5614. pdb prot5614.

Hu, S., Xie, Z., Onishi, A., Yu, X., Jiang, L., Lin, J., … Zhu, H. (2009). Profiling the human protein-DNA interactome reveals ERK2 as a transcriptional repressor of interferon signaling. *Cell*, *139*(3), 610–622. http://dx.doi.org/10.1016/j.cell.2009.08.037.

Huang, J. H., Lin, C. Y., Wu, S. Y., Chen, W. Y., Chu, C. L., Brown, G. D., … Wu-Hsieh, B. A. (2015). CR3 and dectin-1 collaborate in macrophage cytokine response through association on lipid rafts and activation of Syk-JNK-AP-1 pathway. *PLoS Pathogens*, *11*(7), e1004985. http://dx.doi.org/10.1371/journal.ppat.1004985.

Huang, J., Zhu, H., Haggarty, S. J., Spring, D. R., Hwang, H., Jin, F., … Schreiber, S. L. (2004). Finding new components of the target of rapamycin (TOR) signaling network through chemical genetics and proteome chips. *Proceedings of the National Academy of Sciences of the United States of America*, *101*(47), 16594–16599. http://dx.doi.org/10.1073/pnas.0407117101.

Hughes, A. J., Spelke, D. P., Xu, Z., Kang, C. C., Schaffer, D. V., & Herr, A. E. (2014). Single-cell western blotting. *Nature Methods*, *11*(7), 749–755. http://dx.doi.org/10.1038/nmeth.2992.

Hukasova, E., Silva Cascales, H., Kumar, S. R., & Lindqvist, A. (2012). Monitoring kinase and phosphatase activities through the cell cycle by ratiometric FRET. *Journal of Visualized Experiments*, *59*, e3410. http://dx.doi.org/10.3791/3410.

Humphrey, S. J., Azimifar, S. B., & Mann, M. (2015). High-throughput phosphoproteomics reveals in vivo insulin signaling dynamics. *Nature Biotechnology*, *33*(9), 990–995. http://dx.doi.org/10.1038/nbt.3327.

Humphrey, S. J., Yang, G., Yang, P., Fazakerley, D. J., Stockli, J., Yang, J. Y., & James, D. E. (2013). Dynamic adipocyte phosphoproteome reveals that Akt directly regulates mTORC2. *Cell Metabolism*, *17*(6), 1009–1020. http://dx.doi.org/10.1016/j.cmet.2013.04.010.

Jeong, J. S., Rho, H. S., & Zhu, H. (2011). A functional protein microarray approach to characterizing posttranslational modifications on lysine residues. *Methods in Molecular Biology*, *723*, 213–223. http://dx.doi.org/10.1007/978-1-61779-043-0_14.

Jia, W., Andaya, A., & Leary, J. A. (2012). Novel mass spectrometric method for phosphorylation quantification using cerium oxide nanoparticles and tandem mass tags. *Analytical Chemistry*, *84*(5), 2466–2473. http://dx.doi.org/10.1021/ac203248s.

Klarenbeek, J., Goedhart, J., van Batenburg, A., Groenewald, D., & Jalink, K. (2015). Fourth-generation epac-based FRET sensors for cAMP feature exceptional brightness, photostability and dynamic range: Characterization of dedicated sensors for FLIM, for ratiometry and with high affinity. *PloS One*, *10*(4), e0122513. http://dx.doi.org/10.1371/journal.pone.0122513.

Koerber, J. T., Thomsen, N. D., Hannigan, B. T., Degrado, W. F., & Wells, J. A. (2013). Nature-inspired design of motif-specific antibody scaffolds. *Nature Biotechnology*, *31*(10), 916–921. http://dx.doi.org/10.1038/nbt.2672.

Kooij, V., Venkatraman, V., Tra, J., Kirk, J. A., Rowell, J., Blice-Baum, A., … Van Eyk, J. E. (2014). Sizing up models of heart failure: Proteomics from flies to humans. *Proteomics. Clinical Applications*, *8*(9–10), 653–664. http://dx.doi.org/10.1002/prca.201300123.

Kraeutler, M. J., Soltis, A. R., & Saucerman, J. J. (2010). Modeling cardiac beta-adrenergic signaling with normalized-hill differential equations: Comparison with a biochemical model. *BMC Systems Biology*, *4*, 157. http://dx.doi.org/10.1186/1752-0509-4-157.

Kung, L. A., Tao, S. C., Qian, J., Smith, M. G., Snyder, M., & Zhu, H. (2009). Global analysis of the glycoproteome in Saccharomyces cerevisiae reveals new roles for protein glycosylation in eukaryotes. *Molecular Systems Biology*, *5*, 308. http://dx.doi.org/10.1038/msb.2009.64.

Kunkel, M. T., & Newton, A. C. (2009). Spatiotemporal dynamics of kinase signaling visualized by targeted reporters. *Current Protocols in Chemical Biology*, *1*(1), 17–18. http://dx.doi.org/10.1002/9780470559277.ch090106.

Kunkel, M. T., & Newton, A. C. (2014). Imaging kinase activity at protein scaffolds. *Methods in Molecular Biology*, *1071*, 129–137. http://dx.doi.org/10.1007/978-1-62703-622-1_10.

Lam, A. J., St-Pierre, F., Gong, Y., Marshall, J. D., Cranfill, P. J., Baird, M. A., … Lin, M. Z. (2012). Improving FRET dynamic range with bright green and red fluorescent proteins. *Nature Methods*, *9*(10), 1005–1012. http://dx.doi.org/10.1038/nmeth.2171.

Landry, B. D., Clarke, D. C., & Lee, M. J. (2015). Studying cellular signal transduction with OMIC technologies. *Journal of Molecular Biology*, *427*(21), 3416–3440. http://dx.doi.org/10.1016/j.jmb.2015.07.021.

Lin, J. R., Fallahi-Sichani, M., & Sorger, P. K. (2015). Highly multiplexed imaging of single cells using a high-throughput cyclic immunofluorescence method. *Nature Communications*, *6*, 8390. http://dx.doi.org/10.1038/ncomms9390.

Liu, S., He, J., Jin, H., Yang, F., Lu, J., & Yang, J. (2011). Enhanced dynamic range in a genetically encoded Ca2+ sensor. *Biochemical and Biophysical Research Communications*, *412*(1), 155–159. http://dx.doi.org/10.1016/j.bbrc.2011.07.065.

Lu, J. Y., Lin, Y. Y., Boeke, J. D., & Zhu, H. (2013). Using functional proteome microarrays to study protein lysine acetylation. *Methods in Molecular Biology*, *981*, 151–165. http://dx.doi.org/10.1007/978-1-62703-305-3_12.

Maglica, Z., Ozdemir, E., & McKinney, J. D. (2015). Single-cell tracking reveals antibiotic-induced changes in mycobacterial energy metabolism. *MBio*, *6*(1), e02236-02214. http://dx.doi.org/10.1128/mBio.02236-14.

Markoulaki, S., Matson, S., & Ducibella, T. (2004). Fertilization stimulates long-lasting oscillations of CaMKII activity in mouse eggs. *Developmental Biology*, *272*(1), 15–25. http://dx.doi.org/10.1016/j.ydbio.2004.04.008.

Matsunaga-Udagawa, R., Fujita, Y., Yoshiki, S., Terai, K., Kamioka, Y., Kiyokawa, E., … Matsuda, M. (2010). The scaffold protein Shoc2/SUR-8 accelerates the interaction of Ras and Raf. *The Journal of Biological Chemistry, 285*(10), 7818–7826. http://dx.doi.org/10.1074/jbc.M109.053975.

Mehta, S., Aye-Han, N. N., Ganesan, A., Oldach, L., Gorshkov, K., & Zhang, J. (2014). Calmodulin-controlled spatial decoding of oscillatory Ca2+ signals by calcineurin. *eLife, 3*, e03765. http://dx.doi.org/10.7554/eLife.03765.

Mehta, S., & Zhang, J. (2011). Reporting from the field: Genetically encoded fluorescent reporters uncover signaling dynamics in living biological systems. *Annual Review of Biochemistry, 80*, 375–401. http://dx.doi.org/10.1146/annurev-biochem-060409-093259.

Montero, J., Sarosiek, K. A., DeAngelo, J. D., Maertens, O., Ryan, J., Ercan, D., … Letai, A. (2015). Drug-induced death signaling strategy rapidly predicts cancer response to chemotherapy. *Cell, 160*(5), 977–989. http://dx.doi.org/10.1016/j.cell.2015.01.042.

Newman, R. H., Fosbrink, M. D., & Zhang, J. (2011). Genetically encodable fluorescent biosensors for tracking signaling dynamics in living cells. *Chemical Reviews, 111*(5), 3614–3666. http://dx.doi.org/10.1021/cr100002u.

Newman, R. H., Hu, J., Rho, H. S., Xie, Z., Woodard, C., Neiswinger, J., … Qian, J. (2013). Construction of human activity-based phosphorylation networks. *Molecular Systems Biology, 9*, 655. http://dx.doi.org/10.1038/msb.2013.12.

Newman, R. H., & Zhang, J. (2008). Visualization of phosphatase activity in living cells with a FRET-based calcineurin activity sensor. *Molecular BioSystems, 4*(6), 496–501. http://dx.doi.org/10.1039/b720034j.

Newman, R. H., & Zhang, J. (2014). The design and application of genetically encodable biosensors based on fluorescent proteins. *Methods in Molecular Biology, 1071*, 1–16. http://dx.doi.org/10.1007/978-1-62703-622-1_1.

Newman, R. H., Zhang, J., & Zhu, H. (2014). Toward a systems-level view of dynamic phosphorylation networks. *Frontiers in Genetics, 5*, 263. http://dx.doi.org/10.3389/fgene.2014.00263.

Ni, Q., Ganesan, A., Aye-Han, N. N., Gao, X., Allen, M. D., Levchenko, A., & Zhang, J. (2011). Signaling diversity of PKA achieved via a Ca^{2+}-cAMP-PKA oscillatory circuit. *Nature Chemical Biology*, 7(1), 34–40. http://dx.doi.org/10.1038/nchembio.478.

Niino, Y., Hotta, K., & Oka, K. (2010). Blue fluorescent cGMP sensor for multiparameter fluorescence imaging. *PloS One, 5*(2), e9164. http://dx.doi.org/10.1371/journal.pone.0009164.

Nikolaev, V. O., Bunemann, M., Hein, L., Hannawacker, A., & Lohse, M. J. (2004). Novel single chain cAMP sensors for receptor-induced signal propagation. *The Journal of Biological Chemistry, 279*(36), 37215–37218. http://dx.doi.org/10.1074/jbc.C400302200.

Olsen, J. V., Macek, B., Lange, O., Makarov, A., Horning, S., & Mann, M. (2007). Higher-energy C-trap dissociation for peptide modification analysis. *Nature Methods, 4*(9), 709–712. http://dx.doi.org/10.1038/nmeth1060.

Olsen, J. V., & Mann, M. (2013). Status of large-scale analysis of post-translational modifications by mass spectrometry. *Molecular & Cellular Proteomics, 12*(12), 3444–3452. http://dx.doi.org/10.1074/mcp.O113.034181.

Ong, S. E. (2012). The expanding field of SILAC. *Analytical and Bioanalytical Chemistry, 404*(4), 967–976. http://dx.doi.org/10.1007/s00216-012-5998-3.

Ortsater, H., Grankvist, N., Honkanen, R. E., & Sjoholm, A. (2014). Protein phosphatases in pancreatic islets. *The Journal of Endocrinology, 221*(3), R121–R144. http://dx.doi.org/10.1530/JOE-14-0002.

Pearce, L. L., Gandley, R. E., Han, W., Wasserloos, K., Stitt, M., Kanai, A. J., … Levitan, E. S. (2000). Role of metallothionein in nitric oxide signaling as revealed by a green fluorescent fusion protein. *Proceedings of the National Academy of Sciences of the United States of America, 97*(1), 477–482.

Pierobon, M., Wulfkuhle, J., Liotta, L., & Petricoin, E. (2015). Application of molecular technologies for phosphoproteomic analysis of clinical samples. *Oncogene*, *34*(7), 805–814. http://dx.doi.org/10.1038/onc.2014.16.

Popescu, S. C., Popescu, G. V., Bachan, S., Zhang, Z., Seay, M., Gerstein, M., … Dinesh-Kumar, S. P. (2007). Differential binding of calmodulin-related proteins to their targets revealed through high-density Arabidopsis protein microarrays. *Proceedings of the National Academy of Sciences of the United States of America*, *104*(11), 4730–4735. http://dx.doi.org/10.1073/pnas.0611615104.

Roskoski, R., Jr. (2016). Janus kinase (JAK) inhibitors in the treatment of inflammatory and neoplastic diseases. *Pharmacological Research*, *111*, 784–803. http://dx.doi.org/10.1016/j.phrs.2016.07.038.

Ruprecht, B., & Lemeer, S. (2014). Proteomic analysis of phosphorylation in cancer. *Expert Review of Proteomics*, *11*(3), 259–267. http://dx.doi.org/10.1586/14789450.2014.901156.

Russwurm, M., Mullershausen, F., Friebe, A., Jager, R., Russwurm, C., & Koesling, D. (2007). Design of fluorescence resonance energy transfer (FRET)-based cGMP indicators: A systematic approach. *The Biochemical Journal*, *407*(1), 69–77. http://dx.doi.org/10.1042/BJ20070348.

Ryall, K. A., Holland, D. O., Delaney, K. A., Kraeutler, M. J., Parker, A. J., & Saucerman, J. J. (2012). Network reconstruction and systems analysis of cardiac myocyte hypertrophy signaling. *The Journal of Biological Chemistry*, *287*(50), 42259–42268. http://dx.doi.org/10.1074/jbc.M112.382937.

Salzer, E., Santos-Valente, E., Keller, B., Warnatz, K., & Boztug, K. (2016). Protein kinase C delta: A gatekeeper of immune homeostasis. *Journal of Clinical Immunology*, *36*(7), 631–640. http://dx.doi.org/10.1007/s10875-016-0323-0.

Sample, V., Newman, R. H., & Zhang, J. (2009). The structure and function of fluorescent proteins. *Chemical Society Reviews*, *38*(10), 2852–2864. http://dx.doi.org/10.1039/b913033k.

Saucerman, J. J., Zhang, J., Martin, J. C., Peng, L. X., Stenbit, A. E., Tsien, R. Y., & McCulloch, A. D. (2006). Systems analysis of PKA-mediated phosphorylation gradients in live cardiac myocytes. *Proceedings of the National Academy of Sciences of the United States of America*, *103*(34), 12923–12928. http://dx.doi.org/10.1073/pnas.0600137103.

Sevecka, M., & MacBeath, G. (2006). State-based discovery: A multidimensional screen for small-molecule modulators of EGF signaling. *Nature Methods*, *3*(10), 825–831. http://dx.doi.org/10.1038/nmeth931.

Shull, A. Y., Noonepalle, S. K., Awan, F. T., Liu, J., Pei, L., Bollag, R. J., … Shi, H. (2015). RPPA-based protein profiling reveals eIF4G overexpression and 4E–BP1 serine 65 phosphorylation as molecular events that correspond with a pro-survival phenotype in chronic lymphocytic leukemia. *Oncotarget*, *6*(16), 14632–14645. http://dx.doi.org/10.18632/oncotarget.4104.

Spencer, S. L., Gaudet, S., Albeck, J. G., Burke, J. M., & Sorger, P. K. (2009). Non-genetic origins of cell-to-cell variability in TRAIL-induced apoptosis. *Nature*, *459*(7245), 428–432. http://dx.doi.org/10.1038/nature08012.

Stoevesandt, O., & Taussig, M. J. (2013). Phospho-specific antibodies by design. *Nature Biotechnology*, *31*(10), 889–891. http://dx.doi.org/10.1038/nbt.2712.

Sun, Y., Cheng, L., Gu, Y., Xin, A., Wu, B., Zhou, S., … Tao, S. C. (2016). A human lectin microarray for sperm surface glycosylation analysis. *Molecular & Cellular Proteomics*, *15*(9), 2839–2851. http://dx.doi.org/10.1074/mcp.M116.059311.

Sutandy, F. X., Hsiao, F. S., & Chen, C. S. (2016). High throughput platform to explore RNA-protein interactomes. *Critical Reviews in Biotechnology*, *36*(1), 11–19. http://dx.doi.org/10.3109/07388551.2014.922916.

Sutandy, F. X., Qian, J., Chen, C. S., & Zhu, H. (2013). Overview of protein microarrays. *Current Protocols in Protein Science*, 72(27), 1:27.1.1–1:27.1.16. http://dx.doi.org/10.1002/0471140864.ps2701s72. Chapter 27, Unit 27 21.

Tape, C. J. (2016). Systems biology analysis of heterocellular signaling. *Trends in Biotechnology*, *34*(8), 627–637. http://dx.doi.org/10.1016/j.tibtech.2016.02.016.

Terfve, C., & Saez-Rodriguez, J. (2012). Modeling signaling networks using high-throughput phospho-proteomics. *Adv. Exp. Med. Biol.*, *736*, 19–57. http://dx.doi.org/10.1007/978-1-4419-7210-1_2.

Tobias, I. S., & Newton, A. C. (2016). Protein scaffolds control localized protein kinase Czeta activity. *The Journal of Biological Chemistry*, *291*(26), 13809–13822. http://dx.doi.org/10.1074/jbc.M116.729483.

Tsai, L. C., Xie, L., Dore, K., Xie, L., Del Rio, J. C., King, C. C., … Newton, A. C. (2015). Zeta inhibitory peptide disrupts electrostatic interactions that maintain atypical protein kinase C in its active conformation on the scaffold p62. *The Journal of Biological Chemistry*, *290*(36), 21845–21856. http://dx.doi.org/10.1074/jbc.M115.676221.

Violin, J. D., DiPilato, L. M., Yildirim, N., Elston, T. C., Zhang, J., & Lefkowitz, R. J. (2008). beta2-Adrenergic receptor signaling and desensitization elucidated by quantitative modeling of real time cAMP dynamics. *The Journal of Biological Chemistry*, *283*(5), 2949–2961. http://dx.doi.org/10.1074/jbc.M707009200.

Violin, J. D., Zhang, J., Tsien, R. Y., & Newton, A. C. (2003). A genetically encoded fluorescent reporter reveals oscillatory phosphorylation by protein kinase C. *The Journal of Cell Biology*, *161*(5), 899–909. http://dx.doi.org/10.1083/jcb.200302125.

von Stechow, L., Francavilla, C., & Olsen, J. V. (2015). Recent findings and technological advances in phosphoproteomics for cells and tissues. *Expert Review of Proteomics*, *12*(5), 469–487. http://dx.doi.org/10.1586/14789450.2015.1078730.

Woehler, A. (2013). Simultaneous quantitative live cell imaging of multiple FRET-based biosensors. *PloS One*, *8*(4), e61096. http://dx.doi.org/10.1371/journal.pone.0061096.

Woodard, C. L., Goodwin, C. R., Wan, J., Xia, S., Newman, R., Hu, J., … Zhu, H. (2013). Profiling the dynamics of a human phosphorylome reveals new components in HGF/c-Met signaling. *PloS One*, *8*(9), e72671. http://dx.doi.org/10.1371/journal.pone.0072671.

Yang, H., & Yang, L. (2016). Targeting cAMP/PKA pathway for glycemic control and type 2 diabetes therapy. *Journal of Molecular Endocrinology*, *57*(2), R93–R108. http://dx.doi.org/10.1530/JME-15-0316.

Yaniv, Y., Ganesan, A., Yang, D., Ziman, B. D., Lyashkov, A. E., Levchenko, A., … Lakatta, E. G. (2015). Real-time relationship between PKA biochemical signal network dynamics and increased action potential firing rate in heart pacemaker cells: Kinetics of PKA activation in heart pacemaker cells. *Journal of Molecular and Cellular Cardiology*, *86*, 168–178. http://dx.doi.org/10.1016/j.yjmcc.2015.07.024.

Zaytsev, A. V., Segura-Pena, D., Godzi, M., Calderon, A., Ballister, E. R., Stamatov, R., … Grishchuk, E. L. (2016). Bistability of a coupled aurora B kinase-phosphatase system in cell division. *eLife*, *5*, e10644. http://dx.doi.org/10.7554/eLife.10644.

Zeigler, A. C., Richardson, W. J., Holmes, J. W., & Saucerman, J. J. (2016). A computational model of cardiac fibroblast signaling predicts context-dependent drivers of myofibroblast differentiation. *Journal of Molecular and Cellular Cardiology*, *94*, 72–81. http://dx.doi.org/10.1016/j.yjmcc.2016.03.008.

Zhang, J., Ma, Y., Taylor, S. S., & Tsien, R. Y. (2001). Genetically encoded reporters of protein kinase A activity reveal impact of substrate tethering. *Proceedings of the National Academy of Sciences of the United States of America*, *98*(26), 14997–15002. http://dx.doi.org/10.1073/pnas.211566798.

Zhang, H., & Pelech, S. (2012). Using protein microarrays to study phosphorylation-mediated signal transduction. *Seminars in Cell & Developmental Biology*, *23*(8), 872–882. http://dx.doi.org/10.1016/j.semcdb.2012.05.009.

Zhu, H., Bilgin, M., Bangham, R., Hall, D., Casamayor, A., Bertone, P., … Snyder, M. (2001). Global analysis of protein activities using proteome chips. *Science*, *293*(5537), 2101–2105. http://dx.doi.org/10.1126/science.1062191.

CHAPTER SIX

Probing Cdc42 Polarization Dynamics in Budding Yeast Using a Biosensor

Satoshi Okada*,†,1, Mid Eum Lee‡,1,2, Erfei Bi*,3, Hay-Oak Park‡,3
*University of Pennsylvania Perelman School of Medicine, Philadelphia, PA, United States
†Kyushu University Graduate School of Medical Sciences, Fukuoka, Japan
‡The Ohio State University, Columbus, OH, United States
[3]Corresponding authors: e-mail address: ebi@mail.med.upenn.edu; park.294@osu.edu

Contents

Abstract

Cdc42 is a small guanosine triphosphatase (GTPase) that plays a central role in polarity development in diverse cell types. Since the activity of Cdc42 is dynamically controlled in time and space, it is required to develop a biosensor to monitor its activation in vivo. In this chapter, we describe the construction and usage of a simple and robust biosensor for monitoring active Cdc42 in budding yeast. This affinity-based biosensor uses a red fluorescent protein fused to a Cdc42- and Rac-interactive binding motif from one of the Cdc42 effector proteins. Because it binds specifically to the GTP-bound Cdc42, this

[1] These authors contributed equally to this work.
[2] Present address: Department of Biochemistry and Cell Biology, The Geisel School of Medicine at Dartmouth, Hanover, NH 03755, United States.

Methods in Enzymology, Volume 589
ISSN 0076-6879
http://dx.doi.org/10.1016/bs.mie.2017.01.011

biosensor can be used to monitor Cdc42 activation in vivo. This or similar biosensors can be widely used for studying GTPase signaling in other cell types because of the conserved CRIB motif present among GTPase targets.

ABBREVIATIONS

CRIB Cdc42- and Rac-interactive binding
FRAP fluorescence recovery after photobleaching
FRET fluorescence resonance energy transfer
GEF guanine nucleotide exchange factor
GFP green fluorescent protein
GTPase guanosine triphosphatase
PAK p21-activated kinase
PBD p21-binding domain

1. INTRODUCTION

Cdc42 is a highly conserved small guanosine triphosphatase (GTPase) that plays a key role in polarity development in species ranging from yeast to human (Etienne-Manneville, 2004; Park & Bi, 2007). In response to either internal or external cues, Cdc42 cycles between the GDP- and GTP-bound states, and its activity is dynamically controlled in time and space to enable distinct polarization events, including cytoskeletal assembly and membrane trafficking. Since Cdc42 is activated transiently at a specific region, a robust and reliable biosensor for its activation is required to probe the dynamics of Cdc42 activation in live cells.

Earlier use of fluorescence-based biosensors had allowed the determination of the spatiotemporal profile of GTPases' activity, providing mechanistic insights into cell polarization in different model systems. The activation patterns of multiple GTPases have been monitored by several biosensors for more than 10 years. Initially, activation of the endogenous Cdc42 had been monitored using a Wiskott–Aldrich Syndrome protein (WASP) domain that binds only to the active Cdc42. This Cdc42 effector domain labeled with a solvent-sensitive dye produced strong fluorescence signal upon its binding to the activated Cdc42 by placing the dye in a hydrophobic pocket formed by amino acids from both WASP and Cdc42 (Nalbant, Hodgson, Kraynov, Toutchkine, & Hahn, 2004).

Recent polarization studies involved the use of both single and dual chain fluorescence resonance energy transfer (FRET)-based biosensors for small GTPases. The intramolecular FRET-based sensor was designed for monitoring RhoA activation using a fusion construct expressing the Rho effector rhotekin (RBD, aa7–89) tagged with a cyan fluorescent protein, a variant of yellow fluorescent protein, and the full-length RhoA (Pertz, Hodgson, Klemke, & Hahn, 2006). Upon activation, RBD binds to Rho-GTP, which results in changes in the relative position of the two fluorophores, increasing the FRET signal. In budding yeast, a FRET-based biosensor for active Cdc42 consisted of the Cdc42/Rac-interactive binding (CRIB) domain from Cla4, a Cdc42 effector, and Cdc42 that are sandwiched by two fluorophores, green fluorescent protein (GFP), and mCherry (Smith et al., 2013).

In this chapter, we describe the construction and usage of a simple and robust biosensor for monitoring active Cdc42 in the budding yeast *Saccharomyces cerevisiae*. The Cdc42 effector Gic2 contains a conserved CRIB motif, also known as PBD (p21-binding domain), which binds specifically to Cdc42-GTP, and localizes to the sites of polarized growth in a Cdc42-dependent manner (Jaquenoud & Peter, 2000). A fragment of $Gic2^{1-208}$ containing the PBD has been tagged with fluorescent molecules and used for monitoring Cdc42-GTP in live cells (Kang, Lee, & Park, 2014; Lee, Lo, Miller, Chou, & Park, 2015; Lo, Lee, Narayan, Chou, & Park, 2013; Okada et al., 2013; Ozbudak, Becskei, & van Oudenaarden, 2005; Tong et al., 2007). By using this affinity-based biosensor, we can monitor dynamics of Cdc42 activation in specific processes in vivo. This and similar biosensors of GTPases will undoubtedly play key roles in understanding the functions, mechanisms, and spatiotemporal regulation of G protein signaling in diverse cell types.

2. GENERATION OF A BIOSENSOR OF ACTIVE CDC42

2.1 Development of a Cdc42-GTP Biosensor

The Cdc42 effector Gic2 contains a p21-binding domain (PBD, aa133–166) and a CRIB domain (aa134–147) (Fig. 1A), which are required for its interaction with Cdc42-GTP. Jaquenoud and Peter reported that Gic2 localized to the sites of polarized cell growth where Cdc42-GTP was presumably concentrated, and this localization depended on its CRIB domain (Jaquenoud & Peter, 2000). Based on these observations, we developed a Cdc42-GTP biosensor, Gic2-PBD-RFP, by simply fusing the N-terminal portion (aa1–208) of Gic2 to the red fluorescent protein tdTomato (Fig. 1B). As expected, an in vitro interaction assay confirmed that Gic2-PBD-RFP interacted specifically with GTP-bound Cdc42

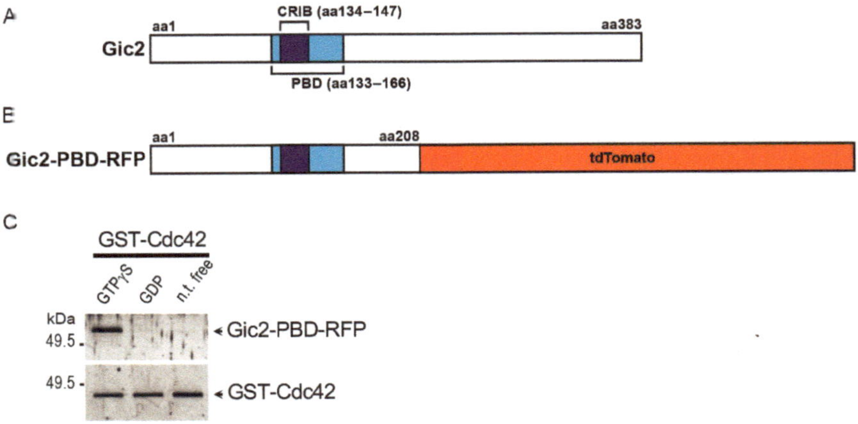

Fig. 1 Schematic representation of Gic2 and Gic2-PBD-RFP and association of PBD-RFP with Cdc42-GTP. (A) Domain structure of Gic2. Gic2-PBD (p21-binding domain, aa133–166) is shown in *cyan*, and CRIB (Cdc42/Rac-interactive binding) domain is shown in *blue*. (B) Domain structure of Gic2-PBD-RFP. A *red* fluorescent protein tdTomato fused to the N-terminal portion of Gic2 is shown in *red*. (C) Association of Gic2-PBD-RFP with GTP-bound Cdc42. GST-Cdc42 purified from the insect cells was immobilized on glutathione Sepharose beads, loaded with GTPγS or GDP, or stripped with EDTA to deplete nucleotides bound to Cdc42 (n.t. free), and then incubated with yeast extracts containing Gic2-PBD-RFP. Gic2-PBD-RFP was detected with polyclonal antibodies against RFP (*upper panel*); GST-Cdc42 was detected with polyclonal antibodies against GST (*lower panel*). *Data provided by Dr. P.J. Kang.*

(Fig. 1C). Gic2-PBD-RFP localized to the sites of polarized cell growth, including the presumptive bud site and bud cortex, but not the bud neck at the onset of cytokinesis (Figs. 2–5) (Kang et al., 2014; Lee et al., 2015; Lo et al., 2013; Okada et al., 2013; Ozbudak et al., 2005; Tong et al., 2007).

To test whether Gic2-PBD-RFP signal genuinely reflected Cdc42 activity (i.e., the level of Cdc42-GTP) in live cells, we examined the fluorescence intensity and distribution of Gic2-PBD-RFP in *cdc24-4* mutant cells, which harbor a temperature-sensitive mutation in *CDC24* that encodes the essential GEF (guanine nucleotide exchange factor) for Cdc42 in budding yeast (Kang et al., 2014; Tong et al., 2007). When the *cdc24-4* mutant cells were cultured at the permissive temperature (25°C), Gic2-PBD-RFP signal localized at the growth sites. After a 20-min treatment at the nonpermissive temperature (37°C), the signal of Gic2-PBD-RFP disappeared from the growth sites. In wild-type cells, the Gic2-PBD-RFP signal remained at the growth sites under the same condition. This result suggested that the loss of Cdc42 GEF (Cdc24) function led to a decrease in Cdc42-GTP (or increase of Cdc42-GDP), which

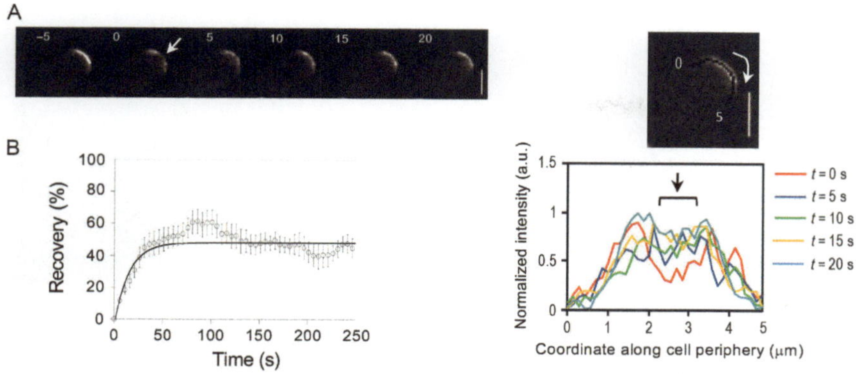

Fig. 2 Gic2-PBD-RFP is highly dynamic. (A) Intensity profile of PBD-RFP cluster along the periphery of **a**/α cell in the G1 phase after bleaching. Numbers indicate time (in s) relative to the bleaching ($t = 0$). The bleached region is marked with a *white arrow* on the image and a *black arrow* with bracket on the graph. Each *colored line* in the graph represents a different time point. Scale bar, 3 μm. (B) FRAP analysis of PBD-RFP cluster in G1 ($n = 10$ cells). Relative intensity is normalized to the prebleach intensity. Mean ± SEM. The curve fit line is drawn in *black-colored line*.

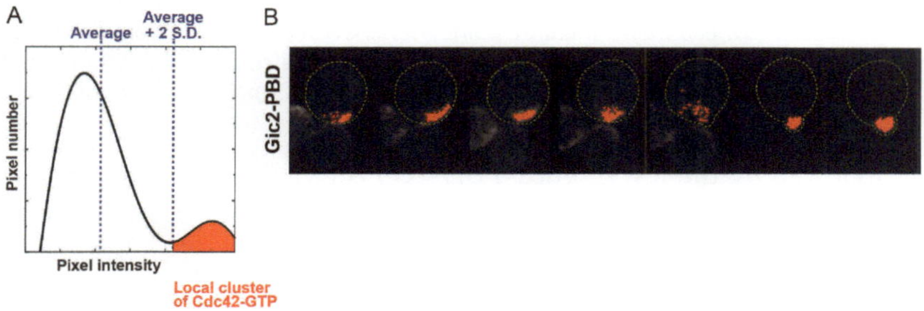

Fig. 3 The threshold method for quantifying Cdc42-GTP signal in live cells. (A) Using the threshold method to assign pixels to a cluster of Cdc42-GTP. After the measurement of the intensities of all pixels inside the cell outline, if a pixel has an intensity higher than the value of the average plus two standard deviations, the pixel is assigned to a cluster of Cdc42-GTP. (B) An example of defining a Cdc42-GTP cluster by the threshold method. Gic2-PBD-RFP signal is shown in grayscale. The pixels calculated as a part of the Cdc42-GTP local cluster are shown in *red*. The cell outlines are marked in *yellow*. *Adapted from Okada, S., Leda, M., Hanna, J., Savage, N. S., Bi, E., & Goryachev, A. B. (2013). Daughter cell identity emerges from the interplay of Cdc42, septins, and exocytosis.* Developmental Cell, 26, *148–161.*

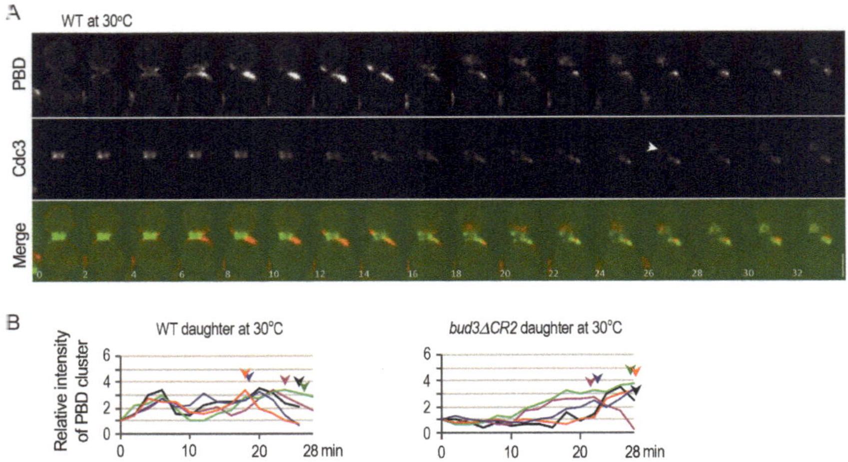

Fig. 4 Dynamic Cdc42-GTP polarization during selection of a new bud site. (A) Time-lapse imaging of PBD-RFP and Cdc3-GFP in wild-type cells at 30°C. *Arrowheads* mark the first appearance of a new septin in a daughter cell. Numbers indicate time (in min) from the onset of cytokinesis ($t = 0$). Scale bar, 3 μm. *Adapted from Kang, P. J., Lee, M. E., Park, H.-O. (2014) Bud3 activates Cdc42 to establish a proper growth site in budding yeast. The Journal of Cell Biology, 206, 19–28.* doi:10.1083/jcb.201402040. (B) PBD-RFP intensity is compared in daughter cells of wild-type and *bud3ΔCR2* cells at 30°C. PBD-RFP intensity was normalized to intensity at the onset of cytokinesis ($t = 0$). Each colored *arrowhead* marks the time point when new septin clouds were first visible (or estimated) in each daughter cell for each corresponding colored line. ©Kang et al., 2014. Originally published in Journal of Cell Biology. doi:10.1083/jcb.201402040.

was followed by the loss of Gic2-PBD-RFP signal from the growth sites. Interestingly, a significant percentage of *cdc24-4* cells carrying a multicopy *BUD3* plasmid exhibited the Gic2-PBD-RFP cluster when arrested as unbudded cells at 37°C. This observation led to the discovery of the second Cdc42 GEF Bud3 (Kang et al., 2014) (see Section 4.1). These studies demonstrated the feasibility of Gic2-PBD-RFP as a Cdc42-GTP biosensor with a sensitivity and responsiveness applicable to experiments involving live cells.

2.2 Further Characterization of the Cdc42 Biosensor by FRAP

We performed fluorescence recovery after photobleaching (FRAP) analysis (see later for the procedure) to further characterize the Cdc42-GTP biosensor in vivo. When the polar cap of PBD-RFP in unbudded cells was bleached, it recovered within 10 s (Fig. 2A), although slightly more slowly than the Gic2-PBD-GFP probe, which was reported to recover its initial intensity in less than 1 s (Ozbudak et al., 2005). Using a 0.83×0.83 μm bleach spot, we found that Gic2-PBD-RFP recovered significantly faster,

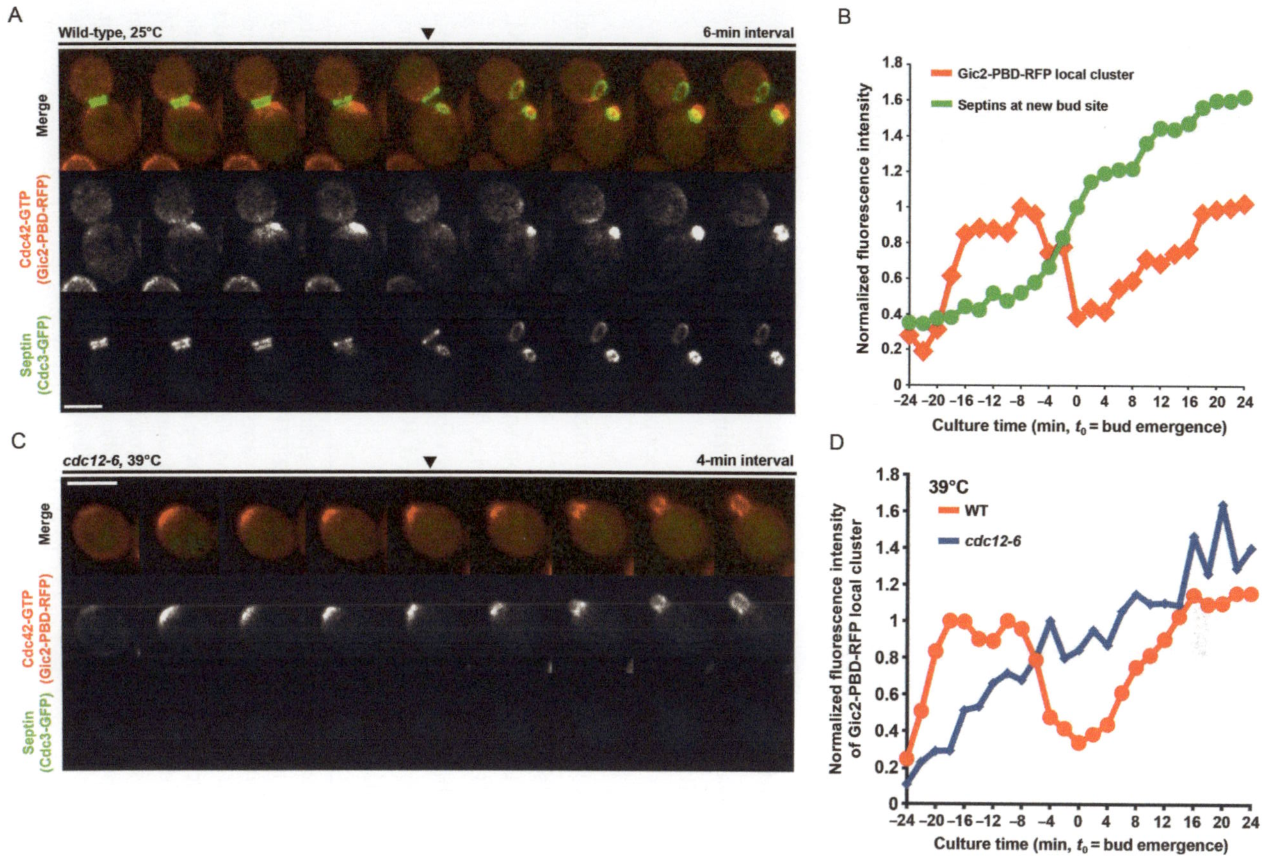

Fig. 5 See legend on next page.

with a recovery half-time of 11.8 ± 4.5 s (Fig. 2B) compared to the GFP-fusion of GTP-locked Cdc42^{Q61L}, albeit somewhat slower than GFP-Cdc42 (Bendezu et al., 2015; Wedlich-Soldner, Wai, Schmidt, & Li, 2004). These data suggest that our biosensor has the specificity and dynamics for monitoring Cdc42 activity in vivo.

3. DETECTION AND QUANTIFICATION OF Cdc42 ACTIVATION IN VIVO

Using the Gic2-PBD-RFP biosensor, we have monitored Cdc42 polarization dynamics throughout the cell cycle. Here, we describe the standard protocols for sample preparation and the equipment used for FRAP and time-lapse imaging at University of Pennsylvania (Penn) and the Ohio State University (OSU).

3.1 Devices

Devices used for time-lapse imaging (performed at the CDB Microscopy Core at PENN)

- Microscope: an inverted microscope (IX71; Olympus, Center Valley, PA, USA)
- Objective lens: a 100 ×, 1.4 (numerical aperture) NA, Plan S-Apo oil immersion (Olympus)
- Camera: a back-thinned electron-multiplying CCD camera (ImagEM, C9100-13; Hamamatsu Photonics, Bridgewater, NJ, USA)
- Light source: diode lasers (488 nm for GFP and 561 nm for RFP) set in a laser integrator (Spectral Applied Research, Richmond Hill, Ontario, Canada)

Fig. 5 Local inhibition of Cdc42 during septin ring formation. (A) An image sequence showing the dynamic behavior of Cdc42-GTP (Gic2-PBD-RFP, *red*) and septins (Cdc3-GFP, *green*) during bud emergence. The image sequence was generated by average projection of 11 *z*-slices with 0.6 μm spacing. *Filled triangle* indicates the timing of bud emergence. Scale bar, 3 μm. (B) Quantitative analysis of Cdc42-GTP cluster and septin accumulation at the site of polarization during bud emergence. The cell shown in (A) was used for the analysis. Normalized total intensity of Cdc42-GTP cluster is shown in *red*, and the normalized septin intensity at the new bud site is shown in *green*. (C) Polarized growth of septin mutant *cdc12-6* strain. At the nonpermissive temperature (39°C), intensity of the Cdc42-GTP cluster does not show a transient decrease around bud emergence. Scale bar, 3 μm. (D) Quantitative analysis of Cdc42-GTP cluster during bud emergence in *cdc12-6* mutant and isogenic wild-type cells cultured at 39°C. The temporal changes of the normalized total intensity of Cdc42-GTP cluster in wild-type and *cdc12-6* mutant cells are shown in *red* and *blue*, respectively.

- Confocal unit: a spinning disk confocal scan head (CSU10; Yokogawa, Sugar Land, TX, USA)
- Environmental chamber: culture dish heater (DH-35; Warner Instruments, Hamden, CT, USA)
- Control software: MetaMorph 7.7 (Molecular Devices, Sunnyvale, CA, USA)

Devices used for FRAP (performed at OSU)

- Microscope: an inverted microscope (Eclipse TE2000-U; Nikon, Melville, NY, USA)
- Objective lens: a 100×/1.4 NA Plan Apo oil immersion (Nikon)
- Light source: a 488-nm argon ion laser and a 568-nm solid-state laser
- Camera: a cooled charge-coupled device camera (ORCA-AG; Hamamatsu Photonics)
- Confocal system: spinning disk confocal microscope (UltraView ERS; Perkin Elmer Life and Analytical Sciences)

Devices used for time-lapse imaging (performed at OSU)

- Microscope: an inverted microscope (Ti-E; Nikon)
- Objective lens: a 100×, 1.4 NA Plan Apo oil immersion (Nikon)
- Light source: 440-, 488-, 515-, and 561-nm solid-state lasers (Modular Laser System 2.0; PerkinElmer, Waltham, MA, USA)
- Camera: a back-thinned electron-multiplying charge-coupled device camera (ImagEM C9100-13; Hamamatsu Photonics)
- Confocal system: spinning disk confocal microscope (Ultra-VIEW VoX CSUX1 system; PerkinElmer)
- Climate chamber: stage top incubator INUB-PPZI2-F1 equipped with UNIV2-D35 dish holder (Tokai Hit, Shizuoka-ken, Japan)

3.2 Sample Preparation

For imaging studies at Penn

1. Inoculate yeast cells in an appropriate medium and continue to culture overnight at 25°C. *Note*: Synthetic complete (SC) medium containing 2% dextrose without buffering material is preferable because the solidity of agarose (used in later steps) will be affected by the pH of the medium. When the fluorescence intensity of the target molecules is expected to be relatively low, it is better to use yeast nitrogen base without riboflavin to reduce the background fluorescence signal from the medium.
2. Dilute the cells into fresh medium (5- to 20-fold dilution) and continue to culture at 25°C until the cells enter the exponential growth phase (OD_{600} = 0.4–0.8).

3. Prepare culture medium containing 1.2% low-melting point agarose and boil the medium for 10 min and keep the melted culture medium in a liquid form at 37°C.
4. Put the yeast liquid culture (500–700 μL) on a polylysine-coated glass-bottom 35-mm dish (P35GC-1.5-14.C; MatTek, Ashland, MA, USA).
5. To settle down the yeast cells to the glass surface, incubate the sample for ~10 min.
6. Using a P1000 micropipette, aspirate the excessive supernatant gently (500–550 μL).
7. Put 1300 μL of the melted culture medium (prepared in step 3) and mix the yeast cells and the agarose-containing medium gently with a P1000 micropipette 3 times.
8. Incubate the dish at room temperature (~25°C) for 15 min to solidify the agarose layer.
9. Put 500–550 μL of liquid medium over the solidified agarose layer.
10. Incubate the sample dish at room temperature (~25°C) for 5–15 min to let the liquid and solid layers reach equilibrium.
11. Set the sample dish into an environmental chamber to maintain a constant condition during time-lapse imaging.

For imaging studies at OSU

Cell growth

1. Grow yeast cells overnight in synthetic media containing appropriate nutrients at 30°C. Since our imaging typically involves fluorescent proteins expressed from their endogenous chromosomal loci, we use SC containing 2% (w/v) dextrose.
2. Dilute overnight culture about 5- to 10-fold in the same fresh media and grow for 4–5 h (to reach $OD_{600}=0.4$–0.8) at 30°C prior to imaging.

Mounting cells

For both time-lapse imaging and FRAP experiments, cells are mounted on agar pads on the glass slide as described in the following steps. Alternatively, cells can be immobilized in glass-bottom dishes (see later).

1. Dissolve 2% media containing low-melting agarose (same media used for growing cells) in a microfuge tube using a heating block.
2. Clean a glass slide with a Kimwipe wiper and then attach two strips of scotch tape on both ends of the glass slide.
3. Place the slide on a cold metal block (covered with a Kimwipe on the top) and drop 30 μL of the agarose mix onto the slide.

4. Place a second slide on the top of the agar mix and then flatten the agar pad by pressing evenly using the thumbs or by putting some weight (for example, placing a glass bottle containing ~50 g of H_2O) on the top.
5. Push apart the two slides using a constant, gentle sliding action. To get a flat agar surface, it is better to detach slides while moisture still remains in the agar pad.
6. Drop ~3 μL of cell culture (spin down cells to concentrate before loading) onto the agarose pad.
7. Place a cover glass on top of the agar pad and seal the cover glass by applying nail polish or 100% pure petroleum jelly (e.g., Vaseline).

   Wait for a couple of minutes for the sealing to harden before using the slide for imaging.

Yeast cells can be immobilized in glass-bottom dishes (35 mm; MatTek), pretreated with concanavalin A (ConA; Type IV, Sigma, St. Louis, MO, USA). *Note*: ConA is a lectin that binds to carbohydrate on the yeast cell surface. ConA stock is prepared in water at 0.5 mg/mL and stored up to a year at –20°C. Alternatively, polylysine-coated glass-bottom dishes can be used. Mounting cells in the dish often improves the spatial resolution of a ring structure at the mother-bud neck (such as the septin ring) by positioning yeast cells vertically (i.e., the mother-bud axis is parallel to the direction of capturing Z-stacks) (Lee et al., 2015).

1. Apply 10 μL of ConA solution into a dish and spread it with the side of a pipette tip, and air dry for about 20 min in a box to prevent dust.
2. Rinse the dish twice with water.
3. Place 100 μL of cell suspension and incubate for 20 min.
4. Wash 3 × with SC medium to remove cells that are not stuck to the glass bottom.
5. Place a thin layer of 0.6% agarose-containing media (cooled down after melting).
6. For long-term imaging, place 500 μL of liquid medium over the solidified agarose layer.

3.3 Imaging Methods

Time-lapse microscopy

1. Set the 35-mm dish sample into an environmental chamber (DH-35; Warner Instruments) mounted on the sample stage of the inverted microscope.
2. Wait for ~10 min until the sample temperature is stabilized.

Note: For imaging of temperature-sensitive mutants such as *cdc24-4*, cells are grown overnight at room temperature, diluted in prewarmed media (37°C), and shaken for 2 h at 37°C before imaging in a temperature-controlled environmental chamber at 37°C.

3. Use a 100 × objective lens with oil immersion. When higher spatial resolution is required, it is preferable to use an objective lens with a higher NA.
4. After the sample temperature is stabilized in the environmental chamber, adjust the focus to the position near the equatorial plane of the cells attached to the glass surface. If the microscope is equipped with an autofocus system, set the focal plane as a reference position at this time.

   Note: When a long time-lapse image acquisition is performed, the effect of focal drift is not negligible. If the microscope system is not equipped with an autofocus system, it is required to readjust the focus whenever needed, usually every 5 min, depending on the time interval settings and the temperature. When a higher temperature is used, the effect of focal drift is larger in general.
5. Set the parameters for image acquisition.

   The duration of imaging depends on the cellular process to be monitored. If a whole cell cycle needs to be covered, 120–180 min duration is required at 25°C. In our laboratory, a 120-min duration time is set typically.

   The interval of image acquisition also depends on the speed of the target process. When Cdc42 activity is monitored, a 1–2-min interval is typically selected.

   Z-stack setting: as Cdc42 molecules can be distributed on the whole cell surface, enough Z sections must be collected to cover the entire cell to yield meaningful data. Typically, 0.6 μm × 11 z-slices (6-μm thickness in total) and 0.7 μm × 11 z-slices (7-μm thickness in total) are used for haploid and diploid strains, respectively. If an abnormally shaped and/or sized strain is monitored, appropriate Z-stack setting is required to accommodate its morphological change.

   Exposure settings: to reduce the photobleaching of fluorescent proteins and to minimize the phototoxicity of the excitation light to the cells, it is essential to choose an appropriate exposure time. In general, it is best to use the shortest exposure time that provides acceptable signal-to-noise ratio. For the Gic2-PBD-RFP biosensor, 100–200 ms exposure time is used for an EMCCD camera (ImagEM, Hamamatsu Photonics) with on-chip gain of 300 and no binning.

6. Start time-lapse microscopy using appropriate imaging software such as MetaMorph to control the microscope and camera.

FRAP

1. Calibrate a microscope for FRAP according to the manufacturer's instruction. We have used a spinning disk confocal microscope Ultra-View ERS or Ultra-VIEW VoX CSUX1 system (PerkinElmer) (see Section 3.1).
2. Place the glass slide on a microscope. Use 100 × objective lens with immersion oil.
3. Focus on yeast cells and place them in the center of the imaging field.

   Note: Positioning cells in the center of the imaging field minimizes variation.
4. Set the imaging conditions. Use 2 × 2 binning for clearer fluorescence imaging if necessary. Selecting a single Z-stack (the middle slice) helps to minimize photobleaching from capturing multiple Z-stacks and also minimize time intervals for fast time-lapse imaging. The time interval can be as small as 1 s for rapid image acquisition. If a cell cycle marker (such as Cdc3-GFP or Whi5-GFP, which we often use to examine cells in the M-G1 phase) is necessary to select cells at a specific cell cycle stage, set the acquisition set-up to capture an initial image with both fluorescent channels (such as GFP and RFP) and subsequent captures with RFP channel only in order to minimize unnecessary photobleaching.
5. Exposure time and gain value can be adjusted. In general, 100–200 ms exposure time was used for imaging Gic2-PBD-RFP. To minimize photobleaching and phototoxicity, shorter exposure times should be used for subsequent time-lapse imaging. Bleaching laser power and cycle can be selected according to the size of ROIs and the method optimized for each microscope system.
6. Select ROIs by drawing a rectangle covering a part of Gic2-PBD-RFP within the fluorescent cap. Remember to record the exact size of ROIs and save ROIs for later data analysis.
7. Start the imaging acquisition including the initial bleaching point.

   Note: Bleaching often occurs even beyond the selected ROIs; thus, bleaching power and time need to be empirically determined for optimization. It is desirable to bleach more than 50% of the original fluorescence signal. Depending on the context of experiments, all conditions need to be optimized including laser intensity, numbers of laser pulse, size of ROIs, etc. In order to minimize phototoxicity, higher numbers of laser pulses with a lower laser power in a smaller ROI will be desirable.

3.4 Methods for Analyzing Cdc42 Activation Dynamics

A threshold method for detecting and quantifying a dynamic Cdc42-GTP cluster

Cdc42 activity changes both spatially and temporally during the cell cycle, which makes the detection and quantification of its local activation challenging. We have developed a "threshold method" to define a Cdc42-GTP cluster objectively using a simple statistical calculation. The specific steps are as follows. Examples of using the threshold method are described in Section 4.

1. We use the image analysis software, Fiji/ImageJ (Schindelin et al., 2012), to import the image data obtained by multidimensional time-lapse imaging experiment(s).
2. Use a 2D projection method (average projection or sum projection) to make a time-series image sequence.
3. Set the cell outline according to cytoplasmic fluorescence signal or a bright field image (if available).
4. Apply a custom-made Fiji/ImageJ macro (available on request) (Okada et al., 2013) to define the Cdc42-GTP cluster. In this macro, the average value and standard deviation of the fluorescence intensity of the pixels inside the cell outline are calculated. If a pixel has fluorescence intensity higher than a value of the average plus two standard deviations, the pixel is defined to be included in a Cdc42-GTP local cluster. For quantitative analysis, the total intensities of the pixels selected by this method are used (Fig. 3A).
5. The capability of the threshold method to detect a Cdc42-GTP cluster was validated (Fig. 3B). Wild-type haploid cells harboring Gic2-PBD-RFP and Cdc3-GFP (septin marker) grown to exponential phase were imaged by dual-color confocal microscopy. The depicted images were averaged projections of 11 z-slices with a 0.6-μm spacing. The pixels calculated as part of a Cdc42-GTP cluster are shown in red. The cell outlines are highlighted in yellow.
6. Measure the average intensity of a region devoid of cells over time as the background signal.
7. Plot the sum of the intensities of the local cluster of Gic2-PBD-RFP after background subtraction and normalization (the peak value before bud emergence set as 100%) over time.

FRAP analysis

1. Import image sequences with ROIs and without ROIs as .tiff files in Image J (NIH).

   Download ImageJ plugin to open image sequences if necessary. Use 16-bit images for analysis.
2. Synchronize all opened windows and draw appropriate shapes along the ROIs in images. For accurate measurement, make sure that the actual bleaching area remains within an ROI shape in the time-lapse series. If cells moved during imaging acquisition, time-series images can be aligned using Image J Plugins such as align stacks, image stabilizer, and others.
3. Use appropriate FRAP template designed with proper formula for analysis. Obtain the total fluorescence intensity from the bleached area (ROIs) by drawing the same size ROIs over images. Subtract background by drawing a shape covering background area devoid of cells. For photobleaching correction, total intensities from the area of unbleached cells can be measured over time. Fluorescence changes should be plotted over time after background subtraction and photobleaching correction using Excel or other statistical programs such as GraphPad or Origin.
4. Draw a recovery graph by normalizing the prebleach intensity set to 100% and the intensity right after bleaching set to 0%. For the curve fit line, you can use the exponential formula $y = m_1 + m_2 \times \exp(-m_3 \times x)$ for the curve fit of fluorescence recovery, in which m_3 is the off-rate (KaleidaGraph; Synergy Software). The half-time of fluorescence recovery was calculated as $t_{1/2} = \ln 2/m_3$ as previously described (Coffman, Wu, Parthun, & Wu, 2011).

As examples of FRAP experiments using Gic2-PBD-RFP, we describe here the dynamic behavior of Cdc42-GTP polarization during the cell cycle in diploid cells. First, the Gic2-PBD polar cap was monitored by live-cell imaging in different stages of the cell cycle. To distinguish cells before and after cytokinesis, we used Cdc3-GFP: a group of cells with a large bud and a single septin ring (hourglass) (right before cytokinesis); and another group of cells with the septin double ring (from the onset of cytokinesis). Once the group of cells was selected based on the bud size and septin ring, a small region in the middle of Gic2-PBD-RFP cap was photobleached and fluorescence recovery was monitored by time-lapse imaging. Fig. 2 shows an example of FRAP images obtained from unbudded cells.

Gic2-PBD-RFP signal recovered right after the bleaching, thus, we measured the recovery of fluorescence signals and plotted the recovery graph over time (Fig. 2B).

4. APPLICATION OF THE CDC42 BIOSENSOR

4.1 Dynamics of Cdc42 Polarization in the G1 Phase

The Cdc42 biosensor Gic2-PBD-RFP was used to examine Cdc42-GTP polarization dynamics by 3D time-lapse microscopy. In addition to the Gic2-PBD-RFP, we coexpressed either Cdc3-GFP as a marker for the bud neck and cell cycle progression, or Whi5-GFP as a cell cycle marker (Kang et al., 2014). The Cdc3-GFP ring splits into two rings at the onset of cytokinesis, and Cdc3-GFP appears at the presumptive bud site in the next G1 phase prior to disappearance of the old Cdc3-GFP ring. Whi5 enters the nucleus late in mitosis and exits the nucleus in mid G1, partitioning G1 phase into T_1 and T_2 (Di Talia, Skotheim, Bean, Siggia, & Cross, 2007). From a series of two-color live-cell images and quantitative measurement of the Cdc42-GTP cluster by a threshold method as described in Section 3.4, we found stepwise activation of Cdc42 in G1 phase (see Fig. 4; Kang et al., 2014). These two steps of Cdc42 activation were observed in T_1 and T_2, and more distinctly separated in daughter cells, which have a longer G1 phase, than mother cells. Imaging cells at a lower temperature (such as 22°C) allowed more clear separation of the two steps than at 30°C because of slower growth. Using a *bud3* mutant that is defective in its GEF activity (*bud3ΔCR2*) and a *cdc24^{ts}* mutant, we showed that Cdc42 activation in the first and second step depends on Bud3 and Cdc24, respectively (see Fig. 4B). Time-lapse imaging of *cdc24^{ts}* cells also indicated that the first Cdc42-GTP peak appeared transiently in early G1 (i.e., prior to arrest in late G1) even after the temperature upshift to 37°C (Kang et al., 2014).

In another recent study, we used this Cdc42 biosensor to examine how a single, proper growth site was determined in haploid **a** or α cells. We imaged Gic2-PBD-RFP and Cdc3-GFP at 1 min time intervals at 22°C. The highest Cdc42-GTP peak appeared at multiple positions near the septin ring in daughter cells until the Cdc42-GTP polar cap became stable at the boundary of T_1 and T_2 (i.e., right before the new septin clouds appear at the new bud site) (Fig. 4; see details in Lee et al., 2015). Thus, the Cdc42 biosensor represents a powerful method for uncovering novel Cdc42 activation dynamics and mechanisms underlying cell polarization.

4.2 Local Inhibition of Cdc42 During Septin Ring Formation

To investigate the spatial and temporal relationship between Cdc42 activity and septin ring assembly during bud emergence, we used a combination of Gic2-PBD-RFP and Cdc3-GFP (Fig. 5) (Okada et al., 2013). The dynamic behaviors of both probes were monitored by 5D time-lapse live-cell imaging, and the spatial and temporal changes of the Cdc42-GTP cluster were analyzed using the threshold method (described in Section 3.4).

During bud emergence, Cdc42-GTP and septins displayed distinct localization dynamics (Fig. 5A and B) (Okada et al., 2013). Cdc42-GTP signal increased gradually and reached a peak as a bright cluster at the presumptive bud site 10–12 min before bud emergence. From the peak time of the cluster intensity to bud emergence, the signals of Cdc42-GTP and septins at the new bud site were inversely correlated. As the septin signal increased, the cluster intensity of Cdc42-GTP decreased. After bud emergence, septin accumulation continued at a slower rate, while a new Cdc42-GTP cluster formed at the bud cortex and intensified over time.

The inverse relationship between Cdc42-GTP and septins during bud emergence suggests that the newly recruited septins might constitute a negative feedback mechanism to inactivate Cdc42 at the presumptive bud site so that Cdc42 can be reactivated within the bud to drive exclusive cell growth within that compartment (Okada et al., 2013). To test this hypothesis, we performed time-lapse imaging of a temperature-sensitive septin mutant *cdc12-6* strain at the nonpermissive temperature. Even at 39°C (the nonpermissive temperature for the *cdc12-6* cells under our imaging conditions), the isogenic wild-type cells showed normal Cdc42 inactivation–reactivation dynamics (Fig. 5D). In contrast, the *cdc12-6* septin mutant cells failed to display the transient inactivation of Cdc42 during bud emergence (Fig. 5C and D). The cluster intensity of Cdc42-GTP increased continuously during bud emergence. As expected, septin accumulation at the presumptive bud site was not observed in the *cdc12-6* mutant at the nonpermissive temperature. Consequently, the bud became elongated instead of ovoid in shape. This result not only provides strong support for our hypothesis but also illustrates the power of the biosensor in uncovering an unexpected interplay between Cdc42 activity and septin ring assembly during bud formation or polarized cell growth.

5. CONCLUDING REMARKS

Small GTPases are cellular switches that play vital roles in diverse processes such as cell polarization, cell migration, membrane trafficking, and cytokinesis (Etienne-Manneville, 2004; Park & Bi, 2007). Their activation and inactivation are exquisitely controlled to ensure that specific cellular events occur only at specific times and places. This also explains why there are far more GEFs and GAPs (GTPase-activating proteins) than their target GTPases (e.g., there are ~70 GEFs and 70 GAPs for 17 Rho GTPases in the human genome) (Bernards, 2003; Bernards & Settleman, 2004). Small GTPases also share two key features in executing specific cellular functions—displaying a unique pattern of individual activation dynamics (e.g., Cdc42) and requiring spatiotemporal coordination with other GTPases (Rac, Rho, Rab, etc.). As we have demonstrated here, the biosensor for Cdc42-GTP in budding yeast has led to the discoveries of new activation and feedback mechanisms during polarization and bud morphogenesis. Strikingly, Cdc42 is inhibited while Rho1 is active during cytokinesis in budding yeast (Atkins et al., 2013; Okada et al., 2013; Okada, Wloka, & Bi, 2017; Onishi, Ko, Nishihama, & Pringle, 2013; Tong et al., 2007). Thus, it will be critically important to explore the potential mutual antagonism between Cdc42 and Rho1 during cytokinesis (Atkins et al., 2013; Onishi et al., 2013). A similar relationship between the GTPases has also been suggested for the budding process (Gao, Caviston, Tcheperegine, & Bi, 2004). Such studies require the development of a rigorously tested biosensor for Rho1 in budding yeast, which has yet to be done. In mammalian cells, the activities of small GTPases such as Ras, Rho, Cdc42, and Rac are coordinately regulated, often displaying mutual antagonism in processes such as cell migration (Machacek et al., 2009; Nobes & Hall, 1999; Tsubouchi et al., 2002), cell–cell adhesion (Arthur, Noren, & Burridge, 2002; Bruewer, Hopkins, Hobert, Nusrat, & Madara, 2004), and chemotaxis (Xu et al., 2003). Biosensors for different small GTPases have played a key role in defining their spatiotemporal relationships in a specific process. These biosensors and newly developed ones will continue to be instrumental for understanding the functions and mechanisms of individual GTPases, discovering their new roles, and understanding their coordination with other GTPases in a given process.

ACKNOWLEDGMENTS

We thank the members of Bi and Park laboratories for discussions, Dr. Pil Jung Kang for providing the PBD-binding figure, Dr. Andrea Stout at the CDB imaging Core, and members of Dr. Jian-Qiu Wu's lab for technical advice. This work is supported by the National Institutes of Health Grants GM115420 (to E.B.) and R01 GM114582 (to H.-O.P.).

REFERENCES

Arthur, W. T., Noren, N. K., & Burridge, K. (2002). Regulation of Rho family GTPases by cell-cell and cell-matrix adhesion. *Biological Research*, *35*, 239–246.

Atkins, B. D., Yoshida, S., Saito, K., Wu, C. F., Lew, D. J., & Pellman, D. (2013). Inhibition of Cdc42 during mitotic exit is required for cytokinesis. *The Journal of Cell Biology*, *202*, 231–240.

Bendezu, F. O., Vincenzetti, V., Vavylonis, D., Wyss, R., Vogel, H., & Martin, S. G. (2015). Spontaneous Cdc42 polarization independent of GDI-mediated extraction and actin-based trafficking. *PLoS Biology*, *13*, e1002097.

Bernards, A. (2003). GAPs galore! A survey of putative Ras superfamily GTPase activating proteins in man and Drosophila. *Biochimica et Biophysica Acta*, *1603*, 47–82.

Bernards, A., & Settleman, J. (2004). GAP control: Regulating the regulators of small GTPases. *Trends in Cell Biology*, *14*, 377–385.

Bruewer, M., Hopkins, A. M., Hobert, M. E., Nusrat, A., & Madara, J. L. (2004). RhoA, Rac1, and Cdc42 exert distinct effects on epithelial barrier via selective structural and biochemical modulation of junctional proteins and F-actin. *American Journal of Physiology. Cell Physiology*, *287*, C327–C335.

Coffman, V. C., Wu, P., Parthun, M. R., & Wu, J. Q. (2011). CENP-A exceeds microtubule attachment sites in centromere clusters of both budding and fission yeast. *The Journal of Cell Biology*, *195*, 563–572.

Di Talia, S., Skotheim, J. M., Bean, J. M., Siggia, E. D., & Cross, F. R. (2007). The effects of molecular noise and size control on variability in the budding yeast cell cycle. *Nature*, *448*, 947–951.

Etienne-Manneville, S. (2004). Cdc42—The centre of polarity. *Journal of Cell Science*, *117*, 1291–1300.

Gao, X. D., Caviston, J. P., Tcheperegine, S. E., & Bi, E. (2004). Pxl1p, a paxillin-like protein in Saccharomyces cerevisiae, may coordinate Cdc42p and Rho1p functions during polarized growth. *Molecular Biology of the Cell*, *15*, 3977–3985.

Jaquenoud, M., & Peter, M. (2000). Gic2p may link activated Cdc42p to components involved in actin polarization, including Bni1p and Bud6p (Aip3p). *Molecular and Cellular Biology*, *20*, 6244–6258.

Kang, P. J., Lee, M. E., & Park, H. O. (2014). Bud3 activates Cdc42 to establish a proper growth site in budding yeast. *The Journal of Cell Biology*, *206*, 19–28. http://dx.doi.org/10.1083/jcb.201402040.

Lee, M. E., Lo, W. C., Miller, K. E., Chou, C. S., & Park, H. O. (2015). Regulation of Cdc42 polarization by the Rsr1 GTPase and Rga1, a Cdc42 GTPase-activating protein, in budding yeast. *Journal of Cell Science*, *128*, 2106–2117.

Lo, W. C., Lee, M. E., Narayan, M., Chou, C. S., & Park, H. O. (2013). Polarization of diploid daughter cells directed by spatial cues and GTP hydrolysis of Cdc42 budding yeast. *PloS One*, *8*, e56665.

Machacek, M., Hodgson, L., Welch, C., Elliott, H., Pertz, O., Nalbant, P., et al. (2009). Coordination of Rho GTPase activities during cell protrusion. *Nature*, *461*, 99–103.

Nalbant, P., Hodgson, L., Kraynov, V., Toutchkine, A., & Hahn, K. M. (2004). Activation of endogenous Cdc42 visualized in living cells. *Science*, *305*, 1615–1619.

Nobes, C. D., & Hall, A. (1999). Rho GTPases control polarity, protrusion, and adhesion during cell movement. *The Journal of Cell Biology*, *144*, 1235–1244.

Okada, S., Leda, M., Hanna, J., Savage, N. S., Bi, E., & Goryachev, A. B. (2013). Daughter cell identity emerges from the interplay of Cdc42, septins, and exocytosis. *Developmental Cell*, *26*, 148–161.

Okada, S., Wloka, C., & Bi, E. (2017). Analysis of protein dynamics during cytokinesis in budding yeast. *Methods in Cell Biology*, *137*, 25–45.

Onishi, M., Ko, N., Nishihama, R., & Pringle, J. R. (2013). Distinct roles of Rho1, Cdc42, and Cyk3 in septum formation and abscission during yeast cytokinesis. *The Journal of Cell Biology*, *202*, 311–329.

Ozbudak, E. M., Becskei, A., & van Oudenaarden, A. (2005). A system of counteracting feedback loops regulates Cdc42p activity during spontaneous cell polarization. *Developmental Cell*, *9*, 565–571.

Park, H.-O., & Bi, E. (2007). Central roles of small GTPases in the development of cell polarity in yeast and beyond. *Microbiology and Molecular Biology Reviews*, *71*, 48–96.

Pertz, O., Hodgson, L., Klemke, R. L., & Hahn, K. M. (2006). Spatiotemporal dynamics of RhoA activity in migrating cells. *Nature*, *440*, 1069–1072.

Schindelin, J., Arganda-Carreras, I., Frise, E., Kaynig, V., Longair, M., Pietzsch, T., et al. (2012). Fiji: An open-source platform for biological-image analysis. *Nature Methods*, *9*, 676–682.

Smith, S. E., Rubinstein, B., Mendes Pinto, I., Slaughter, B. D., Unruh, J. R., & Li, R. (2013). Independence of symmetry breaking on Bem1-mediated autocatalytic activation of Cdc42. *The Journal of Cell Biology*, *202*, 1091–1106.

Tong, Z., Gao, X.-D., Howell, A. S., Bose, I., Lew, D. J., & Bi, E. (2007). Adjacent positioning of cellular structures enabled by a Cdc42 GTPase-activating protein mediated zone of inhibition. *The Journal of Cell Biology*, *179*, 1375–1384.

Tsubouchi, A., Sakakura, J., Yagi, R., Mazaki, Y., Schaefer, E., Yano, H., et al. (2002). Localized suppression of RhoA activity by Tyr31/118-phosphorylated paxillin in cell adhesion and migration. *The Journal of Cell Biology*, *159*, 673–683.

Wedlich-Soldner, R., Wai, S. C., Schmidt, T., & Li, R. (2004). Robust cell polarity is a dynamic state established by coupling transport and GTPase signaling. *The Journal of Cell Biology*, *166*, 889–900.

Xu, J., Wang, F., Van Keymeulen, A., Herzmark, P., Straight, A. F., Kelly, K., et al. (2003). Divergent signals and cytoskeletal assemblies regulate self-organizing polarity in neutrophils. *Cell*, *114*, 201–214.

CHAPTER SEVEN

Novel Fluorescence-Based Biosensors Incorporating Unnatural Amino Acids

Wei Niu[1], Jiantao Guo[1]
University of Nebraska–Lincoln, Lincoln, NE, United States
[1]Corresponding authors: e-mail address: wniu2@unl.edu; jguo4@unl.edu

Contents

Abstract

Fluorescent proteins of different colors are useful probes to study protein structure and function, and to investigate cellular events and conditions. Large efforts have focused on engineering new properties into fluorescent proteins via rational design and directed evolution. In addition to applications in imaging of protein expression level and subcellular localization, fluorescent proteins have been increasingly engineered to act as biosensors to track concentrations of small-molecule metabolites, enzyme activities, and protein conformational changes in living cells. Unlike small-molecule fluorescence biosensors, fluorescent proteins are genetically encodable, and thus can be expressed inside living cells. Attachment of organelle-specific signals to the proteins allows their localization to be specified. Recently, a new class of fluorescent protein biosensors has been developed to include unnatural amino acids as the sensing element.

Methods in Enzymology, Volume 589
ISSN 0076-6879
http://dx.doi.org/10.1016/bs.mie.2017.01.012

The unique chemical and physical properties of the unnatural amino acids enable sensor designs that cannot be realized by using the standard genetic code with the 20 canonical amino acids. In this chapter, we detail the general procedure for the genetic incorporation of unnatural amino acids. We further present two protocols for the in vitro and in vivo detection of hydrogen peroxide (H_2O_2) using a fluorescent protein biosensor that contains an unnatural amino acid, *p*-boronophenylalanine.

1. INTRODUCTION

Fluorescent proteins are useful probes that have greatly facilitated the study of biological molecules and processes (Chudakov, Matz, Lukyanov, & Lukyanov, 2010; Frommer, Davidson, & Campbell, 2009; Newman, Fosbrink, & Zhang, 2011; Tsien, 1998; Wang, Nakata, & Hamachi, 2009; Zimmer, 2002). One important application of fluorescent proteins is to tag and visualize target proteins in living systems under physiological conditions using fluorescence microscopy (Chudakov et al., 2010; Frommer et al., 2009; Newman et al., 2011; Shaner et al., 2004; Tsien, 1998; Wang et al., 2009; Zimmer, 2002). In order to monitor multiple biomolecules simultaneously, many variants of fluorescent proteins with different emission spectral profiles have been isolated or engineered (Matz et al., 1999; Miyawaki, Shcherbakova, & Verkhusha, 2012; Shaner et al., 2004). In addition, fluorescent proteins with photocontrollable properties have been used in superresolution microscopy (Chi, 2009). Fluorescent proteins have also been employed as biosensors to actively track changes in concentrations of small-molecule metabolites, enzyme activities, protein conformations, and signaling pathways in living cells (Frommer et al., 2009; Newman et al., 2011). The beauty of fluorescent protein biosensors lies in their genetic encodability, which allows the easy introduction into biological subjects, including cells, tissues, or whole animals, and the facile targeting into different cell organelles.

The traditional design of fluorescent protein biosensors is restricted to the standard genetic code of the common 20 amino acids. The side-chain functional groups of these amino acids, which are key elements in biosensing, are limited to amine, amide, acid, alcohol, thiol, thioether, and alkyl/aryl groups. Consequently, the genetic incorporation of amino acids containing unnatural functional groups with novel chemical and/or physical properties can potentially enhance one's ability to engineer novel biosensors. This notion is demonstrated by several recent examples in which fluorescent protein engineering is combined with powerful unnatural amino acid mutagenesis methodologies (Ayyadurai et al., 2011; Chen, Chen, Ren, & Ai, 2012; Chen, Ren, Wright, & Ai, 2013; Groff, Wang, Jockusch, Turro, & Schultz, 2010; Liu et al., 2013; Niu & Guo, 2013; Wang, Niu, Guo, & Schultz, 2012).

2. SITE-SPECIFIC INCORPORATION OF UNNATURAL AMINO ACIDS INTO PROTEINS IN LIVE CELLS

A number of methods have been reported for the introduction of unnatural amino acids into proteins in living cells (Budisa et al., 1999; Datta, Wang, Carrico, Mayo, & Tirrell, 2002; Döring et al., 2001; Dougherty, 2000; Hohsaka & Sisido, 2002; Liu & Schultz, 2010). One of the most widely used approaches was developed in the laboratory of Dr. Peter G. Schultz at the Scripps Research Institute (Liu & Schultz, 2010). This general approach has been employed to site-specifically incorporate unnatural amino acids into proteins in living bacterial, yeast, and mammalian cells. The incorporation system features the introduction of two new components into the endogenous protein biosynthetic machinery, including an orthogonal tRNA-codon pair and an aminoacyl-tRNA synthetase that together specifically charge the orthogonal tRNA with the desirable unnatural amino acid.

The orthogonal tRNA must not be recognized by the native aminoacyl-tRNA synthetases of the host cell. In the meantime, after the aminoacylation by its cognate tRNA synthetase, the aminoacylated orthogonal tRNA can function efficiently in translation. This orthogonal tRNA should also be able to decode a noncoding codon (also called "blank codon," which does not encode a natural amino acid). The most utilized noncoding codon in this system is the amber nonsense codon (UAG), which is the least utilized stop codon in both *Escherichia coli* and in *Saccharomyces cerevisiae*. Introduction of the amber suppression system, which consists of an engineered aminoacyl-tRNA synthetase and amber suppressor $tRNA_{CUA}$ pair that can decode the amber codon with an unnatural amino acid, does not significantly affect cell growth. In fact, several common lab *E. coli* strains (e.g., DH5α and XL-1 blue) contain natural amber suppressors. On the other hand, overexpression of amber suppressor $tRNA_{CUA}$ from a plasmid containing multiple copies of $tRNA_{CUA}$ genes could lead to notable slow cell growth and, in some cases, the loss of $tRNA_{CUA}$ genes from the plasmid. In order to improve amber suppression efficiency and to reduce potential adverse effects of $tRNA_{CUA}$ expression, efforts have been made to further engineer $tRNA_{CUA}$ (Guo, Melancon, Lee, Groff, & Schultz, 2009). In recent years, quadruplet codons have also been used to encode unnatural amino acids (Anderson et al., 2004; Neumann, Wang, Davis, Garcia-Alai, & Chin, 2010; Niu, Schultz, & Guo, 2013; Rodriguez, Lester, & Dougherty, 2006; Taki, Matsushita, & Sisido, 2006; Wang et al., 2014). In order to employ quadruplet codons in protein biosynthesis, tRNAs need to be engineered to bear extended anticodon loops (8-nucleotide instead of normal 7-nucleotide loop) and to form apparent quadruplet anticodons (Neumann et al., 2010; Niu et al., 2013; Wang et al., 2014; Wang, Shang, Cerny, Niu, & Guo, 2016).

An engineered orthogonal aminoacyl-tRNA synthetase is the other essential component of the system. The orthogonal synthetase should only aminoacylate the orthogonal tRNA, its cognate partner, but not any of the endogenous tRNAs. In addition, the aminoacylation reaction should only use the desired unnatural amino acid as the substrate, but not any of the 20 common amino acids. Likewise, the unnatural amino acid should not be utilized by any of the endogenous aminoacyl-tRNA synthetases.

The above methodology has enabled high-efficiency incorporations of over 100 unnatural amino acids with a variety of side-chain structures and functions, including spectroscopic probes (Cellitti et al., 2008; Deiters, Geierstanger, & Schultz, 2005; Groff, Thielges, Cellitti, Schultz, & Romesberg, 2009; Jackson, Hammill, & Mehl, 2007; Schultz et al., 2006; Wang, Xie, & Schultz, 2006), metal chelators (Lee, Spraggon, Schultz, & Wang, 2009; Liu et al., 2013, 2012; Xie, Liu, & Schultz, 2007; Ye et al., 2008), posttranslational modification analogs (Guo, Wang, Lee, & Schultz, 2008; Liu & Schultz, 2006; Xie, Supekova, & Schultz, 2007), redox-active groups (Alfonta, Zhang, Uryu, Loo, & Schultz, 2003; Seyedsayamdost, Xie, Chan, Schultz, & Stubbe, 2007; Tippmann & Schultz, 2007), and probes with unique chemical reactivities (Brustad, Bushey, Brock, Chittuluru, & Schultz, 2008; Brustad, Lemke, Schultz, & Deniz, 2008; Chen et al., 2014; Chin et al., 2002; Lang, Davis, Torres-Kolbus et al., 2012; Lang, Davis, Wallace et al., 2012; Wang, Zhang, Brock, & Schultz, 2003; Xiang et al., 2014; Zhang, Wang, Brock, & Schultz, 2002). One example to demonstrate the utility of the incorporated unnatural amino acid is the genetic incorporation of sulfotyrosine into proteins in live cells. In nature, sulfotyrosine is formed through posttranslational modification of tyrosine residues in proteins. The introduction of negatively charged sulfate group plays crucial roles in extracellular biomolecular interactions that dictate various cellular processes including cell adhesion, leukocyte trafficking, hormone activities, and immune responses (Huttner, 1988; Stone, Chuang, Hou, Shoham, & Zhu, 2009). The ability to site-specifically incorporate sulfotyrosine residues into proteins of interest greatly facilitates biochemical studies of protein tyrosine sulfation (Ju et al., 2013; Ju, Niu, & Guo, 2016; Liu, Brustad, Liu, & Schultz, 2007). More commonly, the incorporated unnatural amino acid serves as a chemical handle in protein-labeling experiments. An exciting recent development is to achieve the fluorogenic protein labeling through bioorthogonal reactions enabled by genetically incorporated unnatural amino acids, such as ones bearing alkene groups (Shang et al., 2017; Song, Wang, Qu, Madden, & Lin, 2008). Alkenes can specifically react with either tetrazole (Song et al., 2008) or tetrazine (Shang et al., 2017) to generate fluorescent products. The fluorogenic property of these reactions leads to low background fluorescence interference, which is more desirable in studies of protein folding and function under physiological conditions. In recent years, efforts have also been made to further improve the efficiency and fidelity of unnatural amino acid mutagenesis through optimizing properties of the aminoacyl-tRNA synthetase-$tRNA_{CUA}$ pair (Amiram et al., 2015; Wang, Ju, Niu, & Guo, 2015).

A general scheme for the genetic incorporation of unnatural amino acids into proteins in live *E. coli* cells is shown in Fig. 1. All genetic components are encoded on plasmid(s). Commonly, a two-plasmid system (Fig. 2) is employed. Strains that are able to naturally suppress amber codons should not be used, e.g., XL1-Blue or DH5α. Suitable and nonamber suppressor strains include BL21, BL21(DE3), GeneHogs, DH10B, and TOP10. After transforming the two plasmids into a proper *E. coli* host strain, cells are grown and selected using two compatible antibiotic markers. The protein expression is induced at the exponential phase of cell growth ($OD_{600}=0.6$–1.0). The desirable unnatural amino acid can either be included in the initial culture media or be added to the media together with the induction reagent for protein expression. In general, a chemically and metabolically stable unnatural amino acid is preferably included in initial media in order to suppress the residual misincorporation of natural amino acids by engineered aminoacyl-tRNA synthetase. On the other hand, a relatively unstable unnatural amino acid should be added at the induction step.

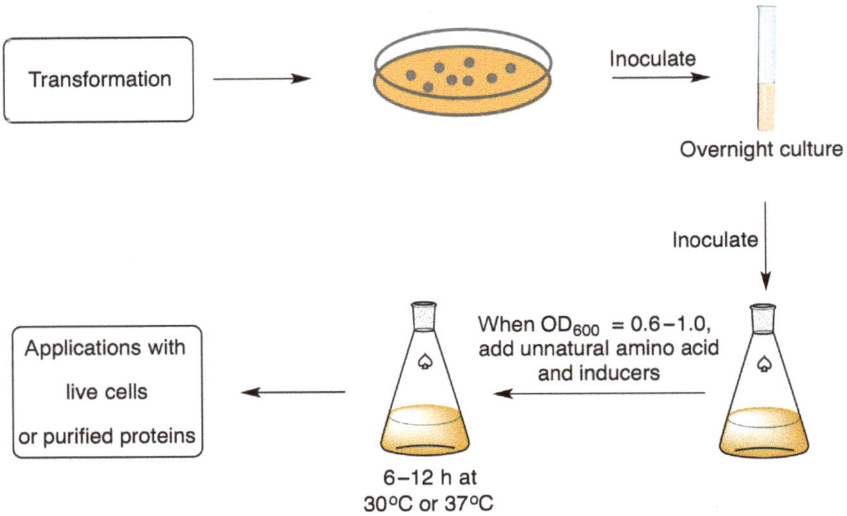

Fig. 1 A general scheme of genetic incorporation of unnatural amino acids into proteins in bacteria.

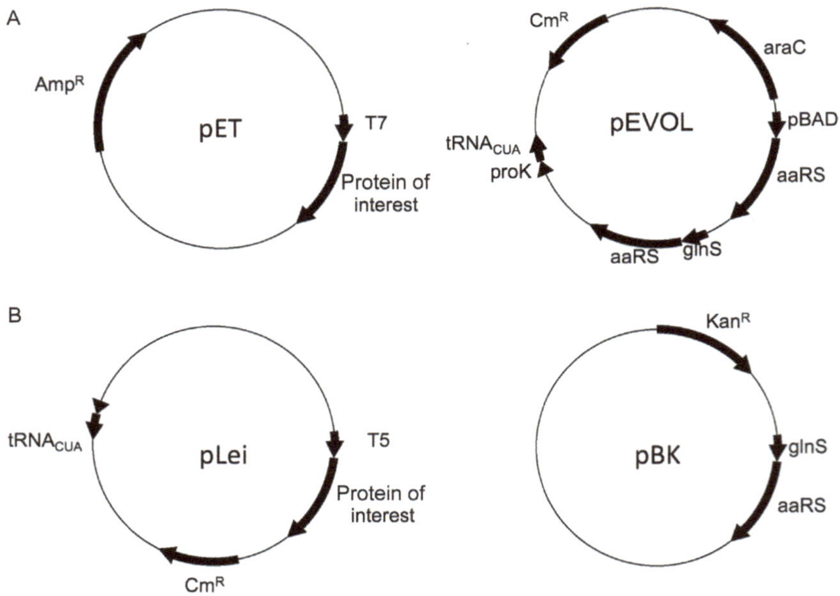

Fig. 2 General plasmid pairs for unnatural amino acid mutagenesis. (A) The pEVOL and pET system. (B) The pBK and pLei system.

2.1 Materials

- 37°C and/or 30°C incubator
- 37°C and/or 30°C shaker
- 42°C water bath
- UV–vis spectrophotometer
- Refrigerated centrifuge
- Autoclave for sterilization
- pEVOL plasmid containing genes encoding amber suppressor tRNA and aminoacyl-tRNA synthetase (available from Professor Peter G. Schultz, The Scripps Research Institute, La Jolla, CA)
- pBK plasmid containing the gene encoding aminoacyl-tRNA synthetase (available from Professor Peter G. Schultz, The Scripps Research Institute, La Jolla, CA)
- pET30 (Invitrogen) or similar expression plasmid containing the encoding gene of the target protein
- pLei plasmid containing the encoding gene of the target protein and the gene of amber suppressor tRNA (Vector is available from Professor Peter G. Schultz, The Scripps Research Institute, La Jolla, CA)
- High-fidelity DNA polymerase (New England Biolabs)
- Antarctic phosphatase (New England Biolabs)
- T4 DNA ligase (New England Biolabs)
- Restriction enzymes (New England Biolabs)
- Chemically competent DH5α (Thermo Fisher Scientific) or similar *E. coli* hosts for plasmid propagation and isolation
- Chemically competent BL21(DE3) (Thermo Fisher Scientific), GeneHogs (Thermo Fisher Scientific), DH10B (Thermo Fisher Scientific), or other similar *E. coli* strains that can be used for the expression of recombinant proteins
- Chloramphenicol (Cm), 34 mg/mL (1000 ×) stock solution in ethanol, stored at –20°C
- Ampicillin, sodium salt (Amp), 100 mg/mL (1000 ×) stock solution in sterile H_2O, stored at –20°C
- Kanamycin sulfate (Kan), 50 mg/mL (1000 ×) stock solution in sterile H_2O, stored at –20°C
- LB liquid medium
- LB Agar plates
- *p*-Azido phenylalanine hydrochloride (pAzPhe) (Bachem)
- *p*-Nitro phenylalanine hydrochloride (pNitroPhe) (Bachem)

- *p*-Boronophenylalanine hydrochloride (pBoPhe) (Sigma-Aldrich)
- Isopropyl β-D-1-thiogalactopyranoside (IPTG), 1 *M* (1000 ×) stock solution in H_2O, sterile filtered through a 0.2-μm membrane, stored at –20°C
- L-(+)-Arabinose, 20 mg/mL (100 ×) stock solution in H_2O, sterile filtered through a 0.2-μm membrane, stored at –20°C
- Ni Sepharose 6 Fast Flow resin (GE Healthcare) and buffers for protein purification
- Fluorometer or fluorescence plate reader
- Home-made or commercial SDS-PAGE gel (e.g., NuPAGE Novex 4%–12% SDS gels from Thermo Fisher Scientific)

2.2 Plasmids for Protein Expression

Two general plasmid systems are commonly used (Fig. 2).

2.2.1 The pEVOL and pET Combination

The pEVOL plasmid contains genes that encode amber suppressor tRNA and aminoacyl-tRNA synthetase (aaRS). There are two copies of aaRS gene on the pEVOL plasmid. One aaRS gene is under the control of the constitutive *glnS* promoter and the second one is under the control of the arabinose-inducible P_{BAD} promoter and the *rrnB* terminator. Simultaneous overexpression of both copies of aaRS gene enhances the production of the full-length target protein (Cellitti et al., 2008). The expression of amber suppressor tRNA is under the control of a *proK* promoter and terminator. The pEVOL plasmid also contains a chloramphenicol acetyltransferase marker (Cm^R), an *araC* repressor gene, and the p15A origin of replication (Fig. 2A).

The target protein is encoded by a pET plasmid. The expression of the target protein is under the control of an inducible *T7* promoter and the *T7* terminator. The pET30 plasmid contains a pBR322 origin of replication and a kanamycin resistance gene (Kan^R). The expressed target protein contains a C-terminal His_6 tag for detection and purification. Other similar vectors can be used to encode the target protein of interest. The target protein vector should have compatible replication origin and selection maker to the pEVOL plasmid so that these two plasmids can be simultaneously maintained in the *E. coli* host. Typical hosts for the pEVOL-pET system are BL21(DE3) or other similar strains that express the T7 RNA polymerase.

A C-terminal His_6 tag is typically employed for unnatural amino acid incorporation using the amber suppression approach. Such design simplifies the purification process, since only full-length target protein, which contains the unnatural amino acid of interest, has a His_6 tag. Inability to decode the amber stop codon leads to translational termination, which produces truncated proteins without a His_6 Tag. On the other hand, if the His_6 tag is fused to the N-terminus of the protein of interest, truncated protein can be copurified with the full-length protein.

2.2.2 The pBK and pLei Combination

The pBK plasmid contains the aaRS gene, under the control of a constitutive *glnS* promoter and terminator. The pBK plasmid also encodes a kanamycin resistance protein (Kan^R) and a ColE1 origin of replication (Fig. 2B).

The target protein is encoded by the pLei plasmid. The expression of the target protein is under the control of an inducible *T5* promoter and a lambda t_0 transcriptional terminator. The pLei plasmid also encodes the amber suppressor tRNA that is under the control of a constitutive *lpp* promoter and *rrnC* terminator. This plasmid contains a p15A origin of replication and a chloramphenicol acetyltransferase marker (Cm^R).

The expressed target protein contains a C-terminal cleavable His_6 tag and S-tag for detection and purification. Other similar vectors can be used to encode the target protein of interest. Such vector should have a compatible replication origin and selection marker to the pBK plasmid so that these two plasmids can be simultaneously maintained in the *E. coli* host. Typical hosts for the pBK-pLei system include DH10B and GeneHogs.

2.3 Expression of Fluorescent Proteins Containing Unnatural Fluorophores

2.3.1 Mutagenesis and Verification

In order to generate an unnatural fluorophore, the fluorophore-forming tyrosine (e.g., Tyr66 in GFP) residue can be replaced with an unnatural amino acid of one's interest. To this end, the codon (TAC) of Tyr66 needs to be mutated into an amber nonsense codon (TAG). Two general methods can be employed for the site-directed mutagenesis. One is the QuikChange (Agilent Technologies) method, which uses a pair of complementary mutagenic primers to amplify the entire plasmid in a thermocycling reaction. A free online program can facilitate the primer design (http://www.genomics.agilent.com/primerDesignProgram.jsp). The second method uses overlapping PCR where a specific primer is designed to contain the desirable amber mutation. After the mutagenesis experiment, the presence of the amber TAG mutation in the gene of interest should be verified by DNA sequencing.

2.3.2 *Cotransformation of Host* E. coli *Cells*

Standard transformation protocols can be used in this step. If the pEVOL and pET combination is used, the transformants are plated onto LB-Agar plates containing Cm (34 μg/mL) and Amp (100 μg/mL). If the pBK and pLei combination is used, the transformants are plated onto LB-Agar plates containing Kan (50 μg/mL) and Cm (34 μg/mL).

2.3.3 *Protein Expression*

A single colony from the Agar plate is used to inoculate 5 mL LB liquid media containing appropriate antibiotics at 37°C. The seed culture (1 mL) is used to inoculate fresh LB media (50 mL) containing appropriate antibiotics at 37°C. For the pBK and pLei system, the protein expression is induced at OD_{600} of 0.6–1.0 by the additions of IPTG (0.2–0.5 m*M*) and unnatural amino acid of interest (1 m*M*). For the pEVOL and pET system, the protein expression is induced at OD_{600} of 0.6–1.0 by the additions of IPTG (0.2–0.5 m*M*), L-arabinose (0.2% (w/v)), and unnatural amino acid of interest (1 m*M*). Following an additional 4–12 h of cultivation, cells are collected by centrifugation at 5000 g and 4°C for 15 min. Harvested cells are resuspended in lysis buffer containing potassium phosphate (20 m*M*, pH 7.4), NaCl (150 m*M*), and imidazole (10 m*M*). Cells are subsequently disrupted by sonication. Cellular debris is removed by centrifugation (21,000 × *g*, 30 min, 4°C). The cell-free lysate is applied to Ni Sepharose 6 Fast Flow resin (GE Healthcare). Protein purification follows the manufacturer's instructions. Protein concentrations are determined by Bradford assay (Bio-Rad). Purified protein can be used directly or desalted prior to next step.

The expression of the target protein using the pLei system is leaky. Significant amounts of protein can be obtained without the addition of IPTG. The optimal IPTG concentration is usually between 0.2 and 0.5 m*M*. If tight control of protein expression is needed (e.g., the target protein is toxic to the host), a host strain that overexpresses the LacI protein can be used. Besides LB, minimal medium, such as GMML (1 × M9, 1 m*M* $MgSO_4$, 0.1 m*M* $CaCl_2$, 8.5 m*M* NaCl, 5 μ*M* $FeSO_4$, 1% glycerol), can be used. If the host is a nutrient auxotroph, appropriate supplements need to be included. In general, higher fidelity can be achieved in minimal medium, but lower protein yield is also observed in some cases due to low cell density. When high cell density is desirable, media that are rich in nutrients, such as 2YT medium, can also be used in the incorporation experiments.

The solubility of some unnatural amino acids is quite low in water. To solve this problem, a proper amount of unnatural amino acid of interest is first dissolved in a small amount of water by adjusting the pH value of the solution. Either base (e.g., 1 *M* NaOH in water) or acid (e.g., 1 *M* HCl in water) can be used depending on the pH stability of the unnatural amino acid of interest. Examples of basic pH-sensitive amino acids include 3-amino-L-tyrosine and 4-borono-L-phenylalanine. An examples of acidic pH-sensitive amino acid is N^{ϵ}-(*tert*-butyloxy-carbonyl)-L-lysine. The unnatural amino acid solution is subsequently added into the culture media, followed by the addition of an equimolar amount of base or acid to adjust the pH back to the original value.

To improve the yield of protein expression, an *E. coli* strain (C321.ΔA) can be used (Lajoie et al., 2013). This strain lacks amber nonsense codons and does not express release factor 1 (RF1; RF1 decodes UAG nonsense codon as the termination signal of protein translation). Such modifications abolish amber nonsense codon-mediated translation termination in this strain. The benefits of using RF1-minus strain in the production of unnatural amino acid-containing proteins have been reported (Chatterjee, Lajoie, Xiao, Church, & Schultz, 2014; Johnson et al., 2011; Wang et al., 2016).

To estimate the fidelity of unnatural amino acid incorporation into a fluorescent protein of interest, two parallel protein expression experiments are commonly conducted. In one experiment, the expression of the target fluorescent protein is carried out in the presence of the unnatural amino acid. In a second experiment, the expression is conducted in the absence of the unnatural amino acid. A good fidelity of incorporation is obtained if full-length fluorescent protein can only be detected when protein expression is conducted in the presence of unnatural amino acid.

2.3.4 Protein Purification and Characterization

Proteins can be purified by commonly used procedures, such as affinity chromatography using Ni-NTA resin. Generally, unnatural amino acids are stable under the affinity purification conditions. If needed, imidazole and salts in the solution of purified protein can be removed using a desalting column or by dialysis. The protein concentration can be quantified by Bradford assay (Bio-Rad). The purified protein can be analyzed by SDS-PAGE. To further confirm the site-specific incorporation of unnatural amino acid into the fluorescent protein of interest, mass spectrometry analysis is commonly conducted. The analysis can be based on the full-length protein or on protein fragments from trypsin digestion. The data of a typical analysis

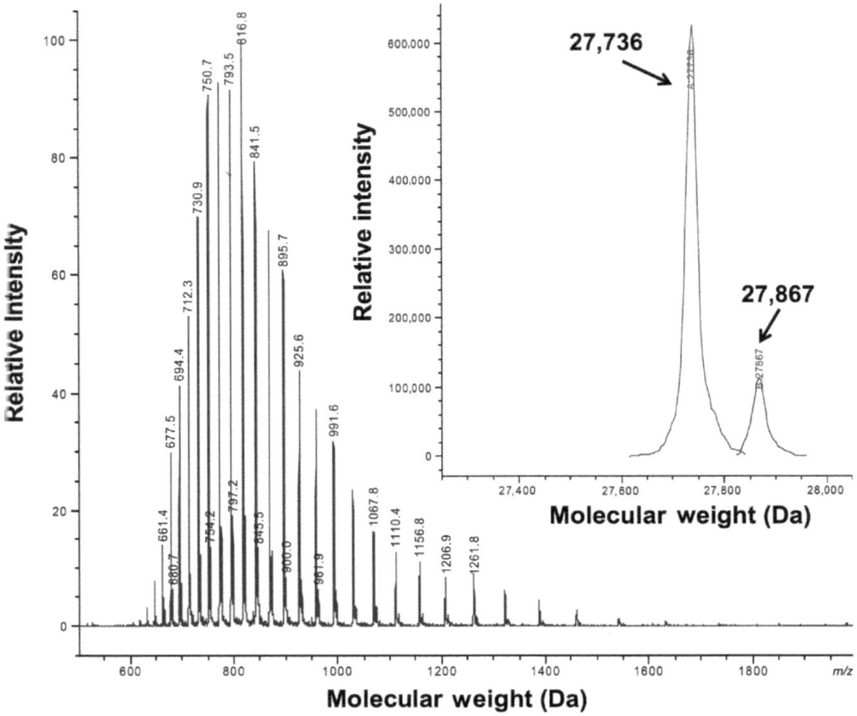

Fig. 3 Mass spectra of GFP mutant that contains *p*-boronophenylalanine (pBoPhe) at position 66. Expected mass: 27,738 Da (without N-terminal methionine) and 27,869 Da (with N-terminal methionine); observed mass: 27,736 Da (without N-terminal methionine) and 27,867 Da (with N-terminal methionine).

of full-length protein are shown in Fig. 3. In a typical analysis of trypsin-digested fragments, the band of the fluorescent protein of interest is cut from SDS-PAGE, then subjected to in-gel digestion with trypsin (in 50 m*M* ammonium bicarbonate, pH 8.0) overnight at 37°C. The resulting peptide fragments are extracted with 0.1% formic acid in 75% acetonitrile, and then analyzed by LC/MS/MS. Database searches are subsequently performed.

The fluorescence property of the unnatural amino acid-containing fluorescent proteins can be examined by a fluorometer or a fluorescence plate reader. The unnatural amino acid-containing fluorescent proteins may have similar or very different emission and excitation wavelength in comparison to their parent fluorescent proteins (Fig. 4).

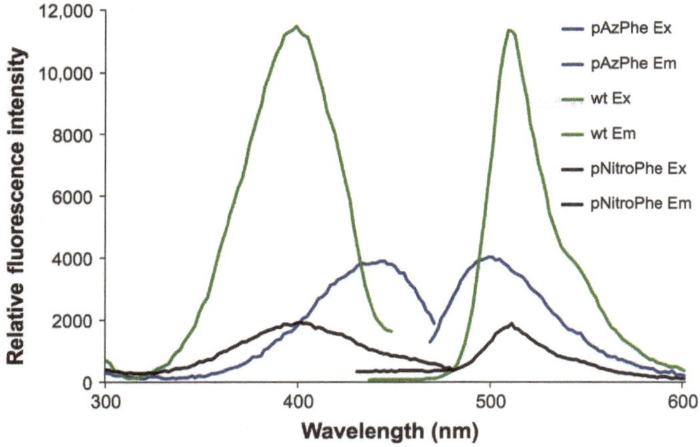

Fig. 4 Fluorescence excitation (Ex) and emission (Em) spectra of wild-type GFP and GFP mutants containing unnatural amino acids at position 66. Fluorescence spectra were measured in phosphate buffer (50 mm, pH 7.4). The emission spectra were recorded with maximum excitation wavelengths of each fluorescent proteins.

3. AN UNNATURAL FLUORESCENT PROTEIN BIOSENSOR FOR HYDROGEN PEROXIDE

3.1 General Design Principles

Since the report of a green fluorescent protein (GFP) biosensor for Ca^{2+} in 1997 (Miyawaki et al., 1997), more than 100 fluorescent protein biosensors have been developed to enable the monitoring of intracellular pH, redox potential, enzyme activities, and concentrations of small molecules (Frommer et al., 2009; Newman et al., 2011).

Several design approaches (Fig. 5) have been developed to engineer fluorescent protein biosensors. The first design (Fig. 5A) focuses on the protonation state of the chromophore. Under different pH, chromophores of fluorescent proteins can adopt altered protonation states, which makes fluorescent proteins naturally sensitive to pH changes (Bizzarri, Serresi, Luin, & Beltram, 2009). While four protonation states (protonated/deprotonated phenol group and alpha nitrogen from Tyr66) can exist in the GFP chromophore, the proton exchange reaction of the phenol group is the main factor in the design of pH sensors that work under physiological conditions. In the model chromophores, the pK_a values of the imidazolinone ring nitrogen (N^{66}; Fig. 5A) and the phenolic hydroxyl group are in the range of

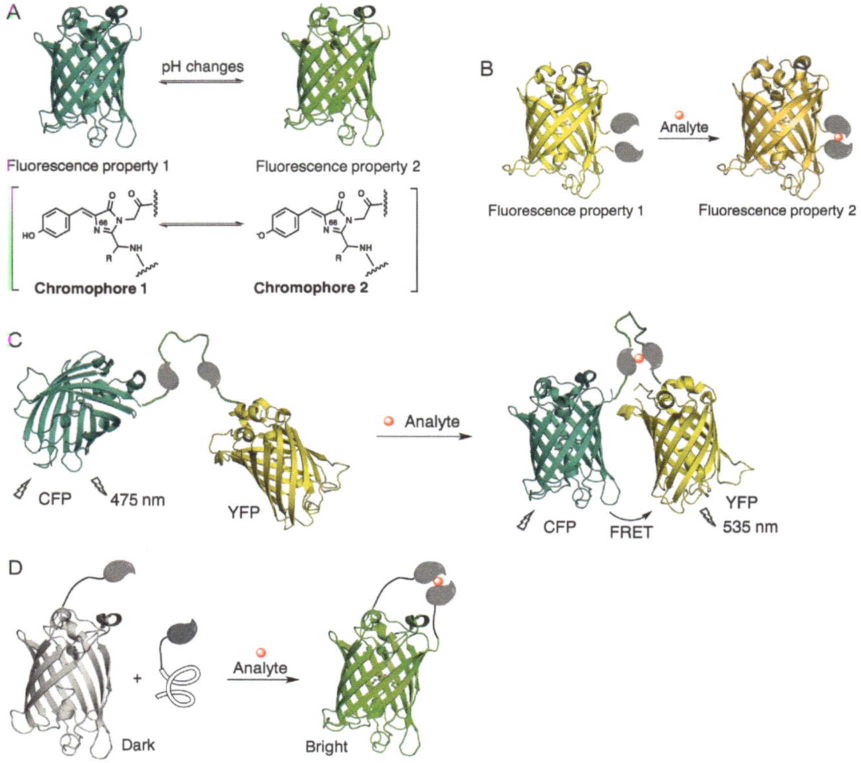

Fig. 5 General design principles of fluorescent protein biosensors. (A) Single fluorescent protein pH sensors that are based on the protonation state of the chromophore. (B) Sensors that are based on conformational change of circularly permuted fluorescent proteins (cpFP). (C) FRET-based fluorescent protein biosensors. (D) Biosensors that are based on bimolecular fluorescence complementation (BiFC).

1.8–2.4 and 8.1–8.5 (Bizzarri et al., 2009), respectively. The estimated pK_a value of the phenolic hydroxyl group of the chromophore in the folded protein is around 5.2–7.5 (Bizzarri et al., 2009). With a protonated phenol group of Tyr66, GFP absorbs at 398 nm and emits at 508 nm. On the other hand, GFP containing a deprotonated phenol group on Tyr66 absorbs at 475 nm and emits at 503 nm (Tsien, 1998).

The second general design is also a class of single fluorescent protein-based biosensors (Fig. 5B) (Frommer et al., 2009; Newman et al., 2011; Wang et al., 2009). It is known that the surrounding molecular environment of a fluorescent protein's chromophore defines its fluorescence signature. Therefore, chemical and/or physical changes to nonchromophore residues can affect the optical properties of a fluorescent protein. For example, the reversible formation of a disulfide bond between two cysteine residues can influence the fluorescence signals of fluorescent proteins. This strategy has been employed to monitor redox potential changes in living cells (Hanson et al., 2004; Ostergaard, Tachibana, & Winther, 2004). In another example, a fluorescent protein, usually a circularly permuted fluorescent protein (cpFP), is fused to a protein scaffold(s) that act as a sensing domain. Upon recognition of the target, the sensing domain causes a conformational change of cpFP and leads to a shift in the fluorescence intensity or the fluorescence hue. This strategy was demonstrated by a genetically encoded hydrogen peroxide (H_2O_2) sensor, Hyper (Belousov et al., 2006), which is based on a cpYFP and the prokaryotic H_2O_2-sensing transcription regulator, OxyR.

In the third design strategy, two fluorescent proteins with overlapping excitation and emission spectra are, respectively, fused to the N- and the C-terminus of a protein scaffold of interest that acts as a sensing domain for the target recognition (Fig. 5C). The sensing domain undergoes a conformational change in response to the target, which leads to the change in the relative distance between the two fused fluorescent proteins. As a consequence, the efficiency of Förster resonance energy transfer (FRET) changed, which often results in the shift in emission (Frommer et al., 2009; Newman et al., 2011; Wang et al., 2009). A well-known example is the "cameleon" class of Ca^{2+} sensors (Mank & Griesbeck, 2008; Miyawaki et al., 1997), in which the fluorescent protein FRET pair is fused to a sensing domain containing calmodulin and a peptide that binds to calmodulin in the presence of Ca^{2+} ions.

The fourth design strategy is based on bimolecular fluorescence complementation (BiFC; Fig. 5D) (Cabantous et al., 2005; Wang et al., 2009), which is commonly known as "split" fluorescent protein. This approach entails the coexpression of two fusion proteins. Each of them contains one fragment of the split fluorescent protein and a part of sensing domains. Association of the sensing domains in the presence of target analyte brings the complementary fragments of the fluorescent protein reporter within proximity, which promotes the reconstitution of the reporter into its native or near native three-dimensional structure and the emission of fluorescence signal. A classic example of such a split fluorescent protein biosensor is the split-pericam (Nagai, Sawano, Park, & Miyawaki, 2001). The association of calmodulin and the M13 peptide in response to Ca^{2+} leads to the formation of an intact GFP.

3.2 General Design Principles for Unnatural Amino Acid-Containing Fluorescent Protein Biosensors

While design principles in the above section have led to the construction of many useful biosensors, significant challenges remain to improve the performance and to expand the diversity of fluorescent protein biosensors. Sensing elements of current fluorescent protein biosensors are either naturally existing proteins, such as transcription factors, or engineered sensing segments derived from natural amino acid (e.g., cysteine residues for disulfide bond formation), which not only limit the specificity and the affinity of fluorescent protein biosensors but also restrict the types of analytes that can be effectively detected. Recently, efforts have been made to develop a new group of fluorescent protein biosensors with expanded chemical/structural space through the incorporation of genetically encoded unnatural amino acids with unique chemical and/or physical properties.

The main focus of above engineering efforts is the chromophore of fluorescent proteins, such as GFP. Since the chemical properties and the structure of the chromophore are the major factors that define the photophysical properties of GFP, a maximum signal output may be achieved by directly modifying the composition of the chromophore. The chromophore of GFP is derived from a tripeptide sequence through the spontaneous post-translational cyclization of the three consecutive amino acid residues (Ser65-Tyr66-Gly67), followed by a dehydration reaction to form an imidazolin-5-one intermediate, and a final dehydrogenation reaction to yield the chromophore (Barondeau, Putnam, Kassmann, Tainer, & Getzoff, 2003; Chalfie, Tu, Euskirchen, Ward, & Prasher, 1994; Cubitt et al., 1995; Pouwels, Zhang, Chan, Dorrestein, & Wachter, 2008; Rao, Kemple, & Prendergast, 1980; Shu, Shaner, Yarbrough, Tsien, & Remington, 2006; Tsien, 1998; Verkhusha & Lukyanov, 2004; Zhang, Patel, Lappe, & Wachter, 2006). This process is self-catalyzed by the protein fold in the presence of molecular oxygen. As Tyr66 forms the major part of the chromophore, the replacement of Tyr66 with an unnatural analog generates an artificial chromophore that cannot be realized in nature. It has been shown previously that these unnatural chromophores displayed altered fluorescent properties from that of the wild-type GFP (Wang et al., 2012; Wang, Xie, et al., 2003).

A general approach to design fluorescent protein biosensors that contain unnatural amino acids is to replace Tyr66 by an unnatural tyrosine analog with a side chain that either selectively reacts with or has high affinity toward the analyte of interest (Fig. 6). The chromophore of the fluorescent protein biosensor changes after the desirable reaction or binding event with the analyte, which leads to fluorescence property changes of the biosensor. These unnatural amino acid-containing fluorescent protein biosensors can generally lead to a large signal output. In order to be highly specific, these unnatural amino acids should only react or bind with the analyte of interest and are chemically orthogonal to other biomolecules in a living cell. Several unnatural amino acid-containing fluorescent protein biosensors have been

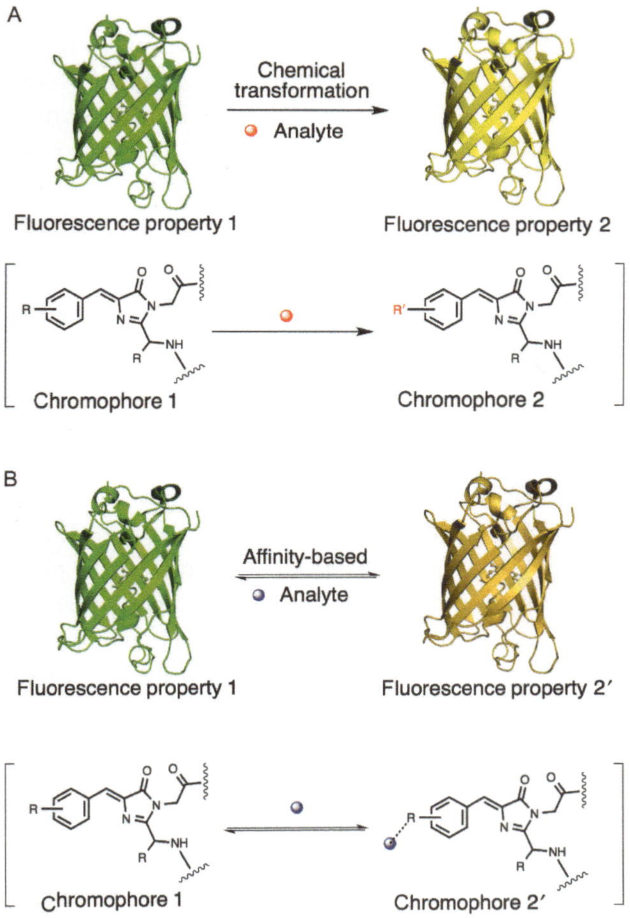

Fig. 6 Basic design principles of green fluorescent protein biosensors containing unnatural amino acid at position 66. (A) Chemical reaction-based biosensors. (B) Affinity/chelation-based biosensors.

developed by following above design principles. These include a transition metal (Cu^{2+} and Zn^{2+}) sensor (Ayyadurai et al., 2011; Liu et al., 2013), a hydrogen peroxide (H_2O_2) sensor (Wang et al., 2012), a peroxynitrite sensor (Chen et al., 2013), a hydrogen sulfide (H_2S) sensor (Chen et al., 2012), and a light sensor (Groff et al., 2010).

To construct this class of fluorescent protein biosensors, an unnatural amino acid of interest is genetically incorporated in response to an amber nonsense codon. Since the chromophore of a fluorescent protein is buried at the center of the β-barrel, it can only be accessed by analytes of small sizes. To enable the analysis of larger analytes and to improve the kinetics of detection, cpFP variants can be employed. Circular permutation is achieved by connecting the original N- and C-termini of fluorescent proteins through flexible peptide linkers and introducing new N- and C-termini at sites in close proximity to the chromophore (Baird, Zacharias, & Tsien, 1999; Nagai et al., 2001). While we discussed GFP specifically here, the above approach can be applied to other fluorescent proteins by replacing functionally equivalent Tyr residues.

3.3 Reaction-Based Fluorescent Protein Biosensor for Hydrogen Peroxide

Reactive oxygen species such as hydrogen peroxide (H_2O_2) acts as a second messenger that selectively modifies cysteine residues on target proteins and, therefore, participates in the control of many intracellular events in response to external stimuli (Paulsen & Carroll, 2010; Poole & Nelson, 2008; Rhee, 2006; Stone & Yang, 2006; Veal, Day, & Morgan, 2007). A prominent example is the reversible inactivation of protein tyrosine phosphatases in receptor tyrosine kinase signaling (Denu & Tanner, 1998; Lee, Kwon, Kim, & Rhee, 1998). Given the importance of H_2O_2 in biology, methods that can be used to selectively image the intracellular concentrations of H_2O_2 and do not interfere with cellular redox events in living systems are highly desirable. A few small-molecule fluorescent probes have been designed based on the chemoselective reaction of arylboronates with H_2O_2 (Lippert, Van de Bittner, & Chang, 2011). These probes are highly specific for H_2O_2 and are bioorthogonal to most cellular processes. In order to generate intracellular H_2O_2 biosensors that have good spatial resolution, a family of genetically encoded H_2O_2 sensor, Hyper (Belousov et al., 2006), Hyper-2 (Markvicheva et al., 2011), and Hyper-3 (Bilan et al., 2012) were constructed by fusing the regulatory domain of bacterial H_2O_2-sensing protein (OxyR) with a circularly permuted yellow fluorescent protein (cpYFP) variant. The output signal is produced through the conformational change of cpYFP, which is induced by the intramolecular disulfide bond formation between two cysteine residues of OxyR in the presence of H_2O_2. A second class of genetically encoded H_2O_2 sensor, roGFP2-Orp1, is based on peroxidase-roGFP2 relay, in which the peroxidase Orp1-mediated oxidation of roGFP2 by H_2O_2 results to signal output (Gutscher et al., 2009). Both the Hyper and roGFP2-Orp1 families of sensors selectively respond to H_2O_2 with very good sensitivity, but lead to less than a fivefold signal enhancement. These hydrogen peroxide sensors are reversible due to the reversibility of disulfide bond formation. Therefore, they can be potentially affected by the redox state of cells, and thus cannot accurately image the accumulated H_2O_2 signal over time.

Here we present a single GFP-based H_2O_2 sensor, in which the Tyr66 is replaced with pBoPhe. The GFP mutant (GFP-Tyr66pBoPhe) is not fluorescent. Based on the crystal structure of GFP-Tyr66pBoPhe (Wang et al., 2012), the $-B(OH)_2$ group forms hydrogen bonds with the backbone nitrogen of His148 and the hydroxyl side chain of Ser205, which affects the planarity of the chromophore. In addition, the bulky $-B(OH)_2$ side chain pushes His148 away from the chromophore (Fig. 3A). The dislocation of His148 (Seifert et al., 2003) may also account for the lack of any observed fluorescence of GFP-Tyr66pBoPhe. In the presence of H_2O_2, arylboronate is oxidized into phenol to restore the native chromophore of GFP (Fig. 7A).

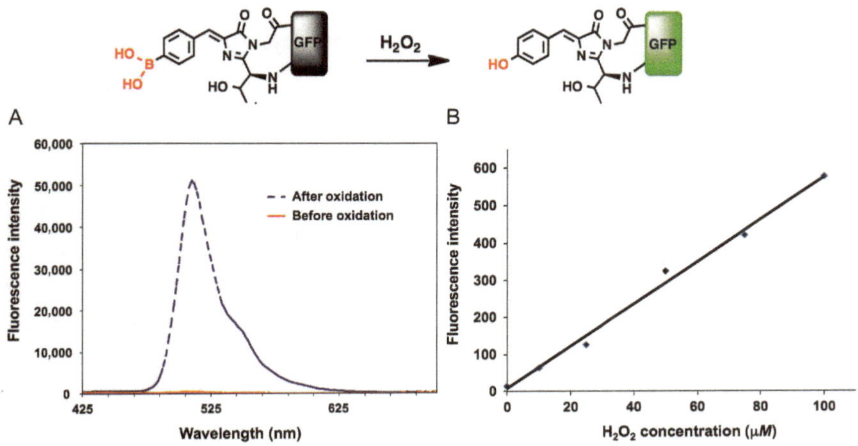

Fig. 7 A fluorescent protein-based H_2O_2 sensor. (A) Fluorescence of GFP-Tyr66pBoPhe before (*red trace*) and after (*blue trace*) H_2O_2 treatment. (B) Changes in fluorescence of GFP-Tyr66pBoPhe after the incubation with different concentrations of H_2O_2.

Therefore, GFP-Tyr66pBoPhe can be used as an H_2O_2 sensor. Since the oxidation of an arylboronate-based fluorescent protein biosensor by H_2O_2 is irreversible, the output signal of the sensor is insensitive to the rapidly recovered cellular redox state after a transient H_2O_2 burst. Therefore, the arylboronate-based fluorescent protein biosensor can be applied to monitoring the cumulative H_2O_2 signal in biological systems.

3.4 Materials

- Fluorescence plate reader
- Inverted confocal fluorescence microscope
- Refrigerated microcentrifuge
- Purified GFP-Tyr66BoF
- Purified wild-type GFP
- *E. coli* cells expressing GFP-Tyr66BoF
- *E. coli* cells expressing wild-type GFP
- 30% H_2O_2 (Sigma)
- PBS buffer
- Assay buffer (50 m*M* phosphate buffer, pH 7.4, containing 10% glycerol)
- 96-Well fluorescence plate (polystyrene plate, black, flat bottomed, and nontransparent; Costar 3915)
- Glass slides for imaging

3.5 Sample Preparation

3.5.1 With Purified Protein

After the purification by affinity chromatography, the isolated protein, GFP-Tyr66BoF (the Tyr66 of GFP is replaced by BoF), is dialyzed against the assay buffer (50 m*M* phosphate buffer, pH 7.4; 10% glycerol).

3.5.2 With Live Cells

Cells (1 mL) expressing GFP-Tyr66BoF are collected using benchtop centrifuge at 5000 × *g* for 8 min. After removal of the supernatant, cells are resuspended in 1 mL PBS buffer and collected by centrifugation. This process is repeated for three times in order to remove residual culture media. After the final washing step, cells are resuspended in 1 mL PBS buffer and are ready for the next step of the experiment.

3.6 Biosensing

3.6.1 With Purified Protein

To a solution of GFP-Tyr66BoF (0.1–1 mg/mL) in assay buffer, a desirable amount of H_2O_2 (0.01–10 m*M*) is added and the resulting mixture is incubated at room temperature for a desirable period of time (3–60 min). As one control, the GFP-Tyr66BoF solution is incubated under the same conditions without the addition of H_2O_2 (replaced with the same amount of water). Fluorescence intensities of the above samples are then measured using a fluorescence plate reader. As shown in Fig. 7A, strong fluorescence can be only observed with the protein sample that is treated with H_2O_2. The fluorescence properties of the H_2O_2-treated protein sample is identical to those of the wild-type GFP $\left(\lambda_{em}^{max} = 508\,nm\right)$. In general, significant fluorescence can be observed after 3 min of incubation with 1 m*M* H_2O_2. Longer time is needed when H_2O_2 concentration is low.

To determine if the fluorescence intensity has a linear correlation with the concentration of H_2O_2, GFP-Tyr66BoF samples are incubated with a range of different concentrations of H_2O_2 (0.01–10 m*M* final concentration). As shown in Fig. 7B, the fluorescence intensities of H_2O_2-treated protein samples correlate well with the concentrations (0.01–0.1 m*M*) of H_2O_2. The linear relationship is observed when H_2O_2 concentration is lower than 10 m*M*. Within a broader range of H_2O_2 concentration (0.01–100 m*M*), a biphasic curve of fluorescence vs H_2O_2 concentration is observed. This likely results from the oxidation of wild-type GFP chromophore at high concentrations of H_2O_2. Indeed, a decrease in fluorescence intensity can be observed when wild-type GFP is incubated with 100 m*M* H_2O_2.

3.6.2 With Live Cells

The fluorescence intensity of washed cells expressing GFP-Tyr66BoF in PBS buffer (100 μL with an OD_{600} of 0.5–1.0) is measured using a fluorescence plate reader. The observed fluorescence intensity is very low (RFU = ~50–100) and is comparable to that of cells that do not express GFP-Tyr66BoF. The same cell sample is incubated with 100 m*M* H_2O_2 at room temperature. The fluorescence intensity is monitored every minute in a 15-min period. As shown in Fig. 8A, the fluorescence intensity of the sample increases as the incubation time increases. In this experiment, a high extracellular concentration of H_2O_2 is needed to attain a sufficient intracellular H_2O_2 concentration. Degradation of H_2O_2 is observed (the appearance of gas bubbles), likely due to the presence of catalase activity in the *E. coli* host. *E. coli* cells are still viable after H_2O_2 treatment at 100 m*M* concentration. As a control, no fluorescence intensity change can be detected with the host *E. coli* cells that are incubated with 100 m*M* H_2O_2 under the same conditions (Fig. 8B).

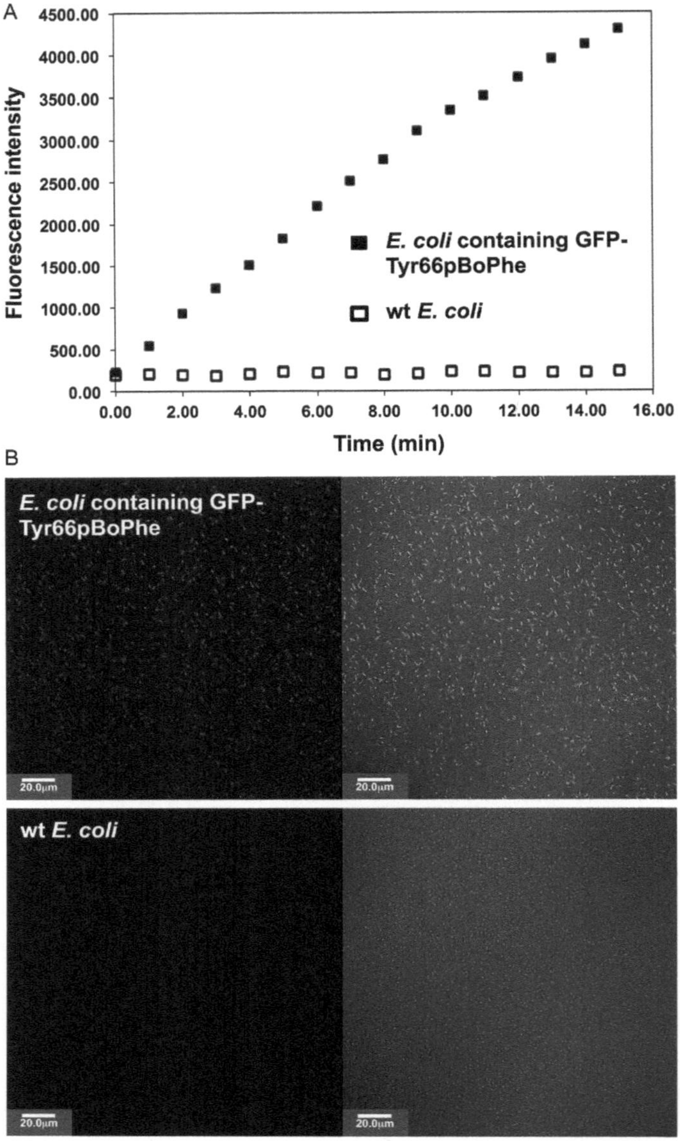

Fig. 8 Sensing H_2O_2 in live cells. (A) Oxidation of wild-type *E. coli* and *E. coli* expressing GFP-Tyr66pBoPhe in the presence of 100 m*M* H_2O_2. (B) Overlapped bright-field and fluorescence images of wild-type *E. coli* and *E. coli* expressing GFP-Tyr66pBoPhe in the presence of 100 m*M* H_2O_2.

In addition to fluorescence intensity measurement, cells can be visualized using a confocal fluorescence microscope. In this set of experiments, a 50 μL of washed cells (with an OD_{600} of one's choice) expressing GFP-Tyr66BoF are resuspended in PBS buffer and added to a glass slide, which is visualized under a confocal fluorescence microscope. As shown in Fig. 8B, no fluorescent cells can be detected. The same batch of cell sample is then incubated with 100 m*M* H_2O_2 at room temperature for 5 min and visualized again. Strong fluorescent signals can be detected. The fluorescence is nicely localized inside cells. Because GFP can be genetically encoded, this approach may allow spatial resolution of redox chemistry when fused to a target protein of interest.

4. CONCLUDING REMARKS

In summary, we report the construction of a H_2O_2 biosensor that combines the desirable properties of the arylboronate-based small-molecule fluorescent probes and fluorescent protein-based biosensors. This new class of H_2O_2 biosensor design aims to reach a maximum signal output by directly modifying the chemical composition of the chromophore. A chemical transformation of the unnatural amino acid (pBoPhe)-containing chromophore leads to dramatic change in fluorescence intensity. The properties of this H_2O_2 biosensor, such as the detection limit and the sensing kinetics, still need to be further optimized in order to be applied to real biological studies. Nevertheless, the construction of this H_2O_2 sensor and other unnatural amino acid-containing fluorescent protein biosensors represent an important step toward the generation of new sensors with high sensitivity and selectivity. Introduction of additional functional groups into fluorescent proteins allows one to explore new designs that are not possible by using the 20 common amino acids. The integration of the two powerful methodologies, fluorescent protein and unnatural amino acid mutagenesis, should lead to the design and construction of novel tools for biological studies in living systems. The design principles of the reaction-based fluorescent protein biosensor for H_2O_2 can be generalized to the construction of other fluorescent protein biosensors. We expect that many well-established aqueous phase chemical reactions can be used for the design of fluorescent protein biosensors and thus may expand the repertoire of analytes that can be detected.

ACKNOWLEDGMENTS

This work was supported by the New Faculty Startup Funds (to J.G. and W.N.) from the University of Nebraska–Lincoln and by an NSF Grant MCB1553041 (to J.G.) and an NIH Grant1R01AI111862 (to J.G. and W.N.). The authors thank Dr. Ronald Cerny in the Nebraska Center for Mass Spectrometry at University of Nebraska–Lincoln for help in protein mass spectrometry. The authors thank Dr. You Zhou in the Microscopy facility at the Center for Biotechnology at University of Nebraska–Lincoln for help in fluorescence microscope analysis.

REFERENCES

Alfonta, L., Zhang, Z., Uryu, S., Loo, J. A., & Schultz, P. G. (2003). Site-specific incorporation of a redox-active amino acid into proteins. *Journal of the American Chemical Society, 125*, 14662–14663.

Amiram, M., Haimovich, A. D., Fan, C., Wang, Y.-S., Aerni, H.-R., Ntai, I., et al. (2015). Evolution of translation machinery in recoded bacteria enables multi-site incorporation of nonstandard amino acids. *Nature Biotechnology, 33*, 1272–1279.

Anderson, J. C., Wu, N., Santoro, S. W., Lakshman, V., King, D. S., & Schultz, P. G. (2004). An expanded genetic code with a functional quadruplet codon. *Proceedings of the National Academy of Sciences of the United States of America, 101*, 7566–7571.

Ayyadurai, N., Prabhu, N. S., Deepankumar, K., Lee, S.-G., Jeong, H.-H., Lee, C.-S., et al. (2011). Development of a selective, sensitive, and reversible biosensor by the genetic incorporation of a metal-binding site into green fluorescent protein. *Angewandte Chemie (International ed. in English), 50*, 6534–6537.

Baird, G. S., Zacharias, D. A., & Tsien, R. Y. (1999). Circular permutation and receptor insertion within green fluorescent proteins. *Proceedings of the National Academy of Sciences of the United States of America, 96*, 11241–11246.

Barondeau, D. P., Putnam, C. D., Kassmann, C. J., Tainer, J. A., & Getzoff, E. D. (2003). Mechanism and energetics of green fluorescent protein chromophore synthesis revealed by trapped intermediate structures. *Proceedings of the National Academy of Sciences of the United States of America, 100*, 12111–12116.

Belousov, V. V., Fradkov, A. F., Lukyanov, K. A., Staroverov, D. B., Shakhbazov, K. S., Terskikh, A. V., et al. (2006). Genetically encoded fluorescent indicator for intracellular hydrogen peroxide. *Nature Methods, 3*, 281–286.

Bilan, D. S., Pase, L., Joosen, L., Gorokhovatsky, A. Y., Ermakova, Y. G., Gadella, T. W. J., et al. (2012). HyPer-3: A genetically encoded H2O2 probe with improved performance for ratiometric and fluorescence lifetime imaging. *ACS Chemical Biology, 8*, 535–542.

Bizzarri, R., Serresi, M., Luin, S., & Beltram, F. (2009). Green fluorescent protein based pH indicators for in vivo use: A review. *Analytical and Bioanalytical Chemistry, 393*, 1107–1122.

Brustad, E., Bushey, M. L., Brock, A., Chittuluru, J., & Schultz, P. G. (2008). A promiscuous aminoacyl-tRNA synthetase that incorporates cysteine, methionine, and alanine homologs into proteins. *Bioorganic and Medicinal Chemistry Letters, 18*, 6004–6006.

Brustad, E. M., Lemke, E. A., Schultz, P. G., & Deniz, A. A. (2008). A general and efficient method for the site-specific dual-labeling of proteins for single molecule fluorescence resonance energy transfer. *Journal of the American Chemical Society, 130*, 17664–17665.

Budisa, N., Minks, C., Alefelder, S., Wenger, W., Dong, F., Moroder, L., et al. (1999). Toward the experimental codon reassignment in vivo: Protein building with an expanded amino acid repertoire. *FASEB Journal, 13*, 41–51.

Cabantous, S., Pedelacq, J.-D., Mark, B. L., Naranjo, C., Terwilliger, T. C., & Waldo, G. S. (2005). Recent advances in GFP folding reporter and split-GFP solubility reporter technologies. Application to improving the folding and solubility of recalcitrant proteins from mycobacterium tuberculosis. *Journal of Structural and Functional Genomics*, *6*, 113–119.

Cellitti, S. E., Jones, D. H., Lagpacan, L., Hao, X., Zhang, Q., Hu, H., et al. (2008). In vivo incorporation of unnatural amino acids to probe structure, dynamics, and ligand binding in a large protein by nuclear magnetic resonance spectroscopy. *Journal of the American Chemical Society*, *130*, 9268–9281.

Chalfie, M., Tu, Y., Euskirchen, G., Ward, W. W., & Prasher, D. C. (1994). Green fluorescent protein as a marker for gene expression. *Science*, *263*, 802–805.

Chatterjee, A., Lajoie, M. J., Xiao, H., Church, G. M., & Schultz, P. G. (2014). A bacterial strain with a unique quadruplet codon specifying non-native amino acids. *Chembiochem: A European Journal of Chemical Biology*, *15*, 1782–1786.

Chen, S., Chen, Z.-J., Ren, W., & Ai, H.-W. (2012). Reaction-based genetically encoded fluorescent hydrogen sulfide sensors. *Journal of the American Chemical Society*, *134*, 9589–9592.

Chen, Z.-J., Ren, W., Wright, Q. E., & Ai, H.-W. (2013). Genetically encoded fluorescent probe for the selective detection of peroxynitrite. *Journal of the American Chemical Society*, *135*, 14940–14943.

Chen, X.-H., Xiang, Z., Hu, Y. S., Lacey, V. K., Cang, H., & Wang, L. (2014). Genetically encoding an electrophilic amino acid for protein stapling and covalent binding to native receptors. *ACS Chemical Biology*, *9*, 1956–1961.

Chi, K. R. (2009). Super-resolution microscopy: Breaking the limits. *Nature Methods*, *6*, 15–18.

Chin, J. W., Santoro, S. W., Martin, A. B., King, D. S., Wang, L., & Schultz, P. G. (2002). Addition of p-azido-L-phenylalanine to the genetic code of Escherichia coli. *Journal of the American Chemical Society*, *124*, 9026–9027.

Chudakov, D. M., Matz, M. V., Lukyanov, S., & Lukyanov, K. A. (2010). Fluorescent proteins and their applications in imaging living cells and tissues. *Physiological Reviews*, *90*, 1103–1163.

Cubitt, A. B., Heim, R., Adams, S. R., Boyd, A. E., Gross, L. A., & Tsien, R. Y. (1995). Understanding, improving and using green fluorescent proteins. *Trends in Biochemical Sciences*, *20*, 448–455.

Datta, D., Wang, P., Carrico, I. S., Mayo, S. L., & Tirrell, D. A. (2002). A designed phenylalanyl-tRNA synthetase variant allows efficient in vivo incorporation of aryl ketone functionality into proteins. *Journal of the American Chemical Society*, *124*, 5652–5653.

Deiters, A., Geierstanger, B. H., & Schultz, P. G. (2005). Site-specific in vivo labeling of proteins for NMR studies. *Chembiochem: A European Journal of Chemical Biology*, *6*, 55–58.

Denu, J. M., & Tanner, K. G. (1998). Specific and reversible inactivation of protein tyrosine phosphatases by hydrogen peroxide: Evidence for a sulfenic acid intermediate and implications for redox regulation. *Biochemistry*, *37*, 5633–5642.

Döring, V., Mootz, H. D., Nangle, L. A., Hendrickson, T. L., de Crécy-Lagard, V., Schimmel, P., et al. (2001). Enlarging the amino acid set of Escherichia coli by infiltration of the valine coding pathway. *Science*, *292*, 501–504.

Dougherty, D. A. (2000). Unnatural amino acids as probes of protein structure and function. *Current Opinion in Chemical Biology*, *4*, 645–652.

Frommer, W. B., Davidson, M. W., & Campbell, R. E. (2009). Genetically encoded biosensors based on engineered fluorescent proteins. *Chemical Society Reviews*, *38*, 2833–2841.

Groff, D., Thielges, M. C., Cellitti, S., Schultz, P. G., & Romesberg, F. E. (2009). Efforts toward the direct experimental characterization of enzyme microenvironments: Tyrosine100 in dihydrofolate reductase. *Angewandte Chemie (International ed. in English)*, *48*, 3478–3481. S3478/3471-S3478/3475.

Groff, D., Wang, F., Jockusch, S., Turro, N. J., & Schultz, P. G. (2010). A new strategy to photoactivate green fluorescent protein. *Angewandte Chemie (International ed. in English), 49*, 7677–7679.

Guo, J., Melancon, C. E., III, Lee, H. S., Groff, D., & Schultz, P. G. (2009). Evolution of amber suppressor tRNAs for efficient bacterial production of proteins containing non-natural amino acids. *Angewandte Chemie (International ed. in English), 48*, 9148–9151. S9148/9141-S9148/9123.

Guo, J., Wang, J., Lee, J. S., & Schultz, P. G. (2008). Site-specific incorporation of methyl- and acetyl-lysine analogues into recombinant proteins. *Angewandte Chemie (International ed. in English), 47*, 6399–6401.

Gutscher, M., Sobotta, M. C., Wabnitz, G. H., Ballikaya, S., Meyer, A. J., Samstag, Y., et al. (2009). Proximity-based protein thiol oxidation by H_2O_2-scavenging peroxidases. *Journal of Biological Chemistry, 284*, 31532–31540.

Hanson, G. T., Aggeler, R., Oglesbee, D., Cannon, M., Capaldi, R. A., Tsien, R. Y., et al. (2004). Investigating mitochondrial redox potential with redox-sensitive green fluorescent protein indicators. *Journal of Biological Chemistry, 279*, 13044–13053.

Hohsaka, T., & Sisido, M. (2002). Incorporation of non-natural amino acids into proteins. *Current Opinion in Chemical Biology, 6*, 809–815.

Huttner, W. B. (1988). Tyrosine sulfation and the secretory pathway. *Annual Review of Physiology, 50*, 363–376.

Jackson, J. C., Hammill, J. T., & Mehl, R. A. (2007). Site-specific incorporation of a 19F-amino acid into proteins as an NMR probe for characterizing protein structure and reactivity. *Journal of the American Chemical Society, 129*, 1160–1166.

Johnson, D. B. F., Xu, J., Shen, Z., Takimoto, J. K., Schultz, M. D., Schmitz, R. J., et al. (2011). RF1 knockout allows ribosomal incorporation of unnatural amino acids at multiple sites. *Nature Chemical Biology, 7*, 779–786.

Ju, T., Niu, W., Cerny, R., Bollman, J., Roy, A., & Guo, J. (2013). Molecular recognition of sulfotyrosine and phosphotyrosine by the Src homology 2 domain. *Molecular BioSystems, 9*, 1829–1832.

Ju, T., Niu, W., & Guo, J. (2016). Evolution of Src homology 2 (SH2) domain to recognize sulfotyrosine. *ACS Chemical Biology, 11*, 2551–2557.

Lajoie, M. J., Rovner, A. J., Goodman, D. B., Aerni, H.-R., Haimovich, A. D., Kuznetsov, G., et al. (2013). Genomically recoded organisms expand biological functions. *Science, 342*, 357–360.

Lang, K., Davis, L., Torres-Kolbus, J., Chou, C., Deiters, A., & Chin, J. W. (2012a). Genetically encoded norbornene directs site-specific cellular protein labelling via a rapid bioorthogonal reaction. *Nature Chemistry, 4*, 298–304.

Lang, K., Davis, L., Wallace, S., Mahesh, M., Cox, D. J., Blackman, M. L., et al. (2012b). Genetic encoding of bicyclononynes and trans-cyclooctenes for site-specific protein labeling in vitro and in live mammalian cells via rapid fluorogenic Diels-alder reactions. *Journal of the American Chemical Society, 134*, 10317–10320.

Lee, S.-R., Kwon, K.-S., Kim, S.-R., & Rhee, S. G. (1998). Reversible inactivation of protein-tyrosine phosphatase 1B in A431 cells stimulated with epidermal growth factor. *Journal of Biological Chemistry, 273*, 15366–15372.

Lee, H. S., Spraggon, G., Schultz, P. G., & Wang, F. (2009). Genetic incorporation of a metal-ion chelating amino acid into proteins as a biophysical probe. *Journal of the American Chemical Society, 131*, 2481–2483.

Lippert, A. R., Van de Bittner, G. C., & Chang, C. J. (2011). Boronate oxidation as a bioorthogonal reaction approach for studying the chemistry of hydrogen peroxide in living systems. *Accounts of Chemical Research, 44*, 793–804.

Liu, C. C., Brustad, E., Liu, W., & Schultz, P. G. (2007). Crystal structure of a biosynthetic sulfo-hirudin complexed to thrombin. *Journal of the American Chemical Society, 129*, 10648–10649.

Liu, X., Li, J., Hu, C., Zhou, Q., Zhang, W., Hu, M., et al. (2013). Significant expansion of the fluorescent protein chromophore through the genetic incorporation of a metal-chelating unnatural amino acid. *Angewandte Chemie (International ed. in English), 52*, 4805–4809.

Liu, C. C., & Schultz, P. G. (2006). Recombinant expression of selectively sulfated proteins in *Escherichia coli*. *Nature Biotechnology, 24*, 1436–1440.

Liu, C. C., & Schultz, P. G. (2010). Adding new chemistries to the genetic code. *Annual Review of Biochemistry, 79*, 413–444.

Liu, X., Yu, Y., Hu, C., Zhang, W., Lu, Y., & Wang, J. (2012). Significant increase of oxidase activity through the genetic incorporation of a tyrosine-histidine cross-link in a myoglobin model of heme-copper oxidase. *Angewandte Chemie (International ed. in English), 51*, 4312–4316.

Mank, M., & Griesbeck, O. (2008). Genetically encoded calcium indicators. *Chemical Reviews, 108*, 1550–1564.

Markvicheva, K. N., Bilan, D. S., Mishina, N. M., Gorokhovatsky, A. Y., Vinokurov, L. M., Lukyanov, S., et al. (2011). A genetically encoded sensor for H2O2 with expanded dynamic range. *Bioorganic and Medicinal Chemistry, 19*, 1079–1084.

Matz, M. V., Fradkov, A. F., Labas, Y. A., Savitsky, A. P., Zaraisky, A. G., Markelov, M. L., et al. (1999). Fluorescent proteins from nonbioluminescent Anthozoa species. *Nature Biotechnology, 17*, 969–973.

Miyawaki, A., Llopis, J., Heim, R., McCaffery, J. M., Adams, J. A., Ikura, M., et al. (1997). Fluorescent indicators for Ca2+ based on green fluorescent proteins and calmodulin. *Nature, 388*, 882–887.

Miyawaki, A., Shcherbakova, D. M., & Verkhusha, V. V. (2012). Red fluorescent proteins: Chromophore formation and cellular applications. *Current Opinion in Structural Biology, 22*, 679–688.

Nagai, T., Sawano, A., Park, E. S., & Miyawaki, A. (2001). Circularly permuted green fluorescent proteins engineered to sense Ca^{2+}. *Proceedings of the National Academy of Sciences of the United States of America, 98*, 3197–3202.

Neumann, H., Wang, K., Davis, L., Garcia-Alai, M., & Chin, J. W. (2010). Encoding multiple unnatural amino acids via evolution of a quadruplet-decoding ribosome. *Nature, 464*, 441–444.

Newman, R. H., Fosbrink, M. D., & Zhang, J. (2011). Genetically encodable fluorescent biosensors for tracking signaling dynamics in living cells. *Chemical Reviews, 111*, 3614–3666.

Niu, W., & Guo, J. (2013). Expanding the chemistry of fluorescent protein biosensors through genetic incorporation of unnatural amino acids. *Molecular BioSystems, 9*, 2961–2970.

Niu, W., Schultz, P. G., & Guo, J. (2013). An expanded genetic code in mammalian cells with a functional quadruplet codon. *ACS Chemical Biology, 8*, 1640–1645.

Ostergaard, H., Tachibana, C., & Winther, J. R. (2004). Monitoring disulfide bond formation in the eukaryotic cytosol. *Journal of Cell Biology, 166*, 337–345.

Paulsen, C. E., & Carroll, K. S. (2010). Orchestrating redox signaling networks through regulatory cysteine switches. *ACS Chemical Biology, 5*, 47–62.

Poole, L. B., & Nelson, K. J. (2008). Discovering mechanisms of signaling-mediated cysteine oxidation. *Current Opinion in Chemical Biology, 12*, 18–24.

Pouwels, L. J., Zhang, L., Chan, N. H., Dorrestein, P. C., & Wachter, R. M. (2008). Kinetic isotope effect studies on the de novo rate of chromophore formation in fast- and slow-maturing GFP variants. *Biochemistry, 47*, 10111–10122.

Rao, B. D., Kemple, M. D., & Prendergast, F. G. (1980). Proton nuclear magnetic resonance and fluorescence spectroscopic studies of segmental mobility in aequorin and a green fluorescent protein from aequorea forskalea. *Biophysical Journal*, *32*, 630–632.

Rhee, S. G. (2006). Cell signaling. H2O2, a necessary evil for cell signaling. *Science*, *312*, 1882–1883.

Rodriguez, E. A., Lester, H. A., & Dougherty, D. A. (2006). In vivo incorporation of multiple unnatural amino acids through nonsense and frameshift suppression. *Proceedings of the National Academy of Sciences of the United States of America*, *103*, 8650–8655.

Schultz, K. C., Supekova, L., Ryu, Y., Xie, J., Perera, R., & Schultz, P. G. (2006). A genetically encoded infrared probe. *Journal of the American Chemical Society*, *128*, 13984–13985.

Seifert, M. H. J., Georgescu, J., Ksiazek, D., Smialowski, P., Rehm, T., Steipe, B., et al. (2003). Backbone dynamics of green fluorescent protein and the effect of histidine 148 substitution. *Biochemistry*, *42*, 2500–2512.

Seyedsayamdost, M. R., Xie, J., Chan, C. T. Y., Schultz, P. G., & Stubbe, J. (2007). Site-specific insertion of 3-aminotyrosine into subunit α2 of E. Coli ribonucleotide reductase: Direct evidence for involvement of Y730 and Y731 in radical propagation. *Journal of the American Chemical Society*, *129*, 15060–15071.

Shaner, N. C., Campbell, R. E., Steinbach, P. A., Giepmans, B. N. G., Palmer, A. E., & Tsien, R. Y. (2004). Improved monomeric red, orange and yellow fluorescent proteins derived from Discosoma sp. red fluorescent protein. *Nature Biotechnology*, *22*, 1567–1572.

Shang, X., Song, X., Faller, C., Lai, R., Li, H., Cerny, R., et al. (2017). Fluorogenic protein labeling using a genetically encoded unstrained alkene. *Chemical Science*, *8*, 1141–1145. http://dx.doi.org/10.1039/C1036SC03635J.

Shu, X., Shaner, N. C., Yarbrough, C. A., Tsien, R. Y., & Remington, S. J. (2006). Novel chromophores and buried charges control color in mFruits. *Biochemistry*, *45*, 9639–9647.

Song, W., Wang, Y., Qu, J., Madden, M. M., & Lin, Q. (2008). A photoinducible 1,3-dipolar cycloaddition reaction for rapid, selective modification of tetrazole-containing proteins. *Angewandte Chemie (International ed. in English)*, *47*, 2832–2835.

Stone, M. J., Chuang, S., Hou, X., Shoham, M., & Zhu, J. Z. (2009). Tyrosine sulfation: An increasingly recognized post-translational modification of secreted proteins. *New Biotechnology*, *25*, 299–317.

Stone, J. R., & Yang, S. (2006). Hydrogen peroxide: A signaling messenger. *Antioxidants & Redox Signaling*, *8*, 243–270.

Taki, M., Matsushita, J., & Sisido, M. (2006). Expanding the genetic code in a mammalian cell line by the introduction of four-base codon/anticodon pairs. *Chembiochem: A European Journal of Chemical Biology*, *7*, 425–428.

Tippmann, E. M., & Schultz, P. G. (2007). A genetically encoded metallocene containing amino acid. *Tetrahedron*, *63*, 6182–6184.

Tsien, R. Y. (1998). The green fluorescent protein. *Annual Review of Biochemistry*, *67*, 509–544.

Veal, E. A., Day, A. M., & Morgan, B. A. (2007). Hydrogen peroxide sensing and signaling. *Molecular Cell*, *26*, 1–14.

Verkhusha, V. V., & Lukyanov, K. A. (2004). The molecular properties and applications of Anthozoa fluorescent proteins and chromoproteins. *Nature Biotechnology*, *22*, 289–296.

Wang, N., Ju, T., Niu, W., & Guo, J. (2015). Fine-tuning interaction between aminoacyl-tRNA synthetase and tRNA for efficient synthesis of proteins containing unnatural amino acids. *ACS Synthetic Biology*, *4*, 207–212.

Wang, H., Nakata, E., & Hamachi, I. (2009). Recent progress in strategies for the creation of protein-based fluorescent biosensors. *Chembiochem: A European Journal of Chemical Biology*, *10*, 2560–2577.

Wang, F., Niu, W., Guo, J., & Schultz, P. G. (2012). Unnatural amino acid mutagenesis of fluorescent proteins. *Angewandte Chemie (International ed. in English)*, *51*, 10132–10135.

Wang, K., Sachdeva, A., Cox, D. J., Wilf, N. M., Lang, K., Wallace, S., et al. (2014). Optimized orthogonal translation of unnatural amino acids enables spontaneous protein double-labeling and FRET. *Nature Chemistry*, *6*, 393–403.

Wang, N., Shang, X., Cerny, R., Niu, W., & Guo, J. (2016). Systematic evolution and study of UAGN decoding tRNAs in a genomically recoded bacteria. *Scientific Reports*, *6*, 21898.

Wang, L., Xie, J., Deniz, A. A., & Schultz, P. G. (2003). Unnatural amino acid mutagenesis of green fluorescent protein. *Journal of Organic Chemistry*, *68*, 174–176.

Wang, J., Xie, J., & Schultz, P. G. (2006). A genetically encoded fluorescent amino acid. *Journal of the American Chemical Society*, *128*, 8738–8739.

Wang, L., Zhang, Z., Brock, A., & Schultz, P. G. (2003). Addition of the keto functional group to the genetic code of Escherichia coli. *Proceedings of the National Academy of Sciences of the United States of America*, *100*, 56–61.

Xiang, Z., Lacey, V. K., Ren, H., Xu, J., Burban, D. J., Jennings, P. A., et al. (2014). Proximity-enabled protein crosslinking through genetically encoding haloalkane unnatural amino acids. *Angewandte Chemie (International ed. in English)*, *53*, 2190–2193.

Xie, J., Liu, W., & Schultz, P. G. (2007). A genetically encoded bidentate, metal-binding amino acid. *Angewandte Chemie (International ed. in English)*, *46*, 9239–9242.

Xie, J., Supekova, L., & Schultz, P. G. (2007). A genetically encoded metabolically stable analog of phosphotyrosine in Escherichia coli. *ACS Chemical Biology*, *2*, 474–478.

Ye, S., Koehrer, C., Huber, T., Kazmi, M., Sachdev, P., Yan, E. C. Y., et al. (2008). Site-specific incorporation of keto amino acids into functional G protein-coupled receptors using unnatural amino acid mutagenesis. *Journal of Biological Chemistry*, *283*, 1525–1533.

Zhang, L., Patel, H. N., Lappe, J. W., & Wachter, R. M. (2006). Reaction progress of chromophore biogenesis in green fluorescent protein. *Journal of the American Chemical Society*, *128*, 4766–4772.

Zhang, Z., Wang, L., Brock, A., & Schultz, P. G. (2002). The selective incorporation of alkenes into proteins in Escherichia coli. *Angewandte Chemie (International ed. in English)*, *41*, 2840–2842.

Zimmer, M. (2002). Green fluorescent protein (GFP): Applications, structure, and related photophysical behavior. *Chemical Reviews*, *102*, 759–781.

CHAPTER EIGHT

Folding- and Dynamics-Based Electrochemical DNA Sensors

Rebecca Y. Lai[1]
University of Nebraska-Lincoln, Lincoln, NE, United States
[1]Corresponding author: e-mail address: rlai2@unl.edu

Contents

Abstract

A number of electrochemical DNA sensors based on the target-induced change in the conformation and/or flexibility of surface-bound oligonucleotides have been developed in recent years. These sensors, which are often termed E-DNA sensors, are comprised of an oligonucleotide probe modified with a redox label (e.g., methylene blue) at one terminus and attached to a gold electrode via a thiol–gold bond at the other. Binding of the target to the DNA probe changes its structure and dynamics, which, in turn, influences the efficiency of electron transfer to the interrogating electrode. Since electrochemically active contaminants are less common, these sensors are resistant to false-positive signals arising from the nonspecific adsorption of contaminants and perform well even when employed directly in serum, whole blood, and other realistically complex sample matrices. Moreover, because all of the sensor components are chemisorbed to the electrode, the E-DNA sensors are essentially label-free and readily reusable. To date, these sensors have achieved state-of-the-art sensitivity, while offering the unprecedented selectivity, reusability, and the operational convenience of direct

Methods in Enzymology, Volume 589
ISSN 0076-6879
http://dx.doi.org/10.1016/bs.mie.2017.02.001

electrochemical detection. This chapter reviews the recent advances in the development of both "signal-off" and "signal-on" E-DNA sensors. Critical aspects that dictate the stability and performance of these sensors are also addressed so as to provide a realistic overview of this oligonucleotide detection platform.

1. INTRODUCTION

One of the major driving forces for the development of DNA and RNA sensing methods is the growing demands for point-of-care medical diagnostics (Gooding, 2002; Peng & Miller, 2011; Sassolas, Leca-Bouvier, & Blum, 2008; Teles & Fonseca, 2008; Wang, 2000). Nucleic acid sensors have also found applications in the agriculture and food industries, as well as in national defense (Mello & Kubota, 2002; Rodriguez-Mozaz, López de Alda, Marco, & Barceló, 2005). However, despite substantial research efforts, very few of these sensors have been successfully commercialized, presumably due to the sensors' inability to simultaneously fulfill a set of stringent criteria that includes high sensitivity, specificity, selectivity, user-friendliness, and cost-effectiveness. While many of the sensing approaches feature impressive sensitivity, a DNA sensing technology that is highly sensitive and selective, does not require exogenous reagents, and is operationally convenient has yet to be realized. Sensing technologies that employ surface plasmon resonance spectroscopy (Castillo et al., 2004; Homola, 2003; Wegner, Lee, & Corn, 2002), quartz crystal microbalance (Cooper, 2003), and microcantilevers (Ziegler, 2004) are known to be rapid and extremely sensitive, but they do rely on binding-induced changes in refractive index or mass at a sensor surface to produce a detectable signal. The analysis could be rather compromised if these techniques were used with real biological samples since all nonspecifically bound contaminants on a sensor will have mass and a polarizable electron density (Homola, 2003; Ziegler, 2004). Electrochemical techniques can potentially circumvent some of the aforementioned issues with optical and mass-based detection approaches. Among the currently available electrochemical DNA sensing platforms, a unique class of DNA sensor, the electrochemical DNA (E-DNA) sensor, has gained substantial popularity since it was first reported by Fan, Plaxco, and Heeger (2003). Merits of this sensor design include high sensitivity and selectivity, in addition to being reagentless and reusable (Fan et al., 2003; Lai, Lagally, et al., 2006; Lubin, Lai, Plaxco, & Heeger, 2006). Various versions of this E-DNA sensor have been described to date (Immoos, Lee, & Grinstaff, 2004; March, Noël, Piro, Reisberg, & Pham, 2008; Rowe et al., 2011; Xiao, Qu, Plaxco, & Heeger, 2007). This book chapter will focus specifically on three types of reagentless and reusable E-DNA sensors. The first type is the original "signal-off" E-DNA sensor (Fan et al., 2003; Lai, Walker, Stormberg, Zaitouna, & Yang, 2013; Yang & Lai, 2011a, 2011b), the second type is the single-probe "signal-on" E-DNA sensor (Immoos et al., 2004; Wu & Lai, 2013, 2014), and the third type is the dual-probe "signal-on" E-DNA sensor (Yu & Lai, 2012, 2013; Yu, Zaitouna, & Lai, 2014). Other topics related to these sensors, including alkanethiol linker length, passivating diluent chain length, electrode substrates, and alternate sensor fabrication approaches, will also be addressed.

2. "SIGNAL-OFF" E-DNA SENSORS

2.1 Stem–Loop Probe and Linear Probe E-DNA Sensors

2.1.1 Design and Signaling Mechanism

The stem–loop probe (SLP) E-DNA sensor is the electrochemical equivalent of the fluorescent molecular beacon assay (Fig. 1A) (Fan et al., 2003; Lai, Lee, Soh, Plaxco, & Heeger, 2006; Lai, Seferos, Heeger, Bazan, & Plaxco, 2006; Lai et al., 2013; Yang & Lai, 2011a, 2011b). The signaling of this sensor originates from the binding-induced change in the conformation of the SLP and the efficiency with which the attached redox label transfers electrons to the electrode. In the absence of the target DNA, the stem–loop structure positions the redox label (e.g., methylene blue (MB)) in close

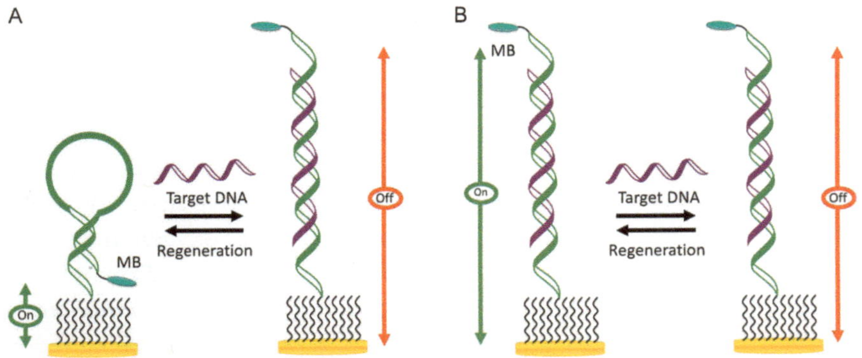

Fig. 1 Design and signaling mechanism of the "signal-off" SLP (A) and LP (B) E-DNA sensors.

proximity to the electrode, enabling efficient electron transfer and the resultant current is high. Upon hybridization to the full complementary target DNA, the enhanced rigidity of the probe–target duplex limits the accessibility of the MB label to the electrode surface, thereby impeding electron transfer, resulting in a detectable reduction in the MB redox current (Farjami, Clima, Gothelf, & Ferapontova, 2011; Lai et al., 2013; Yang & Lai, 2011a, 2011b). An alternate version of the original E-DNA sensor that utilizes a MB-modified linear DNA probe was reported by Ricci, Lai, and Plaxco (2007). The signaling of this linear probe (LP) E-DNA sensor is based on the binding-induced change in the flexibility of the MB-modified probe, which is dampened due to the limited flexibility of the probe–target duplex, leading to a significant reduction in the MB redox current (Fig. 1B). Detailed descriptions of the signaling mechanism can be found in other published works and will not be reviewed here (Anne, Bouchardon, & Moiroux, 2003; Anne & Demaille, 2006; Ricci et al., 2007; White & Plaxco, 2010). It is worth noting that although most reported works focused on the detection of DNA, this class of sensors is well suited for detection of RNA as well. Both SLP and LP sensor designs are highly versatile and can be tailored for detection of essentially any nucleic acid sequence.

2.1.2 Redox Labels

In addition to MB, ferrocene (Fc), thionine (TH), and anthraquinone (AQ) have also been employed as redox labels (Balintová et al., 2011; Mao, Luo, & Ouyang, 2003; Nakayama, Ihara, Nakano, & Maeda, 2002; Yang & Lai, 2012). There are advantages and disadvantages associated with each of them. For example, while Fc has a substantially faster heterogeneous electron transfer rate when compared to MB, in its oxidized form (ferrocenium) it is susceptible to nucleophilic attack (Prins, Korswagen, & Kortbeek, 1972). E-DNA sensors fabricated using Fc-modified DNA probes were found to be less stable when compared to sensors fabricated using MB-modified DNA probes, especially when interrogated in a buffer solution with a high concentration of chloride (unpublished results). The standard reduction potential of AQ (~ -0.4 V vs Ag/AgCl) is more negative than MB and is subject to interference from the reduction of molecular oxygen under certain experimental conditions (Sarapuu, Vaik, Schiffrin, & Tammeveski, 2003) (unpublished results). It is thus more suited to be employed in an oxygen-free environment. A thorough comparison between several redox labels used in this class of folding- and dynamics-based sensors was reported by Kang et al. (2009). Among the known redox labels compatible with the E-DNA sensing platform, MB has emerged as the most versatile one, owing to its stability under a wide range of experimental conditions. In most sensor architectures, the redox label is often located at the distal terminus of the biorecognition probe, but it can be placed in different parts of the probe (Lubin, Vander Stoep Hunt, White, & Plaxco, 2009). The goal is to place the redox label at a location in which target binding could induce the largest change in the redox current; the placement, however, should not interfere with target binding. There are also advantages in labeling the probe with more than one of the same or different redox labels (Yu & Lai, 2012). Modifying a DNA probe with multiple redox labels, especially at different locations, is synthetically challenging; however, the results could be beneficial if the probe coverage is very low or the size of the electrode is extremely small (e.g., nanoelectrodes). This multilabeling approach has found applications in other DNA sensing platforms (Drummond, Hill, & Barton, 2003).

2.1.3 Sensor Fabrication Protocol

Independent of the sensor design and identity of the redox label, the sensor fabrication procedure remains relatively similar. In most reports, a thiol-gold self-assembled monolayer (SAM) was employed (Chaki & Vijayamohanan, 2002). To fabricate either the SLP or LP sensor, the MB-modified DNA probe has to be immobilized onto a cleaned gold electrode surface, followed by surface passivation with an alkanethiol (Lai et al., 2013; Yang & Lai, 2011a, 2011b). The probe immobilization time and concentrations of the reagents should be adjusted according to the required probe coverage and sensor stability. In general, a higher immobilization probe concentration will produce a sensor with a higher probe coverage. The probe coverage requirement differs depending on the specific sensor design and the targeted sensitivity and dynamic range; in general, only 0.05%–5% of the monolayer is composed of DNA probes. A higher probe coverage is essential for other types of DNA sensors, but it is not required or ideal for this class of folding- and dynamics-based E-DNA sensors (Drummond et al., 2003). SAM-based sensors fabricated using a short 4-h protocol are less stable when compared to sensors fabricated using a 24-h protocol. Most of the sensors presented in this book chapter were fabricated using a 2-step sensor fabrication protocol involving direct immobilization of the DNA probe for 1 h, followed by a sensor passivation step in an alkanethiol diluent such as

6-mercapto-1-hexanol (C6-OH) for 16–18 h (Lai et al., 2013; Yang & Lai, 2011a, 2011b). These sensors were found to be stable within the experiment time frame (e.g., 8–24 h).

2.1.4 Alkanethiol Linker Length and Diluent Chain Length

Although most E-DNA sensors developed to date were fabricated using DNA probes modified with a 6-carbon alkanethiol linker (C6-based sensors), there are advantages in employing a longer alkanethiol linker (e.g., 11-carbon) (Lai, Seferos, et al., 2006; Lai et al., 2013; Yang & Lai, 2011a, 2011b). Previous results have shown that C11-based E-DNA sensors were substantially more stable than C6-based sensors; they were functional even after being stored in a buffer under ambient lab conditions for 7 days (Lai, Lee, et al., 2006; Lai, Seferos, et al., 2006). This approach was considered a major improvement since the C6-based E-DNA sensors started to degrade in less than 24 h. In addition to the alkanethiol linker length, the chain length of the passivating diluent could also affect both sensor performance and stability. For most sensors fabricated with a C6-based DNA probe, the preferred diluent is C6-OH (Lai et al., 2013; Yang & Lai, 2011a, 2011b). Because of its length and size, it will not induce steric hindrance to target binding. However, depending on the experimental condition, the use of a longer chain diluent (e.g., 8-mercapto-1-octanol (C8-OH) or 9-mercapto-1-nonanol (C9-OH)) could be beneficial. Yang and Lai reported this effect in a study published in 2011. For example, the SLP sensors fabricated with C8-OH or C9-OH showed larger percent signal suppression (%SS) than sensors fabricated with C6-OH (Yang & Lai, 2011a, 2011b). These sensors showed good stability and DNA sequence mismatch discrimination capability when interrogated at ~47°C, a task that was not achievable with sensors passivated with C6-OH (Fig. 2). E-DNA sensors passivated with C6-OH showed signs of monolayer degradation upon continuous exposure to high temperature solution environments.

2.2 Sensor Response

2.2.1 Alternating Current Voltammetry

A wide range of electrochemical interrogation techniques, including alternating current voltammetry (ACV), cyclic voltammetry (CV), square wave voltammetry (SWV), and differential pulse voltammetry (DPV), have been employed in sensor operation (Fan et al., 2003; Farjami et al., 2011; Lai et al., 2013; White & Plaxco, 2010; Yang & Lai, 2011a, 2011b). ACV is one of the better techniques for both sensor characterization and target analysis, owing to its capability in suppressing the non-Faradaic

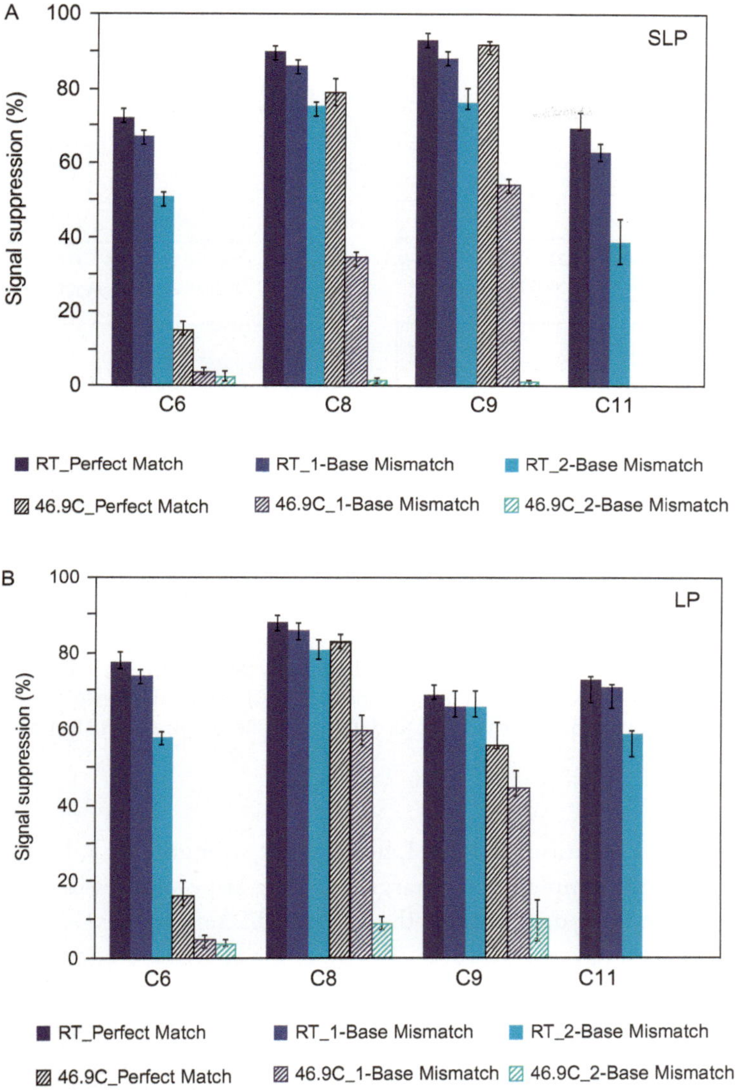

Fig. 2 Mismatch discrimination capability of the SLP (A) and LP (B) sensors fabricated with different passivating diluents at room temperature (RT, $21 \pm 1°C$) and high temperature ($47 \pm 1°C$). All data were obtained in a Phys2 buffer and the concentration of each target DNA was 1.0 μ*M*. Data are averages of three independent experiments.

background current. The responses of the SLP and LP sensors in ACV are shown in Fig. 3A and B. For both sensors, in the absence of the target DNA, a well-defined MB peak was observed at ~−0.27 V (vs Ag/AgCl), a potential consistent with the reduction potential of MB under similar

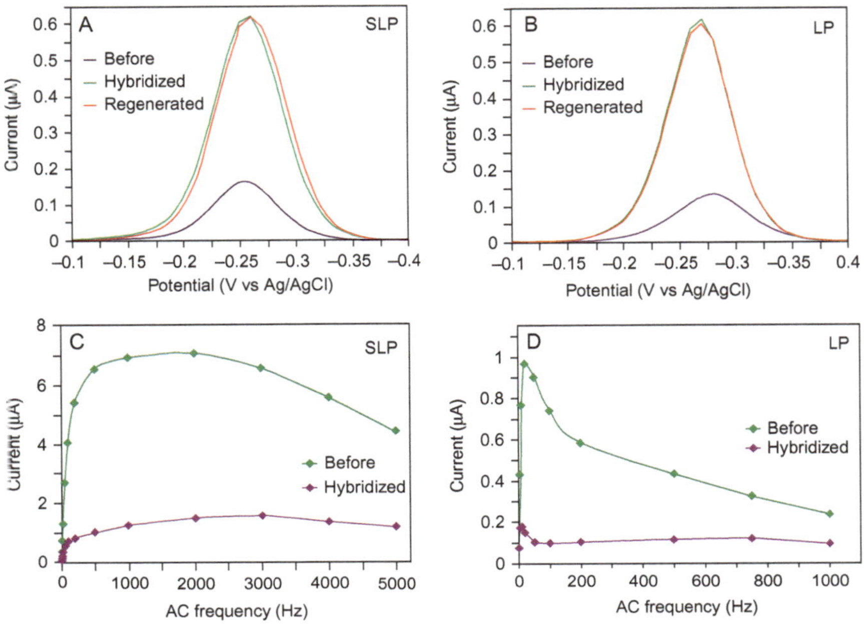

Fig. 3 AC voltammograms of the SLP (A) and LP (B) sensors collected at 10 Hz before and after hybridization with 1.0 μ*M* full complementary target (WT-Gly). The effect of AC frequency on the MB reduction current for both SLP (C) and LP (D) sensors before and after hybridization with 1.0 μ*M* WT-Gly. All data were collected in a Phys2 buffer at room temperature.

experimental conditions (e.g., pH, buffer ionic strength, etc.). In the presence of the full complementary target DNA, a large suppression was registered within 60 and 30 min for the SLP and LP sensors, respectively (Lai et al., 2013; Yang & Lai, 2011a, 2011b).These results are consistent with the design, in which faster binding kinetics is expected for the LP sensor since no unraveling of the hairpin structure is required. These results were obtained from sensors with a probe coverage of $\sim 3 \times 10^{12}$ molecules cm^{-2}. Shorter signal saturation time can be achieved with sensors with a lower probe coverage, albeit with a much smaller %SS. Both sensors were successfully regenerated using a 30-s room temperature deionized (DI) water rinse (Lai et al., 2013; Yang & Lai, 2011a, 2011b) and were reused for no fewer than three times (data not shown).

The large reduction in the MB redox current is attributed to the decrease in flexibility of the probe–target duplex. The LP sensor showed slightly higher %SS than the SLP sensor when interrogated at a lower AC frequency (e.g., 10 Hz). However, at higher AC frequencies, the SLP sensor showed significantly higher %SS than the LP sensor (Fig. 3C and D) (Lai et al., 2013; Yang & Lai, 2011a, 2011b). The difference in sensor behavior between these two "signal-off" E-DNA sensors is expected and can be explained (Creager & Wooster, 1998; Sumner & Creager, 2001; Sumner, Weber, Hockett, & Creager, 2000). For this class of sensors, the MB peak current obtained both before and after target hybridization is highly dependent on the applied AC frequency. For a surface-confined reversible redox system, the peak current should be proportional to the frequency when the frequency is sufficiently lower than the electron transfer rate of the redox label. The peak current, however, decreases at higher AC frequencies. This phenomenon is best explained in terms of the relative time scale of interfacial electron transfer and the potential fluctuation in ACV; as the applied frequency approaches a critical value above which electron transfer can no longer keep up with the rapidly oscillating potential, the peak current of the redox label diminishes relative to the background current (Creager & Wooster, 1998; Sumner & Creager, 2001; Sumner et al., 2000). This "critical" frequency is often an intrinsic characteristic of a specific sensor design. For the SLP sensor, the MB current increased with frequency till the "critical" value of ~2000 Hz. A large decrease in the MB current was evident at frequencies beyond 3000 Hz (Fig. 3C). In comparison, the LP sensor showed a drastically different prehybridization frequency-dependent profile (Fig. 3D). The MB current increased rapidly with frequency, reaching a maximum value at ~20 Hz, followed by a sharp decrease at higher frequencies. The noted difference in the prehybridization "critical" frequencies provides evidence in support of the sensors' signaling mechanism (Lai et al., 2013; Yang & Lai, 2011a, 2011b). The SLP sensor, owing to the SLP structure and the position of the redox label, can support electron transfer at extremely high frequencies; the dynamics-based LP sensor, however, cannot. The posthybridization current responses for both sensors were less frequency dependent; the peak currents were rather smaller throughout the AC frequency range used in the study.

Since the %SS is determined by comparing the pre- and post-hybridization redox currents, it should also be dependent on the applied AC frequency. For the SLP sensor, a sharp increase in %SS was observed between 1 and 200 Hz; only a slight decrease in SS was evident at higher AC frequencies. The LP sensor, however, exhibited a different frequency-dependence profile. A drastic increase in %SS was observed between 1 and 50 Hz, whereas a rapid decrease in %SS was evident at frequencies beyond 100 Hz (Lai et al., 2013; Yang & Lai, 2011a, 2011b). These results not only support the signaling mechanism of the two sensors but also highlight the practical differences between them when interrogated using ACV under specific experimental conditions. As can be seen, the SLP sensor has a much wider usable frequency range when compared to the LP sensor. Thus, depending on the design of the sensor and the experimental conditions, a specific AC frequency should be employed so as to achieve the best sensor response (e.g., %SS).

2.2.2 Differential Pulse Voltammetry

Although ACV has shown itself to be a good sensor interrogation technique, both E-DNA sensors essentially behaved as "signal-off" sensors independent of the applied AC frequency (Lai et al., 2013; Yang & Lai, 2011a, 2011b). In general, "signal-off" sensors are less desirable given that the maximum achievable SS is 100% even under optimal experimental conditions. They are more susceptible to false-positive results since the decrease in the redox current may be attributed to events not related to target binding. For example, monolayer degradation and nonspecific adsorption of matrix contaminants can potentially contribute to the change in the redox current. There are several approaches to convert inherently "signal-off" sensors to "signal-on" sensors. The more complex approach is to redesign the sensors; this aspect will be addressed in later part of this chapter. A more convenient approach is to identify an electrochemical technique that is capable of converting "signal-off" sensors to "signal-on" sensors when different parameters are used. Both SWV and DPV have recently been demonstrated achievement of this rather challenging goal. For SWV, both sensors can be converted into "signal-on" sensors by simply changing the applied frequency (White & Plaxco, 2010), whereas for DPV, the signaling motif can be altered by adjusting the pulse width used to collect the data (Farjami et al., 2011; Lai et al., 2013).

This pulse width-dependent behavior in DPV was found to be generalizable and applicable to both SLP and LP sensors. For example, with a short pulse width such as 1 ms, both SLP and LP sensors behaved as "signal-off" sensors (Fig. 4). The SLP and LP sensors showed 75%SS and 80%SS, respectively, when interrogated with 1.0 μM full complementary target DNA. These values are similar to those obtained using ACV at optimal frequencies. However, when a higher pulse width (e.g., 100 ms) was used, both sensors showed "signal-on" behavior (Lai et al., 2013). For example, ~700%SE (percent signal enhancement) was observed with the SLP sensor at a pulse

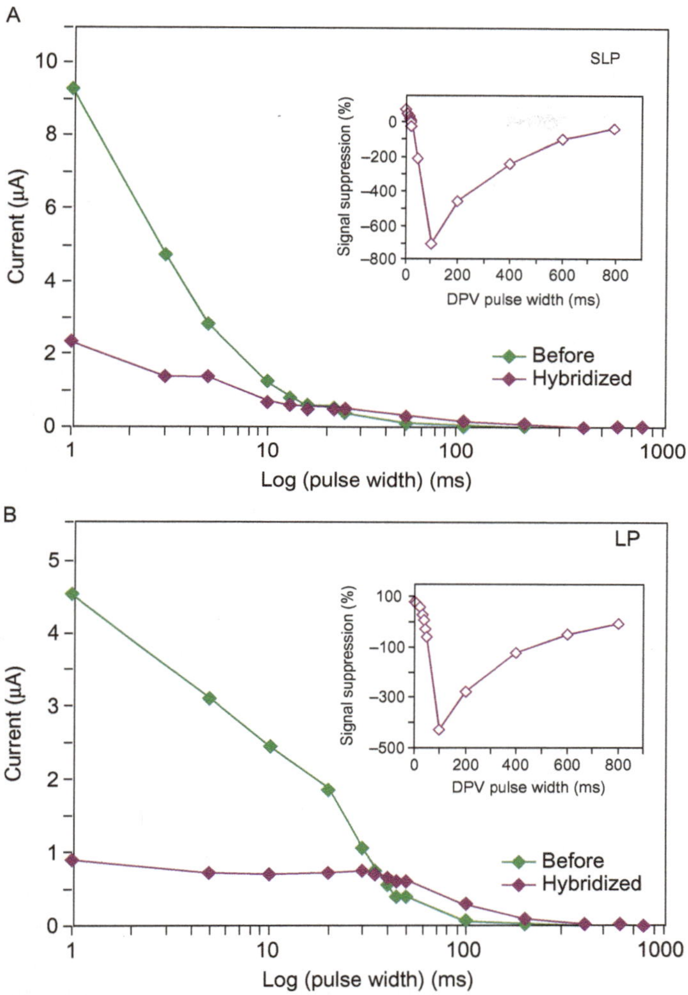

Fig. 4 The effect of DPV pulse width on the MB reduction current for both SLP (A) and LP (B) sensors before and after hybridization with 1.0 μ*M* full complementary target (WT-Gly). Insets: the effect of DPV pulse width on %SS upon target hybridization for both sensors. All data were collected in Phys2 at room temperature.

width of 100 ms, whereas ~430%SE was evident with the LP sensor at the same pulse width. The pulse width where the "switching" occurs was found to be highly dependent on the sensor design and electron transfer kinetics. The "switching" pulse width was ~22 and ~37 ms for the SLP and LP sensors, respectively. The %SE (or −%SS) was substantially higher than the %SS obtained at shorter pulse widths (Fig. 3 insets). Maximum %SE was obtained at 100 ms for both SLP and LP sensors.

These results agree well with the working principles of DPV, a pulse technique that exploits the difference between the rate of the decay of the non-Faradaic and Faradaic currents following a potential step. For the folding- and dynamics-based E-DNA sensors, the rate of current decay is dependent on the electron transfer kinetics. For example, the higher electron transfer rate constant ($k_s = \sim 340\ s^{-1}$) results in a faster decay incurrent for the SLP sensor. The LP sensor, because of the lower k_s ($\sim 38\ s^{-1}$), results in a slower decay in current (Lai et al., 2013; Yang & Lai, 2011a, 2011b). The significantly reduced electron transfer kinetics results in an even slower decay in the MB current after target hybridization. Thus, at some pulse widths, the MB labels on the rigid probe–target duplexes produce more current when compared to the MB labels on either the target-free stem–loop or linear probes. These results highlight the advantages in employing DPV as the sensor interrogation technique, specifically, its versatility. Both SLP and LP sensors can operate as "signal-off" sensors at short pulse widths and "signal-on" sensors at longer pulse widths. Depending on the specific experimental conditions, DPV could be a more attractive sensor interrogation technique than ACV.

3. "SIGNAL-ON" E-DNA SENSORS

3.1 Single-Probe "Signal-On" E-DNA Sensors

3.1.1 Sensor Design and Signaling Mechanism

Although "signal-off" sensors can be readily converted into "signal-on" sensors by simply changing the applied frequency in SWV or pulse width in DPV, there are merits in designing inherently "signal-on" sensors. Since 2003, many research groups have succeeded in developing "signal-on" E-DNA sensors, including one that is based on "inverting" the sensing mechanism of the "signal-off" SLP sensor (Immoos et al., 2004; Wu & Lai, 2013, 2014). This design is the simplest of all "signal-on" E-DNA sensors reported to date (Fig. 5A). The single-probe "signal-on" E-DNA sensor consists of an unstructured MB-labeled DNA probe with a 9-base target recognition region at the 3′-end and an 8-base target recognition region at the 5′-end (Wu & Lai, 2013, 2014). The two regions are connected through an oligothymine (oligo-T) or oligoadenine (oligo-A) spacer. The two target recognition regions are designed to be fully complementary to a 17-base target sequence. In the absence of the target, the unstructured DNA probe is highly flexible; electron transfer between MB and the electrode is thus less efficient. Target hybridization to both recognition regions promotes formation of the "closed" structure, which confines MB close to the electrode surface for efficient electron transfer, resulting in a large increase in the redox current (Wu & Lai, 2013, 2014). Six different DNA probes, each with a different spacer (e.g., A6, A10, A14, T6, T10, T14), were used in the study. The addition of a flexible spacer is deemed necessary, since without it, the formation of the "closed" structure might not be possible due to steric hindrance. It is worth noting that a similar sensor design that employed a poly(ethylene glycol) (PEG) spacer was reported previously (Immoos et al., 2004).

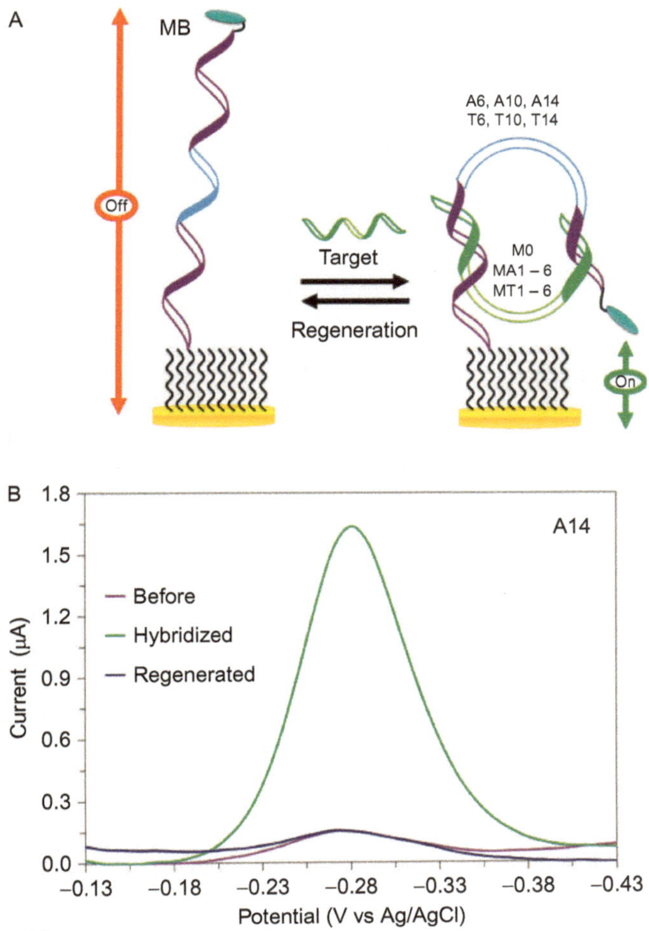

Fig. 5—Cont'd

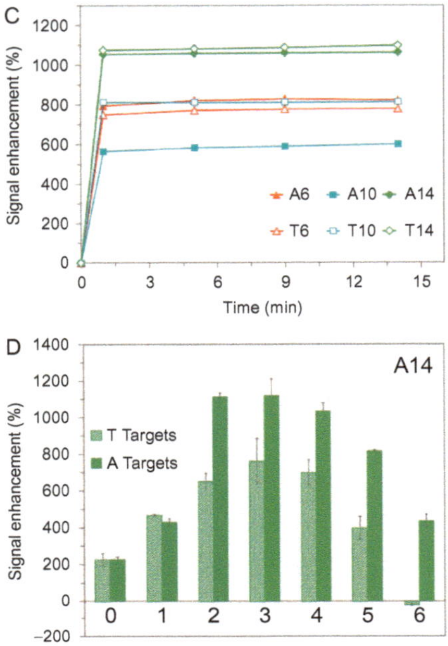

Fig. 5 Sensor design and signaling mechanism of the single-probe "signal-on" E-DNA sensor (A). AC voltammograms of the A14 sensor in the absence, presence of 1 μ*M* full complementary target (MA3), and after sensor regeneration (B). Binding kinetics of the six E-DNA sensors in the presence of 1 μ*M* MA3 (C). %SE recorded for the A14 sensor in the presence of all 13 target sequences (M0, MA1–6, MT1–6). The number indicates the number of bases in the spacer. The concentration of each target was 1 μ*M*.

However, the use of DNA bases as the spacer has several advantages: the tunable design allows better control of the spacer length and flexibility. It is also less synthetically challenging when compared to incorporating PEG into the DNA probe sequence. Steric hindrance to forming the "closed" structure could be alleviated by designing targets with a spacer that does not participate in the hybridization event. A total of 13 target sequences were analyzed in this specific study, with the goal of understanding the effect of spacer length on sensor performance. The design is simple yet versatile and can be tailored for detection of essentially any nucleic acid sequence.

3.1.2 Sensor Response

ACV was used to evaluate all six single-probe "signal-on" E-DNA sensors. Responses of the sensor fabricated with the DNA probe having a 14 adenine (A14) spacer sequence are shown in Fig. 5B (Wu & Lai, 2014). In the absence of the target DNA, a small yet well-defined MB peak was observed at ∼−0.27 V (vs Ag/AgCl), a potential consistent with the reduction potential of MB. When interrogated with the target sequence with a 3 thymine (MA3) spacer sequence, a large increase in the MB current was evident, suggesting the formation of the presumed "closed" structure. The sensor was successfully regenerated via a 30-s room temperature DI water rinse. The sensor regeneration efficiency was close to 100%, as evidenced by the overlapping AC voltammograms shown in Fig. 5B. Among the six sensors, the T14 and A14 sensors showed the highest %SE (∼1250%), whereas the A10 sensor showed the lowest %SE (∼650%) (Wu & Lai, 2014). Overall, DNA probes with a longer spacer appear to be better suited for this sensor design. Despite the differences in %SE, all six sensors showed fast binding kinetics: signal saturation was achieved in ∼60 s (Fig. 5C). This response time is faster than most sensors of this class, including the SLP and LP E-DNA sensors, which require 30–60 min to reach signal saturation. This improvement is likely due to the slightly lower probe coverage, as well as the enhanced probe and target flexibility conferred by the incorporation of a short spacer.

All six sensors' responses to 12 other targets were also analyzed (Fig. 5D). For the T6, T10, and T14 sensors, hybridization to MT3 resulted in the highest %SE, followed by MT2 (Wu & Lai, 2014). The lowest %SE was observed with M0, followed by MT6. These results are not unexpected since hybridization to the target without a spacer is unfavorable because of both steric and electrostatic interferences. Targets with a very long spacer are not suitable either: although the extra bases in the spacer should not affect the hybridization between the 3′ end of the target sequence and the distal end of the probe, it could affect the binding between the 5′ end of the target sequence and the proximal end of the probe. Both interactions are necessary for the formation of the "closed" structure, which gives rise to the increase in the MB current (Fig. 5A). It is worth mentioning that most sensors responded better to the targets with an oligo-A spacer instead of an oligo-T spacer. The responses of the A14 sensor to the 13 targets are shown in Fig. 5D. The sensors' preference for the A targets arises from the fundamental differences between oligo-T and oligo-A. Since T is a smaller base, oligo-T should be relatively more flexible than oligo-A. Rigidity of oligo-T is mostly due to steric restrictions on the backbone (Mills, Vacano, & Hagerman, 1999; Sim, Lipfert, Herschlag, & Doniach, 2012). Unlike oligo-T, oligo-A can form an extensively stacked structure at low temperature (Inners & Felsenfeld, 1970). According to a previous study, even at room temperature two-thirds of the A bases are stacked (Inners & Felsenfeld, 1970). The stacking of the inserted A bases could, in theory, stabilize the interactions between the target and the surface-immobilized DNA probes. However, further studies are required to verify this hypothesis.

3.1.3 Sensor Specificity

Another merit of this sensor design is its ability to discriminate between full complementary and single-base mismatch targets. While a large %SE was observed when the sensor was interrogated with the full complementary target, a large signal reduction was observed when challenged with MA3-1M, a target with an internal C–C mismatch (Wu & Lai, 2014). This "signal-off" behavior is not unexpected since MA3-1M is capable of hybridizing to the 9-base region at the 3′ end, but the single-base mismatch inhibits binding of the target to 8-base region at the 5′ end, the critical step that leads to the formation of the "closed" structure. Formation of the 9-base duplex at the 3′ end reduces the flexibility of the DNA probe, impeding efficient electron transfer to the MB label, resulting in a large reduction in the current. MA3-2M, a target with two internal C–C mismatches was also used in this study. Hybridization to MA3-2M, however, did not result in a substantial change in the MB current for all six sensors (Wu & Lai, 2014). The presence of the two mismatched bases likely impedes hybridization between the 3′ end of the target sequence and the distal end of the probe. Hybridization between the 5′ end of the target sequence to the proximal end of the probe is still possible; this interaction should have limited effect on the MB current since it is located at the distal end of the probe. Only a minor change in the probe flexibility is expected, nevertheless this change could still result in a slight decrease in the MB current.

3.2 Dual-Probe "Signal-On" E-DNA Sensors

3.2.1 Design and Signaling Mechanism

In addition to the aforementioned single-probe "signal-on" E-DNA sensors, displacement-based "signal-on" E-DNA sensors have also been reported in recent years (Fig. 6) (Yu & Lai, 2012, 2013; Yu et al., 2014). To fabricate either the SLP or LP sensors, an unlabeled capture probe and a signaling probe modified with MB have to be simultaneously immobilized onto a cleaned gold electrode. The DNA-modified electrode is then further passivated with C6-OH. The signaling probe is either a SLP

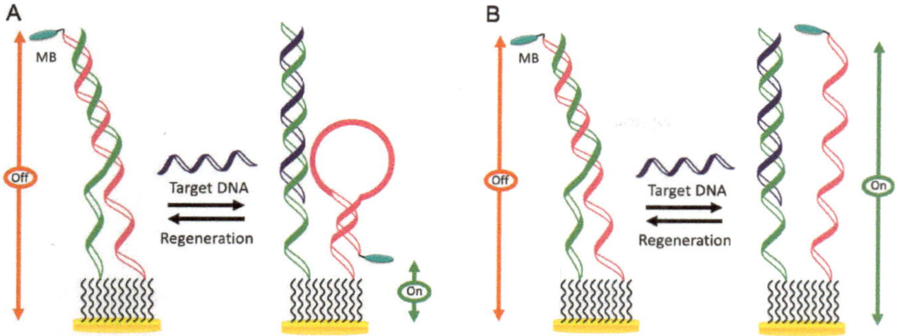

Fig. 6 Design and signaling mechanism of the displacement-based SLP (A) and LP (B) E-DNA sensors.

in which a 15-base loop is flanked by two 5-base self-complementary regions, or a 25-base linear probe. The unlabeled capture probe is designed to possess a 17-base sequence that binds specifically to the target DNA. Eight additional T(s) are added to the 3′ end of capture probe to improve probe flexibility. The surface-immobilized capture probe is designed to be partially complementary to the signaling probe, which allows the formation of a 10-base DNA duplex at the distal end of the probes. For the SLP sensor, formation of this interprobe duplex is thermodynamically more favorable than assuming the intraprobe hairpin structure (Fig. 6A).

3.2.2 Sensor Response

Like most sensors of this class, ACV was used to characterize the sensors. For the SLP sensor, a small yet well-defined MB peak was observed at ~−0.26 V (vs Ag/AgCl), indicating successful immobilization of the probes (Fig. 7A) (Yu & Lai, 2013). Hybridization with the full complementary target DNA (e.g., PM-17) resulted in a large increase in the MB current. These results agree well with the proposed sensing mechanism. In the absence of the target, the 10-base DNA duplex prevents the signaling probe from accessing the electrode surface for efficient electron transfer; formation of the 17-base capture probe/target duplex, however, disrupts the 10-base duplex, liberating the signaling probe and allowing it to assume a stem–loop conformation. Like other sensors of this class, the change in probe conformation and/or flexibility often parallels the change in the electron transfer kinetics and the redox current. The increase in the MB current is likely due to the formation of the stem–loop structure, which confines the MB label closer to the electrode surface for efficient electron transfer. As for the LP sensor, the

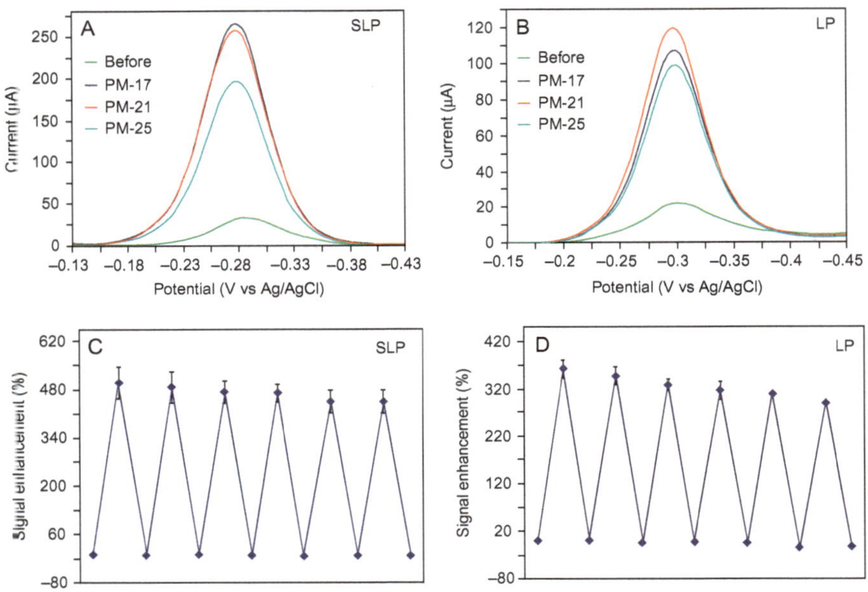

Fig. 7 AC voltammograms of the SLP (A) and LP (B) sensors in the absence and presence of 1 μ*M* PM-17, PM-21, and PM-25. Sensor interrogation and regeneration curves for the SLP (C) and LP (D) sensors. These results were obtained in 50% blood serum and the concentration of the target DNA (PM-17) was 1 μ*M*.

signaling probe regains flexibility after being released from the 10-base interprobe duplex, leading to an increase in the MB current (Fig.7B) (Yu & Lai, 2013).

In addition to PM-17, two longer DNA targets (i.e., PM-21 (21-mer), PM-25 (25-mer)) were used in the study, with the goal of understanding the effect of target length on sensor performance. For the SLP sensor, the %SE was found to be much lower for PM-25; these results suggest that the addition of A(s) to lengthen the target sequence not only does not enhance target binding, but presumably induces steric hindrance, which could inhibit the signaling probe from assuming the stem–loop structure. This effect was not observed with the LP sensor: the %SE was less dependent on the length of the target. It is worth noting that all three interrogations were performed using the same sensor. The sensor was regenerated via a simple 30-s DI water rinse. The sensor regeneration efficiency was close to 100%, as evidenced by the overlapping AC voltammograms shown in Fig. 7A and B (Yu & Lai, 2013). The sensors were not only regenerable after being used in a buffer;

they were reusable even after being exposed to a complex medium such as blood serum. Shown in Fig. 7C and D are the sensor reusability plots obtained for both SLP and LP sensors in 50% serum. The presence of matrix contaminants did not appear to affect the reforming of the 10-base interprobe duplex on the sensor surface. The sensors were stable in 50% serum for over 15 h.

3.2.3 Sensor Specificity

Similar to other displacement-based sensors, these two dual-probe sensors should possess better sensor specificity than conventional E-DNA sensors. Since PM-17 exhibited the best sensor performance, it was used to study the sensors' DNA sequence mismatch discrimination capability. Both the SLP and LP sensors were interrogated with PM-17 and two different mismatched targets (i.e., 17-1M (single-base mismatch) and 17-2M (two-base mismatch)). The four probes used in the study contain either a T3 or T5 linker between the last base of the probe and the MB redox label. It is advantageous to use DNA bases such as T(s) to increase the tether length; both tether length and flexibility can be easily adjusted by simply adding extra bases at the end of the probe sequence, no major synthetic challenges are posed here (Yu et al., 2014). The tuning of the tether length by incorporating extra T(s) or A(s) in the DNA probe sequence to alter sensor performance is widely used and can potentially be used in the design of other E-DNA sensors.

All four sensors (SLP-T3, SLP-T5, LP-T3, and LP-T5) were capable of discriminating between PM-17, 17-1M, and 17-2M in 50% serum. Further experiments were conducted in 100% serum; other than the LP-T3 sensor, the remaining three sensors showed good sensor specificity even in such a complex medium. Both SLP-T5 and LP-T5 sensors showed minimal response to 17-1M and no response to 17-2M (Fig. 8) (Yu et al., 2014). These results are noteworthy; this level of sensor specificity has not been reported for any E-DNA sensor employed in undiluted blood serum. Among the four sensors, the SLP-T5 sensor was found to be the most specific, independent of the interrogation medium. The ability of the SLP sensors to function optimally in 100% serum, especially when compared to the LP sensors, is related to the sensor architecture and signaling mechanism. It is likely that the enhanced probe flexibility enables the MB label to be minimally obstructed by the matrix contaminants adsorbed onto the sensor surface. Furthermore, the added flexibility could promote MB intercalation

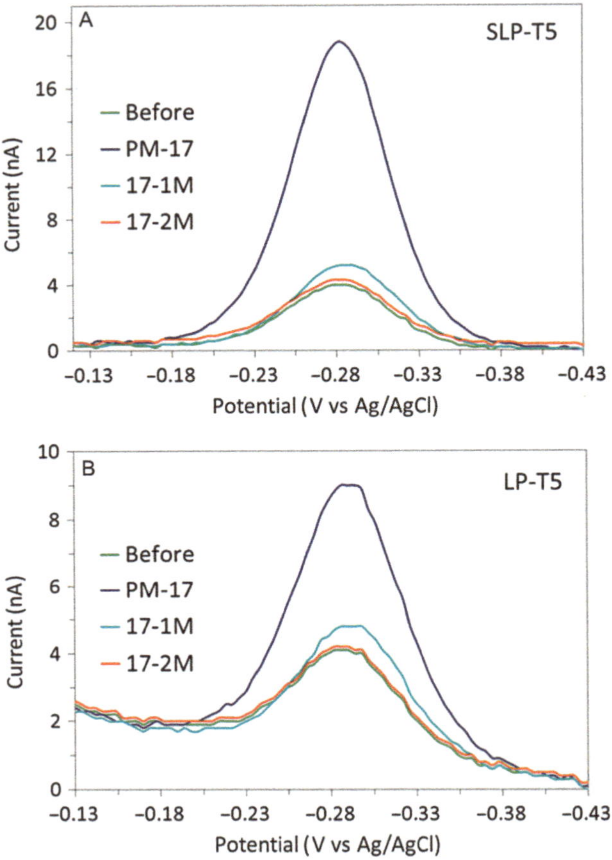

Fig. 8 AC voltammograms of the SLP-T5 (A) and LP-T5 (B) sensors in the absence and presence of 1 μ*M* PM-17, 17-1M, and 17-2M. These results were obtained at room temperature in 100% blood serum.

into the 5-base duplex, stabilizing the stem–loop structure, which could be essential in maintaining good sensor function in a complex medium (Hossain, Giri, & Kumar, 2008).

4. ALTERNATE ELECTRODE SUBSTRATES

4.1 Gold-Plated Screen-Printed Carbon Electrodes

Conventionally sized polycrystalline gold electrodes are the most commonly used electrode substrate for all folding- and dynamics-based E-DNA sensors (Lai et al., 2013; Wu & Lai, 2014; Yang & Lai, 2011a, 2011b; Yu et al., 2014).

The main advantages of this electrode substrate are ruggedness, cost-effectiveness, and user-friendliness; it is ideal for use in the early stages of sensor characterization. However, depending on the specific application, there are needs for other electrode substrates. More recently, many E-DNA sensors have been fabricated on thin film gold electrodes, while ideal for use in all-in-one microfluidic devices, they are, in general, less cost-effective (Lai, Lee, et al., 2006; Pavlovic et al., 2008; Salamifar & Lai, 2014a, 2014b). An equally viable alternative electrode substrate for the E-DNA sensors is the gold-plated screen-printed carbon electrode (SPE) developed by Yang, Gerasimov, and Lai (2009) (Fig. 9A). This cost-effective paper-based substrate is well suited for real world analysis; the sample volume required, when used with a custom-designed electrochemical cell, is minimal (~10 μL) (Fig. 9B). Since its initial development, it has been used in the fabrication of other sensors, including electrochemical aptamer-based sensors and electrochemical metal ion sensors (Guerreiro, Zaitouna, & Lai, 2014; Korshoj, Zaitouna, & Lai, 2015; Zhao, Yang, & Lai, 2011).

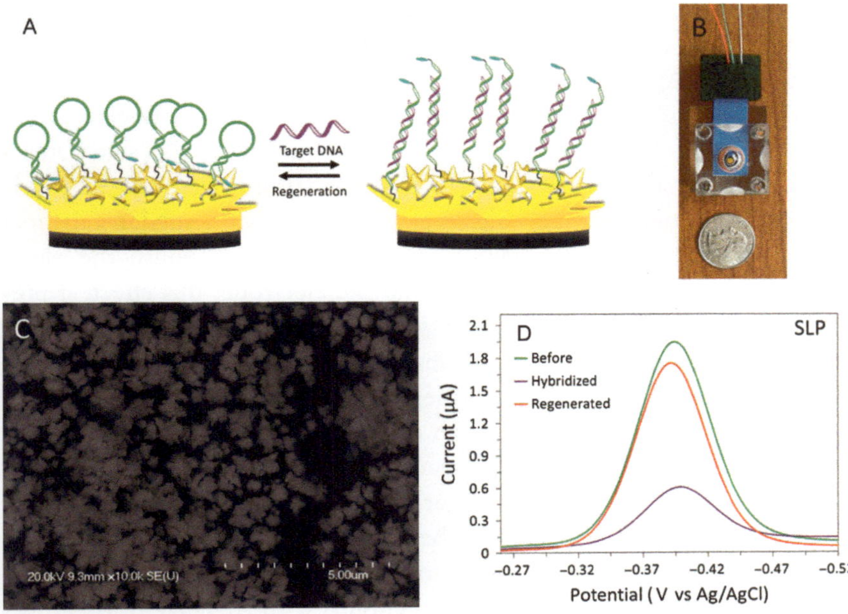

Fig. 9 Design and signaling mechanism of the SLP sensor fabricated on a gold-plated SPE (A). The custom-designed electrochemical cell used with the gold-plated SPEs (B). A representative SEM image of the electrodeposited gold (20 min) on an SPE (C). AC voltammograms of the SLP sensor fabricated on a gold-plated SPE in the absence, presence of 1.0 μ*M* full complementary target (WT-Gly), and after sensor regeneration (D).

A wide range of experimental conditions were employed to identify the most convenient yet effective approach to deposit gold on the SPEs. Parameters evaluated include the concentration of gold, electrochemical deposition method, and deposition time. Among the electrochemical deposition methods tested, chronoamperometry (CA) was found to be ideal for this application (Yang et al., 2009). A microstructured and/or nanostructured gold film was successfully formed by applying −0.4 V (vs Ag/AgCl reference electrode) to the SPE for 20 min in a solution containing 1.2 mg mL^{-1} gold chloride. The scanning electron microscopy (SEM) image of the electrodeposited gold is shown in Fig. 9C. The roughness factor (ρ), which is the ratio between the real surface area and the geometric surface area, was ~4 for the 20-min electrodeposited gold film. The electrodeposited gold adhered relatively well to the SPE; only minimal removal of the gold clusters was evident during the sensor fabrication process. Shorter and longer deposition time did not result in gold films suitable as substrates for the E-DNA sensor.

Similar to the aforementioned E-DNA sensors, ACV was first used to characterize the SLP E-DNA sensor fabricated on a gold-plated SPE (Fig. 9D). A well-defined MB peak was observed at ~−0.34 V; the potential is more negative because the reference electrode employed was the on-chip, screen-printed Ag reference electrode. Despite the difference in the peak location, the sensor responded well to the full complementary target, as indicated by the large %SS (~72%) (Yang et al., 2009). This level of signal attenuation is similar to that obtained on a polycrystalline gold disk electrode (Fig. 3A). Although some DNA probes were probably oriented in a manner where hybridization to the target cannot completely abolish the electron transfer between the redox label and the electrode, this effect, if present, did not drastically alter the sensor behavior. The DNA probe density on this alternate electrode substrate was found to be similar as well ($\sim 3 \times 10^{12}$ molecules cm^{-2}), further supporting the conclusion that E-DNA sensors are compatible with electrode substrates or materials that are microstructured and/or nanostructured. The sensor was regenerated via a 30-s DI water rinse; the sensor regeneration efficiency was slightly lower (~90% instead of ~100%) when compared to the sensor fabricated on a gold disk electrode. This change in sensor regeneration efficiency is presumably attributable to the loss of gold clusters with attached DNA probes during the rinsing process. This loss in DNA probes could be minimized by using other sensor regeneration approaches that do not require direct rinsing of the gold surface (e.g., incubate the sensor in 4 *M* guanidine hydrochloride (GHCl) for 4 min).

The SLP sensor fabricated on this alternate electrode substrate showed good response to the target even when used in 100% serum; the %SS was lower (~64%) but quite reproducible (Yang et al., 2009). The regenerated sensor was reusable as well. Overall, these results highlight the versatility of E-DNA sensors: they are compatible with electrode substrates with a roughness factor as high as 7. With further characterization and optimization, these paper-based E-DNA sensors could find applications in point-of-care medical diagnosis.

4.2 Gold-Plated Recessed Platinum Nanoelectrodes

In addition to gold-plated SPEs, there are other electrode substrates compatible with the folding- and dynamics-based electrochemical sensing platform. Salamifar and Lai recently reported the use of gold-modified platinum (Pt) nanoelectrodes with both "signal-off" SLP and LP E-DNA sensors (Salamifar & Lai, 2014a, 2014b). One of the main advantages of these nanosensors is that they can be used to analyze specific target analytes within the vicinity of cells or even inside the cells if the size of the electrode is sufficiently small (Cox & Zhang, 2012). The E-DNA sensing platform is well suited for this application since it is reagentless and does not require the addition of exogenous reagents to generate the sensor signal. It is also selective enough to be employed in various growth media and inside a living cell.

Gold nanoelectrodes are, in general, extremely challenging to fabricate using a laser puller due to the large difference between the melting points of gold (1064°C) and quartz (1710°C). Thus, laser-pulled Pt nanoelectrodes were used in the study instead. The steps involved in fabricating an E-DNA sensor on a gold-modified Pt nanoelectrode are shown in Fig. 10A (Salamifar & Lai, 2014a, 2014b). In brief, an optimized pulling technique was used to produce the Pt nanoelectrode; the electrode was then mechanically polished using the beveller to expose the Pt wire embedded in the quartz capillary. Next, part of the Pt wire was electrochemically etched to create a recessed electrode; gold was subsequently electrodeposited into the nanopore and used as the substrate for the E-DNA sensor. Both SLP and LP sensors were used in this study, the sensor fabrication procedure was similar to that used with other E-DNA sensors presented in this chapter (Lai et al., 2013; Yang & Lai, 2011a, 2011b). In brief, the thiolated and MB-modified DNA probe was first immobilized on the electrode surface, followed by sensor passivation with C6-OH.

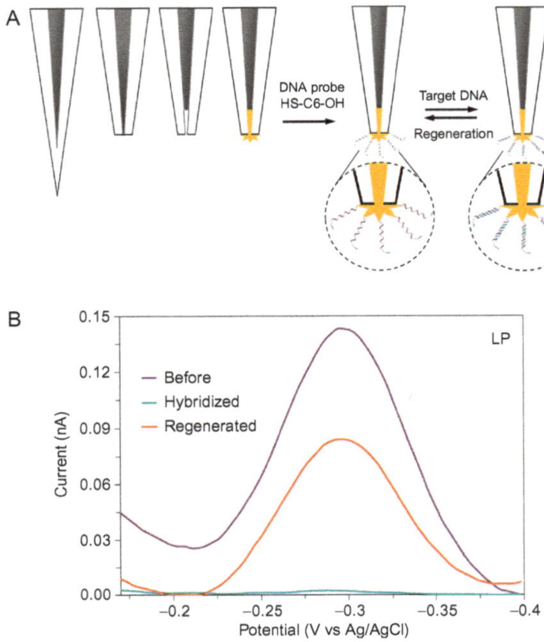

Fig. 10 Steps used in the fabrication of the LP E-DNA sensor on a Pt nanoelectrode. Step 1: fabrication of the electrode using the laser puller; Step 2: electrode polishing using the beveller; Step 3: electrochemical etching of the electrode; Step 4: deposition of gold into the nanopore; Step 5: immobilization of LP DNA probes and passivating diluents onto the gold surface; Step 6: hybridization of the DNA probes with the full complementary target and sensor regeneration (A). AC voltammograms of the LP E-DNA sensor in the absence, presence of 2 μ*M* full complementary target, and after sensor regeneration with 4 M GHCl (B).

The main obstacle in fabricating the nanosensors in a reproducible manner was the amount and quality of electrodeposited gold. Both CV and CA were employed in the gold deposition step, while both techniques resulted in a nanostructured gold film, the structure and porosity of the film were presumed to be different. The gold film produced from CV did not adhere well to the recessed Pt electrode, precluding it as a viable electrode substrate for the E-DNA sensors. Nanostructured gold films produced by applying 0.0 V (vs Ag/AgCl) for 300 s were deemed optimal and this CA method was thus used for the rest of the study (Salamifar & Lai, 2014a, 2014b). The highly porous gold films were found to be suitable substrates for E-DNA sensors.

For both LP and SLP sensors, a small MB peak was observed in the ACVs after the sensor fabrication process, indicating successful immobilization of MB-modified DNA probes. Hybridization to 2 μ*M* full complementary target resulted in a large reduction in the MB current for both sensors. The % SS for the LP sensor, however, was higher than that for the SLP sensor. The lower %SS seen with SLP sensor could be due to incomplete hybridization; it has been hypothesized that not all immobilized stem–loop DNA probes are "hybridizable" (Lai et al., 2013; Yang & Lai, 2011a, 2011b). Another noted difference between the SLP and LP sensors was sensor regenerability; the SLP sensor could not be regenerated via a 4-s incubation in 4 *M* GHCl (data not shown). Other regeneration approaches were employed, but none were successful. It is possible that the nanostructures on the electrode surface obstruct the DNA probes from resuming the original stem–loop conformation even after removal of the target DNA. Unlike the SLP sensor, the regeneration of the LP sensor was partially successful: ~73% of the initial signal was recovered (Fig. 10B). The change in the MB current could be, in part, due to the loss of the electrodeposited gold nanostructures. Despite the difference in the overall sensor performance when compared to the sensors fabricated on conventional gold disk electrodes, these nanosensors can potentially be used for real-time detection of target analytes outside or inside living cells.

5. ALTERNATE SENSOR FABRICATION APPROACHES

5.1 Copper-Catalyzed Azide Alkyne Cycloaddition

The "click" reaction concept introduced by Barry Sharpless in 2001 has stimulated the development of a wide range of efficient, high yield reactions that proceed rapidly under mild conditions (Kolb, Finn, & Sharpless, 2001; McKay & Finn, 2014). The copper-catalyzed azide alkyne cycloaddition is one of the most well-characterized reactions; it is extremely efficient and has demonstrated unprecedented chemoselectivity under experimental conditions compatible with biosensor fabrication. This reaction has been used in the fabrication of various biosensors, including E-DNA sensors. One of the major advantages of this sensor fabrication approach is versatility: a SAM that carries the active azide terminal group can be first immobilized or printed on the substrate electrode, and subsequently modified with the specific alkyne-derivatized DNA probe using either the conventional or potential-assisted "click" chemistry approaches (Fig. 11A) (Cañete & Lai, 2010; Cañete, Yang, & Lai, 2009; Collman, Devaraj, & Chidsey, 2004; Devaraj et al., 2005). This versatile sensor fabrication approach can also be used with other folding- and dynamics-based electrochemical sensors such as the electrochemical peptide-based HIV sensor (Gerasimov & Lai, 2011).

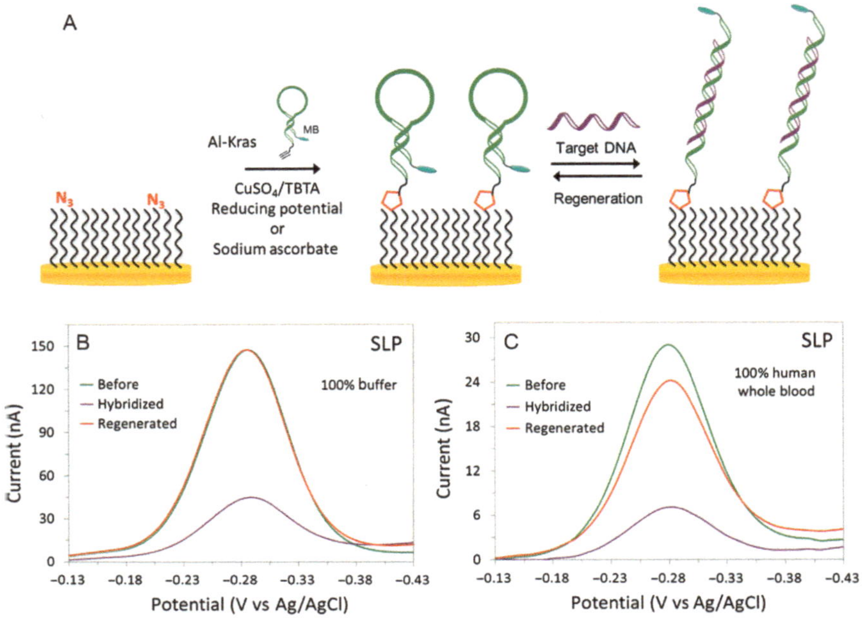

Fig. 11 Steps involved in the fabrication of an E-DNA sensor using Sharpless "click" chemistry (A). AC voltammograms of the SLP sensor in the absence, presence of 1 μ*M* full complementary target, and after sensor regeneration (B and C). The data shown in Panel B were obtained in 100% Phys2 buffer. The data shown in Panel C were obtained in undiluted human whole blood.

5.1.1 Chemical "Click" Reaction

To fabricate the first "click"-based E-DNA sensor, a mixed SAM consisted of 1-azidoundecan-11-thiol (C11-N_3) and 11-mercapto-1-undecanol (C11-OH) was formed on a cleaned gold disk electrode. The alkyne-modified DNA probes were subsequently conjugated to the azide-containing SAM via a short incubation in a solution containing copper sulfate, sodium ascorbate, tris-(benzyltriazolylmethyl)amine (TBTA), and the DNA probes (Cañete et al., 2009). The DNA probe employed in this study was the same stem–loop DNA probe used in previous studies, but the 5′ end of the probe was modified with a C3-alkyne instead of a C6-thiol linker. The resultant sensor responded well to the full complementary target, with a %SS similar to that seen with a sensor fabricated using the conventional 2-step approach (Fig. 11B). More importantly, the sensor performed optimally when interrogated in undiluted human whole blood (Fig. 11C) (Cañete et al., 2009). The sensor was found to be regenerable and reusable even after target hybridization in whole blood. These results show that there are no major differences between the sensors fabricated using this chemical "click" reaction and the more commonly employed approach. This sensor fabrication method could find applications in the manufacturing of large E-DNA sensor arrays for simultaneous detection of diagnostically relevant DNA and RNA sequences.

5.1.2 Potential-Assisted "Click" Reaction

While the chemical "click" reaction is ideal for fabrication of a large sensor array, the potential-assisted "click" reaction is well suited for site-specific immobilization of biorecognition probes on a smaller sensor array (Cañete & Lai, 2010). Prior to sensor fabrication, CV was first used to assess the potential required to electrochemically generate the Cu(I) catalyst across the C11-N_3/C11-OH monolayer in the presence of TBTA. The initial CV results were encouraging: a small, irreversible reduction wave was observed, suggesting the reduction of Cu(II) to Cu(I). The optimal potential for the "click" reaction was later found to be −0.5 V (vs Ag/AgCl); a more negative potential was not used to prevent reductive desorption of the C11-N_3/C11-OH SAM.

To fabricate the 3-pixel sensor array, −0.5 V was applied to Pixel 1 for 30 min, while keeping the other two pixels at open circuit potential (OCP) (i.e., disconnected from the potentiostat). The same procedure was repeated two more times; the resultant 3-pixel E-DNA sensor array was then systematically characterized using ACV (Fig. 12A). Pixel 1 was modified with the *Brac2* probe, whereas Pixels 2 and 3 were modified with the *p53* and *K-ras* probes, respectively (Cañete & Lai, 2010). Sensor interrogation with the full complementary target DNA was performed sequentially to determine possible cross-reactivity (i.e., attachment of DNA probes onto pixels that were held at OCP). In this study, Pixel 3 was the last to be modified with the DNA probe, but no substantial %SS was evident when challenged with target DNA complementary to the other two probes, suggesting minimal cross-reactivity during the functionalization steps (Fig. 12B). All three SLP E-DNA sensors performed well, in that the %SS was not significantly different from the same three sensors fabricated via the conventional 2-step approach. The sensors were regenerated via a 30-s DI water rinse; although the sensor regeneration efficiency was not 100%, the sensors were found to be quite stable within the experimental time frame (e.g., 8 h).

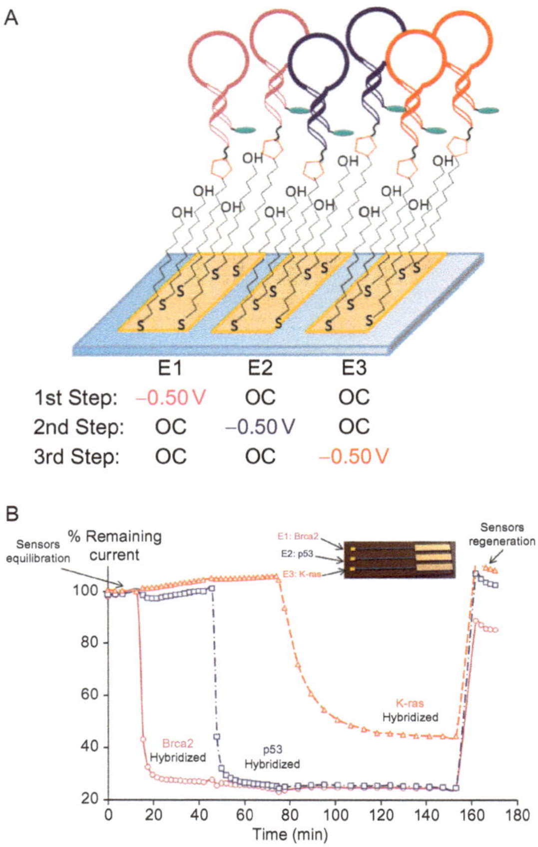

Fig. 12 The steps involved in the fabrication of a 3-pixel E-DNA sensor array using potential-assisted "click" chemistry. A potential of −0.5 V was applied to Pixel 1 for 30 min, Pixels 2 and 3 were disconnected from the potentiostat (open circuit potential or OCP). The same process was repeated twice, but the potential was applied to Pixel 2 and Pixel 3 instead. A 3-pixel sensor array was fabricated after the 90-min process (A). Responses of the three SLP sensors to their corresponding full complementary target DNA as a function of time. These results were obtained in a Phys2 buffer and the concentration of each target DNA was 1 μ*M*. All three sensors were regenerated via a 30-s DI water rinsed at the end of the experiment (B).

6. CONCLUSION AND OUTLOOK

Since its first appearance in 2003, the folding- and dynamics-based E-DNA sensing platform has garnered substantial attention from both academic researchers and industrial scientists. Various types of E-DNA sensors have been developed to date: the two main classes are the "signal-off" and "signal-on" sensors. While many of the newer E-DNA sensors show better overall sensor performance, most are not ready for commercialization. Despite being one of the more promising biosensing platforms developed to date, many aspects of this technology still have to be carefully researched and systematically optimized. Improvements in sensor sensitivity, specificity, and selectivity are needed to enable real world DNA and RNA analysis in complex biological samples. Sensor stability and sensor-to-sensor reproducibility are other aspects that require and warrant more research effort. These challenges provide opportunities for interdisciplinary research teams of chemists, biologists, and engineers to discover innovative solutions that will propel this already promising biosensing platform to a stage where commercialization is within reach. The availability of sensors capable of sensitive, specific, cost-effective, and PCR-free analysis of DNA and RNA could revolutionize pathogen detection and medical diagnosis.

ACKNOWLEDGMENT

This research was supported by the National Science Foundation (CHE-0955439). The author would like to acknowledge the graduate students, postdoctoral researchers, and visiting professors who contributed to the works reviewed in this book chapter.

REFERENCES

Anne, A., Bouchardon, A., & Moiroux, J. (2003). 3′-Ferrocene-labeled oligonucleotide chains end-tethered to gold electrode surfaces: Novel model systems for exploring flexibility of short DNA using cyclic voltammetry. *Journal of the American Chemical Society*, *125*, 1112–1113.

Anne, A., & Demaille, C. (2006). Dynamics of electron transport by elastic bending of short DNA duplexes. Experimental study and quantitative modeling of the cyclic voltammetric behavior of 3′-ferrocenyl DNA end-grafted on gold. *Journal of the American Chemical Society*, *128*, 542–557.

Balintová, J., Pohl, R., Horáková, P., Vidláková, P., Havran, L., Fojta, M., et al. (2011). Anthraquinone as a redox label for DNA: Synthesis, enzymatic incorporation, and electrochemistry of anthraquinone-modified nucleosides, nucleotides, and DNA. *Chemistry—A European Journal*, *17*, 14063–14073.

Cañete, S. J. P., & Lai, R. Y. (2010). Fabrication of an electrochemical DNA sensor array via potential-assisted "click" chemistry. *Chemical Communications*, *46*, 3941–3943.

Cañete, S. J. P., Yang, W., & Lai, R. Y. (2009). Folding-based electrochemical DNA sensor fabricated by "click" chemistry. *Chemical Communications*, *32*, 4835–4837.

Castillo, J., Gaspar, S., Leth, S., Niculescu, M., Mortari, A., Bontidean, I., et al. (2004). Biosensors for life quality design, development and applications. *Sensors and Actuators B: Chemical*, *102*, 179–194.

Chaki, N. K., & Vijayamohanan, K. (2002). Self-assembled monolayers as a tunable platform for biosensor applications. *Biosensors & Bioelectronics*, *17*, 1–12.

Collman, J. P., Devaraj, N. K., & Chidsey, C. E. D. (2004). "Clicking" functionality onto electrode surfaces. *Langmuir*, *20*, 1051–1053.

Cooper, M. A. (2003). Label-free screening of bio-molecular interactions. *Analytical and Bioanalytical Chemistry*, *377*, 834–842.

Cox, J. T., & Zhang, B. (2012). Nanoelectrodes: Recent advances and new directions. *Annual Review of Analytical Chemistry*, *5*, 253–272.

Creager, S. E., & Wooster, T. T. (1998). A new way of using ac voltammetry to study redox kinetics in electroactive monolayers. *Analytical Chemistry*, *70*, 4257–4263.

Devaraj, N. K., Miller, G. P., Ebina, W., Kakaradov, B., Collman, J. P., Kool, E. T., et al. (2005). Chemoselective covalent coupling of oligonucleotide probes to self-assembled monolayers. *Journal of the American Chemical Society*, *127*, 8600–8601.

Drummond, T. G., Hill, M. G., & Barton, J. K. (2003). Electrochemical DNA sensors. *Nature Biotechnology*, *21*, 1192–1199.

Fan, C., Plaxco, K. W., & Heeger, A. J. (2003). Electrochemical interrogation of conformational changes as a reagentless method for sequence specific detection of DNA. *Proceedings of the National Academy of Sciences of the United States of America*, *100*, 9134–9137.

Farjami, E., Clima, L., Gothelf, K., & Ferapontova, E. E. (2011). "Off-on" electrochemical hairpin-DNA-based genosensor for cancer diagnostics. *Analytical Chemistry*, *83*, 1594–1602.

Gerasimov, J. Y., & Lai, R. Y. (2011). Design and characterization of an electrochemical peptide-based sensor fabricated via "click" chemistry. *Chemical Communications*, *2011*(47), 8688–8690.

Gooding, J. J. (2002). Electrochemical DNA hybridization biosensors. *Electroanalysis*, *14*, 1149–1156.

Guerreiro, G. V., Zaitouna, A. J., & Lai, R. Y. (2014). Characterization of an electrochemical mercury sensor using alternating current, cyclic, square wave and differential pulse voltammetry. *Analytica Chimica Acta*, *810*, 79–85.

Homola, J. (2003). Present and future of surface plasmon resonance biosensors. *Analytical and Bioanalytical Chemistry*, *377*, 528–539.

Hossain, M., Giri, P., & Kumar, G. S. (2008). DNA intercalation by quinacrine and methylene blue: A comparative binding and thermodynamic characterization study. *DNA and Cell Biology*, *27*, 81–90.

Immoos, C. E., Lee, S. J., & Grinstaff, M. W. (2004). DNA-PEG-DNA triblock macromolecules for reagentless DNA detection. *Journal of the American Chemical Society*, *126*, 10814–10815.

Inners, L. D., & Felsenfeld, G. J. (1970). Conformation of polyribouridylic acid in solution. *Molecular Biology*, *50*, 373–389.

Kang, D., Zuo, X., Yang, R., Xia, F., Plaxco, K. W., & White, R. J. (2009). Comparing the properties of electrochemical-based DNA sensors employing different redox tags. *Analytical Chemistry*, *81*, 9109–9113.

Kolb, H. C., Finn, M. G., & Sharpless, K. B. (2001). Click chemistry: Diverse chemical function from a few good reactions. *Angewandte Chemie, International Edition*, *40*, 2004–2021.

Korshoj, L. E., Zaitouna, A. J., & Lai, R. Y. (2015). Methylene blue-mediated electrocatalytic detection of hexavalent chromium. *Analytical Chemistry*, *87*, 2560–2564.

Lai, R. Y., Lagally, E. T., Lee, E.-H., Soh, H. T., Plaxco, K. W., & Heeger, A. J. (2006). Rapid, sequence-specific detection of unpurified PCR amplicons via a reusable, electronic sensor. *Proceedings of the National Academy of Sciences*, *103*, 4017–4021.

Lai, R. Y., Lee, S.-H., Soh, H. T., Plaxco, K. W., & Heeger, A. J. (2006). Differential labeling of closely-spaced biosensor electrodes via oxidative desorption. *Langmuir*, *22*, 1932–1936.

Lai, R. Y., Seferos, D. S., Heeger, A. J., Bazan, G. C., & Plaxco, K. W. (2006). A comparison of the signaling and stability of electrochemical DNA sensors fabricated from 6- or 11-carbon self-assembled monolayers. *Langmuir*, *22*, 10796–10800.

Lai, R. Y., Walker, B., Stormberg, K., Zaitouna, A. J., & Yang, W. (2013). Electrochemical techniques for characterization of stem-loop probe and linear probe-based DNA sensors. *Methods*, *64*, 267–275.

Lubin, A. A., Lai, R. Y., Plaxco, K. W., & Heeger, A. J. (2006). Sequence-specific, electronic detection of oligonucleotides in blood, soil and foodstuffs with the reagentless, reusable E-DNA sensor. *Analytical Chemistry*, *78*, 5671–5677.

Lubin, A. A., Vander Stoep Hunt, B., White, R. J., & Plaxco, K. W. (2009). Effects of probe length, probe geometry, and redox-tag placement on the performance of the electrochemical E-DNA sensor. *Analytical Chemistry*, *81*, 2150–2158.

Mao, Y., Luo, C., & Ouyang, Q. (2003). Studies of temperature-dependent electronic transduction on DNA hairpin loop sensor. *Nucleic Acids Research*, *31*, e108.

March, G., Noël, V., Piro, B., Reisberg, S., & Pham, M.-C. (2008). Nanometric layers for direct, signal-on, selective, and sensitive electrochemical detection of oligonucleotides hybridization. *Journal of the American Chemical Society*, *130*, 15752–15753.

McKay, C. S., & Finn, M. G. (2014). Click chemistry in complex mixtures: Bioorthogonal bioconjugation. *Chemistry and Biology*, *21*, 1075–1101.

Mello, L. D., & Kubota, L. T. (2002). Review of the use of biosensors as analytical tools in the food and drink industries. *Food Chemistry*, *77*, 237–256.

Mills, J. B., Vacano, E., & Hagerman, P. J. (1999). Flexibility of single-stranded DNA: Use of gapped duplex helices to determine the persistence lengths of poly(dT) and poly(dA). *Journal of Molecular Biology*, *285*, 245–257.

Nakayama, M., Ihara, T., Nakano, K., & Maeda, M. (2002). DNA sensors using a ferrocene-oligonucleotide conjugate. *Talanta*, *56*, 857–866.

Pavlovic, E., Lai, R. Y., Wu, T. T., Ferguson, B. S., Sun, R., Plaxco, K. W., et al. (2008). Microfluidic device architecture for electrochemical patterning and detection of multiple DNA sequences. *Langmuir*, *24*, 1102–1107.

Peng, H.-I., & Miller, B. L. (2011). Recent advancements in optical DNA biosensors: Exploiting the plasmonic effects of metal nanoparticles. *Analyst*, *136*, 436–447.

Prins, R., Korswagen, A. R., & Kortbeek, A. G. T. G. (1972). Decomposition of the ferricenium cation by nucleophilic reagents. *Journal of Organometallic Chemistry*, *39*, 335–344.

Ricci, F., Lai, R. Y., & Plaxco, K. W. (2007). Linear, redox modified DNA probes as electrochemical DNA sensors. *Chemical Communications*, 3768–3770.

Rodriguez-Mozaz, S., López de Alda, M. J., Marco, M.-P., & Barceló, D. (2005). Biosensors for environmental monitoring: A global perspective. *Talanta*, *65*, 291–297.

Rowe, A. A., Chuh, K. N., Lubin, A. A., Miller, E. A., Cook, B., Hollis, D., et al. (2011). Electrochemical biosensors employing an internal electrode attachment site and achieving reversible, high gain detection of specific nucleic acid sequences. *Analytical Chemistry*, *83*, 9462–9466.

Salamifar, S. E., & Lai, R. Y. (2014a). Fabrication of electrochemical DNA sensors on gold-modified recessed platinum nanoelectrodes. *Analytical Chemistry*, *86*, 2849–2852.

Salamifar, S. E., & Lai, R. Y. (2014b). Application of electrochemical surface plasmon resonance spectroscopy for characterization of electrochemical DNA sensors. *Colloids and Surfaces. B, Biointerfaces*, *122*, 835–839.

Sarapuu, A., Vaik, K., Schiffrin, D. J., & Tammeveski, K. (2003). Electrochemical reduction of oxygen on anthraquinone-modified glassy carbon electrodes in alkaline solution. *Journal of Electroanalytical Chemistry*, *541*, 23–29.

Sassolas, S., Leca-Bouvier, B. D., & Blum, L. J. (2008). DNA biosensors and microarrays. *Chemical Reviews*, *108*, 109–139.

Sim, A. Y. L., Lipfert, J., Herschlag, D., & Doniach, S. (2012). Salt dependence of the radius of gyration and flexibility of single-stranded DNA in solution probed by small-angle x-ray scattering. *Physical Review E*, *86*, 021901.

Sumner, J. J., & Creager, S. E. (2001). Redox kinetics in monolayers on electrodes: Electron transfer is sluggish for ferrocene groups buried within the monolayer interior. *The Journal of Physical Chemistry. B*, *105*, 8739–8745.

Sumner, J. J., Weber, K. S., Hockett, L. A., & Creager, S. E. (2000). Long-range heterogeneous electron transfer between ferrocene and gold mediated by n-alkane and n-alkyl-carboxamide bridges. *The Journal of Physical Chemistry. B*, *104*, 7449–7454.

Teles, F. R. R., & Fonseca, L. P. (2008). Trends in DNA biosensors. *Talanta*, 77(2008), 606–623.

Wang, J. (2000). From DNA biosensors to gene chips. *Nucleic Acids Research*, *28*, 3011–3016.

Wegner, G. J., Lee, H. J., & Corn, R. M. (2002). Characterization and optimization of peptide arrays for the study of epitope-antibody interactions using surface plasmon resonance imaging. *Analytical Chemistry*, *74*, 5161–5168.

White, R. J., & Plaxco, K. W. (2010). Exploiting binding-induced changes in probe flexibility for the optimization of electrochemical biosensors. *Analytical Chemistry*, *82*, 73–76.

Wu, Y., & Lai, R. Y. (2013). Development of a "signal-on" electrochemical DNA sensor with an oligo-thymine spacer for point mutation detection. *Chemical Communications*, *49*, 3422–3424.

Wu, Y., & Lai, R. Y. (2014). Effects of DNA probe and target flexibility on the performance of a "signal-on" electrochemical DNA sensor. *Analytical Chemistry*, *86*, 8888–8895.

Xiao, Y., Qu, X., Plaxco, K. W., & Heeger, A. J. (2007). Label-free electrochemical detection of DNA in blood serum via target-induced resolution of an electrode-bound DNA pseudoknot. *Journal of the American Chemical Society*, *129*, 11896–11897.

Yang, W., Gerasimov, J. Y., & Lai, R. Y. (2009). Folding-based electrochemical DNA sensor fabricated on a gold-plated screen-printed carbon electrode. *Chemical Communications*, *20*, 2902–2904.

Yang, W., & Lai, R. Y. (2011a). Effect of diluent chain length on the performance of the electrochemical DNA sensor at elevated temperature. *Analyst*, *136*, 134–139.

Yang, W., & Lai, R. Y. (2011b). Comparison of the stem-loop and linear probe-based electrochemical DNA sensors by alternating current voltammetry and cyclic voltammetry. *Langmuir*, *27*, 14669–14677.

Yang, W., & Lai, R. Y. (2012). A dual-signaling electrochemical DNA sensor based on target hybridization-induced change in DNA probe flexibility. *Chemical Communications*, *48*, 8703–8705.

Yu, Z., & Lai, R. Y. (2012). A reagentless and reusable electrochemical DNA sensor based on target hybridization-induced stem-loop probe formation. *Chemical Communications*, *48*, 10523–10525.

Yu, Z., & Lai, R. Y. (2013). The effect of signaling probe conformation on sensor performance of a displacement-based electrochemical DNA sensor. *Analytical Chemistry*, *85*, 3340–3346.

Yu, Z., Zaitouna, A. J., & Lai, R. Y. (2014). Effect of redox label tether length and flexibility on sensor performance of displacement-based electrochemical nucleic acid sensors. *Analytica Chimica Acta*, *812*, 176–183.

Zhao, S., Yang, W., & Lai, R. Y. (2011). A folding-based electrochemical aptasensor for detection of vascular endothelial growth factor in human whole blood. *Biosensors & Bioelectronics*, *26*, 2442–2447.

Ziegler, C. (2004). Cantilever-based biosensors. *Analytical and Bioanalytical Chemistry*, *379*, 946–959.

CHAPTER NINE

Construction of Protein-Based Biosensors Using Ligand-Directed Chemistry for Detecting Analyte Binding

Kei Yamaura*, Shigeki Kiyonaka*, Itaru Hamachi*,†,1

*Graduate School of Engineering, Kyoto University, Katsura, Kyoto, Japan

†Core Research for Evolutional Science and Technology (CREST), Japan Science and Technology Agency, Kawaguchi, Saitama, Japan

[1]Corresponding author: e-mail address: ihamachi@sbchem.kyoto-u.ac.jp

Contents

Abstract

Protein-based fluorescent biosensors are powerful tools for quantitative detection of biomolecules or drugs with high sensitivity under physiological conditions. However,

Methods in Enzymology, Volume 589
ISSN 0076-6879
http://dx.doi.org/10.1016/bs.mie.2017.02.006

conventional methods for construction of biosensors require structural data with high resolution or amino acid sequence information in most cases, which hampers applicability of this method to structurally complicated receptor proteins. To sidestep such limitations, we recently developed a new method that employs ligand-directed chemistry coupled with a bimolecular fluorescence quenching and recovery system, which enabled the conversion of various kinds of membrane-bound receptors to "turn-on" type fluorescent sensors. Here, we describe a protocol for construction of "turn-on" type fluorescent biosensors based on the $GABA_A$ receptor which permits quantitative analysis of the ligand affinity.

1. INTRODUCTION

Currently, drug discovery programs are largely dependent on screening large libraries of small chemicals for their effects on target proteins, and new strategies are required for rapid and efficient screening (Loving, Sainlos, & Imperiali, 2010; Mayr & Bojanic, 2009; Prével, Pellerano, Van, & Morris, 2014). One of the potential methods relies on a protein-based fluorescent biosensor, which permits quantitative detection of biomolecules or drugs with high sensitivity in the physiological environment. In most cases, these biosensors detect ligands by transducing the large structural changes upon ligand binding as changes in fluorescence (Tamura & Hamachi, 2014). These are constructed by conjugation of fluorescent proteins with a ligand-binding domain (LBD) (Fehr, Frommer, & Lalonde, 2002; Marvin et al., 2013; Okumoto et al., 2005) or by site-directed modification of an introduced cysteine residue near the ligand-binding site with a small fluorophore (Marvin & Hellinga, 1998; Morii et al., 2002; Simard, Getlik, et al., 2009; Simard, Klüter, et al., 2009; Takikawa et al., 2014).

Recently, a new technique which does not require protein structural changes upon ligand binding has been reported for construction of biosensors (Brun et al., 2012; Brun, Tan, Nakata, Hinner, & Johnsson, 2009; Masharina, Reymond, Maurel, Umezawa, & Johnsson, 2012). These biosensors, termed Snifts (SNAP-tag-based indicator proteins with a fluorescent intramolecular tether), consist of SNAP-tag, CLIP-tag, and LBD, where SNAP-tag is labeled with a small molecular tether bearing a ligand to the LBD and a fluorescent acceptor, and the CLIP-tag is modified with a fluorescent donor. In this method, the covalently attached ligand is displaced from the binding site of the LBD by addition of competitive compounds, resulting in a change in the relative position of the acceptor and the donor, so that the fluorescent signal can be modulated upon the competitor's binding without any need for structural changes in the LBD. While these protein-based biosensors mentioned earlier are very powerful, most of them require structural information on the target protein and strongly rely on genetic manipulation.

We recently reported ligand-directed chemistry for selective labeling of target proteins, in which neither structural information on the LBD nor genetic manipulation are required (Fig. 1) (Fujishima, Yasui, Miki, Ojida, & Hamachi, 2012; Takaoka, Nishikawa, Hashimoto, Sasaki, & Hamachi, 2015; Tsukiji, Miyagawa, Takaoka, Tamura, & Hamachi, 2009). Covalent labeling of proteins using ligand-directed chemistry is driven by selective ligand–protein recognition, which facilitates a chemical reaction of a reactive group having moderate reactivity with nucleophilic amino acid residues located near the LBD with the help of the proximity effect. Importantly, the protein function can be recovered by washing out the excess labeling reagents and the remaining ligand moieties even after the chemical modification. Using ligand-directed acyl imidazole (LDAI) chemistry which is powerful for membrane-protein labeling, we succeeded in chemical labeling of folate receptors on live cells (Fujishima et al., 2012). To our surprise, a fluorescence increase occurred upon addition of ligands such as folic acid and its derivatives, implying that the labeled folate receptors

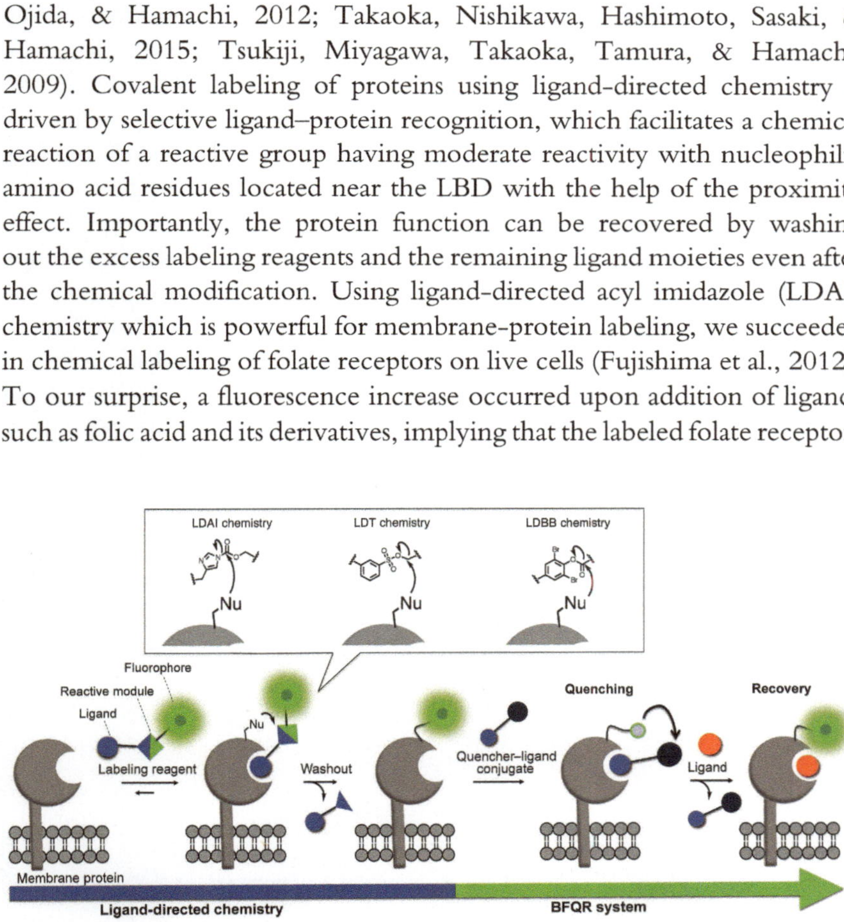

Fig. 1 Construction of protein-based fluorescent biosensors by coupling affinity-based chemical labeling (*left*) with BFQR system (*right*). Schematic illustration for fluorescent labeling using ligand-directed chemistry and the construction of fluorescent biosensors using BFQR system are shown. *Nu*: nucleophile.

can be converted to a fluorescence biosensor. A recent crystal structural analysis reveals there are a few tryptophan residues near the LBD (Chen et al., 2013), which indicates that this fluorescence change would likely be caused by relief of the tryptophan-induced quenching of the labeled fluorophore upon ligand binding. However, such an ideal arrangement of tryptophan residues does not always occur in every LBD of target protein scaffolds, which raises a serious limitation in the applicability of this technology to a wider variety of receptor proteins.

To overcome this limitation, we also developed a method using chemical labeling coupled with bimolecular fluorescence quenching and recovery (BFQR) system. This approach is expected to be more general for constructing fluorescent biosensors since the protein need not provide the quenching species (Fig. 1) (Wang et al., 2011). In this method, after the covalent attachment of a fluorophore in the vicinity of the ligand-binding site, the fluorescence is efficiently quenched by the quencher–ligand conjugate. When the conjugate bound to the ligand-binding site is competitively replaced with small molecules having a sufficient affinity for the same binding site, efficient recovery of the fluorescence (that is, in the turn-on mode) takes place. This method was initially applied to bradykinin receptor 2 and is now extended to various kinds of membrane-surface receptors.

$GABA_A$ receptors ($GABA_ARs$) have been widely recognized as an important drug target over many years (Barnard et al., 1998; Chebib & Johnston, 2000), and a wide variety of orthosteric ligands and allosteric modulators have been reported for $GABA_ARs$ (Fig. 2). Among them,

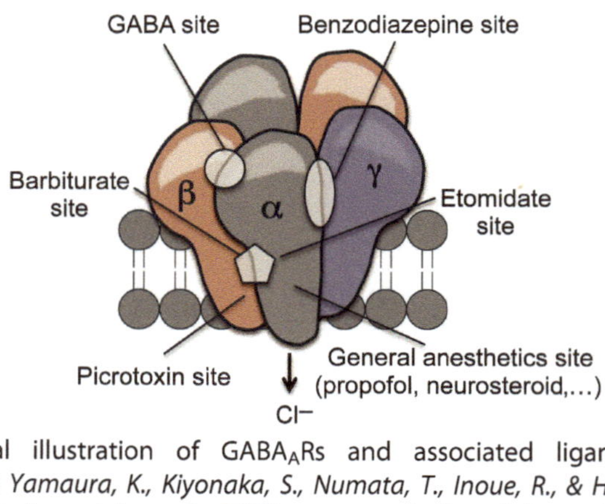

Fig. 2 Structural illustration of $GABA_ARs$ and associated ligand-binding sites. *Reproduced from Yamaura, K., Kiyonaka, S., Numata, T., Inoue, R., & Hamachi, I. (2016). Discovery of allosteric modulators for $GABA_A$ receptors by ligand-directed chemistry.* Nature Chemical Biology, 12, *822–830, with permission from Nature Publishing Group.*

benzodiazepine derivatives exhibit potent sedative/hypnotic and anxiolytic effects (Rudolph & Knoflach, 2011). Barbiturates, etomidate, propofol, and neurosteroids, whose binding sites are located elsewhere on the molecule from that of benzodiazepines, are widely utilized as anesthetics (Sieghart, 2015). Recently, we developed "turn-on" fluorescent biosensors for $GABA_AR$ ligands using LDAI chemistry coupled with the BFQR system (Fig. 1) (Yamaura, Kiyonaka, Numata, Inoue, & Hamachi, 2016). To date, two fluorescent sensors based on $GABA_ARs$ which can respond to orthosteric GABA-site ligands or allosteric benzodiazepine-site ligands have been developed. Here, we provide a step-by-step protocol for researchers who are interested in construction of BFQR-based fluorescent biosensors.

2. EXPRESSION OF $GABA_AR$ CONSISTING OF α/β/γ SUBUNITS IN HEK293T CELLS

$GABA_ARs$ assemble into a heteropentamer consisting of a variety of subunits to form GABA-gated chloride channels. To date, 19 different subunits (α1–α6, β1–β3, γ1–γ3, δ, ε, π, θ, and ρ1–ρ3) have been identified in $GABA_ARs$, which gives rise to a remarkable level of diversity and complexity. Recent studies have demonstrated that physiologically relevant heteromeric $GABA_AR$ forms are mainly composed of two α subunits, two β subunits, and one γ subunit, and that at least 20–30 isoforms consisting of different combinations of these subunits are expressed in the brain (Olsen & Sieghart, 2009). In pharmacological assays using neuronal samples, complicated responses to each drug are observed because each $GABA_AR$ exhibits distinct pharmacological properties. In contrast, a single isoform of $GABA_AR$ consisting of α/β/γ subunits can be expressed in human embryonic kidney (HEK) 293T cells, a mammalian cell line, using a conventional transfection method. Importantly, electrophysiological assays revealed that this expression system gives a functional $GABA_AR$-pentamer consisting of α/β/γ subunits in a 2:2:1 ratio (Tretter, Ehya, Fuchs, & Sieghart, 1997; Yamaura et al., 2016). Here, our protocols for recombinant expression of α/β/γ $GABA_ARs$ in HEK293T cells are described later in detail.

1. Culture HEK293T cells in Dulbecco's modified Eagle's medium (DMEM) containing 10% fetal bovine serum, 30 units mL^{-1} penicillin, and 30 units mL^{-1} streptomycin under 95% air-5% CO_2 atmosphere at 37°C.
2. One day before transfection, seed the cells (2.0×10^5 cells) on 3.5-cm primaria dishes (BD Falcon).

3. Mix 0.7 μg of each expression vector for α, β, and γ subunit (total 2.1 μg) and 0.3 μg of pmCherry-N1 (Clontech) as a transfection marker in a microcentrifuge tube and dilute with 50 μL of Opti-MEM.
4. Dilute 2 μL of Lipofectamine 2000 (Invitrogen) with 50 μL of Opti-MEM in a different tube and incubate for 5 min at room temperature.
5. Mix the plasmid solution with the Lipofectamine solution by pipetting several times and incubate for 20 min at room temperature.
6. Directly add the mixed solution into culture dishes.
7. After 4 h incubation, replace the culture medium with fresh medium.
8. After 42–54 h incubation at 37°C with 5% CO_2, utilize the cells for labeling experiments.

3. DESIGN AND SYNTHESIS OF LABELING REAGENTS FOR GABA$_A$R

In recent years, we have succeeded in chemical labeling of various membrane proteins including carbonic anhydrase 12, folate receptor, bradykinin receptor, NMDA receptor, and GABA$_A$R on live cell surfaces using LDAI chemistry (Fig. 3A, Fujishima et al., 2012; Miki et al., 2014; Yamaura et al., 2016). For selective labeling to a target protein, careful molecular design of LDAI reagents is essential. In this section, we describe the design strategy of LDAI reagents for GABA$_A$R and the synthetic procedure later.

3.1 Design of LDAI Reagents for GABA$_A$R

LDAI reagents consist of four modules: (a) an acyl imidazole moiety as the cleavable reactive group, (b) a ligand moiety having sufficient affinity and selectivity for the target protein, (c) a spacer moiety between acyl imidazole and the ligand, and (d) a fluorophore (or probe) group. For chemical labeling to GABA$_A$R, we prepared three LDAI reagents (**GGAM-Gaba**, **GGAM-Bzp**, or **GGAM-Bbr**), which are targeted for the orthosteric GABA-, allosteric benzodiazepine-, or barbiturate-site, respectively (Fig. 3B and C). As a typical example, we here describe how to design a LDAI reagent for benzodiazepine site (**CGAM-Bzp**) in detail.

1. Appropriate selection of the ligand moiety is critical in LDAI chemistry. A ligand highly selective for the target protein should be chosen. Employing low-affinity ligands would likely cause nonspecific labeling due to the presence of high concentrations of unbound LDAI reagents. So far, we have obtained successful results using highly specific ligands

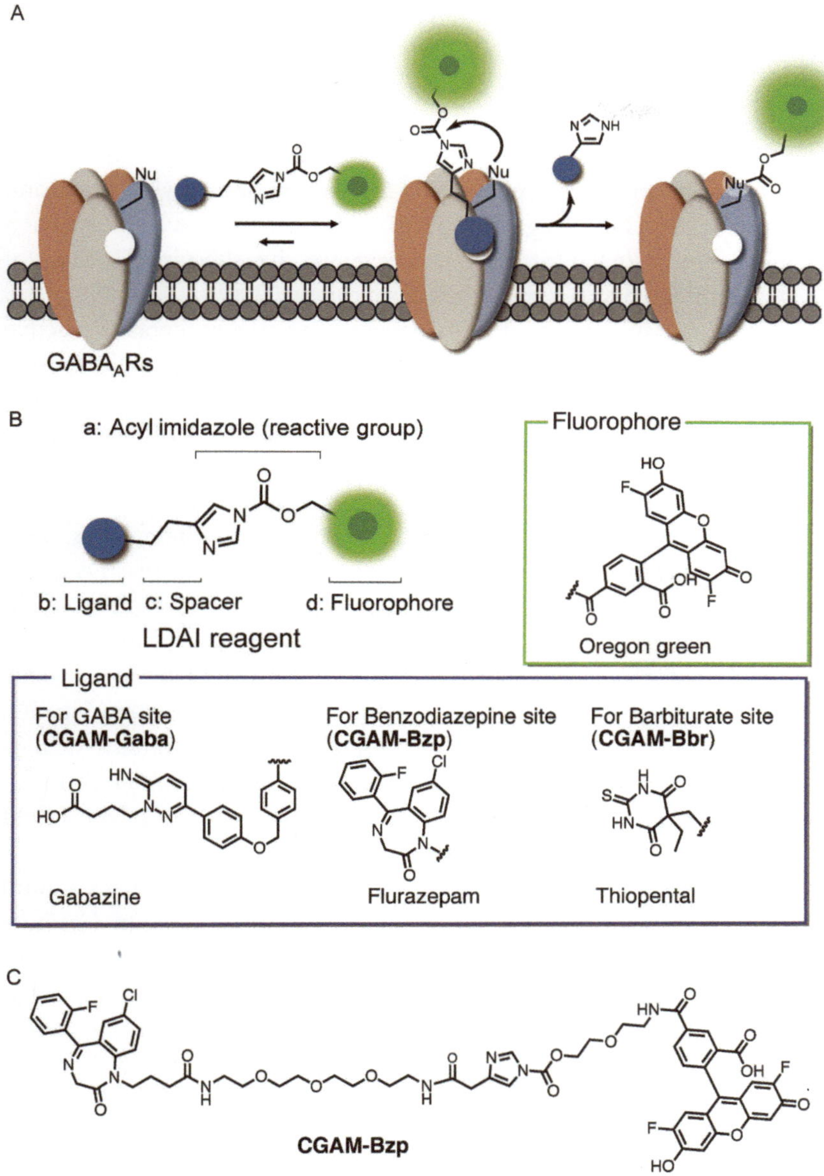

Fig. 3 Chemical labeling and molecular design of GABA$_A$R by LDAI chemistry. (A) Schematic illustration of fluorescent labeling for GABA$_A$R using LDAI chemistry. *Nu*: nucleophile. (B) Basic design configuration of LDAI reagents for GABA$_A$Rs. (C) Chemical structure of LDAI reagent (CGAM-Bzp) for benzodiazepine-site labeling.

with dissociation constants (K_d values) below μ*M* (Fujishima et al., 2012; Miki et al., 2014; Yamaura et al., 2016). It should also be noted that use of a ligand with too high an affinity may perturb the receptor function due to the difficulty of removing the cleaved ligand fragment from the ligand-binding pocket of the target protein after labeling. For labeling the benzodiazepine site, we selected a flurazepam derivative as the ligand moiety for the LDAI reagent (Fig. 3B). This moiety is well known as a highly selective ligand for the benzodiazepine site, showing K_i values of 170 n*M* for $GABA_AR$ (Hanson & Czajkowski, 2008).

2. The acyl imidazole group of LDAI reagents acts as a moderately reactive electrophile. It was demonstrated that positioning this group on the protein surface close to the ligand-binding site facilitates acyl substitution reaction with nucleophilic amino acid residues (Lys, Ser, Thr, or Tyr) (Matsuo et al., 2013; Miki et al., 2016). In other words, the labeling efficiency is largely governed by the proximity between the reactive group of the LDAI reagent and these amino acid side chains on the target protein. Thus, optimization of the spacer length between ligand moiety and the reactive group is crucial for optimizing the labeling efficiency at a high level. Actually, in the case of LDAI labeling at the benzodiazepine site, we found that the labeling yield was dependent on the spacer length (unpublished results), and one of the optimal spacers was triethylene glycol as shown in Fig. 3C.
3. For chemical labeling of surface-exposed membrane proteins such as $GABA_ARs$ on live cells, LDAI reagents should be hydrophilic or negatively charged, which suppresses unfavorable permeation of the LDAI reagents into the live cells. If LDAI reagents permeate into the intracellular region, it would result in nonspecific labeling and high background fluorescence in intracellular spaces. Here, we chose Oregon Green™ (OG) which has a negative charge as the fluorophore group. Compared with fluorescein, OG shows higher photostability and less pH sensitivity.

3.2 Synthesis of LDAI Reagent for $GABA_AR$

Procedures for chemical synthesis of **CGAM-Bzp** (compound 4 in Fig. 4) are described in detail later (Fig. 4).

Synthesis of compound 3

1. Prepare a round-bottom flask containing compound **1** (Yamaura et al., 2016) (18 mg, 0.027 mmol) in anhydrous DMF (270 μL) under N_2 atmosphere.

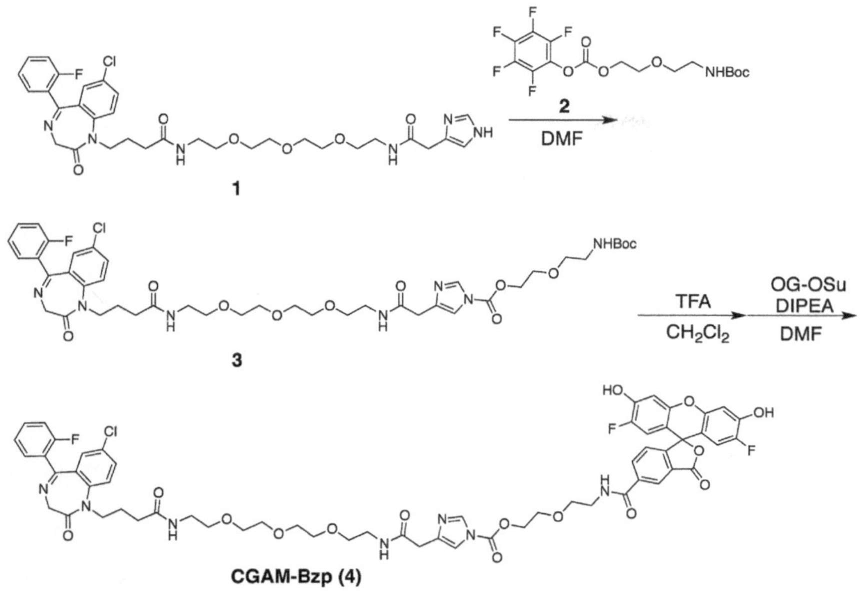

Fig. 4 Synthesis of CGAM-Bzp. *Reproduced from Yamaura, K., Kiyonaka, S., Numata, T., Inoue, R., & Hamachi, I. (2016). Discovery of allosteric modulators for $GABA_A$ receptors by ligand-directed chemistry.* Nature Chemical Biology, 12, *822–830, with permission from Nature Publishing Group.*

2. Add compound **2** (Yamaura et al., 2016) (17 mg, 0.04 mmol).
3. Stir the reaction mixture for 1 h at room temperature.
4. Purify the mixture by silica gel column chromatography ($CHCl_3$ only → $CHCl_3$:MeOH = 10:1) to obtain compound **3** (15 mg, 0.017 mmol, 63%) as a clear solid.
5. *Caution*: compound **3** should be kept in freezer or deep-freezer to prevent decomposition.

Synthesis of CGAM-Bzp (4)

1. Prepare a round-bottom flask containing compound **3** (1.0 mg, 1.1 μmol) in dry CH_2Cl_2 (0.5 mL) under N_2 atmosphere.
2. Add TFA (0.5 mL).
3. Stir the reaction mixture for 1 h at room temperature.
4. Add toluene for azeotropic removal of TFA by rotary evaporation under vacuum. Repeat this step again.
5. Dissolve the residue in dry DMF (100 μL).

6. Add Oregon Green® 488 Carboxylic Acid, Succinimidyl Ester, 5-isomer (Thermo-Fisher, 1 mg, 2.0 μmol), and DIPEA (20 μL, 0.11 mmol).
7. Stir the reaction mixture for 1 h at room temperature.
8. Dilute the reaction mixture with DMSO.
9. Directly purify the mixture by HPLC (mobile phase; CH_3CN (solvent A), 10 m*M* $AcONH_4$ aq. (solvent B)). After purification, go to the next step immediately.
10. Freeze dry the collected fraction to obtain **CGAM-Bzp** (0.8 mg, 0.7 μmol, 64%) as a yellow solid.
11. Dissolve a portion of **CGAM-Bzp** in DMSO to make a stock solution, and determine the **CGAM-Bzp** concentration by UV–vis measurement ($\varepsilon_{495\,nm} = 76{,}000\ M^{-1}\ cm^{-1}$).
12. *Caution*: **CGAM-Bzp** or its DMSO solution should be kept in the deep-freezer to prevent the decomposition.

Caution in the synthesis: The acyl imidazole group is not stable under acidic or basic aqueous conditions. Therefore, compound **3** and **CGAM-Bzp** should be carefully purified by silica gel column chromatography or HPLC under neutral conditions (Miki et al., 2014; Yamaura et al., 2016).

4. CHEMICAL LABELING OF GABA$_A$R USING LDAI REAGENT

Chemical labeling protocols for GABA$_A$R on live cells are described in this section. For confirmation of covalent labeling to target proteins, we generally utilize analytical methods including Western blotting and confocal laser scanning microscopy (CLSM) imaging. In addition, the labeled amino acid residues with LDAI reagents are determined by peptide mass fingerprinting analysis.

4.1 Confirmation of Chemical Labeling of GABA$_A$R by CLSM

CLSM analyses provide valuable information on chemical labeling by LDAI reagents. In addition to chemical labeling of surface-exposed GABA$_A$R, the impermeability of the LDAI reagent, which is also essential for selective visualization of surface-exposed receptors, can be evaluated by CLSM. Although the CLSM observation is performed after washing several times to remove excess LDAI reagent, we cannot exclude the possibility that the observed fluorescence comes from the LDAI reagent noncovalently bound to target proteins. In most cases, however, we confirm that the washing procedure suffices to remove the unreacted labeling reagent and any remaining ligand moiety from target proteins when the ligand moiety has moderate affinity as described in Section 3.1.

To confirm selective labeling of $GABA_AR$, we perform negative control experiments (Fig. 5). While CLSM imaging visualized prominent fluorescence from surface of the cell transfected with $\alpha1/\beta3/\gamma2$ $GABA_ARs$ after treatment with the LDAI reagents, the fluorescence was observed neither in empty vector-transfected cells nor in $\alpha1/\beta3/\gamma2$ $GABA_AR$-transfected cells pretreated with competitive inhibitors. These negative control experiments verify the specific labeling to $GABA_AR$.

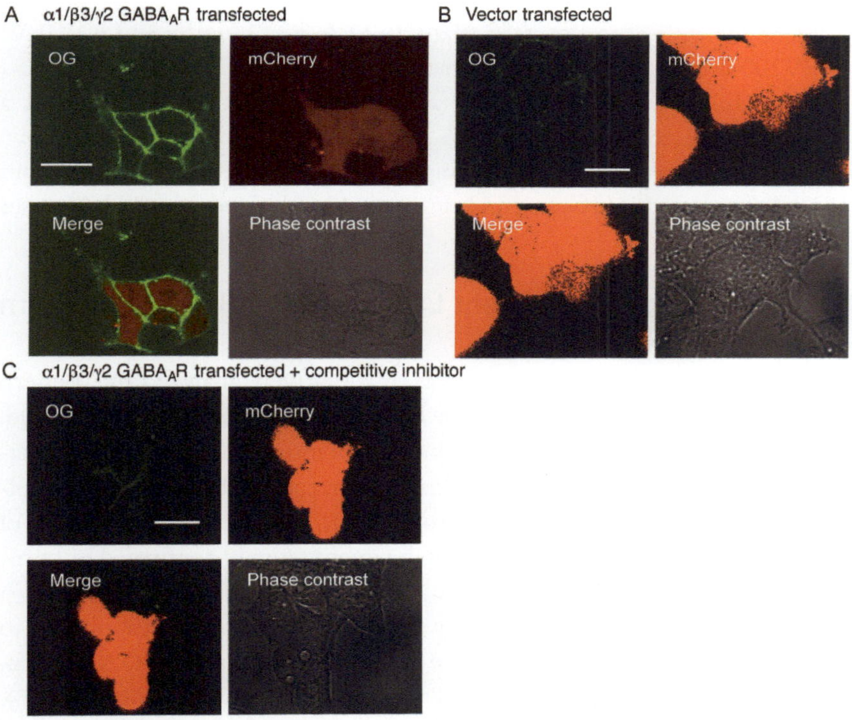

Fig. 5 CLSM imaging visualizes chemical labeling of $GABA_ARs$ using CGAM(Bzp). (A–C) HEK293T cells transfected with $\alpha1/\beta3/\gamma2$ $GABA_ARs$ in (A or C) or control vector in (B) were incubated with 2 μ*M* CGAM-Bzp in serum-free DMEM at 37°C for 3 h in the absence (A or B) or presence of 20 μ*M* flumazenil (competitive inhibitor) in (C). mCherry-N1 was utilized as a transfection marker. Scale bars: 20 μm. *Reproduced from Yamaura, K., Kiyonaka, S., Numata, T., Inoue, R., & Hamachi, I. (2016). Discovery of allosteric modulators for $GABA_A$ receptors by ligand-directed chemistry.* Nature Chemical Biology, 12, *822–830, with permission from Nature Publishing Group.*

The protocols for CLSM imaging of $GABA_AR$ using **CGAM-Bzp** were described later. In a similar manner, we succeeded in selective labeling of $GABA_AR$ using **CGAM-Gaba** by similar procedures.

1. After 24 h of transfection, dissociate transfected cells by pipetting and reseed on glass-bottom dishes (Iwaki) pretreated with poly-L-lysine.
2. 18–30 h after the reseeding, replace the culture medium with serum-free-DMEM medium containing 2 μ*M* **CGAM-Bzp**. *Caution*: **CGAM-Bzp** is not stable in aqueous solution. Thus, after addition of **CGAM-Bzp** in DMEM, the solution should be utilized for the labeling immediately.
3. Incubate the cells at 37°C under 5% CO_2 for 3 h.
4. After removal of the medium, wash the cells twice with serum-free DMEM.
5. Place the dish onto the stage of a CLSM (e.g., Olympus FLUOVIEW FV1000) equipped with an oil immersion objective (60×, NA=1.40), autofocus device (Z-drift compensation, ZDC), and a GaAsP detector.
6. Acquire fluorescence images using the 488 nm excitation derived from an Ar ion laser for OG, and 543 nm excitation derived from a HeNe laser for mCherry.

4.2 Confirmation of Covalent Labeling to $GABA_AR$ by Western Blotting Analysis

Chemical labeling to target proteins is quantitatively analyzed by Western blotting using antibody specific to a labeled fluorophore. In this method, the labeled proteins are analyzed on the basis of molecular weight after denaturation with SDS, which gives us direct information about covalent labeling of the fluorophore to target proteins. In addition, the selectivity of chemical labeling to target proteins can be quantitatively confirmed by comparing the band intensity.

We checked selective labeling by detecting the band corresponding to the molecular weight of $GABA_ARs$ using **CGAM-Gaba**, **CGAM-Bzp**, or **CGAM-Bbr** in cells transfected with α1/β3/γ2 $GABA_ARs$ (Fig. 6). The labeled band was not observed in empty vector-transfected cells or α1/β3/γ2 $GABA_AR$-transfected cells pretreated with the competitive inhibitor, indicating good selectivity of the chemical labeling to $GABA_ARs$. Although $GABA_ARs$ form a heteropentamer consisting of α1-, β3-, and γ2-subunits, we can estimate labeled subunits from the molecular weight difference of the labeled band (Fig. 6B). The protocol is described later.

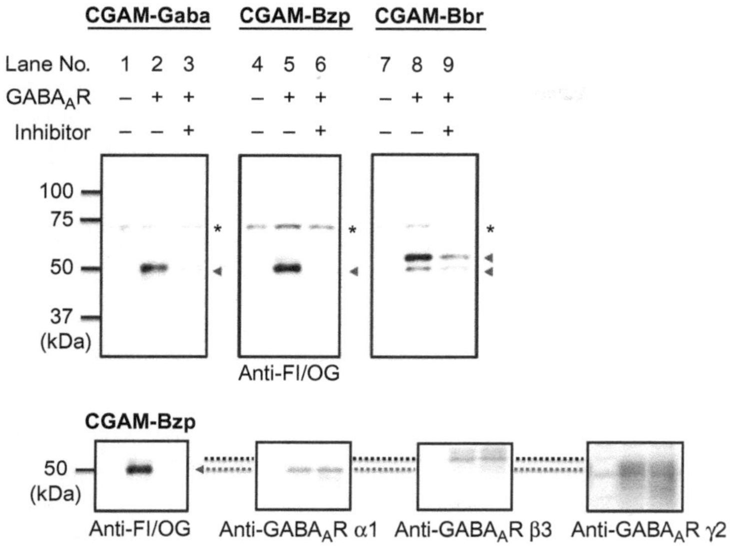

Fig. 6 Chemical labeling of GABA$_A$Rs in HEK293T cells using CGAM reagents. (A) For GABA-site labeling, HEK293T cells transfected with α1/β3/γ2 GABA$_A$Rs (GABA$_A$R(+)) or the control vector (GABA$_A$R(−)) were treated with 1 μ*M* CGAM-Gaba in the presence or absence of 20 μ*M* gabazine. For benzodiazepine-site labeling, the cells were treated with 1 μ*M* CGAM-Bzp in the presence or absence of 20 μ*M* flumazenil. For barbiturate-site labeling, the cells were treated with 10 μ*M* CGAM-Bbr in the presence or absence of 1 m*M* amobarbital. These reactions are conducted in serum-free DMEM at 37°C for 1 h. *Red arrowhead* indicates labeled GABA$_A$Rs. * indicates nonspecific labeling of bovine serum albumin (BSA) included in the culture medium. (B) Comparison of the molecular weight between labeled band detected by anti-Fl/OG antibody and GABA$_A$R subunits detected by anti-GABA$_A$ α1, anti-GABA$_A$ β3, or anti-GABA$_A$ γ2 antibody. *Reproduced from Yamaura, K., Kiyonaka, S., Numata, T., Inoue, R., & Hamachi, I. (2016). Discovery of allosteric modulators for GABA$_A$ receptors by ligand-directed chemistry.* Nature Chemical Biology, *12, 822–830, with permission from Nature Publishing Group.*

1. After 42–54 h of transfection in 3.5 cm dishes, replace the culture medium with serum-free DMEM containing LDAI reagent.
2. Incubate the cells at 37°C under 5% CO_2 for 3–4 h.
3. Remove the medium containing LDAI reagent.
4. Add 1 mL PBS and dissociate labeled cells from the dishes by pipetting.
5. Transfer the suspension to microcentrifuge tubes and centrifuge (600 × *g*, 2 min, 4°C).
6. Remove PBS and wash the cells twice with PBS.
7. Lyse the cells with 50 μL RIPA (25 m*M* Tris/HCl, 150 m*M* NaCl, 1% NP-40, 0.25% deoxycholate (DOC), 0.1% SDS, pH 7.4) buffer

containing protease inhibitor (Calbiochem, 39134) and rotate at 4°C for 30 min.

8. Centrifuge (10,000 × *g*, 10 min, 4°C) and collect the supernatant.
9. Mix the lysate with 5 × SDS-PAGE loading buffer (325 m*M* Tris–HCl, 15% SDS, 20% sucrose, 0.5 *M* DTT, and 0.02% BPB, pH 6.8) and incubate for 1 h at room temperature.
10. Apply to SDS-PAGE and electrotransfer onto immune-blot PVDF membranes (Bio-rad).
11. Block the membrane with 5% nonfat dry milk in TBS containing 0.05% Tween 20 (Sigma-Aldrich).
12. Stain these membranes with rabbit antifluorescein/Oregon Green antibody (Abcam, ab19491, ×2000), rabbit anti-$GABA_AR$ α1 antibody (Millipore, 06-868, ×1000), rabbit anti-$GABA_AR$ β3 antibody (Rockland, 600-401-C79, ×1000), rabbit anti-$GABA_AR$ γ2 antibody (Rockland, 612-401-C81, ×1000).
13. Stain these membranes with an HRP-conjugated goat antirabbit IgG (Santa Cruz, SC-2004, ×5000).
14. Develop these membranes with ECL Prime Western blotting detection system (GE Healthcare) and analyze with a fluorescent imaging system (LAS4000, GE Healthcare).

4.3 Determination of Labeling Yield to $GABA_ARs$ Using Fluorophore-Conjugated Ligand by CLSM

The labeling yield of **CGAM-Bzp** to $GABA_ARs$ on the cell surface can be determined by CLSM imaging under live cell conditions. Here, we utilized OG-conjugated flurazepam (**Bzp-OG**) (Fig. 7A). After incubation of **Bzp-OG** in the cells transfected with α1/β3/γ2 $GABA_ARs$, CLSM visualized the prominent fluorescence from the cell surface, by which intensity we can estimate surface-expressed $GABA_ARs$. On the assumption that the fluorescent properties of **Bzp-OG** noncovalently bound to $GABA_ARs$ are similar to that of OG labeled to $GABA_ARs$ covalently by **CGAM-Bzp**, we can evaluate the labeling yield by comparing the fluorescence intensity in the presence or absence of **Bzp-OG** in the labeled cells (Fig. 7B). Although this method is useful, significant background fluorescence was observed in some cases due to the excess of **Bzp-OG** in the medium, which may seriously interfere with the precise evaluation. To ignore the background fluorescence, we concurrently analyze the fluorescence intensity obtained from

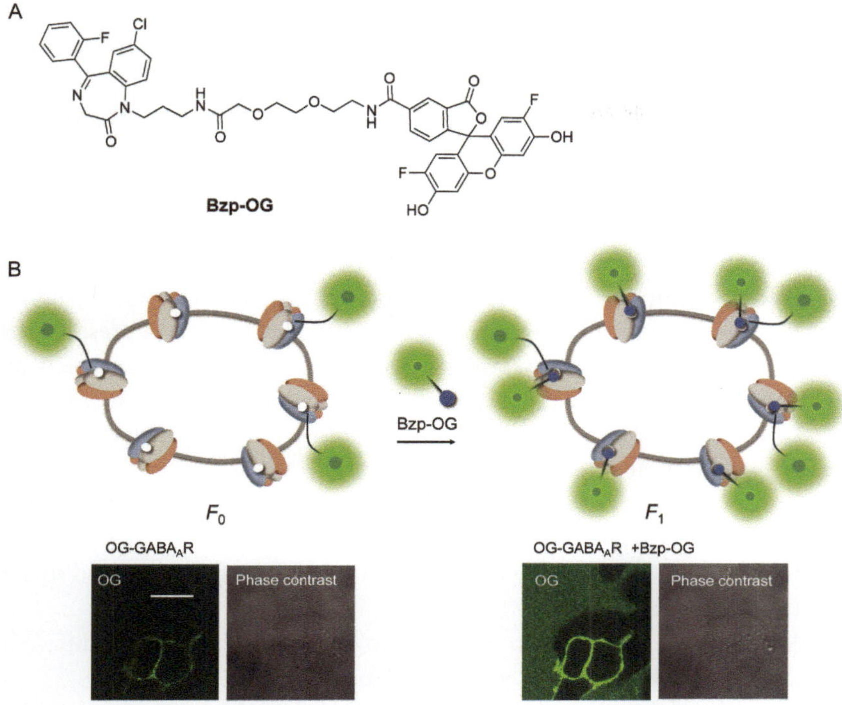

Fig. 7 Determination of the labeling yield for benzodiazepine site on GABA$_A$Rs labeled by CGAM-Bzp. (A) Chemical structure of Bzp-OG. (B) Determination of the labeling yield for OG-labeled GABA$_A$Rs using Bzp-OG. Schematic illustration of the fluorescent changes (*upper*) and representative confocal images (*lower*) are shown. Scale bars: 20 μm. *Reproduced from Yamaura, K., Kiyonaka, S., Numata, T., Inoue, R., & Hamachi, I. (2016). Discovery of allosteric modulators for GABA$_A$ receptors by ligand-directed chemistry.* Nature Chemical Biology, *12, 822–830, with permission from Nature Publishing Group.*

the intercellular interface in CLSM imaging. The detailed protocols were described as follows.

1. Obtain confocal image of labeled GABA$_A$Rs (F_0) as described in Section 4.1.
2. Add 0.1 μ*M* **Bzp-OG**. *Note*: fluorescent titration experiments by CLSM have revealed that **Bzp-OG** binds to most of GABA$_A$Rs on cell surface in this condition (data not shown).
3. After 1 min, obtain confocal image of labeled GABA$_A$Rs (F_1).
4. Quantitatively measure fluorescence intensity from plasma membrane of single labeled cell using the Fluoview software (Olympus).

5. Determine the labeling yield for benzodiazepine site on $GABA_ARs$ based on the following equation: $\text{labeling yield} = F_0/(F_1 - F_0) \times 100$.

4.4 Identification of Labeling Site for Drug Binding Site by Peptide Mass Fingerprinting Analysis

The labeling site of $GABA_ARs$ by **CGAM-Bzp** was determined using peptide mass fingerprinting analysis (Fig. 8). To enrich the OG-labeled $GABA_ARs$, the labeled cell lysate was immunoprecipitated using OG-specific antibody. Then, the immunoprecipitate fraction was separated by SDS-PAGE, and the OG-labeled band was excised using a fluorescent gel imager. After in-gel digestion with trypsin, the peptides were analyzed by LC-MS/MS analysis. The obtained data were further analyzed using Proteome Discoverer and SEQUEST programs to determine the labeled site. The detailed protocols are described later.

1. Collect the supernatant of the cell lysate after labeling (for details, see step 8 in Section 4.2).
2. Add anti-Fl/OG antibody (ab19491, ×400) to the cell lysate and rotate at 4°C for 18 h.
3. Add Dynabeads® Protein A (Thermo Fisher Scientific) to a different tube and wash the beads with RIPA buffer.
4. Add the lysate treated with antibody to the tube containing Dynabeads® Protein A and rotate at 4°C for 2 h.
5. Wash the beads three times with RIPA buffer.
6. Add 2 × SDS-PAGE sample buffer (125 m*M* Tris–HCl, 4% SDS, 20% glycerol, 0.01% bromophenol blue, pH 6.8) and boil in heat block for 5 min.
7. Collect the supernatant and apply to SDS-PAGE.
8. Excise the labeled band corresponding to $GABA_AR$ using fluorescence gel imager (LAS4000, GE Healthcare).
9. Perform in-gel digestion using MS grade Trypsin (Thermo Fisher Scientific).
10. Analyze the peptide by nanoflow reverse phase liquid chromatography followed by tandem MS, using a LTQ Orbitrap XL hybrid mass spectrometer (Thermo Fisher Scientific).
11. Record peptide spectra using Xcalibur 2.1 system (Thermo Fisher Scientific)
12. Analyze the peptide spectra by Proteome Discoverer 1.4.1.14 (Thermo Fisher Scientific).

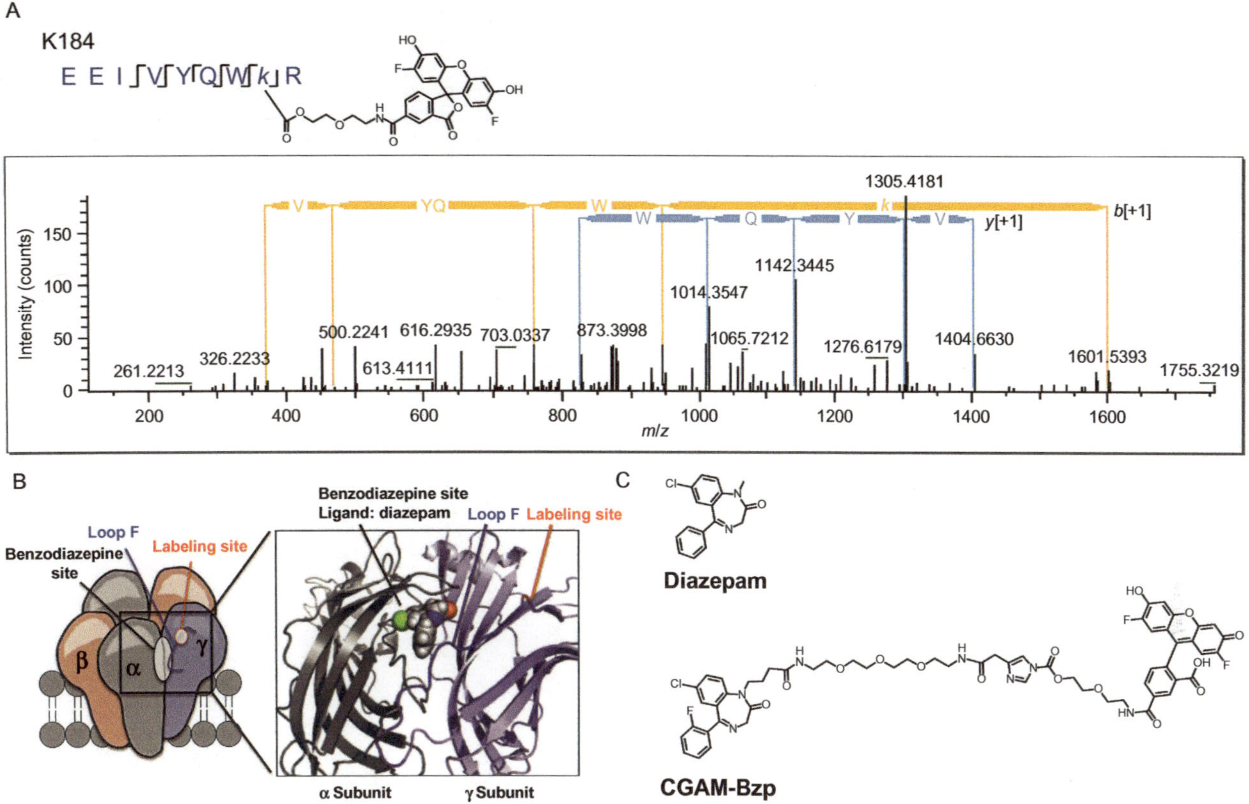

Fig. 8 See legend on next page.

13. Perform the analysis using the SEQUEST HT (Thermo Fisher Scientific) against latest uniprot database for $GABA_AR$ α1 subunit (GBRA1_MOUSE, P62812), $GABA_AR$ β3 subunit (GBRB3_MOUSE, P63080), and $GABA_AR$ γ2 subunit (GBRG2_MOUSE, P22723).

5. CONSTRUCTION OF FLUORESCENT BIOSENSORS USING LABELED $GABA_ARs$

As described in Section 1, we have developed fluorescent biosensors capable of detecting competing ligands using chemical labeling coupled with the BFQR system (Fig. 1, Wang et al., 2011). In this method, molecular design of the quencher–ligand conjugate is crucial for best sensitivity. Thus, we first discuss our design strategy of the quencher–ligand conjugate and subsequently describe how to construct the fluorescent biosensor for $GABA_AR$ ligands using BFQR system in this section.

5.1 Molecular Design of Quencher–Ligand Conjugates for $GABA_ARs$

Quencher–ligand conjugates consist of three modules: (a) a fluorescence quencher moiety, (b) a ligand moiety having affinity to the target protein, (c) a spacer between quencher and the ligand. For $GABA_ARs$, we developed two types of quencher–ligand conjugates (**Gaba-Q** or **Bzp-Q**) binding to the orthosteric GABA site or allosteric benzodiazepine site, respectively (Fig. 9) (Yamaura et al., 2016). Here, we describe the molecular design of the quencher–ligand conjugate for the benzodiazepine site (**Bzp-Q**) as follows.

1. A suitable chemical structure of the quencher moiety is the most crucial for sufficient fluorescence quenching of the labeled fluorophore. In our

Fig. 8 Identification of labeled site for benzodiazepine binding site on $GABA_ARs$ using CGAM-Bzp by peptide mass fingerprinting analysis. (A) MS/MS analysis of the OG-modified peptide detected from γ2 subunit of OG-labeled $GABA_ARs$. (B) Structural illustration and structure model based on homology models and docking studies of $GABA_AR$ (nchembio.917-s2.pdb). Loop F and the amino acid modified by CGAM-Bzp are shown as *blue* and *red* in the cartoon, respectively. Hydrogen, nitrogen, oxygen, and chlorine atom of diazepam ligand are shown as *gray*, *blue*, *red*, and *green spheres*, respectively. (C) Chemical structure of diazepam and CGAM-Bzp. *Reproduced from Yamaura, K., Kiyonaka, S., Numata, T., Inoue, R., & Hamachi, I. (2016). Discovery of allosteric modulators for $GABA_A$ receptors by ligand-directed chemistry.* Nature Chemical Biology, 12, *822–830, with permission from Nature Publishing Group.*

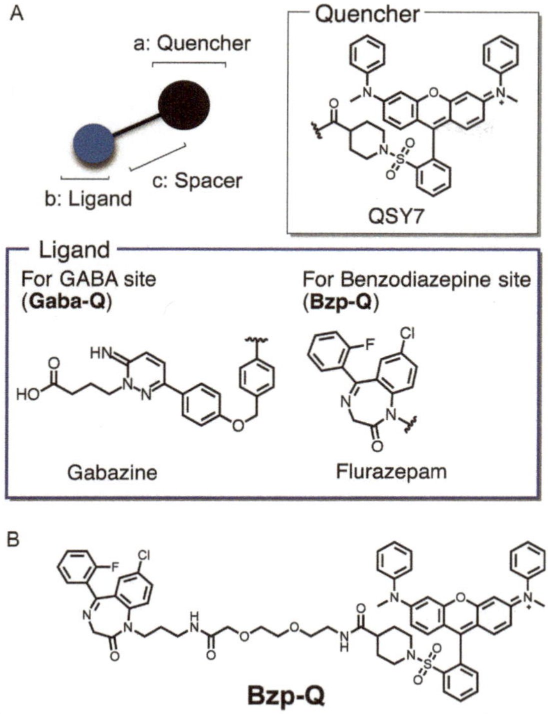

Fig. 9 Molecular design of quencher–ligand conjugates for $GABA_ARs$. (A) Basic design configuration of quencher–ligand conjugates for $GABA_ARs$. (B) Chemical structure of quencher–ligand conjugate (Bzp-Q) for benzodiazepine site of $GABA_ARs$.

BFQR system, the quenching relies on FRET between the fluorescent label and the bound quencher. Thus, the absorption spectrum of the acceptor (quencher) molecule should overlap well with the emission spectrum of the donor (fluorophore) molecule for efficient quenching. In this case, we chose QSY7 as a quencher moiety, exhibiting a high quenching efficiency for labeled OG thanks to the substantial overlap of these spectra (Wruss et al., 2009).

2. The spacer moiety of quencher–ligand conjugate is also important for the fluorescence quenching because FRET is strongly dependent on the distance and relative orientation between the donor (fluorophore) and acceptor (quencher) molecules. In our example, we prepared quencher–ligand conjugates having different spacer lengths for the benzodiazepine site on $GABA_ARs$ (unpublished results); one of the optimized structures is **Bzp-Q** as shown in Fig. 9B.

5.2 Construction of Fluorescent Biosensor for GABA$_A$Rs Employing the BFQR System

A schematic illustration of BFQR system on GABA$_A$Rs is shown in Fig. 10A. The fluorescence intensity following LDAI labeling, where the fluorophore is covalently attached in the vicinity of the ligand-binding site, is defined as F_0. The labeled fluorescence can then be efficiently quenched upon the juxtaposition of a quencher by addition of the quencher–ligand conjugate, and the fluorescence intensity in this step is defined as F_1. When small molecules having a sufficient affinity to the binding site identical with that of the ligand–quencher conjugate are added to the culture medium, the quencher–ligand conjugate bound to the ligand-binding site is competitively released, which results in the fluorescence recovery (F_2). This system can provide the "turn-on" type of fluorescent biosensor (Fig. 10A and B). Using this strategy, we developed two fluorescent sensors, each of which can respond to orthosteric GABA-site ligands or allosteric benzodiazepine-site ligands for GABA$_A$Rs (Yamaura et al., 2016). Importantly, these biosensors discriminated between the target-site ligands and those designed to interact with other sites (Fig. 10C). Moreover, we can quantitatively determine the ligand affinity by the concentration-dependent change of the fluorescence ratio (F_2/F_1), which is discussed in Section 6. Here, the procedures for construction of a fluorescent biosensor for benzodiazepine are described in detail.

1. Label GABA$_A$Rs using **CGAM-Bzp** on glass-bottom dishes as described in Section 4.1.
2. After removal of the medium, add 100 μL of serum-free DMEM and obtain the confocal images (F_0).
3. Add 100 μL of serum-free DMEM including 2 μ*M* **Bzp-Q** (total volume 200 μL) and incubate for 1 min.
4. Obtain the confocal image (F_1).
5. Add 200 μL of serum-free DMEM including the ligand (e.g., 2 μ*M* flumazenil) and 1 μ*M* **Bzp-Q** (total volume 400 μL) and incubate for 1 min.
6. Obtain the confocal image (F_2).
7. Using imageJ software (NIH), subtract the background fluorescence in all images.
8. Measure fluorescence intensity (F_0, F_1, and F_2) from plasma membrane in single cells in these images using imageJ software.
9. *Caution*: The DMSO stock of quencher–ligand conjugates should be stored in the freezer. After diluting the stock solution with serum-free DMEM, immediately utilize for experiment to prevent aggregation in the solution or adsorption to the microtube.

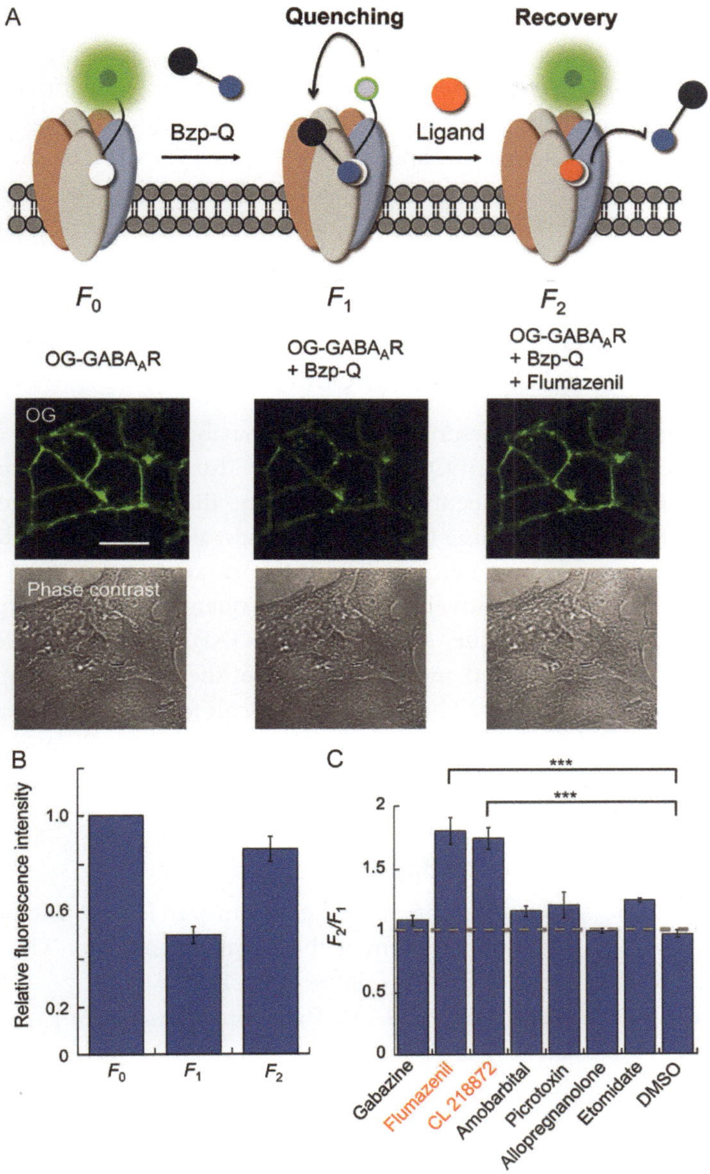

Fig. 10 Construction of GABA$_A$R-based biosensors for benzodiazepine ligands by BFQR system. (A) Fluorescence quenching of OG-labeled GABA$_A$R by 1 μ*M* Bzp-Q and the recovery following addition of 200 μ*M* flumazenil. Schematic illustration (*upper*) and representative confocal images (*lower*) are shown. HEK293T cells transfected with α1/β3/γ2 GABA$_A$Rs were labeled by 2 μ*M* CGAM-Bzp(OG-GABA$_A$R). Scale bars: 20 μm. (B) Relative intensity of labeled (F_0), quenched (F_1), and recovered fluorescence (F_2).

Caution in the fluorescent assay: This BFQR method is powerful for detecting ligands which competitively bind to the quencher–ligand conjugate. However, it was found that these sensors can detect not only competitive ligands but also noncompetitive ligands, which induce structural changes of the receptor upon the ligand binding in some cases (Yamaura et al., 2016). Therefore, the competitive radioligand binding assay should be considered for confirming whether these ligands bind to the receptors competitively or noncompetitively.

6. QUANTITATIVE ANALYSIS OF LIGAND AFFINITY USING BFQR-BASED BIOSENSORS

In this section, we describe quantitative analysis of the ligand affinity on $GABA_A$Rs using the BFQR system. The "turn-on" type fluorescent sensors usually show concentration-dependent fluorescent changes after addition of the corresponding $GABA_A$Rs ligands, and the apparent binding constants (K_{app}) can be determined from the titration (saturation) curve. When the ligand competitively binds to the quencher–ligand conjugate, the real affinity (the K_d value) of the ligand to $GABA_A$Rs can be calculated using the K_{app} of the ligand and the K_a value of the quencher–ligand conjugate (($K_{a(quencher)}$) value). The detailed protocols for the determination of the K_d values of benzodiazepine-site ligands are described as follows; the K_d values of GABA-site ligands can be determined by similar procedures.

6.1 K_d Determination of Bzp-Q

To calculate K_d values of competitive ligands in our BFQR method, we need the $K_{a(quencher)}$ value of the quencher–ligand conjugate. The procedures for **Bzp-Q** are described as follows (Fig. 11).

1. Label $GABA_A$Rs using **CGAM-Bzp** on glass-bottom dishes as described in Section 4.1.

Fig. 10—Cont'd The fluorescence intensity was obtained from confocal imaging, $n=3$. (C) Specificity of the fluorescent biosensor for $GABA_A$R ligands. The F_2/F_1 ratio after addition of each ligand is shown. Each ligand concentration is the following: gabazine (1 m*M*), flumazenil (10 μ*M*), CL217782 (0.05 m*M*), amobarbital (1 m*M*), picrotoxin (1 m*M*), allopregnanolone (0.05 m*M*), and etomidate (1 m*M*), $n=3$. $^{***}P<0.001$. Data represent mean ± SEM. *Reproduced from Yamaura, K., Kiyonaka, S., Numata, T., Inoue, R., & Hamachi, I. (2016). Discovery of allosteric modulators for $GABA_A$ receptors by ligand-directed chemistry.* Nature Chemical Biology, 12, *822–830, with permission from Nature Publishing Group.*

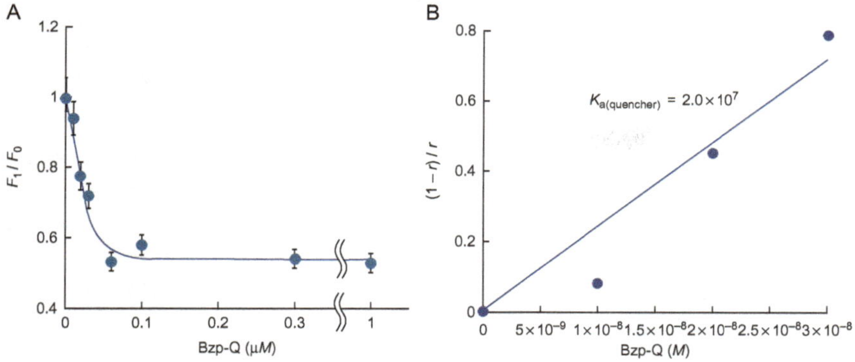

Fig. 11 K_d determination for Bzp-Q. (A) Concentration-dependency of Bzp-Q for the F_1/F_0 ratio of OG-GABA$_A$R, $n=3$. Data represent mean ± SEM. (B) The values determined from the fluorescence titration curve (A) can be plotted to obtain a linear curve with slope. The K_d value (50 n*M*) was directly calculated from the curve by the Hill plot. *Reproduced from Yamaura, K., Kiyonaka, S., Numata, T., Inoue, R., & Hamachi, I. (2016). Discovery of allosteric modulators for GABA$_A$ receptors by ligand-directed chemistry.* Nature Chemical Biology, 12, *822–830, with permission from Nature Publishing Group.*

2. After removal of the medium, add 100 μL of serum-free DMEM and obtain the confocal images (F_0).
3. Add 100 μL of serum-free DMEM including each concentration of **Bzp-Q** (total volume 200 μL) and incubate for 1 min.
4. Obtain the confocal image (F_1) at each concentration of **Bzp-Q**.
5. Using the imageJ software, subtract the background fluorescence in all images.
6. Determine K_d value of **Bzp-Q** by the Hill plot using the following equation.

$$(1-r)/r = K_{a(\text{quencher})} \times [\text{quencher}]$$
$$K_{d(\text{quencher})} = 1/K_{a(\text{quencher})}$$
$$\text{r : relative fluorescence intensity}$$

Note: F_0 and saturated F_1 value are defined as 1 and 0, respectively.

6.2 K_d Determination of Competitive Ligands on Benzodiazepine Site

K_{app} of the each ligand for benzodiazepine site was evaluated by fluorescence titration assay, in which the fluorescence intensity on the cell surfaces was measured upon addition of each concentration of the ligand. The real K_d

value for the ligand was determined using K_{app} of the ligand and $K_{a(quencher)}$ of **Bzp-Q**. As a typical example, procedures for determining the K_d value of flumazenil, a typical benzodiazepine-site ligand, are described as follows (Fig. 12).

1. Label $GABA_A$Rs using **CGAM-Bzp** on glass-bottom dishes as described in Section 4.1.
2. After removal of the medium, add 100 μL of serum-free DMEM and obtain confocal images (F_0).
3. Add 100 μL of serum-free DMEM including 2 μ*M* **Bzp-Q** (total volume 200 μL) and incubate for 1 min.
4. Obtain the confocal image (F_1).
5. Add 200 μL of serum-free DMEM including each concentration of flumazenil and 1 μ*M* **Bzp-Q** (total volume 400 μL) and incubate for 1 min.
6. Obtain the confocal image (F_2) at each concentration of flumazenil.
7. Using the imageJ software, subtract the background fluorescence in all images.
8. To calculate the apparent association constant for flumazenil ($K_{app(ligand)}$), adapt these values to the following equation.

$$r/(1-r) = K_{app} \times [\text{ligand}]$$
$$r: \text{relative fluorescence intensity}$$

Note: F_0 and saturated F_2 values were defined as 0 and 1, respectively.

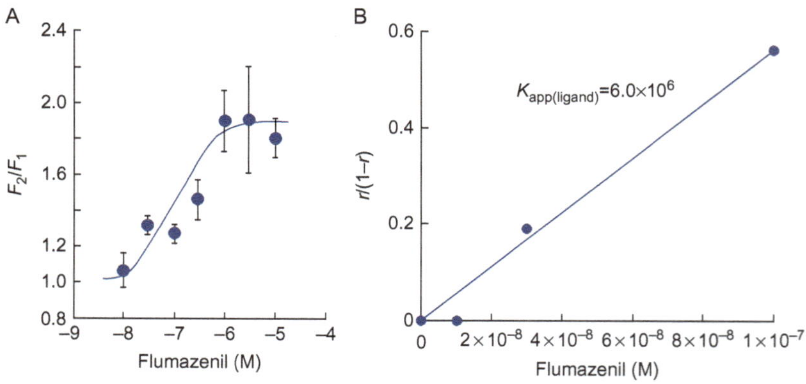

Fig. 12 K_d determination for flumazenil. (A) Concentration-dependency of flumazenil for the F_2/F_1 ratio of OG-$GABA_A$R, $n=3$. Data represent mean ± SEM. (B) The values determined from the fluorescent titration curve (A) can be plotted to obtain a linear curve with slope. The K_{app} value (6.0×10^6 M^{-1}) was calculated from the curve by the Hill plot, which is utilized for calculation of K_d value of flumazenil.

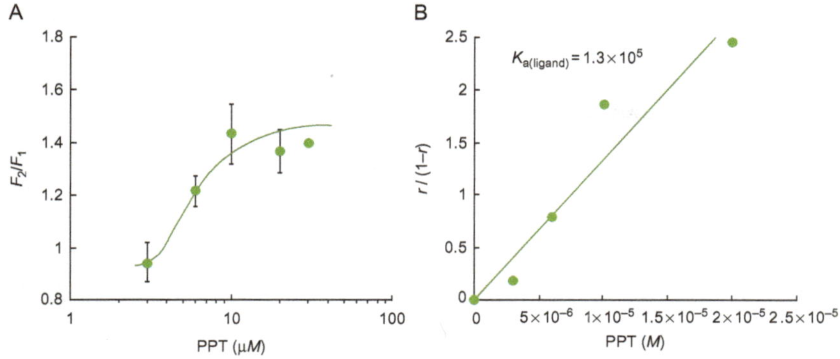

Fig. 13 K_d determination for PPT. (A) Concentration-dependency of PPT for the F_2/F_1 ratio of OG-GABA$_A$R, $n=3$. Data represent mean ± SEM. (B) The values determined from the fluorescent titration curve (A) can be plotted to obtain a linear curve with slope. The K_d value (7.3 μM) was directly calculated from the curve by the Hill plot.

9. To calculate the real K_d value for flumazenil ($K_{d(ligand)}$), adapt calculated K_{app} value and calculated $K_{a(quencher)}$ value to the following equation

$$K_{a(ligand)} = K_{app(ligand)} + \left(K_{app(ligand)} \times K_{a(quencher)} \times [\text{quencher}]\right)$$
$$K_{d(ligand)} = 1/K_{a(ligand)}$$

Caution in calculating the K_d value: In addition to such competitive ligands, it should be noted that our BFQR-based sensors detect noncompetitive ligands which induce structural changes of the receptor upon the ligand binding in some case. In this case, we can directly determine the K_d value of the noncompetitive ligands from the fluorescent titration assay. As a typical example, procedures for determining the K_d value of PPT, a non-competitive benzodiazepine-site ligand, are shown in Fig. 13.

7. SUMMARY AND FUTURE DIRECTIONS

We describe a strategy and protocols for constructing "turn-on" type fluorescent biosensors based on the GABA$_A$Rs scaffold by coupling ligand-directed chemistry with the BFQR fluorescent readout system. It has been demonstrated that our methods allow for quantitative analysis of ligand affinity for GABA$_A$Rs on the live cell surface. Importantly, we can selectively label structurally complicated membrane receptors with fluorophores without any structural information by ligand-directed chemistry. Thus, our biosensor system may be extended to various types of membrane-receptors on live cells, if appropriate small molecule ligands are known.

Recently, protein-based biosensors have been recognized as powerful platforms for high-throughput screening (HTS) of drug candidates. As a successful result, Rauh's group reported a HTS for kinase inhibitors using fluorophore-kinase conjugates, termed FLiK assay that is mostly conducted in the purified test tube system (Simard, Pawar, et al., 2009). It has been demonstrated that our BFQR-based biosensors for $GABA_AR$ ligands are highly amenable to the HTS of a chemical library (Yamaura et al., 2016), which can be conducted with live cell systems. We indeed discovered new small molecules acting as negative allosteric modulators for $GABA_ARs$ by this screening. It is widely recognized that the myriad cell-surface receptors represent the largest family of drug targets. Hence, ongoing drug discovery programs may be accelerated by using our biosensor systems in the near future.

ACKNOWLEDGMENTS

This work was funded by the Japan Science and Technology Agency (JST) Core Research for Evolutional Science and Technology (CREST) of Molecular Technologies to I.H. and by SUNBOR Grant from Suntory Foundation for Life Sciences to S.K.

REFERENCES

Barnard, E. A., Skolnick, P., Olsen, R. W., Mohler, H., Sieghart, W., Biggio, G., et al. (1998). International Union of Pharmacology. XV. Subtypes of γ-aminobutyric acidA receptors: Classification on the basis of subunit structure and receptor function. *Pharmacological Reviews*, *50*, 291–314.

Brun, M. A., Tan, K. T., Griss, R., Kielkowska, A., Reymond, L., & Johnsson, K. (2012). A semisynthetic fluorescent sensor protein for glutamate. *Journal of the American Chemical Society*, *134*, 7676–7678.

Brun, M. A., Tan, K. T., Nakata, E., Hinner, M. J., & Johnsson, K. (2009). Semisynthetic fluorescent sensor proteins based on self-labeling protein tags. *Journal of the American Chemical Society*, *131*, 5873–5884.

Chebib, M., & Johnston, G. A. (2000). GABA-activated ligand gated ion channels: Medicinal chemistry and molecular biology. *Journal of Medicinal Chemistry*, *43*, 1427–1447.

Chen, C., Ke, J., Zhou, X. E., Yi, W., Brunzelle, J. S., Li, J., et al. (2013). Structural basis for molecular recognition of folic acid by folate receptors. *Nature*, *500*, 486–489.

Fehr, M., Frommer, W. B., & Lalonde, S. (2002). Visualization of maltose uptake in living yeast cells by fluorescent nanosensors. *Proceedings of the National Academy of Sciences*, *99*, 9846–9851.

Fujishima, S. H., Yasui, R., Miki, T., Ojida, A., & Hamachi, I. (2012). Ligand-directed acyl imidazole chemistry for labeling of membrane-bound proteins on live cells. *Journal of the American Chemical Society*, *134*, 3961–3964.

Hanson, S. M., & Czajkowski, C. (2008). Structural mechanisms underlying benzodiazepine modulation of the $GABA_A$ receptor. *The Journal of Neuroscience*, *28*, 3490–3499.

Loving, G. S., Sainlos, M., & Imperiali, B. (2010). Monitoring protein interactions and dynamics with solvatochromic fluorophores. *Trends in Biotechnology*, *28*, 73–83.

Marvin, J. S., Borghuis, B. G., Tian, L., Cichon, J., Harnett, M. T., Akerboom, J., et al. (2013). An optimized fluorescent probe for visualizing glutamate neurotransmission. *Nature Methods*, *10*, 162–170.

Marvin, J. S., & Hellinga, H. W. (1998). Engineering biosensors by introducing fluorescent allosteric signal transducers: Construction of a novel glucose sensor. *Journal of the American Chemical Society*, *120*, 7–11.

Masharina, A., Reymond, L., Maurel, D., Umezawa, K., & Johnsson, K. (2012). A fluorescent sensor for GABA and synthetic GABAB receptor ligands. *Journal of the American Chemical Society*, *134*, 19026–19034.

Matsuo, K., Kioi, Y., Yasui, R., Takaoka, Y., Miki, T., Fujishima, S. H., et al. (2013). One-step construction of caged carbonic anhydrase I using a ligand-directed acyl imidazole-based protein labeling method. *Chemical Science*, *4*, 2573–2580.

Mayr, L. M., & Bojanic, D. (2009). Novel trends in high-throughput screening. *Current Opinion in Pharmacology*, *9*, 580–588.

Miki, T., Awa, M., Nishikawa, Y., Kiyonaka, S., Wakabayashi, M., Ishihama, Y., et al. (2016). A conditional proteomics approach to identify proteins involved in zinc homeostasis. *Nature Methods*, *13*, 931–937.

Miki, T., Fujishima, S. H., Komatsu, K., Kuwata, K., Kiyonaka, S., & Hamachi, I. (2014). LDAI-based chemical labeling of intact membrane proteins and its pulse-chase analysis under live cell conditions. *Chemistry & Biology*, *21*, 1013–1022.

Morii, T., Sugimoto, K., Makino, K., Otsuka, M., Imoto, K., & Mori, Y. (2002). A new fluorescent biosensor for inositol trisphosphate. *Journal of the American Chemical Society*, *124*, 1138–1139.

Okumoto, S., Looger, L. L., Micheva, K. D., Reimer, R. J., Smith, S. J., & Frommer, W. B. (2005). Detection of glutamate release from neurons by genetically encoded surface-displayed FRET nanosensors. *Proceedings of the National Academy of Sciences of the United States of America*, *102*, 8740–8745.

Olsen, R. W., & Sieghart, W. (2009). $GABA_A$ receptors: Subtypes provide diversity of function and pharmacology. *Neuropharmacology*, *56*, 141–148.

Prével, C., Pellerano, M., Van, T. N. N., & Morris, M. C. (2014). Fluorescent biosensors for high throughput screening of protein kinase inhibitors. *Biotechnology Journal*, *9*, 253–265.

Rudolph, U., & Knoflach, F. (2011). Beyond classical benzodiazepines: Novel therapeutic potential of $GABA_A$ receptor subtypes. *Nature Reviews. Drug Discovery*, *10*, 685–697.

Sieghart, W. (2015). Allosteric modulation of $GABA_A$ receptors via multiple drug-binding sites. *Advances in Pharmacology*, *72*, 53–96.

Simard, J. R., Getlik, M., Grütter, C., Pawar, V., Wulfert, S., Rabiller, M., et al. (2009). Development of a fluorescent-tagged kinase assay system for the detection and characterization of allosteric kinase inhibitors. *Journal of the American Chemical Society*, *131*, 13286–13296.

Simard, J. R., Klüter, S., Grütter, C., Getlik, M., Rabiller, M., Rode, H. B., et al. (2009). A new screening assay for allosteric inhibitors of cSrc. *Nature Chemical Biology*, *5*, 394–396.

Simard, J. R., Pawar, V., Aust, B., Wolf, A., Rabiller, M., Wulfert, S., et al. (2009). High-throughput screening to identify inhibitors which stabilize inactive kinase conformations in p38α. *Journal of the American Chemical Society*, *131*, 18478–18488.

Takaoka, Y., Nishikawa, Y., Hashimoto, Y., Sasaki, K., & Hamachi, I. (2015). Ligand-directed dibromophenyl benzoate chemistry for rapid and selective acylation of intracellular natural proteins. *Chemical Science*, *6*, 3217–3224.

Takikawa, K., Asanuma, D., Namiki, S., Sakamoto, H., Ariyoshi, T., Kimpara, N., et al. (2014). High-throughput development of a hybrid-type fluorescent glutamate sensor for analysis of synaptic transmission. *Angewandte Chemie, International Edition, 53*, 13439–13443.

Tamura, T., & Hamachi, I. (2014). Recent progress in design of protein-based fluorescent biosensors and their cellular applications. *ACS Chemical Biology, 9*, 2708–2717.

Tretter, V., Ehya, N., Fuchs, K., & Sieghart, W. (1997). Stoichiometry and assembly of a recombinant $GABA_A$ receptor subtype. *The Journal of Neuroscience, 17*, 2728–2737.

Tsukiji, S., Miyagawa, M., Takaoka, Y., Tamura, T., & Hamachi, I. (2009). Ligand-directed tosyl chemistry for protein labeling in vivo. *Nature Chemical Biology, 5*, 341–343.

Wang, H., Koshi, Y., Minato, D., Nonaka, H., Kiyonaka, S., Mori, Y., et al. (2011). Chemical cell-surface receptor engineering using affinity-guided, multivalent organocatalysts. *Journal of the American Chemical Society, 133*, 12220–12228.

Wruss, J., Pollheimer, P. D., Meindl, I., Reichel, A., Schulze, K., Schöfberger, W., et al. (2009). Conformation of receptor adopted upon interaction with virus revealed by site-specific fluorescence quenchers and FRET analysis. *Journal of the American Chemical Society, 131*, 5478–5482.

Yamaura, K., Kiyonaka, S., Numata, T., Inoue, R., & Hamachi, I. (2016). Discovery of allosteric modulators for $GABA_A$ receptors by ligand-directed chemistry. *Nature Chemical Biology, 12*, 822–830.

CHAPTER TEN

Measuring and Imaging Metal Ions With Fluorescence-Based Biosensors: Speciation, Selectivity, Kinetics, and Other Issues

Richard B. Thompson*[,1], Carol A. Fierke†
*University of Maryland School of Medicine, Baltimore, MD, United States
†University of Michigan, Ann Arbor, MI, United States
[1]Corresponding author: e-mail address: rthompson@som.umaryland.edu

Contents

Abstract

Fluorescence-based biosensors have shown themselves to be a powerful tool for the study of a variety of chemical species in biological systems, notably including metal ions. This chapter provides an overview of several important issues in using such sensors to study metallobiochemistry. These issues include selectivity for the analyte over potential interferents, including those that do not themselves induce a signal, the different forms in which metal ions are found (speciation), the utility of metal ion buffers, and the importance of kinetics in studying metal ion binding reactions. Finally, the chapter briefly discusses some of the issues in understanding whole-organism distribution of metal ions and its control.

1. INTRODUCTION

Understanding metallobiochemistry, particularly in cells and living systems, has been greatly aided by the development of indicators (also called

Methods in Enzymology, Volume 589
ISSN 0076-6879
http://dx.doi.org/10.1016/bs.mie.2017.01.003

sensors or probes) that transduce the presence or level of analytes, such as metal ions, as a change in fluorescence (reviewed in Carter, Young, et al., 2014; Que, Domaille, et al., 2008; Thompson, 2006; White & Argauer, 1970). These indicators have been especially useful in biology because of their ready adaptation to microscopy and imaging, correlating the chemical information they provide to the structure of the cell or organism (Negrin & Contag, 2006; Ntziachristos, 2006). A valuable subset of these indicators include those employing biological molecules, to provide high affinity, high selectivity, and other desirable features for these applications; they are the subject of this volume.

However, use of fluorescent biosensors (and other indicators) requires care to obtain accurate, valid data from which biological insight can be derived. Particularly for analytes such as metal ions, consideration of issues such as selectivity, speciation, kinetics, and others are important. This chapter is devoted to these considerations; we and others have previously discussed some of these issues (Bozym, Hurst, et al., 2008; Donat & Bruland, 1990; Stumm & Morgan, 1996). We note that these issues apply in varying degree to measurements with other indicators (such as small molecules), and for other analytes, such as reactive oxygen species. We also note that a considerable proportion of the scientific literature focused on the biology of metal ions neglects these issues to a degree, reducing confidence in observations and conclusions. A key goal of this chapter is not only to provide guidance on the use of fluorescence biosensors to determine metal ions, but also to support valid experiments in this field.

2. SPECIATION OF METAL IONS OF BIOLOGICAL INTEREST

Since a large fraction of biological chemistry takes place in solutions (particularly aqueous solutions) and other condensed phases, metal ions of biological interest rarely are seen in an unliganded form; i.e., without a solvent molecule, counterion, or other ligand molecule closely associated with them. While these ligand associations may be weak or transient, the high concentrations of water, chloride, carbonate/bicarbonate, and other species normally present in biological media assure that virtually all metal ions are bound to one or more such ligands. The identity, number, and distribution of these and other ligands bound to the metal ion are referred to as the metal's "speciation" in solution (Stumm & Morgan, 1996) which depends on the chemical properties of both the metal ion and the ligands. Thus, Zn^{2+} prefers to coordinate nitrogen or sulfur ligands compared to oxygen, frequently in a tetrahedral arrangement around the metal ion; Cu^{2+} prefers sulfur ligands in a trigonal bipyramidyl or octahedral arrangement (Cotton & Wilkinson, 1988). Speciation also refers to the ionization state of the metal for elements that ordinarily exhibit multiple oxidation states like copper or iron. Inside cells, metals such as zinc, copper, and iron are found almost exclusively bound tightly to multidentate ligands such as proteins, where they often serve as enzyme cofactors essential for catalysis or for stabilizing protein structure.

We can distinguish ligands based on the strength (affinity) of their association, expressed as the stability of the complex(es) they make with particular metals. The association reaction in the simplest case of one ligand L binding to the metal M to form a complex LM is expressed as:

$$L + M \leftrightarrow LM \tag{1}$$

The equilibrium dissociation constant K_D (inverse of the association constant, K_A) is expressed in the usual way:

$$K_D = [L][M]/[LM] \tag{2}$$

Since such equilibrium constants depend strongly on pH, ionic strength, and ionization state, they are referred to as "conditional stability constants." If the stability constants, formation constants, concentrations, and pK_a's of the various ligands with the metal ion(s) are known (Smith & Martell, 1973), the proportion of the metal ion(s) in each form can be calculated using computer programs such as MINEQL (Environmental Research Software, Hallowell, Maine). An example is shown in Fig. 1 of the pH dependence of zinc ion speciation in a fresh water (e.g., not physiological) medium (Stumm & Morgan, 1996).

In this fresh water example (total concentrations $[Zn^{2+}] = 10$ nM, $[Mg^{2+}] = 3$ mM, $[Ca^{2+}] = 1$ mM, $[CO_3{}^{2-}] + [HCO_3{}^{-}] = 2$ mM, $[SO_4{}^{2-}] =$ 0:3 mM, $[Na^{+}] = 0.25$ mM, $[Cl^{-}] = 0.25$ mM), it is noteworthy that at pH7.5 more than a third of the zinc ion is bound to carbonate; the physiological concentration of bicarbonate/carbonate (26 mM in serum) is nearly 10-fold higher than in this example. Also noteworthy is the increase in the forms of zinc ion containing hydroxide ligands ($ZnOH^{+}$ and $Zn(OH)_2$)

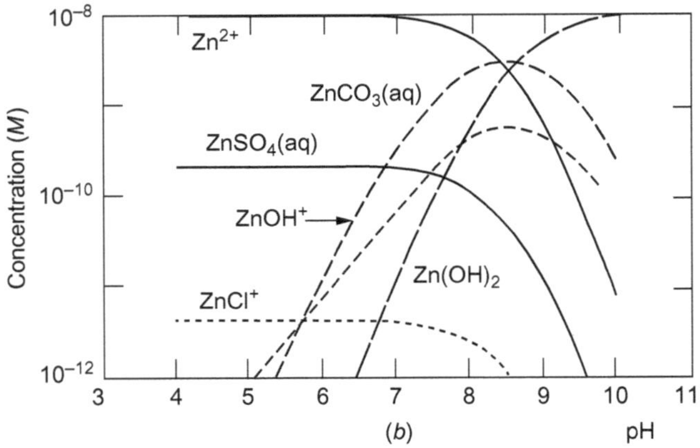

Fig. 1 Zinc ion speciation in fresh water as a function of pH. *Reproduced with permission from Stumm, W., & Morgan J. J. (1996).* Aquatic chemistry: Chemical equilibria and rates in natural waters. *New York: Wiley-Interscience.*

with increasing pH, this contributes to the formation of insoluble precipitates in millimolar zinc solutions at or above neutral pH (Stumm & Morgan, 1996).

As we and others have discussed previously (Bozym et al., 2008; Colvin, Holmes, et al., 2010; Krezel & Maret, 2006), in biological systems, only a small proportion of many of the active, multivalent metal ions such as Zn, Fe, Cu, Mg, Ca, Mn, Co, and Ni, are in complexes with weak, rapidly exchangeable ligands like chloride or water, rather than in high-affinity complexes with multidentate ligands. In the case of zinc at pH 7.5, rapidly exchangeable complexes include at least $ZnHCO_3{}^+$, $ZnCl^+$, $ZnOH^+$, ZnOHCl, and $ZnCl_2$. We term this fraction of the total zinc ion operationally as "free zinc"; the concept is probably similar to what others have described as "labile," "mobile," or "bioavailable." These zinc species are capable of rapidly exchanging ligands to bind to proteins, other biological ligands and fluorescent indicators. In some cases, protein-binding sites recognize and bind metal complexes (Franklin, Ma, et al., 2003); for example, transferrin binds an iron-carbonate complex (Harris, 2012). Although the total zinc concentration in cells is 0.1–0.5 mM (Eide, 2006), the high concentrations of fairly tight metal ion ligands (e.g., ~3 mM reduced glutathione, 2–20 mM glutamic acid, etc.) in addition to proteins that tightly bind zinc (Auld, 2001; Christianson, 1991) decrease the "free" concentration of zinc to the picomolar range (Bozym, Thompson, et al., 2006;

Dittmer, Miranda, et al., 2009; Krezel & Maret, 2006; Van Dongen, Evers, et al., 2007). The pool of zinc ligands in the cell may function as a "zinc buffer" to maintain the free zinc concentration. However, the cellular free zinc concentration can alter dramatically (p*M* to n*M*) with modest changes in the total zinc concentration, as the buffering capacity is overwhelmed by increased zinc (Wang, Hosteen, et al., 2012).

3. METAL ION BUFFERS AND THEIR FORMULATION

Determining the effects in vitro of metal ions present at low concentrations (and particularly low *free* concentrations) in biological media on various processes is challenging, for two major reasons. First, there are many sources of potential contamination with the metal ion of interest, or similar ones. For instance, a physiological buffer containing 125 m*M* NaCl made with "ACS reagent grade" sodium chloride likely contains significant quantities of potential interferents: the American Chemical Society specifies <5 ppm heavy metals (determined as lead) and <2 ppm iron. Thus the buffer containing NaCl might have *several hundred nanomolar* total zinc in addition to whatever else is present, making it difficult to prepare a buffer that contains 5 n*M* total zinc with this much NaCl. We note that other chemical constituents of the media are also likely to be contaminated with zinc (or other interfering metals like copper); borosilicate glassware, some plastic ware, Kimwipe tissues, paper, dust, and human skin are also well-known sources of contaminating metal ions. Second, in most biological media, there are abundant ligands that bind many metal ions with medium to high affinity, effectively lowering the concentration of "free metal" (see Section 2). For these reasons, accurately preparing solutions of metal ions at low concentrations is usually impractical.

A better approach is to prepare metal ion buffers wherein the amount of "free" metal ion changes less following addition or subtraction (by becoming bound to a ligand) of the metal ion than it would in the buffer's absence (Baker, 1988). The effect is closely analogous to the behavior of a pH buffer, where addition of acid or base results in a smaller change in pH than in the buffer's absence. For metal ion buffers, a suitable, fairly strong ligand is added to the solution in excess under conditions where it binds >99% of the metal ion present; importantly, the proportion of metal ion bound by the added ligand is constant under a given set of conditions, such that over a broad range the concentration of free metal ion rises linearly with added metal ion, but much less steeply. For example, suppose we formulate a zinc ion buffer using 2 mM nitrilotriacetic acid (NTA, log $K = 10.4$) as the ligand at pH 7.0, 25°C, and low ionic strength (Bozym et al., 2008). If 0.5 μM total zinc ion is present, the free zinc ion concentration will be 1.06 pM with essentially all of the rest bound to NTA. If additional 1.5 μM total zinc is added as a contaminant, the free concentration only increases to 4.26 pM: i.e., the deleterious effects of the contaminating zinc are minimized. As shown in Fig. 1, metal speciation is pH dependent, so it is important to include a pH buffer in the metal buffer solution. We typically use the sulfonated Good's buffers (HEPES, MOPS, MES, etc.) as pH buffers as they generally have low affinity for metal (Mash, 2003). Other pH buffers, including Tris, phosphate, Bicine, and citrate, have higher metal ion affinity (Dawson, Elliott, et al., 1989); phosphate is additionally problematic due to the low solubility of its zinc salts ($K_{sp} \approx 10^{-33}$). Furthermore, in some cases, the protein can bind a metal–ligand complex (such as Me-NTA) rather than the metal alone. In this case, the concentration of the complex is significantly higher than the concentration of the "free metal ion." Therefore if the metal affinity is calculated using the free metal ion, rather than the complex, the calculated metal affinity will likewise be incorrect. To check for this, measure fraction bound metal at a constant "free metal" but varying metal buffer concentration. Additionally, binding of the metal to the protein should not significantly alter the free metal concentration.

4. SELECTIVITY OF METAL ION BINDING TO INDICATORS AND ITS ASSESSMENT

Due to the many different metal ions normally present in biological media, selectivity is an essential characteristic of indicators designed to quantify them. Thus while an indicator may have high sensitivity for a particular metal ion due to binding it with high affinity, it is equally important that other metal ions present do not compete for that binding site, or they will interfere with the measurement. Thus a sensor useful for determining free zinc ion in cells not only must have picomolar affinity but also must not bind metal ions such as Mg^{2+} or Ca^{2+}, which may be present at billion-fold higher concentrations. A major reason for adapting biological molecules as sensors for metal ions and other analytes is their high selectivity (Thompson, 2006), as they evolved to bind specific metal ions in the presence of other analytes in the cell.

Many fluorescent indicators act by the analyte inducing a change in fluorescence upon binding. It is important to demonstrate both that competing analytes do not induce a change in fluorescence and that the interferent does not bind to the indicator, competing with the analyte binding to the indicator, and inducing a spectral change. Furthermore, a compound or ion can also interfere if it perturbs the affinity of the binding site even if it does not occupy the same site.

To correctly assess interference, one needs to determine whether the compound/ion affects the affinity and/or the signal change upon binding of the analyte to the indicator. The simplest way to detect this is to assess the response of the indicator to a relevant concentration of the analyte (i.e., when [analyte] $= K_D$) in the presence and absence of a relevant concentration of the prospective interferent. The fractional occupancy, ν_i of the indicator binding site by each compound i (analyte and interferent) can be calculated if their concentrations $[M_i]$ and dissociation constants K_i are known:

$$\nu = [M_i]/K_i/1 + \Sigma([M_i]/K_i) \quad (3)$$

For instance, suppose, we wish to know what fraction of human carbonic anhydrase II active sites will be occupied by Zn^{2+} and Cu^{2+} present at free concentrations of 5 and 0.5 nM, respectively; this presumes that binding to the protein does not deplete the free concentration of either metal ion. Since the K_D's for copper and zinc are roughly 10^{-13} and 10^{-12} M, respectively (Huang, Lesburg, et al., 1996; McCall & Fierke, 2004), we can calculate that the binding site is 50% bound with copper and 50% bound with zinc ion under these conditions. Even though its concentration is lower than zinc, the copper ion interferes significantly because its affinity is 10-fold higher. In this example, it is evident that potential interferents with high affinities can interfere dramatically even if their concentrations are quite low and irrespective of whether they induce any spectroscopic change at all. If the interferent does not induce a fluorescence change, it may be necessary to determine its ability to interfere by competing with the analyte; for metal ions this means formulating buffers with the desired range of free interferent metal ions as well as analyte ions.

For example, we wish to determine if calcium or magnesium ions at levels found in serum interfere with zinc determination using the fluorescent zinc indicator FuraZin-1. Upon binding Zn, FuraZin-1 exhibits a change in the ratio of its emission observed at 510 nm when excited at 325 nm, to that

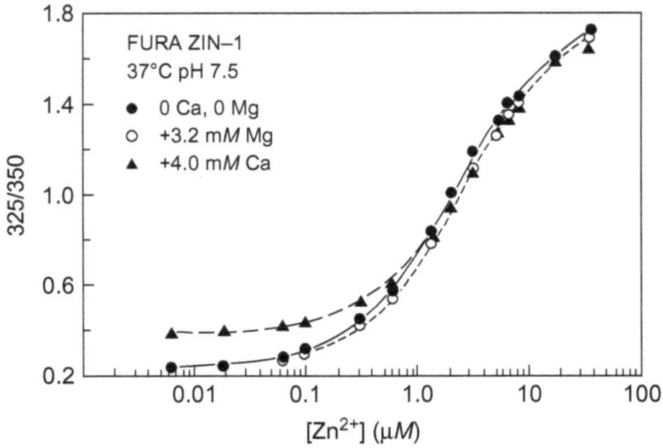

Fig. 2 Zinc-dependent fluorescence excitation ratios (exc 325 nm/350 nm, em 510 nm) for FuraZin-1 in the absence (•) and presence of Mg^{2+} (○) or Ca^{2+} (▲), and best-fit binding isotherms. *Reproduced from Thompson, R. B., Peterson, D., et al. (2002). Fluorescent zinc indicators for neurobiology.* Journal of Neuroscience Methods, 118, *63–75, with permission.*

when excited at 350 nm; this is termed excitation ratiometric (Haugland, 2005). We formulated zinc buffers (124 m*M* NaCl, 1.75 m*M* KCl, 10 m*M* MOPS, 10 m*M* Bicine, pH 7.4 at 37°C) in the absence and presence of relevant calcium (4 m*M*) and magnesium (3.2 m*M*) concentrations (Thompson, Peterson, et al., 2002) having the same range of free zinc concentrations (6 n*M* to 40 μ*M*), and measured the excitation ratio. When the measured ratios were fit to a simple binding isotherm (Fig. 2), the apparent K_D's were 2.1 ± 0.1, 2.9 ± 0.09, and 2.52 ± 0.05 μ*M* for no Ca or Mg, with Ca, and with Mg, respectively. As the figure and derived values make clear, both potential interfering cations exert a modest change on the apparent K_D for zinc that would result in a small error, particularly at lower free zinc concentrations. For most purposes, however, these small errors would be viewed as negligible for measuring free zinc ion in this concentration range. For an example of a substantial interference, see Fig. 4.

5. ADVANTAGES OF BIOMOLECULES FOR METAL ION SENSING

Among the important advantages of biological molecules, such as proteins, as metal ion sensors (e.g., biosensors) is that their metal ion binding properties can be modified and improved either by design, or evolution in vitro. In particular, the binding affinity, selectivity, and binding kinetics of metal ion binding to the biosensor based on human carbonic anhydrase II have all been improved by site-directed mutagenesis (reviewed by Fierke & Thompson, 2001; Hurst, Wang, et al., 2010). Similar optimization has been carried out on other protein-based zinc sensors (Van Dongen et al., 2007), as well as for sensors for other analytes. By comparison, small molecule fluorescent zinc indicators have shown less adaptability in affinity, selectivity, or kinetics since most use variants of one of three zinc-binding moieties: di-(2-picolyl) amine, iminodiacetate, or tetraazacyclododecane (reviewed in Burdette & Lippard, 2001; Kikuchi, Komatsu, et al., 2004; Que et al., 2008). Iminodiacetate has higher affinity for zinc and copper(II) than calcium or magnesium ions, but often levels of the latter two are high enough to interfere. The di-(2-picolyl) amine and tetraazacyclododecane binding moieties give adequate selectivity against Ca and Mg which may be present at micromolar to millimolar ranges, and have nanomolar or tighter affinity for zinc (Aoki, Kaido, et al., 2003; Burdette, Walkup, et al., 2001). Other binding moieties such as 8-aminoquinoline (Nolan, Jaworski, et al., 2005) give faster response and adequate selectivity against intracellular (e.g., high nanomolar) calcium levels, but have an affinity that is too weak for measuring most physiologically relevant (subnanomolar) zinc levels.

For example, the kinetics of zinc ion binding to wild-type apocarbonic anhydrase II (CA II) is slow and it would be desirable to speed up zinc equilibration without significantly decreasing the affinity at the same time. Huang, Lesburg, et al. (1996) used a design process, whereas Hunt and Fierke (1997) used random mutagenesis followed by selection (e.g., evolutionary) process to identify mutants with faster zinc equilibration kinetics. Ultimately a mutant (E117A) was developed with an 800-fold faster zinc association rate constant than the wild type (e.g., nearly diffusion-controlled), which loses less than an order of magnitude in overall zinc affinity. An example of the improved response is shown in Fig. 3, which depicts zinc-binding kinetics to wild type and E117A human apocarbonic anhydrase II in the presence of ABD-N. This fluorophore binds to CA II with a concomitant increase in fluorescence anisotropy (polarization) if and only if zinc is bound to the active site (Thompson, Maliwal, et al., 1998). Under these conditions, equilibration of WT CA II with 0.9 n*M* zinc takes more than 90 min, whereas the E117A variant takes less than 20 min; the differences are more dramatic at lower free zinc concentrations. Since small molecule indicators use only a few different zinc-binding moieties to assure selectivity, it is difficult to see how the structure of the zinc-binding moieties can be altered to enhance kinetics without changing selectivity or affinity.

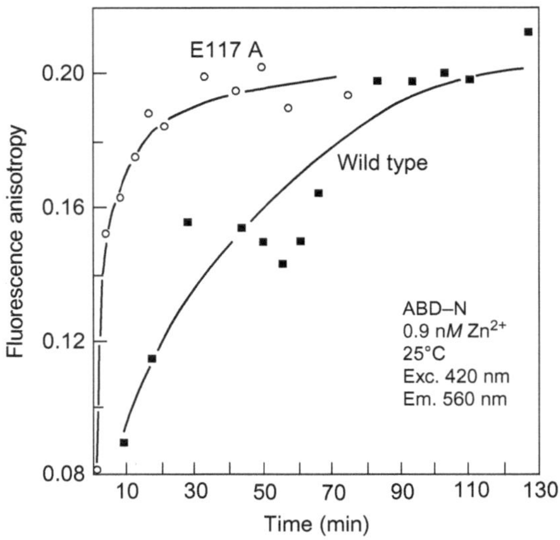

Fig. 3 Fluorescence anisotropy of ABD-N as a function of time after addition of 0.9 n*M* free Zn^{2+} and wild type (■) and E117A (○) apocarbonic anhydrase II.

The Fierke group has also demonstrated that the selectivity of CA-based metal ion sensors can be altered by mutagenesis. In particular, wild-type apocarbonic anhydrase exhibits only about 10-fold higher affinity for Cu(II) than Zn(II) (0.1 p*M* vs 1 p*M*), which makes it difficult to measure the low levels of free Cu(II) predicted to be present in most cells (Rae, Schmidt, et al., 1999) in the presence of physiological free zinc levels. This is illustrated in the upper panel of Fig. 4, which shows a fluorescence lifetime-based determination of Cu(II) in the presence and absence of 1.5 n*M* free zinc ion. It is apparent that the phase angles (inversely proportional to the fluorescence lifetime, or in this case the fraction of the apoprotein with Cu(II) bound) are substantially different in the presence and absence of this concentration of zinc. In particular, the measured phase angle at 10^{-14} *M* free copper in the presence of zinc is close to that measured at 10^{-12} *M* copper in its absence, an error of 100-fold. In this case zinc only competes with Cu(II) for binding to CA II, the bound zinc ion has little effect on the fluorescent signal. However, by use of the Cu(II)-selective Q92A variant, which has similar Cu(II) affinity to the wild type but 17-fold weaker zinc affinity, such that the selectivity is improved to 170-fold, the curves are practically superimposable (lower panel).

There are other key advantages of these protein-based biosensors. Some of ours and others incorporate the fluorescent label by fusing a Fluorescent

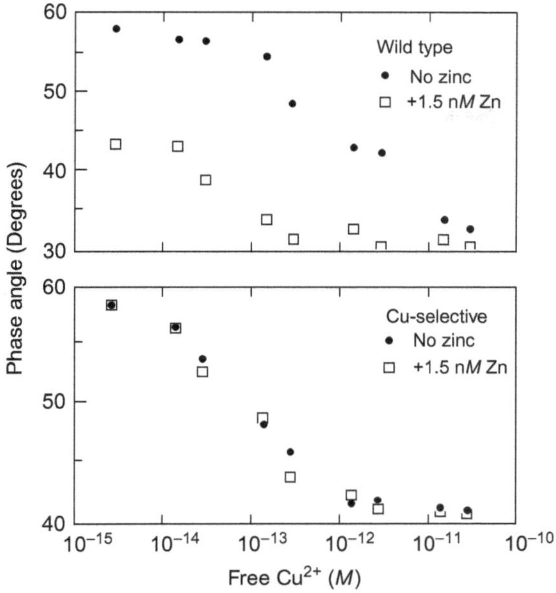

Fig. 4 Cu(II)-dependent phase angles (inversely proportional to the fraction of protein with Cu(II) bound) in the absence (•) and presence (□) of 1.5 n*M* free Zn(II) ion for wild-type (*upper panel*) and Q92A copper-selective variant (*lower panel*). *Reproduced from McCranor, B. J., Szmacinski, H., et al. (2014). Fluorescence lifetime imaging of physiological free Cu(II) levels in live cells with a Cu(II)-selective carbonic anhydrase-based biosensor.* Metallomics *6(5), 1034–1042, with permission.*

Protein variant gene to the carbonic anhydrase variant gene and expressing the fusion protein. First, by expressing the sensor (or using a hybrid approach like ours; Bozym et al., 2006) inside the cell, one is spared the difficulty of chemically labeling the sensor protein with a fluorophore, with the attendant issues of homogeneity, yield, and getting it into the cell. Indeed, one of the key advantages of the expressible approach is the ability to target the sensor to various cell compartments as others (Qin, Dittmer, et al., 2011) and we have demonstrated (McCranor, Bozym, et al., 2012). By comparison, targeting small molecule sensors to organelles (apart from mitochondria with cationic dyes) remains challenging.

6. KINETICS AND MECHANISM OF METAL ION BINDING

An important factor in metallobiochemistry that is frequently overlooked is the kinetics of the metal ion binding reaction, which of course is determined by the microscopic mechanism of the binding reaction. This factor is frequently overlooked in part because of the dearth of available kinetic data (compared with affinities) on many binding reactions of interest, as well as the fact that many such reactions may be thermodynamically feasible but are so slow they are often not observed in realistic time scales. One example is the dissociation of zinc ion from the active site of human carbonic anhydrase II (CA II) (Henkens & Sturtevant, 1968; Hunt & Fierke, 1997). The apparent equilibrium constant K_{EQ} can be expressed as the ratio of the association rate constant k_{ON} and the dissociation rate constant k_{OFF}, assuming a simple one-step binding mechanism:

$$K_{EQ} = k_{ON}/k_{OFF} \tag{4}$$

Thus the observed (net) rate, k_{OBS}, of equilibration to form the enzyme–metal complex is the sum of the off rate constant and the on rate constant times the concentration of metal, since this is a second-order reaction. This analysis assumes that the kinetics are measured under pseudo-first-order conditions where the concentration of the metal ion is higher than the protein and can be assumed to be constant during the reaction:

$$k_{OBS} = k_{OFF} + k_{ON}\left[Zn^{2+}\right] \tag{5}$$

such that the time it takes for the reaction to get halfway to equilibrium ($t_{1/2}$) is

$$t_{1/2} = \ln 2/k_{OBS} = 0.693/k_{OBS} \tag{6}$$

For zinc ion binding to human CA II, the apparent on and off rate constants are $1 \times 10^5\ M^{-1}\ s^{-1}$ and $8.4 \times 10^{-8}\ s^{-1}$, respectively (Hunt & Fierke, 1997). This apparent association rate constant is much slower than a diffusion-controlled second order rate constant for a small ion binding to a protein ($1 \times 10^8\ M^{-1}\ s^{-1}$) (Fersht, 1999), suggesting a two-step binding mechanism. The value for k_{OFF} is measured by adding a high concentration of a high-affinity zinc ligand (like EDTA) that traps zinc upon dissociation from the protein; the half-time for dissociation is limited by the k_{OFF} rate constant, so $t_{1/2}=0.693/8.4 \times 10^{-8}\ s^{-1}$, or 8.25×10^6 s, or about 95 days. Assuming similar binding kinetics under cellular conditions, the off rate is slow enough that carbonic anhydrase will remain metallated for days, even if the zinc concentration is low. However, the association reaction will also be slow so that newly synthesized apo-CA could also be metallated and activated slowly. Note that while carbonic anhydrase II has a slow apparent k_{ON}, even a newly transcribed, zinc-requiring metalloenzyme or transcription factor (Buchsbaum & Berg, 2000) with a picomolar affinity would not bind metal rapidly in the cell due to the low free zinc concentrations. For example, the zinc-binding kinetics for a zinc finger motif in the transcription factor Cp1 have been measured as $k_{ON}=2.8\times10^{7}\ M^{-1}\ s^{-1}$ and $k_{OFF}=1.6\times10^{-4}\ s^{-1}$ (Buchsbaum & Berg, 2000). For a free zinc concentration in cells of 10 p*M* (Bozym et al., 2006; Dittmer et al., 2009; Van Dongen et al., 2007), the observed zinc association rate (k_{OBS}) is estimated as $2.8\times10^{7}\ M^{-1}\ s^{-1}\ (1\times10^{-11}\ M)+(1.6\times10^{-4}\ s^{-1})$ or $4.4\times10^{-4}\ s^{-1}$, and thus $t_{1/2}$ will be ~30 min. If the free zinc concentration is 10-fold higher, the $t_{1/2}$ becomes ~3 min.

There are a variety of mechanisms to avoid slow metallation in cells. One possible mechanism is that zinc associates with carbonic anhydrase as it is folding. Solvent-induced denaturation of CA II in vitro occurs in two steps: native human CA II unfolds to form a stable intermediate that retains much of the secondary structure of the native protein but is inactive; at higher guanidine HCl concentrations, this intermediate further denatures to form an unfolded state which is not well defined but does retain some secondary structure (Maartensson, Jonsson, et al., 1993; Tweedy, Nair, et al., 1993). The zinc affinity of the intermediate state is ~100-fold lower than the native state (Hunt & Fierke, 1997) and the binding kinetics are also likely more rapid. The bound zinc may then be trapped in the active site upon folding to the native state.

A second potential mechanism is that the cellular free zinc concentration is higher in localized regions of the cell or that another zinc species in the cell, other than free zinc, binds to the protein or domain and transfers the metal. Since the total cellular zinc concentration is ~100 μ*M*, this mechanism can potentially substantially increase the concentration of zinc available to bind to the protein and, hence, the apparent association rate constant. Some examples of this mechanism are known: a dipicolinate (DPA)–zinc complex binds to and forms a ternary complex with carbonic anhydrase (Pocker & Fong, 1983). The apparent association rate constant increases since both free zinc and the higher concentration of the DPA–zinc complex bind to CA II. Additionally, DPA bound to CA II facilitates zinc dissociation as a DPA–Zn complex, with a dissociation half-time in the minute range. While the detailed speciation of zinc in the cell remains to be determined, it is estimated that a significant fraction of the 100 μ*M* concentration of total zinc in the cell is bound to small molecules that are relatively abundant like glutathione, glutamate, or histidine. In some cases, a protein chaperone can act as the species that transfers a metal ion to a metalloprotein. For example, copper chaperones like Atx1 selectively transfer Cu(I) to particular apoenzymes, like superoxide dismutase and lysyl oxidase (Boal & Rosenzweig, 2009). Protein chaperones have also been identified for other metal ions, including Fe^{2+} (Shi, Bencze, et al., 2008). To date, no protein chaperones have been well validated for incorporation of zinc into metalloproteins in cells.

7. DISTRIBUTION OF METAL IONS THROUGHOUT THE CELL AND ORGANISM

Recently our knowledge has grown regarding the Zip and Znt families of transporters that are mainly responsible for moving zinc ions into and out of the cell itself and its compartments (reviewed in Eide, 2006; Kambe, Tsuji, et al., 2015; Kambe, Yamaguchi-Iwai, et al., 2004). However, our overall understanding of how zinc is distributed among various tissues and organs is meager. Early views suggested that zinc deficiency presented basically the same phenotype as a broader starvation, but in fact the symptomology of zinc deficiency differs markedly under different circumstances. For instance, the classic work of Prasad et al. (reviewed in Prasad, 1985) showed that young men who were profoundly zinc deficient exhibited stunted growth, underdeveloped external genitalia and secondary sex characteristics, reduced muscular development, and mental deficits. By comparison, infants lacking Zip4 (an intestinal zinc transporter with a major role in dietary uptake) exhibit acrodermatitis enteropathica which manifests as extensive perioral and perianal skin infections: a "super diaper rash." Furthermore, diarrheal disease attributable to zinc deficiency is responsible for a million infant and toddler deaths every year in underdeveloped countries (Walker & Black, 2004). While newborns from Nepalese mothers who were zinc deficient were normal, the mothers were much more prone to death from postpartum infections, suggesting that under conditions of limiting zinc the mother's physiology favors the fetus.

One interpretation of these disparate observations is that there is some level of central control of zinc distribution to various needy tissues and organs. While our understanding is improving on many fronts, such as identification of zinc-dependent transcription factors that control transporter expression (Outten & O'Halloran, 2001) and, hence, zinc levels in individual cells and organelles, the differing symptoms of deficiency suggest a prioritization of jobs for which zinc is required that changes with time and developmental stage. Such coordination of diverting zinc from some tissues or organs to others would imply communication between needy tissues (presumably hormonal) and some central controller. Moreover, there would also seem to be a need for a store of zinc, and/or a zinc-rich tissue which could be catabolized in times of need.

For organismal calcium homeostasis the roles of parathyroid hormone, vitamin D, and calcitonin are widely appreciated, as is that of the bone matrix as a store of calcium. For zinc, there has not been identified a clear, central locus for hormonal communications comparable to the parathyroid for integrated calcium control, nor an inert store from which zinc can be mobilized; we note that the 10–20 μM total zinc in the serum represents 0.1% of the body's total zinc, and thus a very modest buffer for supplying tissue needs (Fig. 5). In principle, bone or skeletal muscle could serve as stores due to the high proportion of bodily zinc they contain, but it is far from clear under what circumstances this might be mobilized.

One of the potential roles of biosensors will be to help elucidate how zinc (and other metal ion) levels are controlled at the organismal level. Such studies will require whole animal studies of zinc levels and fluxes in vivo. Genetically expressed biosensors are well suited for such studies due to their targetability. In many cases, the salient species is likely to be a metal ion complex rather than a free metal ion, the former of which will be easier to detect using a biosensor than a small molecule. Moreover, such in vivo studies require long wavelength fluorescence excitation and emission to avoid background fluorescence, absorption, and scattering so as to be visible in animal models (Ntziachristos, 2006; Thompson, 1994); we and others have made long wavelength fluorescence sensors to meet these needs (Hirayama, Van de Bittner, et al., 2012; Kiyose, Kojima, et al., 2006; Thompson, Maliwal, et al., 2000; Zeng, Matveeva, et al., 2013). Of particular interest are bioluminescent sensors, since they do not require exciting light to penetrate the tissue to reach the sensor, nor are endogenous fluorophores such as flavins, pteridines, and porphyrins potential interferents because their fluorescence is not excited; these advantages have fueled bioluminescent sensors' popularity for in vivo imaging (Negrin & Contag, 2006). We (R. Thompson et al., in preparation; Thompson, Matveeva, et al., 2015) and others (Aper, Dierickx, et al., 2016) have developed bioluminescent sensors for zinc, and bioluminescent sensors for other metals are certain to be developed shortly.

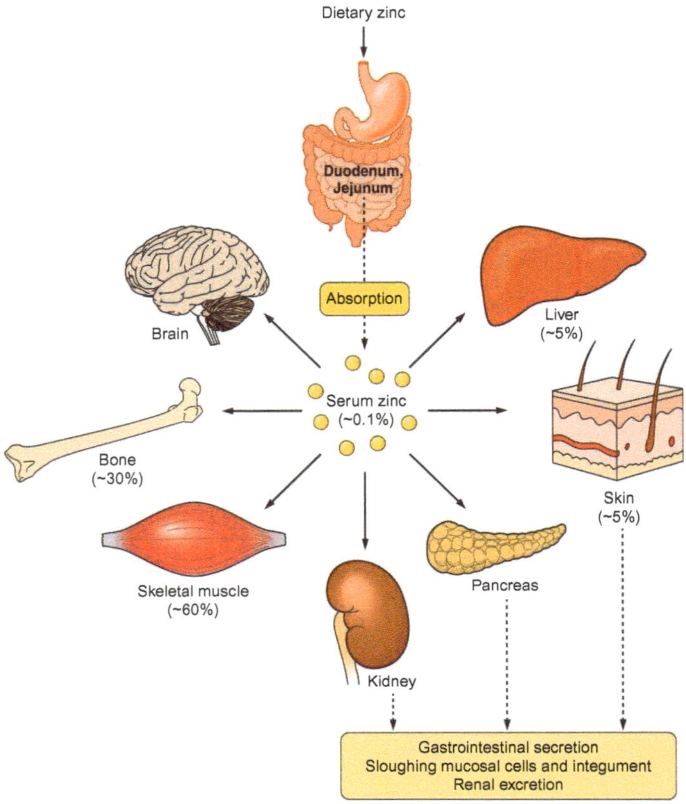

Fig. 5 Distribution of zinc throughout the human body. *Reproduced from Kambe, T., Tsuji, T., et al. (2015). The physiological, biochemical, and molecular roles of zinc transporters in zinc homeostasis and metabolism.* Physiological Reviews, 95*(3), 749, with permission.*

8. CONCLUSION

Overall, biosensors have emerged as important tools in understanding the biology of metal ions, especially those classified historically as being at "trace" levels; with continued innovation by many investigators worldwide, we expect this trend to continue. The evident importance of metal ion fluxes in many disease processes in humans (Nriagu & Skaar, 2015), as also indicated by identification of transporter variants linked to various diseases (Hogstrand, Kille, et al., 2009; Li, Zhang, et al., 2007; Sladek, Rocheleau, et al., 2007; Taylor, Hiscox, et al., 2004), suggests it may be fruitful to track those fluxes in vivo, a job for which biosensors may be very well suited.

REFERENCES

Aoki, S., Kaido, S., et al. (2003). A new zinc(II) fluorophore2-(9-anthrylmethylamino) ethyl-appended 1,4,7,10-tetraazacyclododecane. *Inorganic Chemistry, 42*, 1023–1030.

Aper, S. J. A., Dierickx, P., et al. (2016). Dual readout BRET/FRET sensors for measuring intracellular zinc. *ACS Chemical Biology, 11*(10), 2854–2864.

Auld, D. S. (2001). Zinc coordination sphere in biochemical zinc sites. *Biometals, 14*(3–4), 271–313.

Baker, J. O. (1988). Metal-buffered systems. *Methods in Enzymology, 158*, 33–55.

Boal, A. K., & Rosenzweig, A. C. (2009). Structural biology of copper trafficking. *Chemical Reviews, 109*(10), 4760–4779.

Bozym, R., Hurst, T. K., et al. (2008). Determination of zinc using carbonic anhydrase-based fluorescence biosensors. In L. Brand & M. Johnson (Eds.), *Fluorescence spectroscopy: Vol. 450* (pp. 287–309). San Diego: Academic Press.

Bozym, R. A., Thompson, R. B., et al. (2006). Measuring picomolar intracellular exchangeable zinc in PC-12 cells using a ratiometric fluorescence biosensor. *ACS Chemical Biology, 1*(2), 103–111.

Buchsbaum, J. C., & Berg, J. M. (2000). Kinetics of metal binding by a zinc finger peptide. *Inorganica Chimica Acta, 297*, 217–219.

Burdette, S. C., & Lippard, S. J. (2001). ICCC34-golden edition of coordination chemistry reviews. Coordination chemistry for the neurosciences. *Coordination Chemistry Reviews, 216–217*, 333–361.

Burdette, S. C., Walkup, G. K., et al. (2001). Fluorescent sensors for Zn(2+) based on a fluorescein platform: Synthesis, properties and intracellular distribution. *Journal of the American Chemical Society, 123*(32), 7831–7841.

Carter, K. P., Young, A. M., et al. (2014). Fluorescent sensors for measuring metal ions in living systems. *Chemical Reviews, 114*(8), 4564–4601.

Christianson, D. W. (1991). Structural biology of zinc. *Advances in Protein Chemistry, 42*, 281–355.

Colvin, R. A., Holmes, W. R., et al. (2010). Cytosolic zinc buffering and muffling: Their role in intracellular zinc homeostasis. *Metallomics, 2*(5), 306–317.

Cotton, F. A., & Wilkinson, G. (1988). *Advanced inorganic chemistry*. New York: Wiley-Interscience.

Dawson, R. M. C., & Elliott, D. C., et al.(1989). *Data for biochemical research*. Oxford: Oxford University Press.

Dittmer, P. J., Miranda, J. G., et al. (2009). Genetically encoded sensors to elucidate spatial distribution of cellular zinc. *Journal of Biological Chemistry, 284*, 16289–16297.

Donat, J. R., & Bruland, K. W. (1990). A comparison of two voltammetric techniques for determining zinc speciation in northeast Pacific Ocean waters. *Marine Chemistry, 28*, 301–323.

Eide, D. J. (2006). Zinc transporters and the cellular trafficking of zinc. *Biochimica et Biophysica Acta, 1763*, 711–722.

Fersht, A. R. (1999). *Structure and mechanism in protein science: A guide to enzyme catalysis and protein folding*. New York: W.H. Freeman.

Fierke, C. A., & Thompson, R. B. (2001). Fluorescence-based biosensing of zinc using carbonic anhydrase. *Biometals, 14*(3-4), 205–222.

Franklin, R. B., Ma, J., et al. (2003). Human ZIP1 is a major zinc uptake transporter for the accumulation of zinc in prostate cells. *Journal of Inorganic Biochemistry, 96*, 435–442.

Harris, W. R. (2012). Anion binding properties of the transferrins. Implications for function. *Biochimica et Biophysica Acta (BBA)—General Subjects, 1820*(3), 348–361.

Haugland, R. P. (2005). *The handbook: A guide to fluorescent probes and labeling technologies*. Invitrogen Corp.: Carlsbad, CA.

Henkens, R. W., & Sturtevant, J. M. (1968). The kinetics of the binding of Zn(II) by apocarbonic anhydrase. *Journal of the American Chemical Society, 90*, 2669–2676.

Hirayama, T., Van de Bittner, G. C., et al. (2012). Near-infrared fluorescent sensor for in vivo copper imaging in a murine Wilson disease model. *Proceedings of the National Academy of Sciences of the United States of America, 109*(7), 2228–2233.
Hogstrand, C., Kille, P., et al. (2009). Zinc transporters and cancer: A potential role for ZIP7 as a hub for tyrosine kinase activation. *Trends in Molecular Medicine, 15*(3), 101–111.
Huang, C.-C., Lesburg, C. A., et al. (1996). Reversal of the hydrogen bond to zinc ligand histidine-119 dramatically diminishes catalysis and enhances metal equilibration kinetics in carbonic anhydrase II. *Biochemistry, 35*(11), 3439–3446.
Hunt, J. A., & Fierke, C. A. (1997). Selection of carbonic anhydrase variants displayed on phage: Aromatic residues in zinc binding site enhance metal affinity and equilibration kinetics. *Journal of Biological Chemistry, 272*(33), 20364–20372.
Hurst, T. K., Wang, D., et al. (2010). Carbonic anhydrase II-based metal ion sensing: Advances and new perspectives. *Biochimica et Biophysica Acta (BBA)—Proteins & Proteomics, 1804*(2), 393–403.
Kambe, T., Tsuji, T., et al. (2015). The physiological, biochemical, and molecular roles of zinc transporters in zinc homeostasis and metabolism. *Physiological Reviews, 95*(3), 749.
Kambe, T., Yamaguchi-Iwai, Y., et al. (2004). Overview of mammalian zinc transporters. *Cellular and Molecular Life Sciences, 61*(1), 49–68.
Kikuchi, K., Komatsu, K., et al. (2004). Zinc sensing for cellular application. *Current Opinion in Chemical Biology, 8*(2), 182–191.
Kiyose, K., Kojima, H., et al. (2006). Development of a ratiometric fluorescent zinc ion probe in near-infrared region, based on tricarbocyanine chromophore. *Journal of the American Chemical Society, 128*, 6548–6549.
Krezel, A., & Maret, W. (2006). Zinc-buffering capacity of a eukaryotic cell at physiological pZn. *Journal of Biological Inorganic Chemistry, 11*, 1049–1062.
Li, M., Zhang, Y., et al. (2007). Aberrant expression of zinc transporter ZIP4 (SLC39A4) significantly contributes to human pancreatic cancer pathogenesis and progression. *Proceedings of the National Academy of Sciences of the United States of America, 104*, 18636–18641.
Maartensson, L. G., Jonsson, B. H., et al. (1993). Characterization of folding intermediates of human carbonic anhydrase II: Probing substructure by chemical labeling of sulfhydryl groups introduced by site-directed mutagenesis. *Biochemistry, 32*(1), 224–231.
Mash, H. E. (2003). Complexation of copper by zwitterionic aminosulfonic (good) buffers. *Analytical Chemistry, 75*(3), 671–677.
McCall, K. A., & Fierke, C. A. (2004). Probing determinants of the metal ion selectivity in carbonic anhydrase using mutagenesis. *Biochemistry, 43*, 3979–3986.
McCranor, B., Bozym, R., et al. (2012). Quantitative imaging of mitochondrial and cytosolic free zinc levels in an in vitro model of ischemia/reperfusion. *Journal of Bioenergetics and Biomembranes, 44*(2), 253.
Negrin, R. S., & Contag, C. H. (2006). In vivo imaging using bioluminescence: A tool for probing graft vs. host disease. *Nature Reviews. Immunology, 6*, 484–490.
Nolan, E. M., Jaworski, J., et al. (2005). QZ1 and Qz2: Rapid, reversible quinoline-derivatized fluoresceins for sensing biological Zn(II). *Journal of the American Chemical Society, 127*, 16812–16823.
Nriagu, J. O., & Skaar, E. P. (Eds.), (2015). *Trace metals and infectious diseases. Strungmann forum reports.* Cambridge, MA: MIT Press.
Ntziachristos, V. (2006). Fluorescence molecular imaging. *Annual Review of Biomedical Engineering, 8*, 1–33.
Outten, C. E., & O'Halloran, T. V. (2001). Femtomolar sensitivity of metalloregulatory proteins controlling zinc homeostasis. *Science, 292*, 2488–2492.
Pocker, Y., & Fong, C. T. O. (1983). Inactivation of bovine carbonic anhydrase by dipicolinate: Kinetic studies and mechanistic implications. *Biochemistry, 22*, 813–818.

Prasad, A. S. (1985). Clinical manifestations of zinc deficiency. *Annual Review of Nutrition, 5*, 341–363.
Qin, Y., Dittmer, P. J., et al. (2011). Measuring steady-state and dynamic endoplasmic reticulum and Golgi Zn^{2+} with genetically encoded sensors. *Proceedings of the National Academy of Sciences of the United States of America, 108*, 7351–7356.
Que, E. L., Domaille, D. W., et al. (2008). Metals in neurobiology: Probing their chemistry and biology with molecular imaging. *Chemical Reviews, 108*, 1517–1549.
Rae, T. D., Schmidt, P. J., et al. (1999). Undetectable intracellular free copper: The requirement of a copper chaperone for superoxide dismutase. *Science, 284*, 805–808.
Shi, H., Bencze, K. Z., et al. (2008). A cytosolic iron chaperone that delivers iron to ferritin. *Science, 320*(5880), 1207.
Sladek, R., Rocheleau, G., et al. (2007). A genome-wide association study identifies novel risk loci for type 2 diabetes. *Nature, 445*, 881–885.
Smith, R., & Martell, A. E. (1973). *NIST critical stability constants of metal complexes, standard database: Vol. 46.* Gaithersburg, MD: National Institute of Standards and Technology, U.S. Department of Commerce.
Stumm, W., & Morgan, J. J. (1996). *Aquatic chemistry: Chemical equilibria and rates in natural waters.* New York: Wiley-Interscience.
Taylor, K. M., Hiscox, S., et al. (2004). Zinc transporter LIV-1: A link between cellular development and cancer progression. *Trends in Endocrinology and Metabolism: TEM, 15*(10), 461–463.
Thompson, R. B. (1994). Red and near-infrared fluorometry. Topics in fluorescence spectroscopy. In J. R. Lakowicz (Ed.), *Probe design and chemical sensing: Vol. 4* (pp. 151–181). New York: Plenum Press.
Thompson, R. B. (Ed.), (2006). *Fluorescence sensors and biosensors.* Boca Raton, FL: CRC Press.
Thompson, R. B., Maliwal, B. P., et al. (1998). *High sensitivity determination of Zn(II) and Cu(II) in vitro by fluorescence polarization. Systems and technologies for clinical diagnostics and drug discovery.* San Jose, CA: SPIE.
Thompson, R. B., Maliwal, B. P., et al. (2000). Zinc biosensing with multiphoton excitation using carbonic anhydrase and improved fluorophores. *Journal of Biomedical Optics, 5*(1), 17–22.
Thompson, R. B., Matveeva, E. G., et al. (2015). *Bioluminescent metal ion assay.* USA: U. PTO.
Thompson, R. B., Peterson, D., et al. (2002). Fluorescent zinc indicators for neurobiology. *Journal of Neuroscience Methods, 118*, 63–75.
Tweedy, N. B., Nair, S. K., et al. (1993). Structure and energetics of a non-proline cis-peptidyl linkage in a proline-202 fwdarw alanine carbonic anhydrase II variant. *Biochemistry, 32*(41), 10944–10949.
Van Dongen, E. M. W. M., Evers, T. H., et al. (2007). Variation of linker length in ratiometric fluorescent sensor proteins allows rational tuning of Zn(II) affinity in the picomolar to femtomolar range. *Journal of the American Chemical Society, 129*, 3494–3495.
Walker, C. F., & Black, R. E. (2004). Zinc and the risk for infectious disease. *Annual Review of Nutrition, 24*, 255–275.
Wang, D., Hosteen, O., et al. (2012). ZntR-mediated transcription of zntA responds to nanomolar intracellular free zinc. *Journal of Inorganic Biochemistry, 111*, 173–181.
White, C. E., & Argauer, R. J. (1970). *Fluorescence analysis: A practical approach.* New York: Marcel Dekker, Inc.
Zeng, H., Matveeva, E. G., et al. (2013). Long wavelength fluorescence ratiometric zinc biosensor. *Journal of Fluorescence, 23*, 375–379.

CHAPTER ELEVEN

Bioapplications of Electrochemical Sensors and Biosensors

Eduard Dumitrescu, Silvana Andreescu[1]
Clarkson University, Clarkson, NY, United States
[1]Corresponding author: e-mail address: eandrees@clarkson.edu

Contents

Abstract

Recent progress in the electrochemical field enabled development of miniaturized sensing devices that can be used in biological settings to obtain fundamental and practical biochemically relevant information on physiology, metabolism, and disease states in living systems. Electrochemical sensors and biosensors have demonstrated potential for rapid, real-time measurements of biologically relevant molecules. This chapter provides an overview of the most recent advances in the development of miniaturized sensors for biological investigations in living systems, with focus on the detection of neurotransmitters and oxidative stress markers. The design of electrochemical (bio)sensors, including their detection mechanism and functionality in biological systems, is described as well as their advantages and limitations. Application of these sensors to studies in live cells, embryonic development, and rodent models is discussed.

1. INTRODUCTION

The development of innovative tools is necessary for advancing the understanding of the biochemical processes that are essential for the functioning of living organisms. Electrochemistry is a powerful technique

Methods in Enzymology, Volume 589
ISSN 0076-6879
http://dx.doi.org/10.1016/bs.mie.2017.01.017

enabling real-time in situ detection of biomarkers with high spatial and temporal resolution. The method can be used for studying physiological mechanisms at cell, tissue, and organ levels, as well as helping elucidate health and disease states in otherwise complex systems. Recent developments in this field have demonstrated the potential of electrochemical sensors in clinical and biological settings. Main efforts have been focused on detection of neurotransmitters, such as dopamine, that are electrochemically active and can be conveniently measured at electrode surfaces. In recent years, the technology has been extended to a wider class of analytes that play important roles in live organisms. Neuroscience studies were the first to take advantage of the unique capabilities of electrochemical sensors. Detection of the biological molecules involved in neurological processes is important for the study of normal brain activity, and for understanding the pathophysiology of neurological and neuropsychiatric diseases (Alivisatos et al., 2013). Neurotransmitter monoamines, mainly dopamine, epinephrine, norepinephrine, and serotonin, are implicated in stress response mechanisms in the central nervous system (CNS), in cognitive and memory functions, as well as in brain disorder states (Table 1). The malfunction of normal processes in the CNS is usually indicated by an abnormal level of neurotransmitters present in discrete sections of the brain. To comprehensively study these phenomena, advanced detection tools that can reliably perform in complex conditions and provide measurements at predetermined locations in selected areas of the brain are needed (Alivisatos et al., 2013). Another area of great interest is the fundamental study of neurotransmission for understanding cell signaling and communication between cells. This process is essential for normal cellular operation and intercellular communication in live organisms (Burgoyne & Morgan, 2003), and a fundamental understanding of the basic biological mechanisms involved in cell communication is still needed (Cans & Ewing, 2011). Miniaturized microelectrodes can be extremely useful for studying the release of signal molecules at the single-cell level, a process that requires extremely sensitive methodologies that are able to operate at a very small scale.

A relatively new area of interest in biology is the study of oxidative stress. This topic has gained increasing attention in the study of the antioxidant defense mechanisms against oxidative/nitrosative stress (Finkel & Holbrook, 2000). It is known that, while oxygen is an essential component for living organisms, the biochemical reactions involving oxygen may result in the formation of reactive oxygen metabolites that can challenge the antioxidant defense systems (Buetler, Krauskopf, & Ruegg, 2004). The presence of such reactive species leads to a series of reactions producing free radicals or

Table 1 Role of Neurotransmitters Dopamine, Epinephrine, Norepinephrine, and Serotonin in Normal Cognitive Behavior and Disorders (Selected Examples)

Neurotransmitter	Normal Behavior	Disorders
Dopamine	Reward and learning behavior (Wise, 2004)	Parkinson disease (Lotharius & Brundin, 2002)
	Motor function (Jenkinson & Brown, 2011)	Schizophrenia (Howes & Kapur, 2009)
	Cognitive function (Cropley, Fujita, Innis, & Nathan, 2006)	Depression (Dunlop & Nemeroff, 2007)
		Drug addiction (Di Chiara et al., 2004)
Epinephrine	Stress response (Wurtman, 2002) Memory consolidation (McGaugh & Roozendaal, 2002)	Pheochromocytoma (Zapanti & Ilias, 2006) Hypoglycemia (Desouza, Bolli, & Fonseca, 2010)
Norepinephrine	Stress response (Morilak et al., 2005)	Anxiety and depression (Goddard et al., 2010) Posttraumatic stress disorder (Strawn & Geracioti, 2008)
Serotonin	Stress response (Kaehler, Singewald, Sinner, Thurnher, & Philippu, 2000)	Anxiety (Akimova, Lanzenberger, & Kasper, 2009) Depression (Nemeroff, 2002) Sudden infant death syndrome (Kinney, Richerson, Dymecki, Darnall, & Nattie, 2009)

reactive nonradical species, differentiated in two categories referred to as reactive oxygen species (ROS) and reactive nitrogen species (RNS) (Bauer, 2000; Halliwell, 1996). An example of an ROS/RNS formation pathway is given in Fig. 1. Recent research highlighted the important functions of ROS/RNS in biochemical processes. However, due to the high reactivity and short lifetime of these species, their study is difficult. Reactive species have both beneficial and damaging effects. The overproduction of ROS/RNS is usually associated with their destructive effects, when the

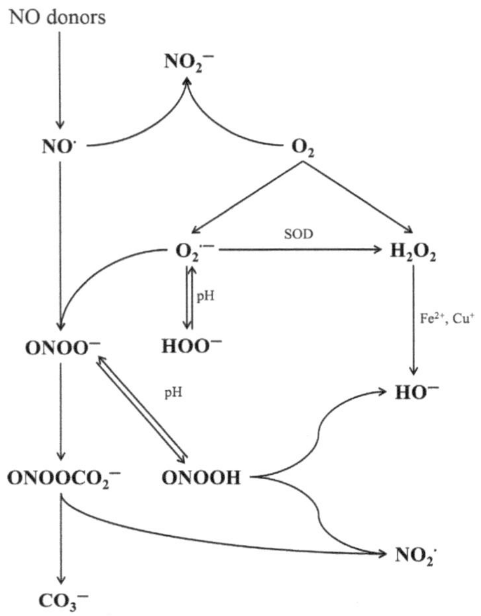

Fig. 1 Examples of ROS/RNS formed as a result of oxygen (O_2) metabolism and NO· formation in live organisms. *Reprinted with permission from Fang, Y.-Z., Yang, S., & Wu, G. (2002). Free radicals, antioxidants, and nutrition.* Nutrition, 18*(10), 872–879. http://dx.doi.org/10.1016/S0899-9007(02)00916-4.*

innate antioxidant protection mechanism of cells is overwhelmed. Due to their high reactivity, ROS/RNS can induce structural alterations of DNA bases and decrease the activity of DNA repair enzymes (Wiseman & Halliwell, 1996). Several recent works reviewed the current understating of the detrimental effects of ROS/RNS against DNA and the different mechanisms behind their action (Franco, Schoneveld, Georgakilas, & Panayiotidis, 2008; Jena, 2012; Kryston, Georgiev, Pissis, & Georgakilas, 2011). Peroxynitrite ions ($ONOO^-$) and hydroxyl radicals ($HO^{\cdot}$) are more likely to be involved due to their higher reactivity in comparison with the other ROS/RNS. $ONOO^-$ is formed by the fast reaction between superoxide ($O_2^{\cdot-}$) and nitric oxide ($NO^{\cdot}$) radicals (Radi, Peluffo, Alvarez, Naviliat, & Cayota, 2001). Due to its oxidative and nitrating nature, $ONOO^-$ can easily react with a wide array of biological molecules (Beckman & Koppenol, 1996). Formed either as a result of $ONOO^-$ decomposition (Beckman, Beckman, Chen, Marshall, & Freeman, 1990) or through the Fenton reaction of hydrogen peroxide (H_2O_2) (Imlay, Chin, & Linn, 1988), $HO^{\cdot}$ is extremely short-lived, virtually reacting with the first molecule that it encounters (Genestra, 2007). When reactive species are present in small concentrations, they can play a protective or signaling role in normal physiological processes in organisms. $O_2^{\cdot -}$, formed as a result of oxygen reduction, is implicated in cell growth and proliferation (Bloomfield & Pears, 2003; Pimentel et al., 2001), as well as in signaling pathways in the CNS (Zimmerman et al., 2002). $O_2^{\cdot -}$ can be further reduced to H_2O_2. Its presence is associated with cell signaling processes (Rhee, 1999) and cellular growth (Arnold et al., 2001), differentiation (Sauer, Wartenberg, & Hescheler, 2001), and apoptosis (Cai, 2005). Generated by the enzymatic reaction of L-arginine in the presence of nitric oxide synthase isoforms (Ghafourifar & Cadenas, 2005), $NO^{\cdot}$ has gained increased significance in recent years. The main functions of $NO^{\cdot}$ are neurotransmission (Calabrese et al., 2007), cardiovascular regulation (Massion, Feron, Dessy, & Balligand, 2003), and smooth muscle relaxant factor action (Ignarro, 2002). The ROS/RNS interplay has also been suggested to play a role in disease states, in aging (Finkel & Holbrook, 2000), neurodegenerative disorders (Lin & Beal, 2006), cancer (Valko, Rhodes, Moncol, Izakovic, & Mazur, 2006), diabetic complications (Giacco & Brownlee, 2010), vascular disease (Madamanchi, Vendrov, & Runge, 2005), and other health issues (Aruoma, 1998; Dalle-Donne, Rossi, Colombo, Giustarini, & Milzani, 2006; Pham-Huy, He, & Pham-Huy, 2008).

This review chapter provides an overview of recent developments (2010 to present) in the electrochemical detection field, with a focus on measurements in biological systems, in cellular environments, tissues, and whole organisms. Fabrication, design characteristics, and performance of electrochemical sensors for detection of several neurotransmitters (dopamine, epinephrine, norepinephrine, and serotonin) and ROS/RNS species ($O_2^{\cdot -}$, $NO^{\cdot}$, H_2O_2, $ONOO^-$, and $HO^{\cdot}$) will be discussed in detail, with examples of applications in the fundamental understanding of biological mechanisms in these systems and monitoring disease status. Challenges associated with the functionality of these probes in complex matrixes, as well as dealing with small sample volumes and/or size constraints, are examined and potential solutions to these problems are provided. Opportunities for future development of this technology are also discussed at the end of the chapter.

2. ELECTROCHEMICAL SENSING IN BIOLOGICAL SYSTEMS

2.1 Electrochemical Measurements in Cell Environments

The study of living cell biology is essential for gaining fundamental knowledge about cellular mechanisms at the smallest level (Xie, Yu, & Yang, 2006).

Electrochemical sensors have the ability to solve some of the inherent challenges of working in small spatial dimensions, at the micro- or even nanometer level. The development of reliable methodologies for measuring biological analytes and elucidating their mechanisms in cell populations, single cells, and ultimately inside live cells has been the focus of recent research (Yum, Wang, & Yu, 2010). Thus, these methods are applied to test relevant hypotheses in biological and clinical research.

2.1.1 Measurements in Cell Populations

Compared with the necessary setup and conditions for single-cell studies (e.g., patch-clamping) or in vivo studies, measurements in cell cultures are a more accessible option for validation of sensors' performance. Different biological processes can be easily modulated or triggered in live cells, such as the production of an oxidative response, which may then be studied using electrochemical sensors. Several studies report electrochemical ROS/RNS measurements in cell cultures. The species that are most widely investigated are H_2O_2, $NO^{\bullet}$, and $O_2^{\bullet -}$. As bare carbon or platinum-based electrodes suffer from poor sensitivity and selectivity, most sensor designs involve surface modification with conductive or catalytic layers, as well as polymeric materials to increase surface area, enhance sensor signals, and minimize interference effects. Graphene-based materials have been widely used in recent years to coat electrodes for detection of H_2O_2 in cell cultures, either as a simple nitrogen-doped material (Wu, Cai, Gao, Zhang, & Cai, 2011) or by modification with more suitable catalysts, such as nanoparticles (NPs) (Maji, Sreejith, Mandal, Ma, & Zhao, 2014; Wang, Zhang, Cheng, et al., 2015) or recognition molecules (Xi, Zhao, Wang, & Chen, 2013). In some cases, the detection limit for these sensors is relatively high and the validation is done in high-density cell cultures to quantify a measurable signal (Wang, Zhang, Cheng, et al., 2015). The use of platinum-based materials is another common approach for detection of H_2O_2 (Marcu et al., 2012; Zhang, Bai, et al., 2014). Functionalized graphene oxide–carbon nanotube (CNT) hybrids with Pt NPs used as electrode material enabled fabrication of a sensitive H_2O_2 sensor with a nanomolar detection limit and demonstrated functionality in live macrophages (Sun, He, Zhang, Zhou, & Duan, 2015). Wang et al. reported a platform modified with MoS_2 ultrasmall NPs for detection of H_2O_2 in the nanomolar range (Wang, Zhu, et al., 2013). The application of enzyme-based configurations was also reported. Horseradish peroxidase immobilized on inorganic nanomaterials was applied for measurements of H_2O_2 flux from macrophages (Li, Zhang, et al., 2011;

Wu, Cai, Chen, Zhang, & Cai, 2011). However, the performance of this sensor was limited to detection in the micromolar range.

Measurement of cellular evolved $NO^{\cdot}$ was reported using NP-modified graphene sheets (Li, Xie, Gao, & Li, 2015; Zan et al., 2013). Zan et al. highlighted the potential application of graphene as flexible sensor (Zan et al., 2013). As similar materials were also used for H_2O_2 detection, a rigorous assessment of the electrochemical performance and selectivity assessment in the presence of different ROS/RNS species is required. It is well known that the stimulation of cells triggers oxidative bursts containing different reactive species, which may interfere with the measurement (Amatore et al., 2000b). A potential solution was reported by Abdelwahab et al. through the use of multienzymes that work synergistically to eliminate interferences (Abdelwahab, Koh, Noh, & Shim, 2010). Microperoxidase was used to facilitate the detection of $NO^{\cdot}$, while catalase and superoxide dismutase (SOD) were employed to prevent interferences from H_2O_2 and $O_2^{\cdot-}$, respectively.

Selective detection of $O_2^{\cdot-}$ in cell cultures was demonstrated through the use of SOD-modified sensors (Zhu et al., 2016). Due to its low stability in solution, development of probes capable of detecting submicromolar concentrations of $O_2^{\cdot-}$ is required. Such performance was reported with electrodes functionalized with a biocatalytic layer of SOD entrapped within a sodium alginate sol-gel film (Wang, Han, Bao, Tu, & Dai, 2012). The sensor was used for detection of $O_2^{\cdot-}$ released by macrophages. Thiol-based immobilization of cytochrome *c* (Cyt *c*) on a nanoporous gold material is a more widely used approach to develop a selective $O_2^{\cdot-}$ detection system with low detection limit (Sadeghian et al., 2016). Nanomolar concentrations of $O_2^{\cdot-}$ were measured using this approach from drug-exposed C_2C_{12} myoblasts and differentiated myotubes. The validity of the sensor response was confirmed by the use of a $O_2^{\cdot-}$-specific fluorescent probe. Similar electrochemical performance was reported for the use of a MnSOD mimic supported on an MWCNT-formylstyrylpyridine matrix (Yuan et al., 2014).

$ONOO^-$ detection in cell cultures was also reported using electrodes modified with electrocatalytic mediators such as a manganese ion-conductive polymer complex (Koh, Son, Choe, & Shim, 2010). The synthesized compound allowed detection of $ONOO^-$ in the nanomolar range. The performance of the sensor was confirmed by measurements done in stimulated rat glioma YPEN-1 cells. A singular example of $HO^{\cdot}$ detection was reported, using electrochemical impedance measurements at a thiol-modified gold electrode (Zhu, Liu, Rui, & Tian, 2011). While real-time detection is not possible, this study demonstrates the potential of this methodology for live cell monitoring, thus opening new directions for future development of $HO^{\bullet}$ electrochemical sensing devices.

In spite of all successful applications, most studies measure one species at a time. As an emerging technology, measurements are used as a validation of sensor responses. Future sensors should be developed to simultaneously detect multiple ROS/RNS species.

2.1.2 Single-Cell Electrochemistry

Meaningful information can be achieved by performing electrochemistry at the single-cell level. The implications of applying electrochemistry at small scales have been treated in detail elsewhere (Amatore, Arbault, Guille, & Lemaître, 2008; Lemaître, Guille Collignon, & Amatore, 2014; Trouillon, Passarelli, Wang, Kurczy, & Ewing, 2013). In particular, the study of exocytosis from single cells through electrochemical detection of neurotransmitters, mainly catecholamines, is of great importance (Ge, Koseoglu, & Haynes, 2010). Small-size carbon fiber electrodes are the main tool used for this type of measurement. Neurotransmitter release can be measured by amperometry at a working potential where all contents released during the stimulated exocytosis are oxidized. The proximity between the cell and electrode, and their comparable size ensures that all species are detected. The temporal resolution of the method allows for identification of single exocytotic events in the form of amperometric peaks. The graphical representation of an exocytosis process and a typical exocytotic peak with its quantifiable features measured by amperometry at a carbon fiber microelectrode is shown in Fig. 2. These measurements contribute toward achieving a complete understanding of the exocytotic release events including the fusion process of neurotransmitter-containing vesicles with the cell membrane (Amatore, Oleinick, & Svir, 2010), assessment of the catecholamine content in single vesicles (Dunevall et al., 2015), and the influence of cellular membrane structure in determining the kinetics of exocytotic release (Trouillon & Ewing, 2013). Amperometric investigation of exocytosis at BON cells reconfirmed the performance of the single-cell electrochemistry approach (Meunier et al., 2014). The reduced exocytotic activity of BON cells was shown to be well within the sensitivity limits achieved at carbon fiber electrodes. Other groups have applied a similar approach to study medically relevant issues. Insights into the pharmacological effects of β-blockers in hypertension treatment were revealed by the use of single-cell amperometry (Montesinos et al., 2010). Exocytosis at single

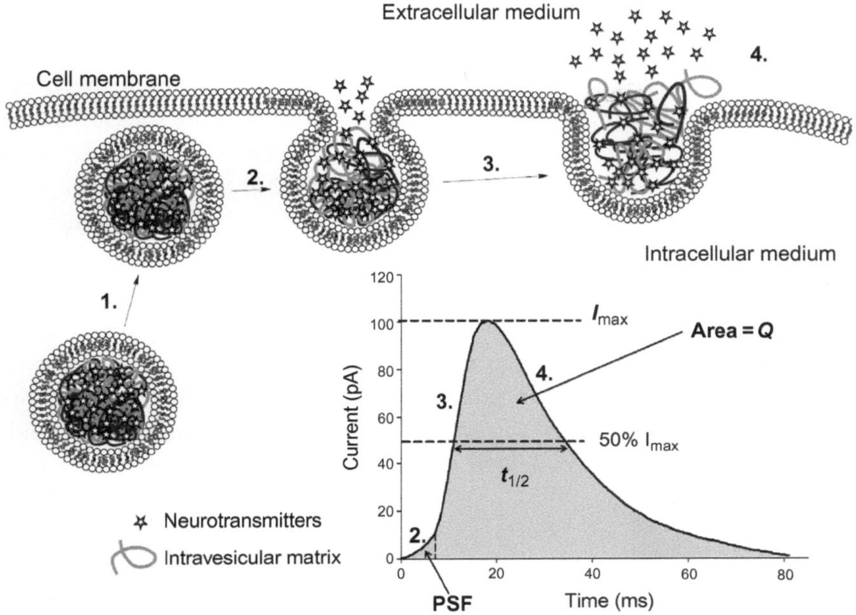

Fig. 2 *Top*: Schematic representation of the successive stages involved in vesicular exocytosis: (1) vesicle displacement up to vesicle membrane docking, (2) formation of the initial fusion pore, and (3) expansion of the pore and (4) massive release. *Bottom*: Schematic amperometric spike indicating the definition of the usual parameters extracted for analysis of the exocytotic release (I_{max}, Q, $t_{1/2}$) and the corresponding steps of the exocytotic mechanism. A typical prespike feature (PSF) is also indicated. *Reprinted with permission from Meunier, A., Bretou, M., Darchen, F., Guille Collignon, M., Lemaître, F., & Amatore, C. (2014). Amperometric detection of vesicular exocytosis from BON cells at carbon fiber microelectrodes.* Electrochimica Acta, 126, *74–80. http://dx.doi.org/10.1016/j.electacta.2013.07.11.*

cells was studied using gold-modified carbon fiber electrodes (Adams, Jena, Percival, & Zhang, 2011), planar gold nanoelectrodes (Liu, Li, Zhang, Zhu, & Shi, 2015), or conductive polymer-based nanoelectrodes (Yang et al., 2011). These modified electrodes provide improved performance as compared to bare carbon fiber electrodes, such as longer shelf life and increased resistance to fouling.

Amatore's group has made significant advancements in the development of electrochemical probes for ROS/RNS detection in the vicinity of single cells (Amatore, Arbault, Guille, et al., 2008). Their methodology is based exclusively on the use of platinized carbon fiber microelectrodes. Electro-deposited platinum black provides a powerful catalytic system for detection of $ONOO^-$, H_2O_2, $NO^\bullet$, and nitrite ions, the only four species that can be measured at a stimulated single cell. The amperometric monitoring of these species is done successively at four different potentials. In-depth validation work provided a complete description of the transition from the primordial species formed inside the cell, $NO^\bullet$ and $O_2^{\bullet -}$, to the mix of species detected at the small surface of the electrode positioned close to a single cell. The major feature of this methodology is its capability to help reconstruct the complete ROS/RNS flux, and offer a qualitative and quantitative perspective of the complex oxidative response at individual cells. The typical single-cell amperometry setup is shown in Fig. 3 for a measurement done at a single MG63 osteosarcoma cell submitted to mechanical stress. The most recent work focuses on characterization of the oxidative stress response using different cell lines (Hu et al., 2010), using single-cell electrochemistry as a tool. The antioxidant activity of two SOD mimics was studied at single

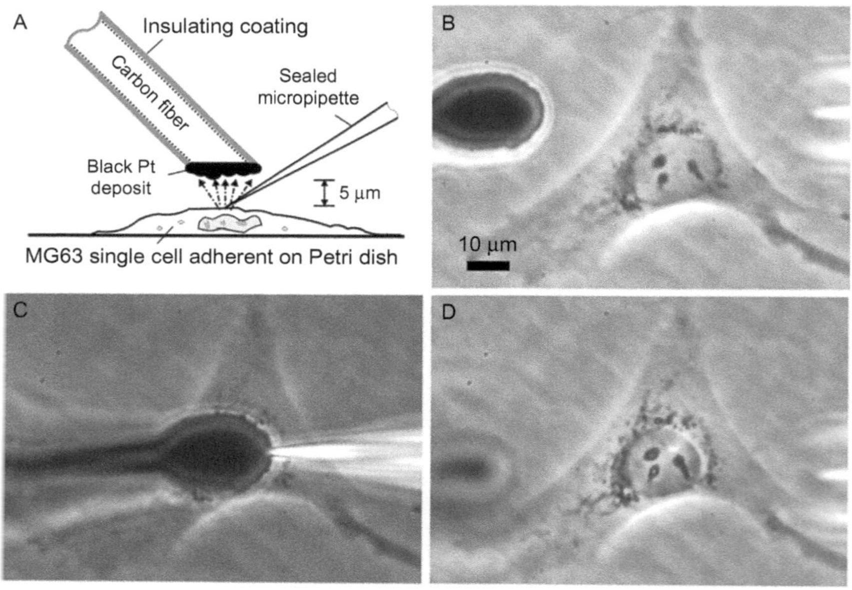

Fig. 3 (A) Schematic view of the "artificial synapse" configuration showing the placement of the platinized carbon fiber electrode and of the micropipette sealed tip used to induce a brief mechanical stress onto an adherent single MG63 osteosarcoma cell. (B–D) Optical microphotographs illustrating different phases of the experimental protocol. *Reprinted with permission from Hu, R., Guille, M., Arbault, S., Lin, C. J., & Amatore, C. (2010). In situ electrochemical monitoring of reactive oxygen and nitrogen species released by single MG63 osteosarcoma cell submitted to a mechanical stress.* Physical Chemistry Chemical Physics, 12*(34), 10048–10054. doi: 10.1039/c0cp00398.*

macrophages (Bernard et al., 2012; Filipović et al., 2010). These studies provided a characterization protocol for studying the antioxidant effect of compounds with potential therapeutic value (Filipović et al., 2010). Moreover, the mechanism of action of therapeutic drugs can also be revealed by electrochemistry at the single-cell level, as shown by the study of zidovudine (Amatore, Arbault, Jaouen, et al., 2010) and ferrocifens (Lu et al., 2014). This work demonstrated that it is possible to establish a connection between the structure of a drug and its effects, based on the measured oxidative stress response when administered to relevant cell lines. The results obtained by the application of single-cell electrochemistry are encouraging for future application of this methodology in biological and clinical research. The main drawback of ROS/RNS monitoring at single cells is the large amount of data necessary to provide statistically significant results. In order to simplify the measurement protocol, triple potential-step chronoamperometry was developed as a powerful alternative to the sequential detection of the different reactive species (Amatore, Arbault, & Koh, 2010). While further work and optimization are required for better reliability and precision, these preliminary results are promising. Malinski's group has also reported electrochemical studies at a single cell, using nanosensors with surface modification for selective detection of $NO^{\bullet}$ or $ONOO^{-}$ (Burewicz, Dawoud, Jiang, & Malinski, 2013; Malinski, 2015).

The next big challenge for single-cell electrochemistry is development of feasible approaches for intracellular monitoring of species of interest. The use of nanotip electrodes was demonstrated for transmitter measurements inside PC12 cells (Li, Majdi, Dunevall, Fathali, & Ewing, 2015) and in single synapses (Li et al., 2014). Similarly, nanosized platinized platinum electrodes were applied for ROS/RNS detection inside macrophages (Wang, Noël, et al., 2012). By employing both internal and external detection of these species, it is possible to gain an ever more complete view of the release process of reactive species, and therefore of fundamental biological mechanisms taking place at single cells.

2.2 In Vivo Electrochemistry

2.2.1 Neurotransmitter Detection

The very early research that built the foundation for the development of applied electrochemistry can be traced back to the work of Ralph Adams and coworkers (Johnson, 2013). The first in vivo application demonstrating measurement of catecholamines in brain tissue used a carbon paste electrode (Kissinger, Hart, & Adams, 1973). Significant improvements have been made regarding selectivity between different electrochemically active biological molecules, as well as sensitivity, temporal resolution, and data acquisition (Troyer, Heien, Venton, & Wightman, 2002). Recent reviews have also acknowledged the progress made in the field (Eckert et al., 2013; Trouillon, Svensson, Berglund, Cans, & Ewing, 2012; Zhang, Wang, & Tian, 2014). In particular, the development of this methodology has had an impact in neuroscience. The majority of applications for measurement of brain signaling molecules are based on the use of fast-scan cyclic voltammetry (FSCV) at bare carbon fiber microelectrodes (Lama, Charlson, Anantharam, & Hashemi, 2012; Maina, Khalid, Apawu, & Mathews, 2012). Wightman's group has made significant advancements in the development of this technique for neurotransmitter detection (Kuhr & Wightman, 1986; Wipf, Kristensen, Deakin, & Wightman, 1988). It was demonstrated that, by controlling the parameters of the FSCV, it is possible to distinguish and quantify the electrochemically active molecules present in the brain based on the voltammogram shape (Heien, Phillips, Stuber, Seipel, & Wightman, 2003). It was also shown that the sensitivity and temporal resolution of the technique can be improved by employing different waveforms, scan rates, and scan frequencies (Keithley et al., 2011; Kile et al., 2012). Most recently, they have observed the release of serotonin (Hashemi, Dankoski, Wood, Ambrose, & Wightman, 2011) and norepinephrine (Fox, Bucher, Johnson, & Wightman, 2016; Herr et al., 2012) at different locations in the brains of rodent models. The technique enabled investigations of the simultaneous release dynamics of dopamine and norepinephrine at two different locations in the brain of anesthetized rats (Park, Takmakov, & Wightman, 2011). For instance, Fig. 4 shows the recorded voltammetric data and its transformation into time-dependent release traces for the two analytes, presented comparatively. The applicability of FSCV measurements goes past characterization of neurotransmitter dynamics in the brain. The brain reward pathways, commonly activated by the natural reward-seeking behavior or by drugs, have been studied by measuring the differentiated profiles of dopamine and norepinephrine in the nucleus accumbens of rat brain under the influence of drugs (Park, Aragona, Kile, Carelli, & Wightman, 2010). Similarly, the dopamine dynamics in the nucleus accumbens of freely moving rats were assessed during intracranial self-stimulation, as a model for drug-of-abuse exposure (Rodeberg, Johnson, Bucher, & Wightman, 2016). A variation of the FSCV, fast-scan controlled-adsorption voltammetry, has enabled measurement of absolute concentrations of neurotransmitters in vivo, revealing

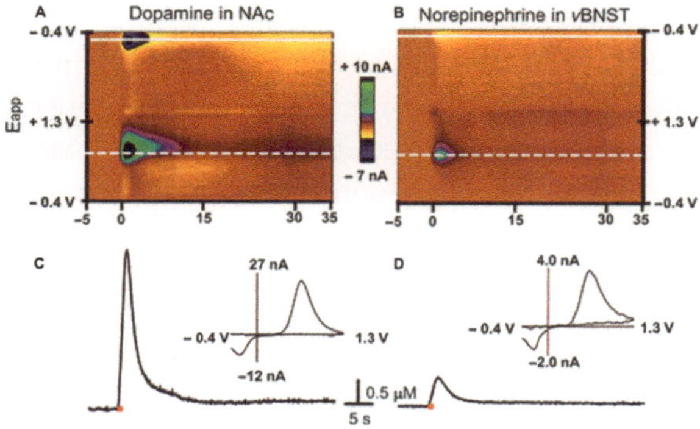

Fig. 4 Signal identification of voltammetric data recorded in the anterior nucleus accumbens (NAc) and ventral bed nucleus of the stria terminalis (vBNST). *Upper panels*: *Color plots* for the voltammetric data, with current changes encoded in false color for dopamine in NAc (A) and norepinephrine in the vBNST (B). The sets are composed of all background subtracted cyclic voltammograms recorded for 40 s before and after electrical stimulation of the ventral tegmental area/substantia nigra, and the ventral noradrenergic bundle (60 Hz, 60 pulses; delivered at 0 s). Catecholamine concentration changes are apparent in the *color plots* at the potential for their oxidation (~0.65 V, *dotted line*) and reduction (~0.23 V, *solid line*). The traces of dopamine in NAc (C) and norepinephrine in vBNST (D) evoked by the electrical stimulation were measured; the traces (*dotted line)* are shown at the potential at which catecholamine is oxidized. Electrical stimulation is indicated with the *solid red bars* under the traces. *Insets*: Background-subtracted cyclic voltammagrams recorded at the maximum of the evoked release. *Reprinted with permission from Park, J., Takmakov, P., & Wightman, R. M. (2011). In vivo comparison of norepinephrine and dopamine release in rat brain by simultaneous measurements with fast-scan cyclic voltammetry.* Journal of Neurochemistry, 119*(5), 932–944. doi: 10.1111/j.1471-4159.2011.07494.x.*

information about their local diffusion kinetics (Atcherley, Wood, Parent, Hashemi, & Heien, 2015; Burrell, Atcherley, Heien, & Lipski, 2015). Improvement in selectivity, sensitivity, and/or fouling resistance of the carbon fiber electrodes during FSCV was reported using surface modification with Nafion–CNTs or overoxidized polypyrrole–CNTs (Peairs, Ross, & Venton, 2011), as well as Nafion, base-hydrolyzed cellulose acetate, or fibronectin (Singh, Sawarynski, Dabiri, Choi, & Andrews, 2011), or polyethylenedioxythiophene (PEDOT):Nafion composite (Vreeland et al., 2015). Improved electrochemical characteristics were observed when CNT yarn electrodes (Schmidt, Wang, Zhu, & Sombers, 2013) and ultrananocrystalline diamond microsensors (Arumugam et al., 2013) were used as compared to single carbon fiber microelectrodes. FSCV at carbon fiber microelectrodes has been employed for development of a Wireless

Instantaneous Neurochemical Concentration Sensing (WINCS) system (Chang et al., 2013; Griessenauer et al., 2010). The wireless device was recently tested on human patients using both a carbon fiber electrode and a diamond-based electrode (Bennet et al., 2016). For long-term repeated measurements in vivo, the boron-doped diamond electrode shows better resistance with no obvious surface morphology change after 6 days. On the other hand, the carbon fiber electrode started to get eroded after less than 3 days of continuous use. The setup presented shows great promise for basic research for the investigation of long-term neurochemical responses in human subjects. Short-term human brain measurements were previously reported using a carbon fiber electrode (Kishida et al., 2011). In this study, dopamine release was studied in the context of human decision-making behavior. Therefore, it can be stated that the use of microelectrodes as an established clinical tool for neurobiological studies is within reach and progress is currently being made in that direction (Kasasbeh, Lee, Bieber, Bennet, & Chang, 2013).

Amperometry, as an alternative to the use of FSCV, has been shown as a viable approach for in vivo studies. Dopamine detection in rat striatum was reported using a molecularly imprinted overoxidized polypyrrole platinum sensor (Tsai et al., 2012) and a Au NP-sheathed glass capillary nanoelectrode (Liu, Yao, et al., 2015). Our group has reported an enzymatic sensor for dopamine detection employing tyrosinase immobilized at the surface of a carbon wire electrode in a chitosan matrix containing oxygen-rich NPs (Njagi, Chernov, Leiter, & Andreescu, 2010). The biosensor was used for detection of dopamine in anesthetized rats under electrical stimulation. This work demonstrates a viable approach for implementation of NPs in sensor designs for in vivo studies in a low oxygen environment (Sardesai, Ganesana, Karimi, Leiter, & Andreescu, 2015). Detection of neurotransmitters in smaller vertebrates is also possible and opens up new directions for applications of electrochemistry in vivo. Recent work in our group was focused on the use of zebrafish embryos as a screening model for nanotoxicity (Bowman & Zon, 2010). In vivo detection of serotonin in the intestine of embryos was demonstrated at a Nafion–CNTs-modified carbon fiber microelectrode (Njagi, Ball, Best, Wallace, & Andreescu, 2010) and at a chitosan-modified carbon fiber microelectrode (Özel, Wallace, & Andreescu, 2011). The latter was successfully applied to study the impact of NP exposure in the zebrafish intestine (Ozel, Hayat, Wallace, & Andreescu, 2013; Ozel, Wallace, & Andreescu, 2014). These results suggest that in vivo detection of neurotransmitters at carbon fiber microelectrodes can be used as a methodology for nanotoxicity screening (Özel, Liu, Alkasir, & Andreescu, 2014) and study of neurotransmission in developmental biology.

2.2.2 ROS/RNS Detection

The detection of ROS/RNS in biological models has been reviewed recently (Calas-Blanchard, Catanante, & Noguer, 2014; Chen, Ren, Yang, Wen, & Zhao, 2012; Liu, Dumitrescu, & Andreescu, 2015). Overall, few studies have reported successful detection of reactive species in vivo due to the short lifetimes of these species in biological tissue, cross-reactivity, and interferences. The latest reports of in vivo measurements of $NO^{\cdot}$ almost exclusively use platinized platinum wire electrodes. Jo et al. have reported the in-depth profiling of $NO^{\cdot}$ in the cortical and hippocampal brain areas of a rodent model using a glass nanopore-platinized platinum electrode (Jo, Do, Jhon, Suh, & Lee, 2011). Dual-sensor configurations have been reported for simultaneous detection of $NO^{\cdot}$ and oxygen (Park, Hong, et al., 2010), carbon monoxide (Park et al., 2015), or potassium ions (Moon et al., 2016). These sensors were employed in the brain of a rat model to study the real-time concentration changes and signaling role of these species. The concentration–time profile of these analytes was studied in relation to a model of epileptic seizure. Recent work in our group has been focused on development of sensors for detection of reactive species and their application in relevant biological systems, as tools for studying oxidative stress mechanisms (Fig. 5). Detection of $NO^{\cdot}$ was reported using a Meldola Blue-modified platinum electrode (Njagi, Erlichman, Aston, Leiter, & Andreescu, 2010). The combination of electrochemical mediator and protective membranes ensured the selectivity of the sensor, a low detection limit, as well as robustness during the implantation process in brain slices. Similarly, we have demonstrated the detection of $NO^{\cdot}$ in zebrafish embryo intestine using a carbon fiber microelectrode modified with selective membranes (Özel, Alkasir, Ray, Wallace, & Andreescu, 2013). The sensor was used as a tool to investigate the effects of oxidative stress injury in relation with NP exposure. $O_2^{\cdot -}$ detection in rodent brain was reported for sensor configurations employing immobilized Cyt *c* (Rahman, Kothalam, Choe, Won, & Shim, 2012) or SOD (Wang, Liu, et al., 2013) at the surface of microelectrodes. The immobilization of Cyt *c* on a thiol-modified gold wire electrode and its use for probing the $O_2^{\cdot -}$ dynamics in a brain ischemia model was also reported (Ganesana et al., 2012). The Cyt *c*-based biosensor

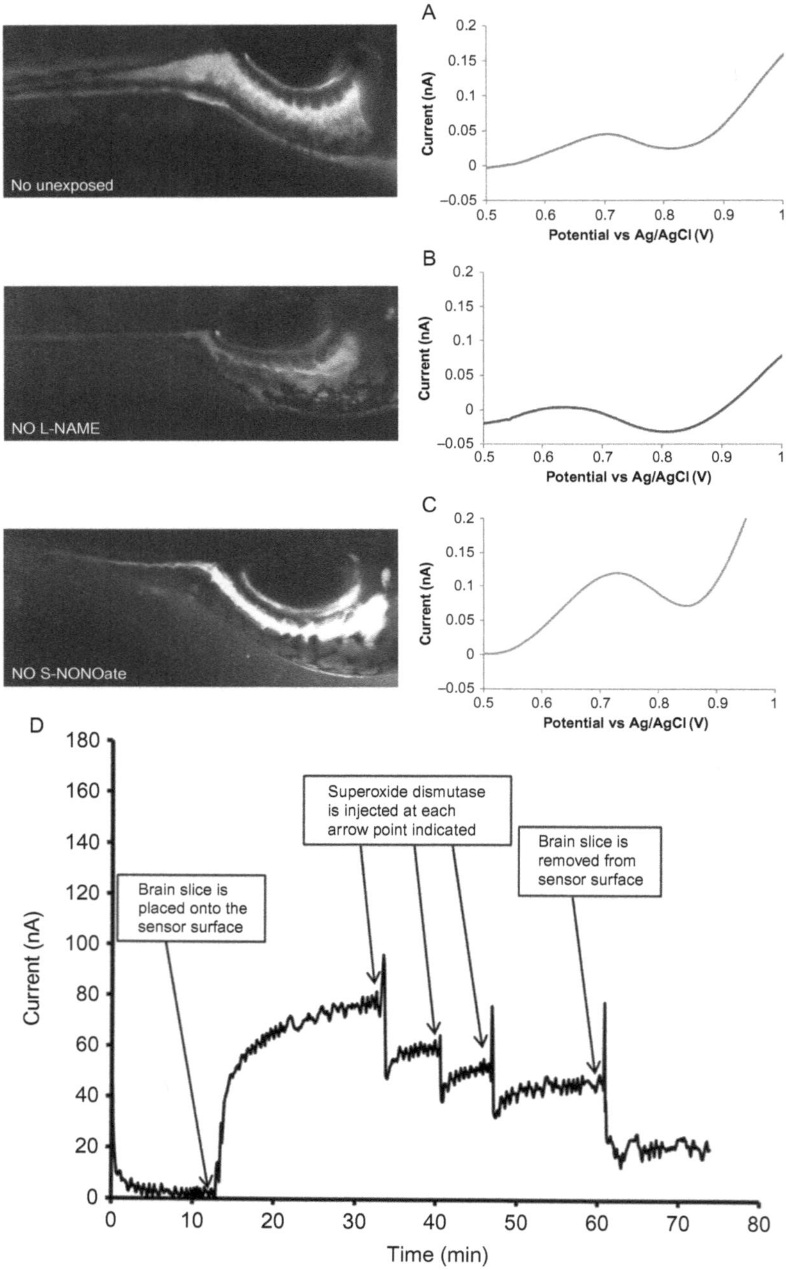

Fig. 5 *Upper panels (A–C)*: Electrochemical measurements in the live zebrafish intestine showing normal physiological NO concentration and altered levels as a result of pharmacological manipulation in 5 dpf zebrafish embryos. Fluorescence microscopy images (on the *left panel*) and their corresponding DPV (on the *right panel*) measured

was also used in bacterial cultures, demonstrating its robustness for varied biological applications (Liu, Marrakchi, Jahne, Rogers, & Andreescu, 2016).

3. RECENT ADVANCEMENTS IN ELECTRODE DESIGNS FOR BIOLOGICAL MONITORING

The increasing interest in applying electrochemistry for studies in biological models determined recent progress in the development of electrochemical probes to improve detection characteristics and enable measurements in varying sample types. Most sensor configurations are still in the proof of concept stage, being used primarily for fundamental studies in standard laboratory conditions. More recent studies explore the use of this technology for practical tools in biomedical research. Two primary factors contribute to the choice of the sensor design: the target analyte and the biological model. Fig. 6 summarizes characteristics and targeted performance metrics in the design of electrochemical sensors. Most recent advances in electrode design and the potential of these new approaches to address limitations of existing sensors are discussed herein.

(A) for control embryos, (B) for embryos exposed to a NO synthase inhibitor, 3 m*M* L-NAME for 24 h, and (C) an NO donor, 100 μ*M* Spermine NONOate (S-NONOate) for 1 h. Application of L-NAME to 5 dpf wild-type embryos drastically lowers NO concentrations throughout the digestive tract. Spermine NONOate treatment, however, significantly increases physiological NO levels within the digestive system. All embryos were incubated with 5 μ*M* DAF-FM-DA for 2 h before fluorescence imaging and were alive during imaging. Electrochemical recordings were performed on live embryos without any additional preparation beyond electrode insertion. *Lower panel (D)*: Specificity of a Cyt *c*-based biosensor response toward superoxide released by a hippocampal brain slice under normal physiological conditions. With each addition of 500 U/mL SOD, there is a decrease in the amperometric response, indicating specific inactivation of the produced superoxide radicals. The *arrows* indicate the times of SOD addition and the exposure of the brain to/from the electrode surface. *Panel (A)–(C): Reprinted with permission from Özel, R. E., Alkasir, R. S. J., Ray, K., Wallace, K. N., & Andreescu, S. (2013). Comparative evaluation of intestinal nitric oxide in embryonic zebrafish exposed to metal oxide nanoparticles.* Small, *9(24), 4250–4261. http://dx.doi.org/10.1002/smll.201301087. Panel (D): Reprinted with permission from Ganesana, M., Erlichman, J. S., & Andreescu, S. (2012). Real-time monitoring of superoxide accumulation and antioxidant activity in a brain slice model using an electrochemical cytochrome c biosensor.* Free Radical Biology and Medicine, *53(12), 2240–2249. doi: http://dx.doi.org/10.1016/j.freeradbiomed.2012.10.540.*

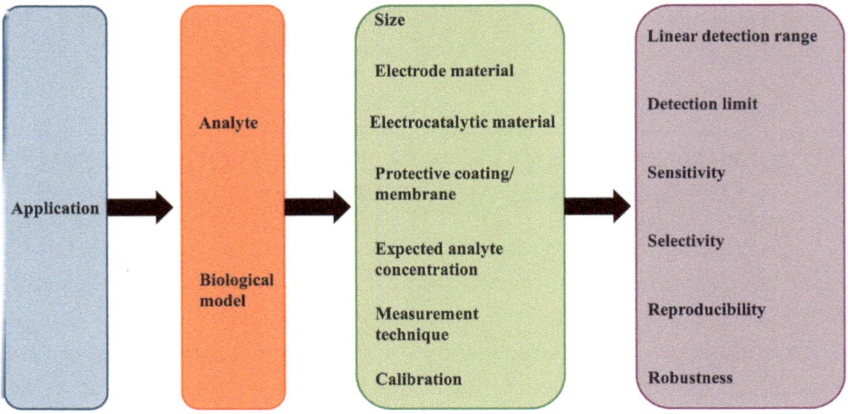

Fig. 6 Approach in the development of biological sensors and their targeted performance characteristics.

3.1 Electrode Fabrication Methodology

One of the most important parameters to consider is the size of the sensor matching the biological model under study. The design of implantable sensors poses significantly greater challenges due to difficulties with in vivo calibration and the potential damage associated with the implantation (Bucur, 2012; Polikov, Tresco, & Reichert, 2005). When nanometer-sized electrodes were used in individual cells, it was shown that penetration of cells does not affect the cell viability (Yum et al., 2010). Significant efforts are oriented toward miniaturization of electrochemical probes. Recent developments are concerned with fabrication of nanoelectrodes (Yeh & Shi, 2010). Flame etching of micrometer-sized carbon fibers was shown to yield a nanotip with a diameter of 50–100 nm at the top and a length of 30–100 μm (Li, Majdi, et al., 2015). The morphology of the etched carbon fiber tip is shown in Fig. 7. These electrodes were used for intracellular measurement of catecholamines. However, the use of carbon fibers as starting material does not always allow for size control during fabrication of nanoelectrodes. Procedures for direct deposition of carbon inside of pulled capillaries with nanosize openings were reported and shown to be more reliable approaches. The filling of nanopipettes with carbon can be done by pyrolitic decomposition of butane (Takahashi et al., 2011) or through chemical vapor deposition of methane (Hu et al., 2013). Carbon nanodisk electrodes were fabricated by the pyrolitic deposition of carbon within prepulled nanopipettes with openings varying from 5 to 200 nm in diameter

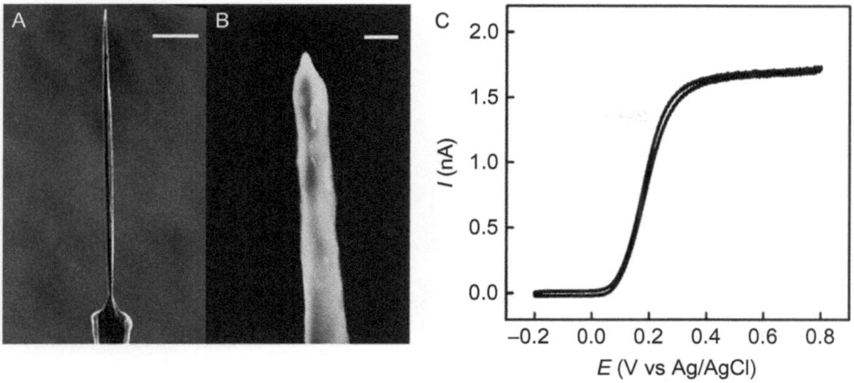

Fig. 7 Characterization of a flame-etched carbon fiber nanotip electrode used for intracellular measurements. (A) Global view (scanning electron microscopy) of a nanotip conical carbon fiber microelectrode (scale bar: 5 mm). (B) Magnified view of the tip of another nanotip conical carbon fiber microelectrode (scale bar: 400 nm). (C) Representative cyclic voltammogram of 0.10 m*M* dopamine at a nanotip conical carbon fiber microelectrode (scan rate: 0.10 V/s). *Reprinted with permission from Li, X., Majdi, S., Dunevall, J., Fathali, H., & Ewing, A. G. (2015). Quantitative measurement of transmitters in individual vesicles in the cytoplasm of single cells with nanotip electrodes.* Angewandte Chemie International Edition, *54(41), 11978–11982. http://dx.doi.org/10.1002/anie.201504839.*

(Actis et al., 2014). Further etching of the carbon at the tip of the nanoelectrode produces a nanocavity that can be electrochemically filled with catalytic materials (Clausmeyer, Actis, López Córdoba, Korchev, & Schuhmann, 2014). This procedure is aimed to increase the mechanical and chemical stability of electroactive films at the surface of small-size electrodes. The concept was demonstrated for deposition of Prussian Blue and subsequent detection of H_2O_2. The fabrication of a gold-based nanotip electrode with a diameter down to 6 nm was also reported, showing that fine size control of metallic wires can be achieved (Liu, Li, et al., 2015). Trikantzopoulos et al. developed a procedure for fabrication of carbon fiber electrodes using a 3D printed mold (Trikantzopoulos, Yang, Ganesana, Wang, & Venton, 2016). The design of the 3D printed mold and the step-by-step procedure for fabrication of carbon fiber microelectrodes is shown in Fig. 8. This work and others suggest that printing technology can improve the fabrication process of sensors due to advantages, such as fast high volume production, reproducibility, and control over the dimensions of the electrodes (Othman, Karimi, & Andreescu, 2016).

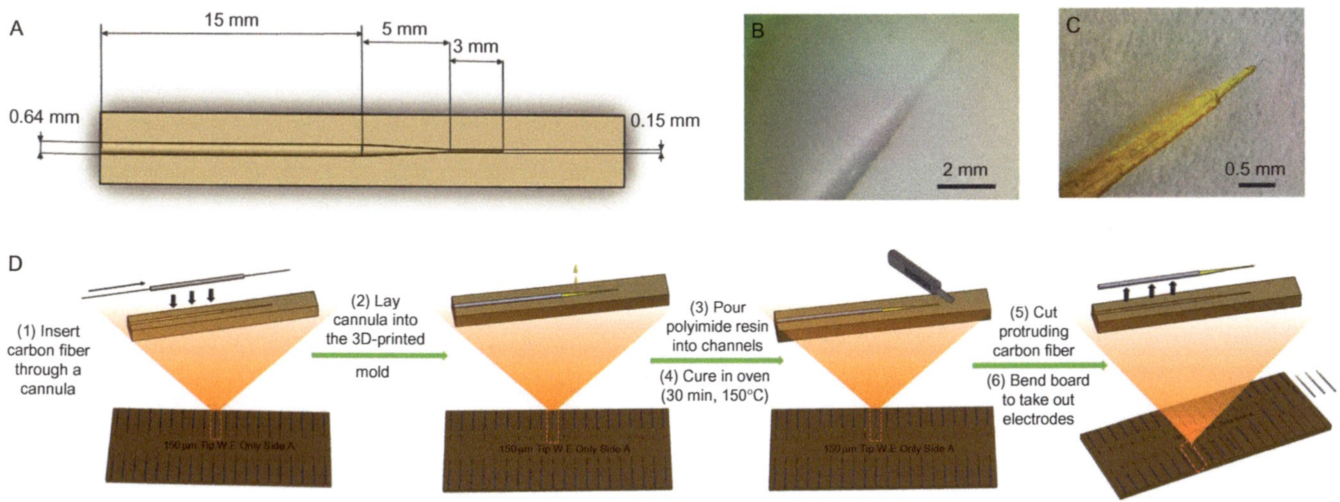

Fig. 8 3D-printed mold design and polyimide-insulated carbon fiber microelectrodes fabrication process. (A) Close-up of the design of one channel for 3D-printed microelectrode fabrication. (B) Image of the tip from a 3D-printed mold, scale bar: 2 mm. (C) Image of a polyimide-insulated carbon fiber microelectrode fabricated using the mold, scale bar: 0.5 mm. (D) Scheme of the process of microelectrode fabrication. *Reprinted with permission from Trikantzopoulos, E., Yang, C., Ganesana, M., Wang, Y., & Venton, B. J. (2016). Novel carbon-fiber microelectrode batch fabrication using a 3D-printed mold and polyimide resin.* Analyst, 141, *5256–5260. http://dx.doi.org/10.1039/c6an01469k.*

3.2 Emerging Materials for Improving Electrochemical Detection

In recent years, a wide variety of materials have been used for improving characteristics of electrodes for detection of neurotransmitters and ROS/RNS (Fig. 9). In this section, we discuss properties of emerging materials used to modify electrode surfaces for improving electrochemical detection and their suitability for potential use in biological applications.

3.2.1 Materials for Neurotransmitters Sensing

Several different materials have been used to improve the detection of neurotransmitters. Most applications are reported with electrodes that incorporate carbon nanomaterials and metallic NPs. There are also reports employing other materials, such as electrochemical mediators (Jiang et al., 2015; Reddaiah, Reddy, Mallikarjuna, & Narasimha, 2013), conductive polymers (Harley, Rooney, & Breslin, 2010; Marti, Fabregat, Estrany, Aleman, & Armelin, 2010), self-assembled monolayers (Shervedani, Siadat-Barzoki, & Bagherzadeh, 2010), layered double hydroxides (Zhu et al., 2010), and aptamers (Farjami et al., 2013).

The incorporation of carbon nanomaterials in electrochemical devices is closely related to the discovery of their special properties such as high surface area, high conductivity, exceptional strength, and stiffness. Most applications are with CNTs (Merkoçi et al., 2005), followed closely by graphene materials (Pumera, Ambrosi, Bonanni, Chng, & Poh, 2010). Their relatively recent discovery did not allow for the exploration of their full potential for sensing applications (Yang et al., 2010). The attractiveness of carbon materials consists in their ability to substantially increase the active surface of the electrode due to their innate high specific surface area (Shao et al., 2010; Swamy & Venton, 2007). This is particularly useful for development of small-size probes with increased electrochemically active surface (Jacobs, Ivanov, Nguyen, Zestos, & Venton, 2014; Mercante et al., 2015). By architectural modeling of the 3D structure and nitrogen doping, the electrochemical properties of graphene can be further improved, which can be advantageous for sensing applications (Feng et al., 2015). Graphene can also exhibit electrocatalytic properties. Valentini et al. showed that oxidized graphene nanoribbons demonstrate notable electroanalytical performance, considering the shifted detection potential at modified electrodes used for measurement of catecholamines and serotonin as well as for other biological molecules of interest (Valentini, Romanazzo, Carbone, & Palleschi, 2012).

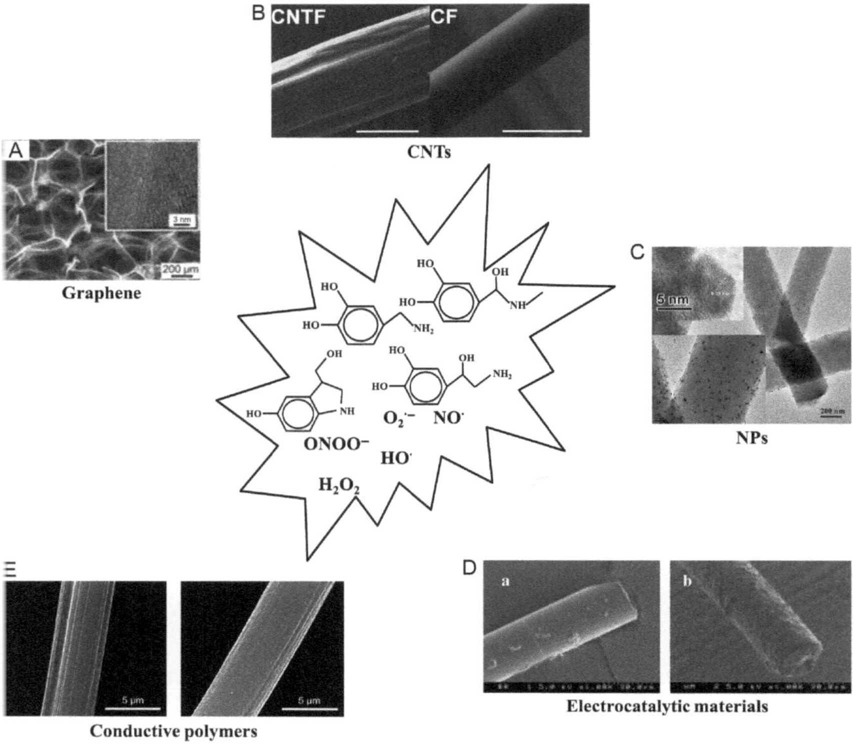

Fig. 9 Examples of emerging materials often used in new systems for detection of neurotransmitters and ROS/RNS species. (A) SEM image of 3D graphene nanosheets. (B) SEM images of the body of a CNT-modified carbon fiber and a bare carbon fiber; the bar indicates 50 µm. (C) TEM images of Ag–Pt bimetallic NP-modified electrospun carbon nanofibers. (D) Typical surface morphology of bare and hemin-modified carbon fibers as analyzed by FESEM. (E) Electron micrograph of uncoated and PEDOT:Nafion-coated carbon fiber microelectrodes. *Panel (A): Reprinted with permission from Cao, X., Zeng, Z., Shi, W., Yep, P., Yan, Q., & Zhang, H. (2013). Three-dimensional graphene network composites for detection of hydrogen peroxide.* Small, *9(9–10), 1703–1707. http://dx.doi.org/10.1002/smll.201200683. Panel (B): Reprinted with permission from Harreither, W., Trouillon, R., Poulin, P., Neri, W., Ewing, A. G., & Safina, G. (2013). Carbon nanotube fiber microelectrodes show a higher resistance to dopamine fouling.* Analytical Chemistry, *85(15), 7447–7453. http://dx.doi.org/10.1021/ac401399s. Panel (C): Reprinted with permission from Huang, Y., Miao, Y.-E., Ji, S., Tjiu, W. W., & Liu, T. (2014). Electrospun carbon nanofibers decorated with Ag–Pt bimetallic nanoparticles for selective detection of dopamine.* ACS Applied Materials & Interfaces, *6(15), 12449–12456. http://dx.doi.org/10.1021/am502344p. Panel (D): Reprinted with permission from Peteu, S. F., Whitman, B. W., Galligan, J. J., & Swain, G. M. (2016). Electrochemical detection of peroxynitrite using hemin-PEDOT functionalized boron-doped diamond microelectrode.* Analyst, 141*(5), 1796–1806. http://dx.doi.org/10.1039/c5an02587g. Panel (E): Reprinted with permission from Vreeland, R. F., Atcherley, C. W., Russell, W. S., Xie, J. Y., Lu, D., Laude, N. D., . . . Heien, M. L. (2015). Biocompatible PEDOT:Nafion composite electrode coatings for selective detection of neurotransmitters in vivo.* Analytical Chemistry, 87*(5), 2600–2607. http://dx.doi.org/10.1021/ac502165f.*

A common difficulty in translating electrochemical sensors to measurements in "real" biological environments is the possible interfering effects of coexisting compounds in the complex biological matrix. Improved detection was observed with electrodes modified with electrochemically pretreated CNTs, which were used for measurement of dopamine in the presence of common interferents (Alwarappan, Liu, & Li, 2010). Graphene materials can be functionalized to enhance interaction with the analyte of interest, while rejecting interfering species (Bagherzadeh & Heydari, 2013). The sulfonation of graphene was shown to improve the detection of dopamine with no interference from uric acid and ascorbic acid, demonstrating the ability of graphene to be tailored based on the specific application requirements (Li, He, et al., 2013). A major problem for electrochemical detection of dopamine is the fouling of the electrode due to formation of by-products during the oxidation process (Hawley, Tatawawadi, Piekarski, & Adams, 1967). It was shown that the preparation of fiber electrodes with carbon nanotubes improves resistance to fouling (Harreither et al., 2013). CNTs and graphene can also be employed in composite materials, which exhibit an overall improved electrochemical performance. Other carbon architectures can be grown in different geometries and still exhibit improved electrochemical properties. Arrays of aligned carbon nanofibers (Rand et al., 2013) or nitrogen-doped porous carbon nanorods (Yuan, Yuan, Zhou, Zou, & Zhou, 2012) are possible alternatives to CNTs or graphene, yielding comparable performance for detection of dopamine and serotonin. Charged carbon NPs can also be used to provide selectivity for dopamine oxidation in the presence of common interferents (Kundys et al., 2014). The synthesis of carbon nanofibers modified with metallic NPs was highlighted by Huang, Miao, Ji, Tjiu, and Liu (2014). The in situ preparation of bimetallic Au–Pt NPs represents a practical methodology for sensor preparation as an alternative to the more commonly used drop-casting method.

In addition to carbon-based nanostructures, several types of inorganic NPs have been used to modify the surface of electrochemical sensors for neurotransmitter detection. Common strategies report the combined use of carbon nanomaterials with NPs to provide a synergistic effect for enlarging surface area and increase of the electrocatalytic current. Several reports also highlight their use as main electrocatalytic element in the sensing design. The use of metallic NPs is related to their high specific surface area and good electrocatalytic activity, similar to carbon nanomaterials (Yu, Zeng, Qi, Zhou, & Shi, 2012). Surface functionalization of a mesoporous silica nanomaterial enabled selective detection of dopamine and serotonin in human serum samples (Hasanzadeh, Shadjou, & Omidinia, 2013). In spite of the highlighted advantages, the existing methodology for employing NPs in electrochemical sensor design does not allow their use for more advanced applications with micro- or nanosensors (Khoobi, Ghoreishi, Behpour, & Masoum, 2014). Structures similar to the ones obtained by drop casting of NPs at the surface of electrodes can be formed by electrochemical approaches. A particular example is the electrodeposition of gold. The electrochemistry of common neurotransmitters at electrodes modified by gold electrodeposition was reported (Atta, Galal, Abu-Attia, & Azab, 2011).

3.2.2 Materials for ROS/RNS Sensing

Most electrochemical sensors for ROS/RNS detection focus on H_2O_2 as a representative analyte. The detection of H_2O_2 is more often reported due to its accessibility and stability in comparison with the other ROS/RNS, as well as the interest in developing enzymatic sensors (Grieshaber, MacKenzie, Vörös, & Reimhult, 2008). This section describes the most relevant examples of electrochemical sensors for H_2O_2 detection. In addition to chemical detection based on carbon, NPs, or electrocatalytic materials, a few enzymatic approaches using enzymes entrapped in porous nanodiamond materials (Gopalan, Komathi, Sai Anand, & Lee, 2013), metal-loaded porous inorganic nanomaterials (Jin, Sakthivel, Gandhi, Huy, & Lee, 2016), or porous metal architectures (Meng, Yan, Liu, Gu, & Zou, 2011; Zheng et al., 2012) were reported.

The applicability of carbon materials for H_2O_2 detection is mainly driven by their good electrical conductivity and high specific surface area, which allows loading of a greater quantity of catalytic materials on their surface, thus improving the sensing performance for H_2O_2 detection (Cao et al., 2013). Catalytic materials used in electrode design are generally metallic-based nanomaterials, such as NPs, or polymerized electrocatalytic mediators. Applications employing CNTs include loading of the nanotubes with Pd NPs (You et al., 2011), Ag NPs (Bui et al., 2010), or Co_3O_4 NPs (Kaçar, Dalkiran, Erden, & Kiliç, 2014). Similarly, graphene materials can be modified with NPs as catalytic materials for H_2O_2 sensing. Examples of nanomaterials often used are Pt NPs (Xu et al., 2011), Pd NPs (Chen et al., 2013; You et al., 2013), Au NPs (Liu et al., 2012), Ag NPs (Liu, Tian, Wang, & Sun, 2011; Zhao, Liu, Fu, & Yang, 2013), CuO NPs (Liu, Liu, & Chen, 2013; Xu, Deng, Li, Chen, & Wang, 2013), SnO_2 NPs (Adam et al., 2016), and MnO_2 NPs (Li, Du, et al., 2010). Functionalization of graphene with electrochemical mediators, such as Prussian Blue (Yang, Myoung, & Hong, 2012) and hemin (Chen, Zhao, Bai, & Shi, 2011; Guo, Li, & Dong, 2011), is also possible based on interactions of a physical nature (e.g., π–π interactions) between the two structures. High-surface carbon architectures can be synthesized and loaded with Pd (Bian et al., 2012) or with Pt NPs (Bo, Bai, Qi, & Guo, 2011) for H_2O_2 sensing. In particular, the in situ synthesis of ultrafine Pt NPs with the help of ionic liquids, and using mesoporous carbons as substrate could be potentially applied on other, more practical substrates (Bo et al., 2011). A similar example is provided by the work of Liu et al. where the Pt NPs are synthesized directly on carbon nanofibers (Liu, Wang, Xu, Hou, & You, 2011).

For sensing applications, NPs can be loaded on silicon-based nanomaterials, in the form of porous materials (Ensafi, Abarghoui, & Rezaei, 2014) or nanowires (Yin et al., 2011), and on conductive polymer-based materials (Bian et al., 2010). Bimetallic (Wang, Han, Xu, Gao, & Cao, 2015), metal–nonmetal (Li, Zhang, Xuan, Jiang, & Zhu, 2010), or metal–metal oxide (Liu et al., 2010; Miao et al., 2013) nanomaterials are prevalent, and widely employed as electrocatalytic materials for H_2O_2 detection. CuO-based nanomaterials show particularly good performance for electrochemical sensing of H_2O_2 (Ping, Ru, Fan, Wu, & Ying, 2010). An advantage is the possibility to form in situ nano-CuO films on copper substrates, which can be used for the development of miniaturized sensors (Song, Hwang, & Whang, 2010).

The use of electrocatalytic mediators is historically the main approach for H_2O_2 detection since the emergence of well-known materials such as Prussian Blue (Karyakin, 2001), Meldola Blue (Wring & Hart, 1992), and others. Most applications are still based on iron-based complexes. Prussian Blue was deposited on the nanoporous gold substrate for improvement of sensitivity toward H_2O_2 reduction (Ghaderi & Mehrgardi, 2014). Yagati et al. reported an alternative option to the use of Prussian Blue by employing ferredoxin, an iron(III) complex, covalently bound on a thiol-based self-assembled monolayer (Yagati, Lee, Min, & Choi, 2011). Biological molecules containing iron centers, such as myoglobin, are also used as biocatalytic material candidates; however, their activity depends greatly on the immobilization method (Safavi & Farjami, 2010). Other recognition molecules for H_2O_2 detection (e.g., thionine) are also studied (Wang, Zhang, & Tian, 2015).

Recently reported $NO^{\cdot}$ detection systems involve the use of similar materials to the ones discussed for H_2O_2 detection. Composites of carbon nanomaterials, CNTs, and graphene with ionic liquids (Ng, Guo, & Li, 2011) and NPs (Li, Liu, et al., 2013; Maluta, Canevari, & Machado, 2014) are widely used. Liu et al. demonstrated that the catalytic behavior of $NO^{\cdot}$ at graphene-modified electrodes can be controlled and improved by differing functionalization of the material, the main determinant parameter being the lattice defects and the C/O ratio in the synthesized graphene (Liu, Punckt, Pope, Gelperin, & Aksay, 2013). Detection of $NO^{\bullet}$ using metallic complex mediators is another relevant research direction for discovery of new or improved catalytic materials (Santos et al., 2013). Functionalization and modification of the metallic phthalocyanine structure were reported and it was shown that the metallic center used can influence the sensitivity of the material toward $NO^{\bullet}$ detection (Yap, Xu, & Ang, 2013). These materials are of great applicability mostly due to the possibility of electropolymerization on different electrode substrates.

The detection of $O_2^{\bullet -}$ is more difficult compared with H_2O_2 or $NO^{\bullet}$ due to the short life and instability of $O_2^{\bullet -}$ in normal conditions. The validation of $O_2^{\bullet -}$ sensors is a big challenge, as there are few verified methods for $O_2^{\bullet -}$ production in vitro and their use is limited by solvent choice, pH, temperature, and electrode fouling (in particular for the xanthine oxidase-based calibration method) (Ge & Lisdat, 2002). As a result, most research is based on the use of the two well-known $O_2^{\bullet -}$ recognition molecules, Cyt *c* (McNeil, Smith, Bellavite, & Bannister, 1989) and SOD (Tanaka & Muto, 1992). The use of Cyt *c* was demonstrated as part of a composite material with CNTs and ionic liquid, and its performance for $O_2^{\bullet -}$ detection was reported (Rahimi, Ghourchian, & Rafiee-Pour, 2011). Materials with a high surface area and improved electrochemical performance, such as CNTs (Li, Wang, Xu, & Chen, 2011; Rajesh et al., 2010), graphene (Tang et al., 2015), NPs (Rafiee-Pour, Noorbakhsh, Salimi, & Ghourchian, 2010; Thandavan, Gandhi, Sethuraman, Rayappan, & Krishnan, 2013), and carbon nanofiber composites (Santhosh et al., 2011), are often employed to increase the load of SOD. A matrix based on the chitosan–ionic liquid composite was also reported, taking advantage of the special electrochemical properties of ionic liquids and the biocompatibility of chitosan (Wang, Wen, et al., 2013). As the use and activity of biologically derived sensor elements are limited by pH, temperature, or denaturation, the discovery of inorganic materials functioning as SOD mimics is of great importance. The SOD activity of bimetallic Cu–Zn NPs was explored by Derkus et al., who reported good sensing performance for $O_2^{\bullet -}$ detection (Derkus, Emregul, & Emregul, 2015). Manganese-based materials have also been reported as SOD mimics. A manganese phosphate silica-supported nanomaterial was shown to offer good sensitivity and a low detection limit for electrochemical determination of $O_2^{\bullet -}$, thus demonstrating its SOD mimetic properties (Shen et al., 2016). Similarly, a manganese-based porphyrin supported on polypyrrole films has been reported as an SOD mimetic material and used for detection of $O_2^{\bullet -}$ (Rajesh et al., 2011).

The development of electrochemical sensors for $ONOO^-$ detection is a big challenge due to its instability and short lifetime, as well as issues related to the preparation and handling of $ONOO^-$ in vitro. Similar to $O_2^{\cdot -}$, the validation of $ONOO^-$ sensors is often questionable. Because of a lack of selective catalysts for its detection, only a few publications report electrode modifications for $ONOO^-$ sensing, and the existing work is based exclusively on electrochemical mediators. Hemin, an iron-containing porphyrin, was shown to possess electrocatalytic activity for $ONOO^-$ (Peteu, Bose, & Bayachou, 2013). The sensing performance of hemin toward $ONOO^-$ was improved by synthesis of composite materials with graphene (Oprea et al., 2013) or PEDOT (Peteu, Whitman, Galligan, & Swain, 2016). The hemin-PEDOT material is particularly attractive for use in biological studies since its deposition on carbon fiber electrodes was already reported (Peteu, Peiris, Gebremichael, & Bayachou, 2010). Cobalt-containing mediators have also been used for $ONOO^-$ detection. The performance of a cobalt porphyrin–graphene composite material has been demonstrated for simultaneous detection of $ONOO^-$ and H_2O_2 (Hosu et al., 2015). Catalytic activity for $ONOO^-$detection was also reported for a cobalt center-containing complex, cyanocobalamin (Wang & Chen, 2010). Its use is also advantageous due to the possibility of forming thin films at the electrode surface by electrochemical polymerization from monomer solutions.

$HO^{\cdot}$ has the lowest lifetime in solution among the ROS/RNS discussed herein. At the same time, the synthesis of $HO^{\cdot}$ for in vitro testing is limited to the use of the Fenton reaction. As a result, the only methods reported for the detection of $HO^{\cdot}$ are of an indirect nature and therefore not suitable for real-time measurement of the analyte. One study reported detection of $HO^{\cdot}$ based on its strong oxidation properties when in contact with polypyrrole films (Gualandi, Guadagnini, Zappoli, & Tonelli, 2014). The quantitative determination was related to the different degree of polypyrrole overoxidation observed after exposure to $HO^{\cdot}$ during the Fenton reaction. Similarly, another reported configuration is based on the oxidation properties of $HO^{\cdot}$ toward DNA strands (Wu et al., 2012). The oxidation process was quantified through the use of an electrochemical mediator, methylene blue, and amplified through the use of gold NPs.

3.3 Limitations of Current Sensing Technologies

3.3.1 Expected Analyte Concentration

Currently, there are a considerable number of in vivo studies that can help in estimating the expected concentrations of analytes in different biological models. It was shown that the concentration of neurotransmitters measured in different models is usually lower than 1 μM (Liu, Yao, et al., 2015; Ozel et al., 2014). The ROS/RNS concentration in vivo is harder to estimate due to their reduced stability and potential for cross-reactivity. The concentration level of reactive species can reach a maximum in the lower micromolar range, depending on the model used and measurement conditions (Ganesana et al., 2012; Njagi, Erlichman, et al., 2010; Özel et al., 2013). For cell models, the expected production of ROS/RNS can be determined based on the numbers provided by single-cell studies. For one cell, the flux for each of the detected species ($ONOO^-$, H_2O_2, and $NO^\bullet$) varies in between 0 and 30 fM per oxidative burst, depending on the cell type and stimulation method (Amatore et al., 2006, 2000a; Amatore, Arbault, Bouton, et al., 2008; Hu et al., 2010). When translating these numbers to cell cultures, several other factors can affect the detected concentration, such as cell population, distance between sensor and cells, half-life and diffusion coefficient for each species, and cross-reactivity.

3.3.2 Calibration and Data Interpretation

The transition of electrodes from ideal standard solutions or in vitro conditions to the in vivo environment is often difficult. The biological matrix is complex and cannot be reproduced for external calibration purposes (Bucur, 2012; Chen & Rice, 1999). Also, the behavior of the analyte could be different in the biological environment than during calibration done in ideal conditions (Oleinick et al., 2006). From this perspective, an in situ calibration strategy was proposed for the use of FSCV in biological samples, which eliminates the need for tracking the sensitivity of the electrodes during experiments (Roberts, Toups, Eyualem, McCarty, & Sombers, 2013). For ROS/RNS sensors, the calibration procedures are unreliable due to low stability of the analytes in normal conditions, in particular for $NO^\bullet$ (Bedioui & Villeneuve, 2003), $O_2^{\bullet -}$ (Ge & Lisdat, 2002), and $ONOO^-$ (Peteu, Boukherroub, & Szunerits, 2014). For these reasons, validation of results using alternative methods is recommended, either during calibration or during measurements in biological models (Özel et al., 2013). These issues have been discussed specifically for the detection of $NO^\bullet$ in biological media (Hunter, Storm, Coneski, & Schoenfisch, 2013), but it is expected that sensing of ROS/RNS in general is limited by these effects. The calibration limitations can be avoided, however, for certain applications. Electrochemical measurements at single cells do not require calibration. It is assumed that, due to the close positioning of the electrode to the cell and the comparable size between the two, all species released are quantitatively oxidized at the electrode surface and no signal is lost (Amatore, Arbault, Guille, et al., 2008).

3.3.3 Biocompatibility

The implications of electrode insertion into cell cultures or tissue are not always considered during the development of sensors for biological applications. The implantation of a foreign object into a live organism induces a response, "foreign body reaction" (Ratner & Bryant, 2004). The response is time dependent, and for long-term sensor implantation, it becomes especially problematic. The foreign body reaction usually starts with protein adsorption on the surface of the electrode, which takes place in the first hour of insertion, and can continue with acute inflammation at the insertion site, when mediators of inflammation and reactive species can be formed. All these phenomena may affect the physical integrity of the sensor and its performance, especially the sensitivity. Even though generally it is considered that biocompatibility can be assured only by the use of biocompatible materials, the characteristics of the electrode, such as size and shape, are also important, and their effect can only be tested by long-term implantation in biological models (Wilson & Johnson, 2008). Only a few recent studies have considered the biocompatibility issue when reporting sensing systems for detection of neurotransmitters and ROS/RNS. Guo et al. reported a biomimetic sensor for detection of $NO^{\cdot}$ in cell cultures (Guo et al., 2012). RGD peptide covalently bound on functionalized graphene was used to promote selective cell adhesion properties. The applicability of the sensor was shown for $NO^{\cdot}$ sensing in human umbilical vein endothelial cells under drug stimulation. The preparation steps are schematically shown in Fig. 10. Clark et al. studied the biocompatibility of carbon fiber microsensors during long-term implantation in rodent models (Clark et al., 2010). It was shown that 4 months of continuous use in vivo does not elicit overt reactive gliosis, encapsulation of the carbon fiber, or a loss of tyrosine hydroxylase expression. These results have been credited to the small size of the implanted sensor (Seymour & Kipke, 2007). Strategies for improving biocompatibility were reviewed recently for the use of other implantable sensors, especially glucose sensors (Lin, Lin, Mansour, & Gu, 2013; Nichols, Koh, Storm, Shin, & Schoenfisch, 2013). These already available results may be considered for the future development of sensors designed for in vivo measurements.

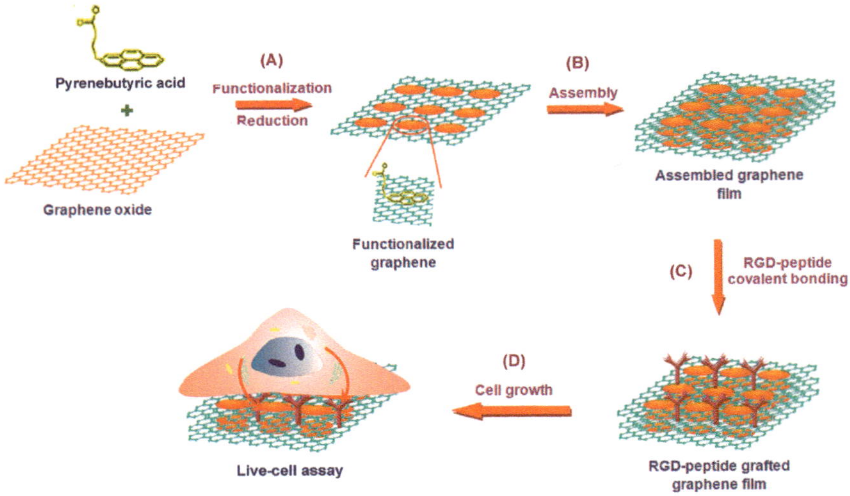

Fig. 10 Schematic illustration of preparation of a free-standing biomimetic film and its use in a live-cell assay for detection of NO·. (A) Chemical reduction of graphene oxide and pyrenebutyric acid functionalization; (B) preparation of pyrenebutyric acid functionalized graphene film by filtration of the solution of pyrenebutyric acid functionalized graphene; (C) fabrication of the biomimetic functional film by covalently bonding RGD peptide on the surface of the pyrenebutyric acid functionalized graphene film via EDC/NHS coupling; (D) the functional biofilm as cell adhesion and sensing matrix. *Reprinted with permission from Guo, C. X., Ng, S. R., Khoo, S. Y., Zheng, X., Chen, P., & Li, C. M. (2012). RGD-peptide functionalized graphene biomimetic live-cell sensor for real-time detection of nitric oxide molecules.* ACS Nano, *6(8), 6944–6951. http://dx.doi.org/10.1021/nn301974u.*

4. CONCLUSIONS AND FUTURE TRENDS

Current research in the development of electrochemical sensors demonstrates the potential of these devices as valuable investigative tools for studying complex biochemical mechanisms in biological and clinical research. Electrochemical sensors can provide rapid, sensitive, and real-time detection of analytes with essential activity in live organisms. This chapter summarizes the current status of electrochemical methodologies for detection of neurotransmitters and ROS/RNS as examples of relevant targets. Real-time in situ monitoring of these species enables detailed investigation of biochemical mechanisms in complex organs, such as the brain, and at a fundamental level, in cell populations or at single cells. These studies can help elucidate chemical and biological processes that take place inside cells or living organisms, as well as the study of disease or injury states. Electrochemical sensors can provide information that is not currently obtainable through other methods, like real-time observation of neurotransmitter release in situ. Although the potential of these methods has been demonstrated, several challenges still exist which prevents their broad adoption by the biological and biomedical community. These include effective miniaturization of the probes, quantitative data interpretation and calibration, selectivity, antifouling, and biocompatibility. Development of new sensing designs and materials could improve sensor performance and address these challenges in the future. In addition to innovation in sensor design, progress is expected in the implementation of this technology to study fundamental biological mechanisms in relevant model organisms and in realistic biomedical settings. The investigation of exocytosis and oxidative stress at single cells has brought important insights regarding biochemical processes that take place at a fundamental level. Single-cell electrochemistry can be applied in the future in clinical research for studying drug action and antioxidative mechanisms. Work focusing on intracellular measurements is also expected and can make significant contributions to this field. Development of methodologies for simultaneous detection of multiple species, especially for ROS/RNS, is underway and successful implementation can help decipher complex oxidative stress mechanisms and interrelated processes. Detection of neurotransmitters by FSCV in brain is the other successful application of electrochemistry in vivo. It was shown that voltammetry at carbon fiber electrodes can be used to study normal behavior and neurological disorders in the brain of rodent models. The applicability of this approach for measurements in living organisms is unquestionable, and recent studies have emphasized the potential of this technology for detection in human subjects. Thus, it is expected that progress will be made in this direction in the next few years. On the other hand, few examples of ROS/RNS detection in more complex organisms than single cells have been reported. Better approaches are required in order to selectively and reliably detect such reactive species in biological conditions. The developmental work performed in ideal conditions should aid in designing future electrochemical methodologies for detection of ROS/RNS and other species in vivo. These studies will provide a wealth of information on the biochemistry of living organisms and advance biological and clinical research and enable the development of next-generation biological and medical devices.

ACKNOWLEDGMENTS

This work was supported by NSF Grants #1336493 and 1561491. Any opinions, findings, and conclusions or recommendations expressed in this material are those of the author(s) and do not necessarily reflect the views of the funding agencies.

REFERENCES

Abdelwahab, A. A., Koh, W. C. A., Noh, H.-B., & Shim, Y.-B. (2010). A selective nitric oxide nanocomposite biosensor based on direct electron transfer of microperoxidase: Removal of interferences by co-immobilized enzymes. *Biosensors and Bioelectronics*, *26*(3), 1080–1086. http://dx.doi.org/10.1016/j.bios.2010.08.070.

Actis, P., Tokar, S., Clausmeyer, J., Babakinejad, B., Mikhaleva, S., Cornut, R., … Korchev, Y. E. (2014). Electrochemical nanoprobes for single-cell analysis. *ACS Nano*, *8*(1), 875–884. http://dx.doi.org/10.1021/nn405612q.

Adam, J., Adamova, D., Aggarwal, M. M., Aglieri Rinella, G., Agnello, M., Agrawal, N., … Collaboration, A. (2016). Anisotropic flow of charged particles in Pb-Pb collisions at radical[s{NN}] = 5.02 TeV. *Physical Review Letters*, *116*(13), 132302. http://dx.doi.org/10.1103/PhysRevLett.116.132302.

Adams, K. L., Jena, B. K., Percival, S. J., & Zhang, B. (2011). Highly sensitive detection of exocytotic dopamine release using a gold-nanoparticle-network microelectrode. *Analytical Chemistry*, *83*(3), 920–927. http://dx.doi.org/10.1021/ac102599s.

Akimova, E., Lanzenberger, R., & Kasper, S. (2009). The serotonin-1A receptor in anxiety disorders. *Biological Psychiatry*, *66*(7), 627–635. http://dx.doi.org/10.1016/j.biopsych.2009.03.012.

Alivisatos, A. P., Andrews, A. M., Boyden, E. S., Chun, M., Church, G. M., Deisseroth, K., … Zhuang, X. (2013). Nanotools for neuroscience and brain activity mapping. *ACS Nano*, 7(3), 1850–1866. http://dx.doi.org/10.1021/nn4012847.

Alwarappan, S., Liu, G., & Li, C.-Z. (2010). Simultaneous detection of dopamine, ascorbic acid, and uric acid at electrochemically pretreated carbon nanotube biosensors. *Nanomedicine: Nanotechnology, Biology and Medicine*, *6*(1), 52–57. http://dx.doi.org/10.1016/j.nano.2009.06.003.

Amatore, C., Arbault, S., Bouton, C., Coffi, K., Drapier, J.-C., Ghandour, H., & Tong, Y. (2006). Monitoring in real time with a microelectrode the release of reactive oxygen and nitrogen species by a single macrophage stimulated by its membrane mechanical depolarization. *ChemBioChem*, 7(4), 653–661. http://dx.doi.org/10.1002/cbic.200500359.

Amatore, C., Arbault, S., Bouton, C., Drapier, J.-C., Ghandour, H., & Koh, A. C. W. (2008). Real-time amperometric analysis of reactive oxygen and nitrogen species released by single immunostimulated macrophages. *ChemBioChem*, *9*(9), 1472–1480. http://dx.doi.org/10.1002/cbic.200700746.

Amatore, C., Arbault, S., Bruce, D., De Oliveira, P., Erard, M., Sojic, N., & Vuillaume, M. (2000a). Nitrogen monoxide and oxidative stress: Composition and intensity of cellular oxidative bursts cocktail. A study through artificial electrochemical synapses on single human fibroblasts. *Analusis*, *28*(6), 506–517.

Amatore, C., Arbault, S., Bruce, D., de Oliveira, P., Erard, M., & Vuillaume, M. (2000b). Analysis of individual biochemical events based on artificial synapses using ultramicroelectrodes: Cellular oxidative burst. *Faraday Discussions*, *116*, 319–333. http://dx.doi.org/10.1039/b001448f.

Amatore, C., Arbault, S., Guille, M., & Lemaître, F. (2008b). Electrochemical monitoring of single cell secretion: Vesicular exocytosis and oxidative stress. *Chemical Reviews*, *108*(7), 2585–2621. http://dx.doi.org/10.1021/cr068062g.

Amatore, C., Arbault, S., Jaouen, G., Koh, A. C. W., Leong, W. K., Top, S., ... Woo, C. H. (2010a). Pro-oxidant properties of AZT and other thymidine analogues in macrophages: Implication of the azido moiety in oxidative stress. *ChemMedChem*, *5*(2), 296–301. http://dx.doi.org/10.1002/cmdc.200900464.

Amatore, C., Arbault, S., & Koh, A. C. W. (2010b). Simultaneous detection of reactive oxygen and nitrogen species released by a single macrophage by triple potential-step chronoamperometry. *Analytical Chemistry*, *82*(4), 1411–1419. http://dx.doi.org/10.1021/ac902486x.

Amatore, C., Oleinick, A. I., & Svir, I. (2010). Reconstruction of aperture functions during full fusion in vesicular exocytosis of neurotransmitters. *ChemPhysChem*, *11*(1), 159–174. http://dx.doi.org/10.1002/cphc.200900647.

Arnold, R. S., Shi, J., Murad, E., Whalen, A. M., Sun, C. Q., Polavarapu, R., ... Lambeth, J. D. (2001). Hydrogen peroxide mediates the cell growth and transformation caused by the mitogenic oxidase Nox1. *Proceedings of the National Academy of Sciences of the United States of America*, *98*(10), 5550–5555. http://dx.doi.org/10.1073/pnas.101505898.

Arumugam, P. U., Zeng, H., Siddiqui, S., Covey, D. P., Carlisle, J. A., & Garris, P. A. (2013). Characterization of ultrananocrystalline diamond microsensors for in vivo dopamine detection. *Applied Physics Letters*, *102*(25), 253107. http://dx.doi.org/10.1063/1.4811785.

Aruoma, O. I. (1998). Free radicals, oxidative stress, and antioxidants in human health and disease. *Journal of the American Oil Chemists' Society*, *75*(2), 199–212. http://dx.doi.org/10.1007/s11746-998-0032-9.

Atcherley, C. W., Wood, K. M., Parent, K. L., Hashemi, P., & Heien, M. L. (2015). The coaction of tonic and phasic dopamine dynamics. *Chemical Communications*, *51*(12), 2235–2238. http://dx.doi.org/10.1039/c4cc06165a.

Atta, N. F., Galal, A., Abu-Attia, F. M., & Azab, S. M. (2011). Simultaneous determination of paracetamol and neurotransmitters in biological fluids using a carbon paste sensor modified with gold nanoparticles. *Journal of Materials Chemistry*, *21*(34), 13015–13024. http://dx.doi.org/10.1039/c1jm11795e.

Bagherzadeh, M., & Heydari, M. (2013). Electrochemical detection of dopamine based on pre-concentration by graphene nanosheets. *Analyst*, *138*(20), 6044–6051. http://dx.doi.org/10.1039/c3an01318a.

Bauer, G. (2000). Reactive oxygen and nitrogen species: Efficient, selective, and interactive signals during intercellular induction of apoptosis. *Anticancer Research*, *20*(6B), 4115–4139.

Beckman, J. S., Beckman, T. W., Chen, J., Marshall, P. A., & Freeman, B. A. (1990). Apparent hydroxyl radical production by peroxynitrite: Implications for endothelial injury from nitric oxide and superoxide. *Proceedings of the National Academy of Sciences of the United States of America*, *87*(4), 1620–1624. http://dx.doi.org/10.1073/pnas.87.4.1620.

Beckman, J. S., & Koppenol, W. H. (1996). Nitric oxide, superoxide, and peroxynitrite: The good, the bad, and ugly. *The American Journal of Physiology*, *271*(5 Pt. 1), C1424–C1437.

Bedioui, F., & Villeneuve, N. (2003). Electrochemical nitric oxide sensors for biological samples—Principle, selected examples and applications. *Electroanalysis*, *15*(1), 5–18. http://dx.doi.org/10.1002/elan.200390006.

Bennet, K. E., Tomshine, J. R., Min, H.-K., Manciu, F. S., Marsh, M. P., Paek, S. B., ... Lee, K. H. (2016). A diamond-based electrode for detection of neurochemicals in the human brain. *Frontiers in Human Neuroscience*, *10*, 102. http://dx.doi.org/10.3389/fnhum.2016.00102.

Bernard, A.-S., Giroud, C., Ching, H. Y. V., Meunier, A., Ambike, V., Amatore, C., ... Policar, C. (2012). Evaluation of the anti-oxidant properties of a SOD-mimic

Mn-complex in activated macrophages. *Dalton Transactions*, *41*(21), 6399–6403. http://dx.doi.org/10.1039/c2dt12479c.
Bian, X., Guo, K., Liao, L., Xiao, J., Kong, J., Ji, C., & Liu, B. (2012). Nanocomposites of palladium nanoparticle-loaded mesoporous carbon nanospheres for the electrochemical determination of hydrogen peroxide. *Talanta*, *99*, 256–261. http://dx.doi.org/10.1016/j.talanta.2012.05.048.
Bian, X., Lu, X., Jin, E., Kong, L., Zhang, W., & Wang, C. (2010). Fabrication of Pt/polypyrrole hybrid hollow microspheres and their application in electrochemical biosensing towards hydrogen peroxide. *Talanta*, *81*(3), 813–818. http://dx.doi.org/10.1016/j.talanta.2010.01.020.
Bloomfield, G., & Pears, C. (2003). Superoxide signalling required for multicellular development of Dictyostelium. *Journal of Cell Science*, *116*(16), 3387–3397. http://dx.doi.org/10.1242/jcs.00649.
Bo, X., Bai, J., Qi, B., & Guo, L. (2011). Ultra-fine Pt nanoparticles supported on ionic liquid polymer-functionalized ordered mesoporous carbons for nonenzymatic hydrogen peroxide detection. *Biosensors and Bioelectronics*, *28*(1), 77–83. http://dx.doi.org/10.1016/j.bios.2011.07.001.
Bowman, T. V., & Zon, L. I. (2010). Swimming into the future of drug discovery: In vivo chemical screens in zebrafish. *ACS Chemical Biology*, *5*(2), 159–161. http://dx.doi.org/10.1021/cb100029t.
Bucur, B. (2012). Technological barriers in the use of electrochemical microsensors and microbiosensors for in vivo analysis of neurological relevant substances. *Current Neuropharmacology*, *10*(3), 197–211. http://dx.doi.org/10.2174/157015912803217350.
Buetler, T. M., Krauskopf, A., & Ruegg, U. T. (2004). Role of superoxide as a signaling molecule. *Physiology*, *19*(3), 120–123. http://dx.doi.org/10.1152/nips.01514.2003.
Bui, M.-P. N., Pham, X.-H., Han, K. N., Li, C. A., Kim, Y. S., & Seong, G. H. (2010). Electrocatalytic reduction of hydrogen peroxide by silver particles patterned on single-walled carbon nanotubes. *Sensors and Actuators B: Chemical*, *150*(1), 436–441. http://dx.doi.org/10.1016/j.snb.2010.06.019.
Burewicz, A., Dawoud, H., Jiang, L.-L., & Malinski, T. (2013). Nitric oxide/peroxynitrite redox imbalance in endothelial cells measured with amperometric nanosensors. *American Journal of Analytical Chemistry*, *04*(10), 7. http://dx.doi.org/10.4236/ajac.2013.410A1004.
Burgoyne, R. D., & Morgan, A. (2003). Secretory granule exocytosis. *Physiological Reviews*, *83*(2), 581–632. http://dx.doi.org/10.1152/physrev.00031.2002.
Burrell, M. H., Atcherley, C. W., Heien, M. L., & Lipski, J. (2015). A novel electrochemical approach for prolonged measurement of absolute levels of extracellular dopamine in brain slices. *ACS Chemical Neuroscience*, *6*(11), 1802–1812. http://dx.doi.org/10.1021/acschemneuro.5b00120.
Cai, H. (2005). Hydrogen peroxide regulation of endothelial function: Origins, mechanisms, and consequences. *Cardiovascular Research*, *68*(1), 26–36. http://dx.doi.org/10.1016/j.cardiores.2005.06.021.
Calabrese, V., Mancuso, C., Calvani, M., Rizzarelli, E., Butterfield, D. A., & Giuffrida Stella, A. M. (2007). Nitric oxide in the central nervous system: Neuroprotection versus neurotoxicity. *Nature Reviews. Neuroscience*, *8*(10), 766–775.
Calas-Blanchard, C., Catanante, G., & Noguer, T. (2014). Electrochemical sensor and biosensor strategies for ROS/RNS detection in biological systems. *Electroanalysis*, *26*(6), 1277–1286. http://dx.doi.org/10.1002/elan.201400083.
Cans, A.-S., & Ewing, A. G. (2011). Highlights of 20 years of electrochemical measurements of exocytosis at cells and artificial cells. *Journal of Solid State Electrochemistry*, *15*(7), 1437–1450. http://dx.doi.org/10.1007/s10008-011-1369-9.

Cao, X., Zeng, Z., Shi, W., Yep, P., Yan, Q., & Zhang, H. (2013). Three-dimensional graphene network composites for detection of hydrogen peroxide. *Small*, *9*(9–10), 1703–1707. http://dx.doi.org/10.1002/smll.201200683.

Chang, S.-Y., Kimble, C. J., Kim, I., Paek, S. B., Kressin, K. R., Boesche, J. B., … Lee, K. H. (2013). Development of the Mayo Investigational Neuromodulation Control System: Toward a closed-loop electrochemical feedback system for deep brain stimulation: Laboratory investigation. *Journal of Neurosurgery*, *119*(6), 1556–1565. http://dx.doi.org/10.3171/2013.8.jns122142.

Chen, X.-m., Cai, Z.-x., Huang, Z.-y., Oyama, M., Jiang, Y.-q., & Chen, X. (2013). Ultrafine palladium nanoparticles grown on graphene nanosheets for enhanced electrochemical sensing of hydrogen peroxide. *Electrochimica Acta*, *97*, 398–403. http://dx.doi.org/10.1016/j.electacta.2013.02.047.

Chen, W., Ren, Q.-Q., Yang, Q., Wen, W., & Zhao, Y.-D. (2012). In vivo electrochemical biosensors for reactive oxygen species detection: A mini-review. *Analytical Letters*, *45*(2–3), 156–167. http://dx.doi.org/10.1080/00032719.2011.633185.

Chen, B. T., & Rice, M. E. (1999). Calibration factors for cationic and anionic neurochemicals at carbon-fiber microelectrodes are oppositely affected by the presence of Ca2+ and Mg2+. *Electroanalysis*, *11*(5), 344–348. http://dx.doi.org/10.1002/(sici)1521-4109(199905)11:5<344::aid-elan344>3.0.co;2-k.

Chen, J., Zhao, L., Bai, H., & Shi, G. (2011). Electrochemical detection of dioxygen and hydrogen peroxide by hemin immobilized on chemically converted graphene. *Journal of Electroanalytical Chemistry*, *657*(1–2), 34–38. http://dx.doi.org/10.1016/j.jelechem.2011.03.005.

Clark, J. J., Sandberg, S. G., Wanat, M. J., Gan, J. O., Horne, E. A., Hart, A. S., … Phillips, P. E. M. (2010). Chronic microsensors for longitudinal, subsecond dopamine detection in behaving animals. *Nature Methods*, 7(2), 126–129. http://dx.doi.org/10.1038/nmeth.1412. http://www.nature.com/nmeth/journal/v7/n2/suppinfo/nmeth.1412_S1.html.

Clausmeyer, J., Actis, P., López Córdoba, A., Korchev, Y., & Schuhmann, W. (2014). Nanosensors for the detection of hydrogen peroxide. *Electrochemistry Communications*, *40*, 28–30. http://dx.doi.org/10.1016/j.elecom.2013.12.015.

Cropley, V. L., Fujita, M., Innis, R. B., & Nathan, P. J. (2006). Molecular imaging of the dopaminergic system and its association with human cognitive function. *Biological Psychiatry*, *59*(10), 898–907. http://dx.doi.org/10.1016/j.biopsych.2006.03.004.

Dalle-Donne, I., Rossi, R., Colombo, R., Giustarini, D., & Milzani, A. (2006). Biomarkers of oxidative damage in human disease. *Clinical Chemistry*, *52*(4), 601–623. http://dx.doi.org/10.1373/clinchem.2005.061408.

Derkus, B., Emregul, E., & Emregul, K. C. (2015). Copper–zinc alloy nanoparticle based enzyme-free superoxide radical sensing on a screen-printed electrode. *Talanta*, *134*, 206–214. http://dx.doi.org/10.1016/j.talanta.2014.11.003.

Desouza, C. V., Bolli, G. B., & Fonseca, V. (2010). Hypoglycemia, diabetes, and cardiovascular events. *Diabetes Care*, *33*(6), 1389–1394. http://dx.doi.org/10.2337/dc09-2082.

Di Chiara, G., Bassareo, V., Fenu, S., De Luca, M. A., Spina, L., Cadoni, C., … Lecca, D. (2004). Dopamine and drug addiction: The nucleus accumbens shell connection. *Neuropharmacology*, *47*(Suppl. 1), 227–241. http://dx.doi.org/10.1016/j.neuropharm.2004.06.032.

Dunevall, J., Fathali, H., Najafinobar, N., Lovric, J., Wigström, J., Cans, A.-S., & Ewing, A. G. (2015). Characterizing the catecholamine content of single mammalian vesicles by collision–adsorption events at an electrode. *Journal of the American Chemical Society*, *137*(13), 4344–4346. http://dx.doi.org/10.1021/ja512972f.

Dunlop, B. W., & Nemeroff, C. B. (2007). The role of dopamine in the pathophysiology of depression. *Archives of General Psychiatry*, *64*(3), 327–337. http://dx.doi.org/10.1001/archpsyc.64.3.327.

Eckert, M. A., Vu, P. Q., Zhang, K., Kang, D., Ali, M. M., Xu, C., & Zhao, W. (2013). Novel molecular and nanosensors for in vivo sensing. *Theranostics*, *3*(8), 583–594. http://dx.doi.org/10.7150/thno.6584.

Ensafi, A. A., Abarghoui, M. M., & Rezaei, B. (2014). Electrochemical determination of hydrogen peroxide using copper/porous silicon based non-enzymatic sensor. *Sensors and Actuators B: Chemical*, *196*, 398–405. http://dx.doi.org/10.1016/j.snb.2014.02.028.

Farjami, E., Campos, R., Nielsen, J. S., Gothelf, K. V., Kjems, J., & Ferapontova, E. E. (2013). RNA aptamer-based electrochemical biosensor for selective and label-free analysis of dopamine. *Analytical Chemistry*, *85*(1), 121–128. http://dx.doi.org/10.1021/ac302134s.

Feng, X., Zhang, Y., Zhou, J., Li, Y., Chen, S., Zhang, L., … Yan, X. (2015). Three-dimensional nitrogen-doped graphene as an ultrasensitive electrochemical sensor for the detection of dopamine. *Nanoscale*, 7(6), 2427–2432. http://dx.doi.org/10.1039/c4nr06623e.

Filipović, M. R., Koh, A. C. W., Arbault, S., Niketić, V., Debus, A., Schleicher, U., … Ivanović-Burmazović, I. (2010). Striking inflammation from both sides: Manganese(II) pentaazamacrocyclic SOD mimics act also as nitric oxide dismutases: A single-cell study. *Angewandte Chemie International Edition*, *49*(25), 4228–4232. http://dx.doi.org/10.1002/anie.200905936.

Finkel, T., & Holbrook, N. J. (2000). Oxidants, oxidative stress and the biology of ageing. *Nature*, *408*(6809), 239–247. http://dx.doi.org/10.1038/35041687.

Fox, M. E., Bucher, E. S., Johnson, J. A., & Wightman, R. M. (2016). Medullary norepinephrine projections release norepinephrine into the contralateral bed nucleus of the stria terminalis. *ACS Chemical Neuroscience*, 7(12), 1681–1689. http://dx.doi.org/10.1021/acschemneuro.6b00210.

Franco, R., Schoneveld, O., Georgakilas, A. G., & Panayiotidis, M. I. (2008). Oxidative stress, DNA methylation and carcinogenesis. *Cancer Letters*, *266*(1), 6–11. http://dx.doi.org/10.1016/j.canlet.2008.02.026.

Ganesana, M., Erlichman, J. S., & Andreescu, S. (2012). Real-time monitoring of superoxide accumulation and antioxidant activity in a brain slice model using an electrochemical cytochrome c biosensor. *Free Radical Biology and Medicine*, *53*(12), 2240–2249. http://dx.doi.org/10.1016/j.freeradbiomed.2012.10.540.

Ge, S., Koseoglu, S., & Haynes, C. L. (2010). Bioanalytical tools for single-cell study of exocytosis. *Analytical and Bioanalytical Chemistry*, *397*(8), 3281–3304. http://dx.doi.org/10.1007/s00216-010-3843-0.

Ge, B., & Lisdat, F. (2002). Superoxide sensor based on cytochrome c immobilized on mixed-thiol SAM with a new calibration method. *Analytica Chimica Acta*, *454*(1), 53–64. http://dx.doi.org/10.1016/S0003-2670(01)01545-8.

Genestra, M. (2007). Oxyl radicals, redox-sensitive signalling cascades and antioxidants. *Cellular Signalling*, *19*(9), 1807–1819. http://dx.doi.org/10.1016/j.cellsig.2007.04.009.

Ghaderi, S., & Mehrgardi, M. A. (2014). Prussian blue-modified nanoporous gold film electrode for amperometric determination of hydrogen peroxide. *Bioelectrochemistry*, *98*, 64–69. http://dx.doi.org/10.1016/j.bioelechem.2014.03.007.

Ghafourifar, P., & Cadenas, E. (2005). Mitochondrial nitric oxide synthase. *Trends in Pharmacological Sciences*, *26*(4), 190–195. http://dx.doi.org/10.1016/j.tips.2005.02.005.

Giacco, F., & Brownlee, M. (2010). Oxidative stress and diabetic complications. *Circulation Research*, *107*(9), 1058–1070. http://dx.doi.org/10.1161/circresaha.110.223545.

Goddard, A. W., Ball, S. G., Martinez, J., Robinson, M. J., Yang, C. R., Russell, J. M., & Shekhar, A. (2010). Current perspectives of the roles of the central norepinephrine system in anxiety and depression. *Depression and Anxiety*, *27*(4), 339–350. http://dx.doi.org/10.1002/da.20642.

Gopalan, A. I., Komathi, S., Sai Anand, G., & Lee, K.-P. (2013). Nanodiamond based sponges with entrapped enzyme: A novel electrochemical probe for hydrogen peroxide. *Biosensors and Bioelectronics*, *46*, 136–141. http://dx.doi.org/10.1016/j.bios.2013.02.036.

Grieshaber, D., MacKenzie, R., Vörös, J., & Reimhult, E. (2008). Electrochemical biosensors—Sensor principles and architectures. *Sensors (Basel, Switzerland)*, *8*(3), 1400–1458.

Griessenauer, C. J., Chang, S.-Y., Tye, S. J., Kimble, C. J., Bennet, K. E., Garris, P. A., & Lee, K. H. (2010). Wireless instantaneous neurotransmitter concentration system: Electrochemical monitoring of serotonin using fast-scan cyclic voltammetry—A proof-of-principle study. *Journal of Neurosurgery*, *113*(3), 656–665. http://dx.doi.org/10.3171/2010.3.jns091627.

Gualandi, I., Guadagnini, L., Zappoli, S., & Tonelli, D. (2014). A polypyrrole based sensor for the electrochemical detection of OH radicals. *Electroanalysis*, *26*(7), 1544–1550. http://dx.doi.org/10.1002/elan.201400054.

Guo, Y., Li, J., & Dong, S. (2011). Hemin functionalized graphene nanosheets-based dual biosensor platforms for hydrogen peroxide and glucose. *Sensors and Actuators B: Chemical*, *160*(1), 295–300. http://dx.doi.org/10.1016/j.snb.2011.07.050.

Guo, C. X., Ng, S. R., Khoo, S. Y., Zheng, X., Chen, P., & Li, C. M. (2012). RGD-peptide functionalized graphene biomimetic live-cell sensor for real-time detection of nitric oxide molecules. *ACS Nano*, *6*(8), 6944–6951. http://dx.doi.org/10.1021/nn301974u.

Halliwell, B. (1996). Free radicals, proteins and DNA: Oxidative damage versus redox regulation. *Biochemical Society Transactions*, *24*(4), 1023–1027. http://dx.doi.org/10.1042/bst0241023.

Harley, C. C., Rooney, A. D., & Breslin, C. B. (2010). The selective detection of dopamine at a polypyrrole film doped with sulfonated β-cyclodextrins. *Sensors and Actuators B: Chemical*, *150*(2), 498–504. http://dx.doi.org/10.1016/j.snb.2010.09.012.

Harreither, W., Trouillon, R., Poulin, P., Neri, W., Ewing, A. G., & Safina, G. (2013). Carbon nanotube fiber microelectrodes show a higher resistance to dopamine fouling. *Analytical Chemistry*, *85*(15), 7447–7453. http://dx.doi.org/10.1021/ac401399s.

Hasanzadeh, M., Shadjou, N., & Omidinia, E. (2013). A novel electroanalytical method for simultaneous detection of two neurotransmitter dopamine and serotonin in human serum. *Journal of Neuroscience Methods*, *219*(1), 52–60. http://dx.doi.org/10.1016/j.jneumeth.2013.07.007.

Hashemi, P., Dankoski, E. C., Wood, K. M., Ambrose, R. E., & Wightman, R. M. (2011). In vivo electrochemical evidence for simultaneous 5-HT and histamine release in the rat substantia nigra pars reticulata following medial forebrain bundle stimulation. *Journal of Neurochemistry*, *118*(5), 749–759. http://dx.doi.org/10.1111/j.1471-4159.2011.07352.x.

Hawley, M. D., Tatawawadi, S. V., Piekarski, S., & Adams, R. N. (1967). Electrochemical studies of the oxidation pathways of catecholamines. *Journal of the American Chemical Society*, *89*(2), 447–450. http://dx.doi.org/10.1021/ja00978a051.

Heien, M. L. A. V., Phillips, P. E. M., Stuber, G. D., Seipel, A. T., & Wightman, R. M. (2003). Overoxidation of carbon-fiber microelectrodes enhances dopamine adsorption and increases sensitivity. *Analyst*, *128*(12), 1413–1419. http://dx.doi.org/10.1039/b307024g.

Herr, N. R., Park, J., McElligott, Z. A., Belle, A. M., Carelli, R. M., & Wightman, R. M. (2012). In vivo voltammetry monitoring of electrically evoked extracellular norepinephrine in subregions of the bed nucleus of the stria terminalis. *Journal of Neurophysiology*, *107*(6), 1731–1737. http://dx.doi.org/10.1152/jn.00620.2011.

Hosu, I. S., Wang, Q., Vasilescu, A., Peteu, S. F., Raditoiu, V., Railian, S., … Szunerits, S. (2015). Cobalt phthalocyanine tetracarboxylic acid modified reduced graphene oxide:

A sensitive matrix for the electrocatalytic detection of peroxynitrite and hydrogen peroxide. *RSC Advances*, *5*(2), 1474–1484. http://dx.doi.org/10.1039/c4ra09781e.

Howes, O. D., & Kapur, S. (2009). The dopamine hypothesis of schizophrenia: Version III—The final common pathway. *Schizophrenia Bulletin*, *35*(3), 549–562. http://dx.doi.org/10.1093/schbul/sbp006.

Hu, K., Gao, Y., Wang, Y., Yu, Y., Zhao, X., Rotenberg, S. A., … Gogotsi, Y. (2013). Platinized carbon nanoelectrodes as potentiometric and amperometric SECM probes. *Journal of Solid State Electrochemistry*, *17*(12), 2971–2977. http://dx.doi.org/10.1007/s10008-013-2173-5.

Hu, R., Guille, M., Arbault, S., Lin, C. J., & Amatore, C. (2010). In situ electrochemical monitoring of reactive oxygen and nitrogen species released by single MG63 osteosarcoma cell submitted to a mechanical stress. *Physical Chemistry Chemical Physics*, *12*(34), 10048–10054. http://dx.doi.org/10.1039/c0cp00398k.

Huang, Y., Miao, Y.-E., Ji, S., Tjiu, W. W., & Liu, T. (2014). Electrospun carbon nanofibers decorated with Ag–Pt bimetallic nanoparticles for selective detection of dopamine. *ACS Applied Materials & Interfaces*, *6*(15), 12449–12456. http://dx.doi.org/10.1021/am502344p.

Hunter, R. A., Storm, W. L., Coneski, P. N., & Schoenfisch, M. H. (2013). Inaccuracies of nitric oxide measurement methods in biological media. *Analytical Chemistry*, *85*(3), 1957–1963. http://dx.doi.org/10.1021/ac303787p.

Ignarro, L. J. (2002). Nitric oxide as a unique signaling molecule in the vascular system: A historical overview. *Journal of Physiology and Pharmacology: An Official Journal of the Polish Physiological Society*, *53*(4 Pt. 1), 503–514.

Imlay, J., Chin, S., & Linn, S. (1988). Toxic DNA damage by hydrogen peroxide through the Fenton reaction in vivo and in vitro. *Science*, *240*(4852), 640–642. http://dx.doi.org/10.1126/science.2834821.

Jacobs, C. B., Ivanov, I. N., Nguyen, M. D., Zestos, A. G., & Venton, B. J. (2014). High temporal resolution measurements of dopamine with carbon nanotube yarn microelectrodes. *Analytical Chemistry*, *86*(12), 5721–5727. http://dx.doi.org/10.1021/ac404050t.

Jena, N. R. (2012). DNA damage by reactive species: Mechanisms, mutation and repair. *Journal of Biosciences*, *37*(3), 503–517. http://dx.doi.org/10.1007/s12038-012-9218-2.

Jenkinson, N., & Brown, P. (2011). New insights into the relationship between dopamine, beta oscillations and motor function. *Trends in Neurosciences*, *34*(12), 611–618. http://dx.doi.org/10.1016/j.tins.2011.09.003.

Jiang, G., Gu, X., Jiang, G., Chen, T., Zhan, W., & Tian, S. (2015). Application of a mercapto-terminated binuclear Cu(II) complex modified Au electrode to improve the sensitivity and selectivity for dopamine detection. *Sensors and Actuators B: Chemical*, *209*, 122–130. http://dx.doi.org/10.1016/j.snb.2014.11.109.

Jin, D., Sakthivel, K., Gandhi, S., Huy, B. T., & Lee, Y.-I. (2016). An improved non-enzymatic hydrogen peroxide sensor based on europium functionalized inorganic hybrid material—Evaluation of optical and electrochemical properties. *Sensors and Actuators B: Chemical*, *237*, 81–89. http://dx.doi.org/10.1016/j.snb.2016.06.078.

Jo, A., Do, H., Jhon, G.-J., Suh, M., & Lee, Y. (2011). Real-time evaluation of nitric oxide (NO) levels in cortical and hippocampal areas with a nanopore-based electrochemical NO sensor. *Neuroscience Letters*, *498*(1), 22–25. http://dx.doi.org/10.1016/j.neulet.2011.04.053.

Johnson, M. A. (2013). In vivo electrochemical measurements: Past, present and future. *Bioanalysis*, *5*(2), 119–122. http://dx.doi.org/10.4155/bio.12.322.

Kaçar, C., Dalkiran, B., Erden, P. E., & Kiliç, E. (2014). An amperometric hydrogen peroxide biosensor based on Co_3O_4 nanoparticles and multiwalled carbon nanotube modified glassy carbon electrode. *Applied Surface Science*, *311*, 139–146. http://dx.doi.org/10.1016/j.apsusc.2014.05.028.

Kaehler, S. T., Singewald, N., Sinner, C., Thurnher, C., & Philippu, A. (2000). Conditioned fear and inescapable shock modify the release of serotonin in the locus coeruleus. *Brain Research*, *859*(2), 249–254. http://dx.doi.org/10.1016/S0006-8993(00)01967-3.

Karyakin, A. A. (2001). Prussian blue and its analogues: Electrochemistry and analytical applications. *Electroanalysis*, *13*(10), 813–819. http://dx.doi.org/10.1002/1521-4109(200106)13:10<813::aid-elan813>3.0.co;2-z.

Kasasbeh, A., Lee, K., Bieber, A., Bennet, K., & Chang, S. Y. (2013). Wireless neurochemical monitoring in humans. *Stereotactic and Functional Neurosurgery*, *91*(3), 141–147.

Keithley, R. B., Takmakov, P., Bucher, E. S., Belle, A. M., Owesson-White, C. A., Park, J., & Wightman, R. M. (2011). Higher sensitivity dopamine measurements with faster-scan cyclic voltammetry. *Analytical Chemistry*, *83*(9), 3563–3571. http://dx.doi.org/10.1021/ac200143v.

Khoobi, A., Ghoreishi, S. M., Behpour, M., & Masoum, S. (2014). Three-dimensional voltammetry: A chemometrical analysis of electrochemical data for determination of dopamine in the presence of unexpected interference by a biosensor based on gold nanoparticles. *Analytical Chemistry*, *86*(18), 8967–8973. http://dx.doi.org/10.1021/ac501238a.

Kile, B. M., Walsh, P. L., McElligott, Z. A., Bucher, E. S., Guillot, T. S., Salahpour, A., … Wightman, R. M. (2012). Optimizing the temporal resolution of fast-scan cyclic voltammetry. *ACS Chemical Neuroscience*, *3*(4), 285–292. http://dx.doi.org/10.1021/cn200119u.

Kinney, H. C., Richerson, G. B., Dymecki, S. M., Darnall, R. A., & Nattie, E. E. (2009). The brainstem and serotonin in the sudden infant death syndrome. *Annual Review of Pathology*, *4*, 517–550. http://dx.doi.org/10.1146/annurev.pathol.4.110807.092322.

Kishida, K. T., Sandberg, S. G., Lohrenz, T., Comair, Y. G., Sáez, I., Phillips, P. E. M., & Montague, P. R. (2011). Sub-second dopamine detection in human striatum. *PLoS One*, *6*(8), e23291. http://dx.doi.org/10.1371/journal.pone.0023291.

Kissinger, P. T., Hart, J. B., & Adams, R. N. (1973). Voltammetry in brain tissue—A new neurophysiological measurement. *Brain Research*, *55*(1), 209–213. http://dx.doi.org/10.1016/0006-8993(73)90503-9.

Koh, W. C. A., Son, J. I., Choe, E. S., & Shim, Y.-B. (2010). Electrochemical detection of peroxynitrite using a biosensor based on a conducting polymer–manganese ion complex. *Analytical Chemistry*, *82*(24), 10075–10082. http://dx.doi.org/10.1021/ac102041u.

Kryston, T. B., Georgiev, A. B., Pissis, P., & Georgakilas, A. G. (2011). Role of oxidative stress and DNA damage in human carcinogenesis. *Mutation Research/Fundamental and Molecular Mechanisms of Mutagenesis*, *711*(1–2), 193–201. http://dx.doi.org/10.1016/j.mrfmmm.2010.12.016.

Kuhr, W. G., & Wightman, R. M. (1986). Real-time measurement of dopamine release in rat brain. *Brain Research*, *381*(1), 168–171. http://dx.doi.org/10.1016/0006-8993(86)90707-9.

Kundys, M., Szot, K., Rozniecka, E., Jonsson-Niedziolka, M., Lawrence, R., Bull, S. D., … Opallo, M. (2014). Electrochemical determination of selected neurotransmitters at electrodes modified with oppositely charged carbon nanoparticles. *Analytical Methods*, *6*(18), 7532–7539. http://dx.doi.org/10.1039/c4ay01344a.

Lama, R. D., Charlson, K., Anantharam, A., & Hashemi, P. (2012). Ultrafast detection and quantification of brain signaling molecules with carbon fiber microelectrodes. *Analytical Chemistry*, *84*(19), 8096–8101. http://dx.doi.org/10.1021/ac301670h.

Lemaître, F., Guille Collignon, M., & Amatore, C. (2014). Recent advances in electrochemical detection of exocytosis. *Electrochimica Acta*, *140*, 457–466. http://dx.doi.org/10.1016/j.electacta.2014.02.059.

Li, L., Du, Z., Liu, S., Hao, Q., Wang, Y., Li, Q., & Wang, T. (2010). A novel nonenzymatic hydrogen peroxide sensor based on MnO_2/graphene oxide nanocomposite. *Talanta*, *82*(5), 1637–1641. http://dx.doi.org/10.1016/j.talanta.2010.07.020.

Li, S.-J., He, J.-Z., Zhang, M.-J., Zhang, R.-X., Lv, X.-L., Li, S.-H., & Pang, H. (2013). Electrochemical detection of dopamine using water-soluble sulfonated graphene. *Electrochimica Acta*, *102*, 58–65. http://dx.doi.org/10.1016/j.electacta.2013.03.176.

Li, X., Liu, Y., Zheng, L., Dong, M., Xue, Z., Lu, X., & Liu, X. (2013). A novel nonenzymatic hydrogen peroxide sensor based on silver nanoparticles and ionic liquid functionalized multiwalled carbon nanotube composite modified electrode. *Electrochimica Acta*, *113*, 170–175. http://dx.doi.org/10.1016/j.electacta.2013.09.049.

Li, X., Majdi, S., Dunevall, J., Fathali, H., & Ewing, A. G. (2015). Quantitative measurement of transmitters in individual vesicles in the cytoplasm of single cells with nanotip electrodes. *Angewandte Chemie International Edition*, *54*(41), 11978–11982. http://dx.doi.org/10.1002/anie.201504839.

Li, X.-R., Wang, B., Xu, J.-J., & Chen, H.-Y. (2011). In vitro detection of superoxide anions released from cancer cells based on potassium-doped carbon nanotubes-ionic liquid composite gels. *Nanoscale*, *3*(12), 5026–5033. http://dx.doi.org/10.1039/c1nr11025j.

Li, J., Xie, J., Gao, L., & Li, C. M. (2015). Au nanoparticles–3D graphene hydrogel nanocomposite to boost synergistically in situ detection sensitivity toward cell-released nitric oxide. *ACS Applied Materials & Interfaces*, 7(4), 2726–2734. http://dx.doi.org/10.1021/am5077777.

Li, Y.-T., Zhang, S.-H., Wang, L., Xiao, R.-R., Liu, W., Zhang, X.-W., … Huang, W.-H. (2014). Nanoelectrode for amperometric monitoring of individual vesicular exocytosis inside single synapses. *Angewandte Chemie International Edition*, *53*(46), 12456–12460. http://dx.doi.org/10.1002/anie.201404744.

Li, C., Zhang, H., Wu, P., Gong, Z., Xu, G., & Cai, C. (2011). Electrochemical detection of extracellular hydrogen peroxide released from RAW 264.7 murine macrophage cells based on horseradish peroxidase-hydroxyapatite nanohybrids. *Analyst*, *136*(6), 1116–1123. http://dx.doi.org/10.1039/c0an00825g.

Li, Y., Zhang, J.-J., Xuan, J., Jiang, L.-P., & Zhu, J.-J. (2010). Fabrication of a novel nonenzymatic hydrogen peroxide sensor based on Se/Pt nanocomposites. *Electrochemistry Communications*, *12*(6), 777–780. http://dx.doi.org/10.1016/j.elecom.2010.03.031.

Lin, M. T., & Beal, M. F. (2006). Mitochondrial dysfunction and oxidative stress in neurodegenerative diseases. *Nature*, *443*(7113), 787–795.

Lin, P., Lin, C.-W., Mansour, R., & Gu, F. (2013). Improving biocompatibility by surface modification techniques on implantable bioelectronics. *Biosensors and Bioelectronics*, *47*, 451–460. http://dx.doi.org/10.1016/j.bios.2013.01.071.

Liu, X., Dumitrescu, E., & Andreescu, S. (2015). Electrochemical biosensors for real-time monitoring of reactive oxygen and nitrogen species. In M. Hepel & S. Andreescu (Eds.), *Oxidative Stress: Diagnostics, Prevention, and Therapy Volume 2: Vol. 1200* (pp. 301–327). Washington, DC: American Chemical Society.

Liu, R., Li, S., Yu, X., Zhang, G., Zhang, S., Yao, J., … Zhi, L. (2012). Facile synthesis of Au-nanoparticle/polyoxometalate/graphene tricomponent nanohybrids: An enzyme-free electrochemical biosensor for hydrogen peroxide. *Small*, *8*(9), 1398–1406. http://dx.doi.org/10.1002/smll.201102298.

Liu, Y., Li, M., Zhang, F., Zhu, A., & Shi, G. (2015). Development of Au disk nanoelectrode down to 3 nm in radius for detection of dopamine release from a single cell. *Analytical Chemistry*, *87*(11), 5531–5538. http://dx.doi.org/10.1021/ac5042999.

Liu, M., Liu, R., & Chen, W. (2013). Graphene wrapped Cu_2O nanocubes: Non-enzymatic electrochemical sensors for the detection of glucose and hydrogen peroxide with

enhanced stability. *Biosensors and Bioelectronics*, *45*, 206–212. http://dx.doi.org/10.1016/j.bios.2013.02.010.
Liu, X., Marrakchi, M., Jahne, M., Rogers, S., & Andreescu, S. (2016). Real-time investigation of antibiotics-induced oxidative stress and superoxide release in bacteria using an electrochemical biosensor. *Free Radical Biology and Medicine*, *91*, 25–33. http://dx.doi.org/10.1016/j.freeradbiomed.2015.12.001.
Liu, Y. M., Punckt, C., Pope, M. A., Gelperin, A., & Aksay, I. A. (2013). Electrochemical sensing of nitric oxide with functionalized graphene electrodes. *ACS Applied Materials & Interfaces*, *5*(23), 12624–12630. http://dx.doi.org/10.1021/am403983g.
Liu, S., Tian, J., Wang, L., & Sun, X. (2011). A method for the production of reduced graphene oxide using benzylamine as a reducing and stabilizing agent and its subsequent decoration with Ag nanoparticles for enzymeless hydrogen peroxide detection. *Carbon*, *49*(10), 3158–3164. http://dx.doi.org/10.1016/j.carbon.2011.03.036.
Liu, Y., Wang, D., Xu, L., Hou, H., & You, T. (2011). A novel and simple route to prepare a Pt nanoparticle-loaded carbon nanofiber electrode for hydrogen peroxide sensing. *Biosensors and Bioelectronics*, *26*(11), 4585–4590. http://dx.doi.org/10.1016/j.bios.2011.05.034.
Liu, Y., Yao, Q., Zhang, X., Li, M., Zhu, A., & Shi, G. (2015). Development of gold nanoparticle-sheathed glass capillary nanoelectrodes for sensitive detection of cerebral dopamine. *Biosensors and Bioelectronics*, *63*, 262–268. http://dx.doi.org/10.1016/j.bios.2014.07.040.
Liu, Z., Zhao, B., Shi, Y., Guo, C., Yang, H., & Li, Z. (2010). Novel nonenzymatic hydrogen peroxide sensor based on iron oxide–silver hybrid submicrospheres. *Talanta*, *81*(4–5), 1650–1654. http://dx.doi.org/10.1016/j.talanta.2010.03.019.
Lotharius, J., & Brundin, P. (2002). Pathogenesis of Parkinson's disease: Dopamine, vesicles and [alpha]-synuclein. *Nature Reviews. Neuroscience*, *3*(12), 932–942. http://dx.doi.org/10.1038/nrn983.
Lu, C., Heldt, J.-M., Guille-Collignon, M., Lemaître, F., Jaouen, G., Vessières, A., & Amatore, C. (2014). Quantitative analyses of ROS and RNS production in breast cancer cell lines incubated with ferrocifens. *ChemMedChem*, *9*(6), 1286–1293. http://dx.doi.org/10.1002/cmdc.201402016.
Madamanchi, N. R., Vendrov, A., & Runge, M. S. (2005). Oxidative stress and vascular disease. *Arteriosclerosis, Thrombosis, and Vascular Biology*, *25*(1), 29–38. http://dx.doi.org/10.1161/01.atv.0000150649.39934.13.
Maina, F. K., Khalid, M., Apawu, A. K., & Mathews, T. A. (2012). Presynaptic dopamine dynamics in Striatal brain slices with fast-scan cyclic voltammetry. *Journal of Visualized Experiments: JoVE*, (59), e3464. http://dx.doi.org/10.3791/3464.
Maji, S. K., Sreejith, S., Mandal, A. K., Ma, X., & Zhao, Y. (2014). Immobilizing gold nanoparticles in mesoporous silica covered reduced graphene oxide: A hybrid material for cancer cell detection through hydrogen peroxide sensing. *ACS Applied Materials & Interfaces*, *6*(16), 13648–13656. http://dx.doi.org/10.1021/am503110s.
Malinski, T. (2015). Using nanosensors for in situ monitoring and measurement of nitric oxide and peroxynitrite in a single cell. In D. Armstrong (Ed.), *Advanced protocols in oxidative stress III* (pp. 139–155). New York, NY: Springer New York.
Maluta, J. R., Canevari, T. C., & Machado, S. A. S. (2014). Sensitive determination of nitric oxide using an electrochemical sensor based on MWCNTs decorated with spherical Au nanoparticles. *Journal of Solid State Electrochemistry*, *18*(9), 2497–2504. http://dx.doi.org/10.1007/s10008-014-2505-0.
Marcu, R., Rapino, S., Trinei, M., Valenti, G., Marcaccio, M., Pelicci, P. G., ... Giorgio, M. (2012). Electrochemical study of hydrogen peroxide formation in isolated mitochondria. *Bioelectrochemistry*, *85*, 21–28. http://dx.doi.org/10.1016/j.bioelechem.2011.11.005.

Marti, M., Fabregat, G., Estrany, F., Aleman, C., & Armelin, E. (2010). Nanostructured conducting polymer for dopamine detection. *Journal of Materials Chemistry*, *20*(47), 10652–10660. http://dx.doi.org/10.1039/c0jm01364a.

Massion, P. B., Feron, O., Dessy, C., & Balligand, J.-L. (2003). Nitric oxide and cardiac function: Ten years after, and continuing. *Circulation Research*, *93*(5), 388–398. http://dx.doi.org/10.1161/01.res.0000088351.58510.21.

McGaugh, J. L., & Roozendaal, B. (2002). Role of adrenal stress hormones in forming lasting memories in the brain. *Current Opinion in Neurobiology*, *12*(2), 205–210. http://dx.doi.org/10.1016/S0959-4388(02)00306-9.

McNeil, C. J., Smith, K. A., Bellavite, P., & Bannister, J. V. (1989). Application of the electrochemistry of cytochrome C to the measurement of superoxide radical production. *Free Radical Research Communications*, 7(2), 89–96. http://dx.doi.org/10.3109/10715768909087928.

Meng, F., Yan, X., Liu, J., Gu, J., & Zou, Z. (2011). Nanoporous gold as non-enzymatic sensor for hydrogen peroxide. *Electrochimica Acta*, *56*(12), 4657–4662. http://dx.doi.org/10.1016/j.electacta.2011.02.105.

Mercante, L. A., Pavinatto, A., Iwaki, L. E. O., Scagion, V. P., Zucolotto, V., Oliveira, O. N., … Correa, D. S. (2015). Electrospun polyamide 6/poly(allylamine hydrochloride) nanofibers functionalized with carbon nanotubes for electrochemical detection of dopamine. *ACS Applied Materials & Interfaces*, 7(8), 4784–4790. http://dx.doi.org/10.1021/am508709c.

Merkoçi, A., Pumera, M., Llopis, X., Pérez, B., del Valle, M., & Alegret, S. (2005). New materials for electrochemical sensing. VI: Carbon nanotubes. *TrAC Trends in Analytical Chemistry*, *24*(9), 826–838. http://dx.doi.org/10.1016/j.trac.2005.03.019.

Meunier, A., Bretou, M., Darchen, F., Guille Collignon, M., Lemaître, F., & Amatore, C. (2014). Amperometric detection of vesicular exocytosis from BON cells at carbon fiber microelectrodes. *Electrochimica Acta*, *126*, 74–80. http://dx.doi.org/10.1016/j.electacta.2013.07.110.

Miao, Y.-E., He, S., Zhong, Y., Yang, Z., Tjiu, W. W., & Liu, T. (2013). A novel hydrogen peroxide sensor based on Ag/SnO_2 composite nanotubes by electrospinning. *Electrochimica Acta*, *99*, 117–123. http://dx.doi.org/10.1016/j.electacta.2013.03.063.

Montesinos, M. S., Camacho, M., Machado, J. D., Viveros, O. H., Beltrán, B., & Borges, R. (2010). The quantal secretion of catecholamines is impaired by the accumulation of β-adrenoceptor antagonists into chromaffin cell vesicles. *British Journal of Pharmacology*, *159*(7), 1548–1556. http://dx.doi.org/10.1111/j.1476-5381.2010.00650.x.

Moon, J., Ha, Y., Kim, M., Sim, J., Lee, Y., & Suh, M. (2016). Dual electrochemical microsensor for real-time simultaneous monitoring of nitric oxide and potassium ion changes in a rat brain during spontaneous neocortical epileptic seizure. *Analytical Chemistry*. http://dx.doi.org/10.1021/acs.analchem.6b02396.

Morilak, D. A., Barrera, G., Echevarria, D. J., Garcia, A. S., Hernandez, A., Ma, S., & Petre, C. O. (2005). Role of brain norepinephrine in the behavioral response to stress. *Progress in Neuro-Psychopharmacology and Biological Psychiatry*, *29*(8), 1214–1224. http://dx.doi.org/10.1016/j.pnpbp.2005.08.007.

Nemeroff, C. B. (2002). Recent advances in the neurobiology of depression. *Psychopharmacology Bulletin*, *36*(Suppl. 2), 6–23.

Ng, S. R., Guo, C. X., & Li, C. M. (2011). Highly sensitive nitric oxide sensing using three-dimensional graphene/ionic liquid nanocomposite. *Electroanalysis*, *23*(2), 442–448. http://dx.doi.org/10.1002/elan.201000344.

Nichols, S. P., Koh, A., Storm, W. L., Shin, J. H., & Schoenfisch, M. H. (2013). Biocompatible materials for continuous glucose monitoring devices. *Chemical Reviews*, *113*(4), 2528–2549. http://dx.doi.org/10.1021/cr300387j.

Njagi, J., Ball, M., Best, M., Wallace, K. N., & Andreescu, S. (2010). Electrochemical quantification of serotonin in the live embryonic zebrafish intestine. *Analytical Chemistry, 82*(5), 1822–1830. http://dx.doi.org/10.1021/ac902465v.

Njagi, J., Chernov, M. M., Leiter, J. C., & Andreescu, S. (2010). Amperometric detection of dopamine in vivo with an enzyme based carbon fiber microbiosensor. *Analytical Chemistry, 82*(3), 989–996. http://dx.doi.org/10.1021/ac9022605.

Njagi, J., Erlichman, J. S., Aston, J. W., Leiter, J. C., & Andreescu, S. (2010). A sensitive electrochemical sensor based on chitosan and electropolymerized Meldola blue for monitoring NO in brain slices. *Sensors and Actuators B: Chemical, 143*(2), 673–680. http://dx.doi.org/10.1016/j.snb.2009.10.029.

Oleinick, A. I., Amatore, C., Guille, M., Arbault, S., Klymenko, O. V., & Svir, I. (2006). Modelling release of nitric oxide in a slice of rat's brain: Describing stimulated functional hyperemia with diffusion-reaction equations. *Mathematical Medicine and Biology, 23*(1), 27–44. http://dx.doi.org/10.1093/imammb/dql001.

Oprea, R., Peteu, S. F., Subramanian, P., Qi, W., Pichonat, E., Happy, H., ... Szunerits, S. (2013). Peroxynitrite activity of hemin-functionalized reduced graphene oxide. *Analyst, 138*(15), 4345–4352. http://dx.doi.org/10.1039/c3an00678f.

Othman, A., Karimi, A., & Andreescu, S. (2016). Functional nanostructures for enzyme based biosensors: Properties, fabrication and applications. *Journal of Materials Chemistry B, 4*(45), 7178–7203. http://dx.doi.org/10.1039/c6tb02009g.

Özel, R. E., Alkasir, R. S. J., Ray, K., Wallace, K. N., & Andreescu, S. (2013). Comparative evaluation of intestinal nitric oxide in embryonic zebrafish exposed to metal oxide nanoparticles. *Small, 9*(24), 4250–4261. http://dx.doi.org/10.1002/smll.201301087.

Ozel, R. E., Hayat, A., Wallace, K. N., & Andreescu, S. (2013). Effect of cerium oxide nanoparticles on intestinal serotonin in zebrafish. *RSC Advances, 3*(35), 15298–15309. http://dx.doi.org/10.1039/c3ra41739e.

Özel, R. E., Liu, X., Alkasir, R. S. J., & Andreescu, S. (2014). Electrochemical methods for nanotoxicity assessment. *TrAC Trends in Analytical Chemistry, 59*, 112–120. http://dx.doi.org/10.1016/j.trac.2014.04.006.

Özel, R. E., Wallace, K. N., & Andreescu, S. (2011). Chitosan coated carbon fiber microelectrode for selective in vivo detection of neurotransmitters in live zebrafish embryos. *Analytica Chimica Acta, 695*(1–2), 89–95. http://dx.doi.org/10.1016/j.aca.2011.03.057.

Ozel, R. E., Wallace, K. N., & Andreescu, S. (2014). Alterations of intestinal serotonin following nanoparticle exposure in embryonic zebrafish. *Environmental Science: Nano, 1*(1), 27–36. http://dx.doi.org/10.1039/c3en00001j.

Park, J., Aragona, B. J., Kile, B. M., Carelli, R. M., & Wightman, R. M. (2010). In vivo voltammetric monitoring of catecholamine release in subterritories of the nucleus accumbens shell. *Neuroscience, 169*(1), 132–142. http://dx.doi.org/10.1016/j.neuroscience.2010.04.076.

Park, S. S., Hong, M., Ha, Y., Sim, J., Jhon, G.-J., Lee, Y., & Suh, M. (2015). The real-time in vivo electrochemical measurement of nitric oxide and carbon monoxide release upon direct epidural electrical stimulation of the rat neocortex. *Analyst, 140*(10), 3415–3421. http://dx.doi.org/10.1039/c5an00122f.

Park, S. S., Hong, M., Song, C.-K., Jhon, G.-J., Lee, Y., & Suh, M. (2010). Real-time in vivo simultaneous measurements of nitric oxide and oxygen using an amperometric dual microsensor. *Analytical Chemistry, 82*(18), 7618–7624. http://dx.doi.org/10.1021/ac1013496.

Park, J., Takmakov, P., & Wightman, R. M. (2011). In vivo comparison of norepinephrine and dopamine release in rat brain by simultaneous measurements with fast-scan cyclic voltammetry. *Journal of Neurochemistry, 119*(5), 932–944. http://dx.doi.org/10.1111/j.1471-4159.2011.07494.x.

Feairs, M. J., Ross, A. E., & Venton, B. J. (2011). Comparison of Nafion- and overoxidized polypyrrole-carbon nanotube electrodes for neurotransmitter detection. *Analytical Methods*, *3*(10), 2379–2386. http://dx.doi.org/10.1039/c1ay05348e.

Feteu, S. F., Bose, T., & Bayachou, M. (2013). Polymerized hemin as an electrocatalytic platform for peroxynitrite's oxidation and detection. *Analytica Chimica Acta*, *780*, 81–88. http://dx.doi.org/10.1016/j.aca.2013.03.057.

Peteu, S. F., Boukherroub, R., & Szunerits, S. (2014). Nitro-oxidative species in vivo biosensing: Challenges and advances with focus on peroxynitrite quantification. *Biosensors and Bioelectronics*, *58*, 359–373. http://dx.doi.org/10.1016/j.bios.2014.02.025.

Peteu, S., Peiris, P., Gebremichael, E., & Bayachou, M. (2010). Nanostructured poly(3,4-ethylenedioxythiophene)–metalloporphyrin films: Improved catalytic detection of peroxynitrite. *Biosensors and Bioelectronics*, *25*(8), 1914–1921. http://dx.doi.org/10.1016/j.bios.2010.01.008.

Peteu, S. F., Whitman, B. W., Galligan, J. J., & Swain, G. M. (2016). Electrochemical detection of peroxynitrite using hemin-PEDOT functionalized boron-doped diamond microelectrode. *Analyst*, *141*(5), 1796–1806. http://dx.doi.org/10.1039/c5an02587g.

Pham-Huy, L. A., He, H., & Pham-Huy, C. (2008). Free radicals, antioxidants in disease and health. *International Journal of Biomedical Science: IJBS*, *4*(2), 89–96.

Pimentel, D. R., Amin, J. K., Xiao, L., Miller, T., Viereck, J., Oliver-Krasinski, J., ... Sawyer, D. B. (2001). Reactive oxygen species mediate amplitude-dependent hypertrophic and apoptotic responses to mechanical stretch in cardiac myocytes. *Circulation Research*, *89*(5), 453–460. http://dx.doi.org/10.1161/hh1701.096615.

Ping, J., Ru, S., Fan, K., Wu, J., & Ying, Y. (2010). Copper oxide nanoparticles and ionic liquid modified carbon electrode for the non-enzymatic electrochemical sensing of hydrogen peroxide. *Microchimica Acta*, *171*(1), 117–123. http://dx.doi.org/10.1007/s00604-010-0420-3.

Polikov, V. S., Tresco, P. A., & Reichert, W. M. (2005). Response of brain tissue to chronically implanted neural electrodes. *Journal of Neuroscience Methods*, *148*(1), 1–18. http://dx.doi.org/10.1016/j.jneumeth.2005.08.015.

Pumera, M., Ambrosi, A., Bonanni, A., Chng, E. L. K., & Poh, H. L. (2010). Graphene for electrochemical sensing and biosensing. *TrAC Trends in Analytical Chemistry*, *29*(9), 954–965. http://dx.doi.org/10.1016/j.trac.2010.05.011.

Radi, R., Peluffo, G., Alvarez, M. N., Naviliat, M., & Cayota, A. (2001). Unraveling peroxynitrite formation in biological systems. *Free Radical Biology and Medicine*, *30*(5), 463–488. http://dx.doi.org/10.1016/S0891-5849(00)00373-7.

Rafiee-Pour, H.-A., Noorbakhsh, A., Salimi, A., & Ghourchian, H. (2010). Sensitive superoxide biosensor based on silicon carbide nanoparticles. *Electroanalysis*, *22*(14), 1599–1606. http://dx.doi.org/10.1002/elan.200900577.

Rahimi, P., Ghourchian, H., & Rafiee-Pour, H.-A. (2011). Superoxide radical biosensor based on a nano-composite containing cytochrome c. *Analyst*, *136*(18), 3803–3808. http://dx.doi.org/10.1039/c1an15380c.

Rahman, M. A., Kothalam, A., Choe, E. S., Won, M.-S., & Shim, Y.-B. (2012). Stability and sensitivity enhanced electrochemical in vivo superoxide microbiosensor based on covalently co-immobilized lipid and cytochrome c. *Analytical Chemistry*, *84*(15), 6654–6660. http://dx.doi.org/10.1021/ac301086m.

Rajesh, S., Kanugula, A. K., Bhargava, K., Ilavazhagan, G., Kotamraju, S., & Karunakaran, C. (2010). Simultaneous electrochemical determination of superoxide anion radical and nitrite using Cu, ZnSOD immobilized on carbon nanotube in polypyrrole matrix. *Biosensors and Bioelectronics*, *26*(2), 689–695. http://dx.doi.org/10.1016/j.bios.2010.06.063.

Rajesh, S., Sethy, N. K., Bhargava, K., Ilavazhagan, G., Singh, S. K., & Karunakaran, C. (2011). Electrochemical sensor for simultaneous measurement of nitrite and superoxide anion radical using superoxide dismutase-mimetic manganese(III) tetrakis(1-methyl-4-pyridyl) porphyrin on polypyrrole matrix. *Sensor Letters*, *9*(5), 1682–1688. http://dx.doi.org/10.1166/sl.2011.1733.

Rand, E., Periyakaruppan, A., Tanaka, Z., Zhang, D. A., Marsh, M. P., Andrews, R. J., … Koehne, J. E. (2013). A carbon nanofiber based biosensor for simultaneous detection of dopamine and serotonin in the presence of ascorbic acid. *Biosensors and Bioelectronics*, *42*, 434–438. http://dx.doi.org/10.1016/j.bios.2012.10.080.

Ratner, B. D., & Bryant, S. J. (2004). Biomaterials: Where we have been and where we are going. *Annual Review of Biomedical Engineering*, *6*(1), 41–75. http://dx.doi.org/10.1146/annurev.bioeng.6.040803.140027.

Reddaiah, K., Reddy, T. M., Mallikarjuna, K., & Narasimha, G. (2013). Electrochemical detection of dopamine at poly(solochrome cyanine)/Pd nanoparticles doped modified carbon paste electrode and simultaneous resolution in the presence of ascorbic acid and uric acid: A voltammetric method. *Analytical Methods*, *5*(20), 5627–5636. http://dx.doi.org/10.1039/c3ay41039k.

Rhee, S. G. (1999). Redox signaling: Hydrogen peroxide as intracellular messenger. *Experimental & Molecular Medicine*, *31*(2), 53–59.

Roberts, J. G., Toups, J. V., Eyualem, E., McCarty, G. S., & Sombers, L. A. (2013). In situ electrode calibration strategy for voltammetric measurements in vivo. *Analytical Chemistry*, *85*(23), 11568–11575. http://dx.doi.org/10.1021/ac402884n.

Rodeberg, N. T., Johnson, J. A., Bucher, E. S., & Wightman, R. M. (2016). Dopamine dynamics during continuous intracranial self-stimulation: Effect of waveform on fast-scan cyclic voltammetry data. *ACS Chemical Neuroscience*, 7(11), 1508–1518. http://dx.doi.org/10.1021/acschemneuro.6b00142.

Sadeghian, R. B., Ostrovidov, S., Han, J., Salehi, S., Bahraminejad, B., Bae, H., … Khademhosseini, A. (2016). Online monitoring of superoxide anions released from skeletal muscle cells using an electrochemical biosensor based on thick-film nanoporous gold. *ACS Sensors*, *1*(7), 921–928. http://dx.doi.org/10.1021/acssensors.6b00325.

Safavi, A., & Farjami, F. (2010). Hydrogen peroxide biosensor based on a myoglobin/hydrophilic room temperature ionic liquid film. *Analytical Biochemistry*, *402*(1), 20–25. http://dx.doi.org/10.1016/j.ab.2010.03.013.

Santhosh, P., Manesh, K. M., Lee, S.-H., Uthayakumar, S., Gopalan, A. I., & Lee, K.-P. (2011). Sensitive electrochemical detection of superoxide anion using gold nanoparticles distributed poly(methyl methacrylate)-polyaniline core-shell electrospun composite electrode. *Analyst*, *136*(8), 1557–1561. http://dx.doi.org/10.1039/c0an00616e.

Santos, V. N., Mendonça, G. L. F., Freire, V. N., Holanda, A. K. M., Sousa, J. R., Lopes, L. G. F., … de Lima-Neto, P. (2013). Electrochemical and Monte Carlo studies of self-assembled trans-[Fe(cyclam)(NCS)2]+ complex ion on gold surface as electrochemical sensor for nitric oxide. *Electrochimica Acta*, *91*, 1–10. http://dx.doi.org/10.1016/j.electacta.2012.11.132.

Sardesai, N. P., Ganesana, M., Karimi, A., Leiter, J. C., & Andreescu, S. (2015). Platinum-doped ceria based biosensor for in vitro and in vivo monitoring of lactate during hypoxia. *Analytical Chemistry*, *87*(5), 2996–3003. http://dx.doi.org/10.1021/ac5047455.

Sauer, H., Wartenberg, M., & Hescheler, J. (2001). Reactive oxygen species as intracellular messengers during cell growth and differentiation. *Cellular Physiology and Biochemistry*, *11*(4), 173–186.

Schmidt, A. C., Wang, X., Zhu, Y., & Sombers, L. A. (2013). Carbon nanotube yarn electrodes for enhanced detection of neurotransmitter dynamics in live brain tissue. *ACS Nano*, 7(9), 7864–7873. http://dx.doi.org/10.1021/nn402857u.

Seymour, J. P., & Kipke, D. R. (2007). Neural probe design for reduced tissue encapsulation in CNS. *Biomaterials*, *28*(25), 3594–3607. http://dx.doi.org/10.1016/j.biomaterials.2007.03.024.

Shao, Y., Wang, J., Wu, H., Liu, J., Aksay, I. A., & Lin, Y. (2010). Graphene based electrochemical sensors and biosensors: A review. *Electroanalysis*, *22*(10), 1027–1036. http://dx.doi.org/10.1002/elan.200900571.

Shen, X., Wang, Q., Liu, Y., Xue, W., Ma, L., Feng, S., … Mao, C. (2016). Manganese phosphate self-assembled nanoparticle surface and its application for superoxide anion detection. *Scientific Reports*, *6*, 28989. http://dx.doi.org/10.1038/srep28989. http://www.nature.com/articles/srep28989#supplementary-information.

Shervedani, R. K., Siadat-Barzoki, S. M., & Bagherzadeh, M. (2010). Electrochemical characterization of gold 6-amino-2-mercaptobenzothiazole self-assembled monolayer for dopamine detection in pharmaceutical samples. *Electroanalysis*, *22*(9), 969–977. http://dx.doi.org/10.1002/elan.200900327.

Singh, Y. S., Sawarynski, L. E., Dabiri, P. D., Choi, W. R., & Andrews, A. M. (2011). Head-to-head comparisons of carbon fiber microelectrode coatings for sensitive and selective neurotransmitter detection by voltammetry. *Analytical Chemistry*, *83*(17), 6658–6666. http://dx.doi.org/10.1021/ac2011729.

Song, M.-J., Hwang, S. W., & Whang, D. (2010). Non-enzymatic electrochemical CuO nanoflowers sensor for hydrogen peroxide detection. *Talanta*, *80*(5), 1648–1652. http://dx.doi.org/10.1016/j.talanta.2009.09.061.

Strawn, J. R., & Geracioti, T. D. (2008). Noradrenergic dysfunction and the psychopharmacology of posttraumatic stress disorder. *Depression and Anxiety*, *25*(3), 260–271. http://dx.doi.org/10.1002/da.20292.

Sun, Y., He, K., Zhang, Z., Zhou, A., & Duan, H. (2015). Real-time electrochemical detection of hydrogen peroxide secretion in live cells by Pt nanoparticles decorated graphene–carbon nanotube hybrid paper electrode. *Biosensors and Bioelectronics*, *68*, 358–364. http://dx.doi.org/10.1016/j.bios.2015.01.017.

Swamy, B. E. K., & Venton, B. J. (2007). Carbon nanotube-modified microelectrodes for simultaneous detection of dopamine and serotonin in vivo. *Analyst*, *132*(9), 876–884. http://dx.doi.org/10.1039/b705552h.

Takahashi, Y., Shevchuk, A. I., Novak, P., Zhang, Y., Ebejer, N., Macpherson, J. V., … Korchev, Y. E. (2011). Multifunctional nanoprobes for nanoscale chemical imaging and localized chemical delivery at surfaces and interfaces. *Angewandte Chemie International Edition*, *50*(41), 9638–9642. http://dx.doi.org/10.1002/anie.201102796.

Tanaka, K., & Muto, Y. (1992). Amperometric determination of superoxide anions generated from phytoplankton Chattonella antiqua. *Bioelectrochemistry and Bioenergetics*, *29*(1), 143–147. http://dx.doi.org/10.1016/0302-4598(92)80062-L.

Tang, J., Zhu, X., Niu, X., Liu, T., Zhao, H., & Lan, M. (2015). An amperometric superoxide anion radical biosensor based on SOD/PtPd-PDARGO modified electrode. *Talanta*, *137*, 18–24. http://dx.doi.org/10.1016/j.talanta.2015.01.009.

Thandavan, K., Gandhi, S., Sethuraman, S., Rayappan, J. B. B., & Krishnan, U. M. (2013). A novel nano-interfaced superoxide biosensor. *Sensors and Actuators B: Chemical*, *176*, 884–892. http://dx.doi.org/10.1016/j.snb.2012.09.031.

Trikantzopoulos, E., Yang, C., Ganesana, M., Wang, Y., & Venton, B. J. (2016). Novel carbon-fiber microelectrode batch fabrication using a 3D-printed mold and polyimide resin. *Analyst*, *141*(18), 5256–5260. http://dx.doi.org/10.1039/c6an01469k.

Trouillon, R., & Ewing, A. G. (2013). Single cell amperometry reveals glycocalyx hinders the release of neurotransmitters during exocytosis. *Analytical Chemistry*, *85*(9), 4822–4828. http://dx.doi.org/10.1021/ac4008682.

Trouillon, R., Passarelli, M. K., Wang, J., Kurczy, M. E., & Ewing, A. G. (2013). Chemical analysis of single cells. *Analytical Chemistry*, *85*(2), 522–542. http://dx.doi.org/10.1021/ac303290s.

Trouillon, R., Svensson, M. I., Berglund, E. C., Cans, A.-S., & Ewing, A. G. (2012). Highlights of selected recent electrochemical measurements in living systems. *Electrochimica Acta*, *84*, 84–95. http://dx.doi.org/10.1016/j.electacta.2012.04.164.

Troyer, K. P., Heien, M. L. A. V., Venton, B. J., & Wightman, R. M. (2002). Neurochemistry and electroanalytical probes. *Current Opinion in Chemical Biology*, *6*(5), 696–703. http://dx.doi.org/10.1016/S1367-5931(02)00374-5.

Tsai, T.-C., Han, H.-Z., Cheng, C.-C., Chen, L.-C., Chang, H.-C., & Chen, J.-J. J. (2012). Modification of platinum microelectrode with molecularly imprinted over-oxidized polypyrrole for dopamine measurement in rat striatum. *Sensors and Actuators B: Chemical*, *171–172*, 93–101. http://dx.doi.org/10.1016/j.snb.2011.07.052.

Valentini, F., Romanazzo, D., Carbone, M., & Palleschi, G. (2012). Modified screen-printed electrodes based on oxidized graphene nanoribbons for the selective electrochemical detection of several molecules. *Electroanalysis*, *24*(4), 872–881. http://dx.doi.org/10.1002/elan.201100415.

Valko, M., Rhodes, C. J., Moncol, J., Izakovic, M., & Mazur, M. (2006). Free radicals, metals and antioxidants in oxidative stress-induced cancer. *Chemico-Biological Interactions*, *160*(1), 1–40. http://dx.doi.org/10.1016/j.cbi.2005.12.009.

Vreeland, R. F., Atcherley, C. W., Russell, W. S., Xie, J. Y., Lu, D., Laude, N. D., … Heien, M. L. (2015). Biocompatible PEDOT:Nafion composite electrode coatings for selective detection of neurotransmitters in vivo. *Analytical Chemistry*, *87*(5), 2600–2607. http://dx.doi.org/10.1021/ac502165f.

Wang, Y., & Chen, Z.-z. (2010). A novel poly(cyanocobalamin) modified glassy carbon electrode as electrochemical sensor for voltammetric determination of peroxynitrite. *Talanta*, *82*(2), 534–539. http://dx.doi.org/10.1016/j.talanta.2010.05.020.

Wang, X., Han, M., Bao, J., Tu, W., & Dai, Z. (2012). A superoxide anion biosensor based on direct electron transfer of superoxide dismutase on sodium alginate sol–gel film and its application to monitoring of living cells. *Analytica Chimica Acta*, *717*, 61–66. http://dx.doi.org/10.1016/j.aca.2011.12.045.

Wang, N., Han, Y., Xu, Y., Gao, C., & Cao, X. (2015). Detection of H2O2 at the nanomolar level by electrode modified with ultrathin AuCu nanowires. *Analytical Chemistry*, *87*(1), 457–463. http://dx.doi.org/10.1021/ac502682n.

Wang, Z., Liu, D., Gu, H., Zhu, A., Tian, Y., & Shi, G. (2013). NTA-modified carbon electrode as a general relaying substrate to facilitate electron transfer of SOD: Application to in vivo monitoring of O2− in a rat brain. *Biosensors and Bioelectronics*, *43*, 101–107. http://dx.doi.org/10.1016/j.bios.2012.10.071.

Wang, Y., Noël, J.-M., Velmurugan, J., Nogala, W., Mirkin, M. V., Lu, C., … Amatore, C. (2012). Nanoelectrodes for determination of reactive oxygen and nitrogen species inside murine macrophages. *Proceedings of the National Academy of Sciences of the United States of America*, *109*(29), 11534–11539. http://dx.doi.org/10.1073/pnas.1201552109.

Wang, L., Wen, W., Xiong, H., Zhang, X., Gu, H., & Wang, S. (2013). A novel amperometric biosensor for superoxide anion based on superoxide dismutase immobilized on gold nanoparticle-chitosan-ionic liquid biocomposite film. *Analytica Chimica Acta*, *758*, 66–71. http://dx.doi.org/10.1016/j.aca.2012.10.050.

Wang, L., Zhang, Y., Cheng, C., Liu, X., Jiang, H., & Wang, X. (2015a). Highly sensitive electrochemical biosensor for evaluation of oxidative stress based on the nanointerface of graphene nanocomposites blended with gold, Fe3O4, and platinum nanoparticles. *ACS Applied Materials & Interfaces*, 7(33), 18441–18449. http://dx.doi.org/10.1021/acsami.5b04553.

Wang, Z., Zhang, L., & Tian, Y. (2015b). A durable non-enzymatic electrochemical sensor for monitoring H2O2 in rat brain microdialysates based on one-step fabrication of hydrogels. *Analyst*, *140*(11), 3788–3793. http://dx.doi.org/10.1039/c4an02003k.

Wang, T., Zhu, H., Zhuo, J., Zhu, Z., Papakonstantinou, P., Lubarsky, G., … Li, M. (2013). Biosensor based on ultrasmall MoS2 nanoparticles for electrochemical detection of H2O2 released by cells at the nanomolar level. *Analytical Chemistry*, *85*(21), 10289–10295. http://dx.doi.org/10.1021/ac402114c.

Wilson, G. S., & Johnson, M. A. (2008). In-vivo electrochemistry: What can we learn about living systems? *Chemical Reviews*, *108*(7), 2462–2481. http://dx.doi.org/10.1021/cr068082i.

Wipf, D. O., Kristensen, E. W., Deakin, M. R., & Wightman, R. M. (1988). Fast-scan cyclic voltammetry as a method to measure rapid heterogeneous electron-transfer kinetics. *Analytical Chemistry*, *60*(4), 306–310. http://dx.doi.org/10.1021/ac00155a006.

Wise, R. A. (2004). Dopamine, learning and motivation. *Nature Reviews. Neuroscience*, *5*(6), 483–494.

Wiseman, H., & Halliwell, B. (1996). Damage to DNA by reactive oxygen and nitrogen species: Role in inflammatory disease and progression to cancer. *Biochemical Journal*, *313*(Pt. 1), 17–29.

Wring, S. A., & Hart, J. P. (1992). Chemically modified, carbon-based electrodes and their application as electrochemical sensors for the analysis of biologically important compounds. A review. *Analyst*, *117*(8), 1215–1229. http://dx.doi.org/10.1039/an9921701215.

Wu, P., Cai, Z., Chen, J., Zhang, H., & Cai, C. (2011). Electrochemical measurement of the flux of hydrogen peroxide releasing from RAW 264.7 macrophage cells based on enzyme-attapulgite clay nanohybrids. *Biosensors and Bioelectronics*, *26*(10), 4012–4017. http://dx.doi.org/10.1016/j.bios.2011.03.018.

Wu, P., Cai, Z., Gao, Y., Zhang, H., & Cai, C. (2011). Enhancing the electrochemical reduction of hydrogen peroxide based on nitrogen-doped graphene for measurement of its releasing process from living cells. *Chemical Communications*, *47*(40), 11327–11329. http://dx.doi.org/10.1039/c1cc14419g.

Wu, L., Yang, Y., Zhang, H., Zhu, G., Zhang, X., & Chen, J. (2012). Sensitive electrochemical detection of hydroxyl radical with biobarcode amplification. *Analytica Chimica Acta*, *756*, 1–6. http://dx.doi.org/10.1016/j.aca.2012.10.039.

Wurtman, R. J. (2002). Stress and the adrenocortical control of epinephrine synthesis. *Metabolism*, *51*(6), 11–14. http://dx.doi.org/10.1053/meta.2002.33185.

Xi, F., Zhao, D., Wang, X., & Chen, P. (2013). Non-enzymatic detection of hydrogen peroxide using a functionalized three-dimensional graphene electrode. *Electrochemistry Communications*, *26*, 81–84. http://dx.doi.org/10.1016/j.elecom.2012.10.017.

Xie, X. S., Yu, J., & Yang, W. Y. (2006). Living cells as test tubes. *Science*, *312*(5771), 228–230. http://dx.doi.org/10.1126/science.1127566.

Xu, F., Deng, M., Li, G., Chen, S., & Wang, L. (2013). Electrochemical behavior of cuprous oxide-reduced graphene oxide nanocomposites and their application in nonenzymatic hydrogen peroxide sensing. *Electrochimica Acta*, *88*, 59–65. http://dx.doi.org/10.1016/j.electacta.2012.10.070.

Xu, F., Sun, Y., Zhang, Y., Shi, Y., Wen, Z., & Li, Z. (2011). Graphene–Pt nanocomposite for nonenzymatic detection of hydrogen peroxide with enhanced sensitivity. *Electrochemistry Communications*, *13*(10), 1131–1134. http://dx.doi.org/10.1016/j.elecom.2011.07.017.

Yagati, A. K., Lee, T., Min, J., & Choi, J.-W. (2011). Amperometric sensor for hydrogen peroxide based on direct electron transfer of spinach ferredoxin on Au electrode. *Bioelectrochemistry*, *80*(2), 169–174. http://dx.doi.org/10.1016/j.bioelechem.2010.08.002.

Yang, S. Y., Kim, B. N., Zakhidov, A. A., Taylor, P. G., Lee, J.-K., Ober, C. K., … Malliaras, G. G. (2011). Detection of transmitter release from single living cells using conducting polymer microelectrodes. *Advanced Materials*, *23*(24), H184–H188. http://dx.doi.org/10.1002/adma.201100035.

Yang, J.-H., Myoung, N., & Hong, H.-G. (2012). Facile and controllable synthesis of Prussian blue on chitosan-functionalized graphene nanosheets for the electrochemical detection of hydrogen peroxide. *Electrochimica Acta*, *81*, 37–43. http://dx.doi.org/10.1016/j.electacta.2012.07.038.

Yang, W., Ratinac, K. R., Ringer, S. P., Thordarson, P., Gooding, J. J., & Braet, F. (2010). Carbon nanomaterials in biosensors: Should you use nanotubes or graphene? *Angewandte Chemie International Edition*, *49*(12), 2114–2138. http://dx.doi.org/10.1002/anie.200903463.

Yap, C. M., Xu, G. Q., & Ang, S. G. (2013). Amperometric nitric oxide sensor based on nanoporous platinum phthalocyanine modified electrodes. *Analytical Chemistry*, *85*(1), 107–113. http://dx.doi.org/10.1021/ac302081h.

Yeh, J. I., & Shi, H. (2010). Nanoelectrodes for biological measurements. *Wiley Interdisciplinary Reviews: Nanomedicine and Nanobiotechnology*, *2*(2), 176–188. http://dx.doi.org/10.1002/wnan.70.

Yin, J., Qi, X., Yang, L., Hao, G., Li, J., & Zhong, J. (2011). A hydrogen peroxide electrochemical sensor based on silver nanoparticles decorated silicon nanowire arrays. *Electrochimica Acta*, *56*(11), 3884–3889. http://dx.doi.org/10.1016/j.electacta.2011.02.033.

You, J.-M., Jeong, Y. N., Ahmed, M. S., Kim, S. K., Choi, H. C., & Jeon, S. (2011). Reductive determination of hydrogen peroxide with MWCNTs-Pd nanoparticles on a modified glassy carbon electrode. *Biosensors and Bioelectronics*, *26*(5), 2287–2291. http://dx.doi.org/10.1016/j.bios.2010.09.053.

You, J.-M., Kim, D., Kim, S. K., Kim, M.-S., Han, H. S., & Jeon, S. (2013). Novel determination of hydrogen peroxide by electrochemically reduced graphene oxide grafted with aminothiophenol–Pd nanoparticles. *Sensors and Actuators B: Chemical*, *178*, 450–457. http://dx.doi.org/10.1016/j.snb.2013.01.006.

Yu, D., Zeng, Y., Qi, Y., Zhou, T., & Shi, G. (2012). A novel electrochemical sensor for determination of dopamine based on AuNPs@SiO2 core-shell imprinted composite. *Biosensors and Bioelectronics*, *38*(1), 270–277. http://dx.doi.org/10.1016/j.bios.2012.05.045.

Yuan, L., Liu, S., Tu, W., Zhang, Z., Bao, J., & Dai, Z. (2014). Biomimetic superoxide dismutase stabilized by photopolymerization for superoxide anions biosensing and cell monitoring. *Analytical Chemistry*, *86*(10), 4783–4790. http://dx.doi.org/10.1021/ac403920q.

Yuan, D., Yuan, X., Zhou, S., Zou, W., & Zhou, T. (2012). N-Doped carbon nanorods as ultrasensitive electrochemical sensors for the determination of dopamine. *RSC Advances*, *2*(21), 8157–8163. http://dx.doi.org/10.1039/c2ra21041j.

Yum, K., Wang, N., & Yu, M.-F. (2010). Nanoneedle: A multifunctional tool for biological studies in living cells. *Nanoscale*, *2*(3), 363–372. http://dx.doi.org/10.1039/b9nr00231f.

Zan, X., Fang, Z., Wu, J., Xiao, F., Huo, F., & Duan, H. (2013). Freestanding graphene paper decorated with 2D-assembly of Au@Pt nanoparticles as flexible biosensors to monitor live cell secretion of nitric oxide. *Biosensors and Bioelectronics*, *49*, 71–78. http://dx.doi.org/10.1016/j.bios.2013.05.006.

Zapanti, E., & Ilias, I. (2006). Pheochromocytoma. *Annals of the New York Academy of Sciences*, *1088*(1), 346–360. http://dx.doi.org/10.1196/annals.1366.022.

Zhang, Y., Bai, X., Wang, X., Shiu, K.-K., Zhu, Y., & Jiang, H. (2014). Highly sensitive graphene–Pt nanocomposites amperometric biosensor and its application in living cell H2O2 detection. *Analytical Chemistry*, *86*(19), 9459–9465. http://dx.doi.org/10.1021/ac5009699.

Zhang, L., Wang, J., & Tian, Y. (2014). Electrochemical in-vivo sensors using nanomaterials made from carbon species, noble metals, or semiconductors. *Microchimica Acta*, *181*(13), 1471–1484. http://dx.doi.org/10.1007/s00604-014-1203-z.

Zhao, B., Liu, Z., Fu, W., & Yang, H. (2013). Construction of 3D electrochemically reduced graphene oxide–silver nanocomposite film and application as nonenzymatic hydrogen peroxide sensor. *Electrochemistry Communications*, *27*, 1–4. http://dx.doi.org/10.1016/j.elecom.2012.10.040.

Zheng, M., Li, P., Yang, C., Zhu, H., Chen, Y., Tang, Y., … Lu, T. (2012). Ferric ion immobilized on three-dimensional nanoporous gold films modified with self-assembled monolayers for electrochemical detection of hydrogen peroxide. *Analyst*, *137*(5), 1182–1189. http://dx.doi.org/10.1039/c2an15957k.

Zhu, A., Liu, Y., Rui, Q., & Tian, Y. (2011). Selective and sensitive determination of hydroxyl radicals generated from living cells through an electrochemical impedance method. *Chemical Communications*, *47*(14), 4279–4281. http://dx.doi.org/10.1039/c0cc05821a.

Zhu, X., Liu, T., Zhao, H., Shi, L., Li, X., & Lan, M. (2016). Ultrasensitive detection of superoxide anion released from living cells using a porous Pt–Pd decorated enzymatic sensor. *Biosensors and Bioelectronics*, *79*, 449–456. http://dx.doi.org/10.1016/j.bios.2015.12.061.

Zhu, Z., Qu, L., Guo, Y., Zeng, Y., Sun, W., & Huang, X. (2010). Electrochemical detection of dopamine on a Ni/Al layered double hydroxide modified carbon ionic liquid electrode. *Sensors and Actuators B: Chemical*, *151*(1), 146–152. http://dx.doi.org/10.1016/j.snb.2010.09.032.

Zimmerman, M. C., Lazartigues, E., Lang, J. A., Sinnayah, P., Ahmad, I. M., Spitz, D. R., & Davisson, R. L. (2002). Superoxide mediates the actions of angiotensin II in the central nervous system. *Circulation Research*, *91*(11), 1038–1045. http://dx.doi.org/10.1161/01.res.0000043501.47934.fa.

CHAPTER TWELVE

A Highly Sensitive Biosensor for ATP Using a Chimeric Firefly Luciferase

Bruce R. Branchini[1], Tara L. Southworth
Connecticut College, New London, CT, United States
[1]Corresponding author: e-mail address: brbra@conncoll.edu

Contents

Abstract

Firefly luciferases, which emit visible light in a highly specific ATP-dependent process, have been adapted for a variety of applications based on the detection of the enzymes or using the proteins to measure ATP levels. Based on studies of chimeric luciferases, we engineered a novel luciferase called PLG2 that has enhanced specific activity, and thermal and pH stability compared to the commonly used *Photinus pyralis* luciferase. We present here protocols for preparing a single assay mixture containing PLG2 that can be used to readily detect femtomole levels of ATP. Our methodology can be used with a variety of samples, including human and bacterial cells, where measurements of ATP can be used as a biosensor for the detection of viable cells.

1. INTRODUCTION

Bioluminescence, the emission of visible light by living organisms, is broadly distributed in the marine and terrestrial environments. Typically,

Methods in Enzymology, Volume 589
ISSN 0076-6879
http://dx.doi.org/10.1016/bs.mie.2017.01.004

light is produced by enzymes (luciferases) and substrates (luciferins and O_2) in biochemical reactions that often require additional cofactors. Methods for analyte detection based on a variety of bioluminescence materials are important research tools in cell biology, molecular biology, and analytical chemistry (Roda, 2011; Widder & Falls, 2014). The general advantages of bioluminescence methods include: relative ease of performance, high sensitivity due to low background, wide dynamic range, and low cost of photomultiplier tube (PMT) and charge-coupled device detectors. Numerous applications of beetle luciferases, especially those from fireflies, include whole-cell-based biosensors (Binkowski et al., 2011; Bottcher, Pollet, Delius, & Niehrs, 2004; Etzov, Cohen, & Marks, 2015; Hakkila, Maksimow, Karp, & Virta, 2002; Roda et al., 2011; Shi et al., 2014; Urban et al., 2007), gene expression assays (Gould & Subramani, 1988; Leclerc, Boockfor, Faught, & Frawley, 2000; Paddison, Caudy, & Hannon, 2002; Vannini et al., 2012), and imaging disease states in live mammals (Cevenini et al., 2014; Kuo, Auriau, Pierre-Eugene, & Issad, 2014; Lewis et al., 2014; Taylor & Kelly, 2014). Many of these applications can be applied in high-throughput screening format in support of drug development (Afshari, Uhde-Stone, & Lu, 2015; Che, Cui, Kutsch, Cui, & Li, 2012; Hattori & Ozawa, 2015; Roda, Guardigli, Michelini, & Mirasoli, 2009; Thorne et al., 2012; Yoon, Namkung, & Lee, 2013). In addition to the general advantages of bioluminescence methods, beetle bioluminescence methods take advantage of the stringent requirement for ATP, the universal biological energy source. Beetle luciferase (abbreviated here as Luc) assays for the production, consumption, or presence of ATP are based on the role of the metabolite in the activation of beetle luciferin (LH_2), the first step (Eq. 1) in the production of light (Eqs. 1–3).

$$\text{Luc} + \text{LH}_2 + \text{ATP} \overset{\text{Mg}^{2+}}{\leftrightarrows} \text{Luc} \cdot \text{LH}_2 - \text{AMP} + \text{PP}_i \tag{1}$$

$$\text{Luc} \cdot \text{LH}_2 - \text{AMP} + \text{O}_2 \rightarrow \text{Luc} \cdot \text{AMP} \cdot \text{Oxyluciferin}^* + \text{CO}_2 \tag{2}$$

$$\text{Luc} \cdot \text{AMP} \cdot \text{Oxyluciferin}^* \rightarrow \text{Luc} + \text{Oxyluciferin} + \text{AMP} + h\nu \tag{3}$$

The *Photinus pyralis* Luc bioluminescence system is highly efficient with a quantum yield of 41.0% ± 7.4% (Ando et al., 2008). It has been estimated that attomole levels of Luc can be quantified (Lundin, 1993), and unlike with fluorescence systems that require an external light source,

Table 1 Properties of Luciferases

Enzyme	Relative Integrated Specific Activity[a]	K_m (μ*M*)		Bioluminescence λ_{max}[b] (nm)		Thermal Inactivation[c] (h) at 37°C
		LH$_2$	Mg-ATP	pH 7.8	pH 6.5	
PpyWT[d]	100 ± 2	15 ± 2	86 ± 7	560 (70)	609, 560 sh (92)	0.3 ± 0.03
PLG2	425 ± 2	52 ± 6	79 ± 5	559 (67)	561 (71)	>24 ± 5

[a]Specific activities are based on total light emission in 15 min, were obtained at pH 7.8 with LH$_2$ (400 μ*M*) and Mg-ATP (2 m*M*), and are expressed relative to the PpyWT value, which is defined as 100.
[b]Bioluminescence emission maxima were obtained from two trials and the standard deviation is ±1 nm. The bandwidth at half-maximal intensity is shown in parentheses and sh indicates a shoulder.
[c]Time for the maximum initial activity to decay to 50% at 37°C. All activity values were obtained from at least three trials and are reported as means ± standard deviation.
[d]Recombinant *P. pyralis* luciferase containing the additional N-terminal peptide GlyProLeuGlySer (Branchini et al., 2014).

there is virtually no nonspecific light production induced in the cellular environment.

We have developed improved biosensor methods for the detection of ATP by taking advantage of the highly specific and sensitive detection of ATP generally afforded by beetle Lucs. The key component is a highly engineered version of *P. pyralis* Luc called PLG2 (Table 1) (Branchini et al., 2015). The design of this enzyme was inspired by earlier studies of the enhanced specific activity of a chimeric Luc (Branchini et al., 2014), which contains the N-domain of *P. pyralis* Luc joined to the C-domain of the *Luciola italica* enzyme (Branchini, Southworth, DeAngelis, Roda, & Michelini, 2006). We present procedures using PLG2 in in vitro assays designed to detect femtomole levels of ATP from biological samples including living cells.

2. ATP DETECTION PROTOCOL

2.1 Sample Preparation

The assay protocol described here can be used to determine levels of cells and microorganisms in a variety of products including foodstuffs, water, and surfaces based on their ATP content. Methods of sample preparation, including extraction procedures when necessary, will vary according to the specific materials being analyzed. An excellent guide to sample preparation can be found in the technical bulletin of the Promega ENLITEN® ATP Assay

System (ENLITEN ATP Assay System Bioluminescence Detection Kit for ATP Measurement, https://www.promega.com/-/media/files/resources/protocols/technical-bulletins/0/enliten-atp-assay-system-bioluminescence-detection-kit-for-atp-protocl.pdf).

2.2 Light Measurement

1. Luminometer. Microwell plate reading and single tube luminometers are available from several manufacturers and can be purchased from Promega, Thermo Scientific, Berthold Technologies, BioTek, and others. Although not essential, it is preferable to use an instrument with a single automatic injector. Assays can be performed at room temperature, but standard curves should be made at the temperature where unknown assays will be performed. The detection limit of ATP assays will ultimately depend on the "dark background," the signal produced in the absence of ATP. For the protocols described here, we used a Thermo Luminoskan Ascent instrument equipped with a PMT operated at 900 V and a single automatic injector. The reagent lines and injector were cleaned prior to and after use with 70% ethanol. If more stringent cleaning procedures are required, consult the technical bulletin of the Promega ENLITEN® ATP Assay System (ENLIGHTEN ATP Assay System Bioluminescence Detection Kit for ATP Measurement).
2. Measurement of relative light units (RLUs). Luciferase activity is commonly expressed in RLUs that are defined for the equipment being employed. The RLUs in the protocols presented here have been related to the definition described in the Sigma-Aldrich catalog for recombinant *P. pyralis* luciferase. We prepared a stock solution of Sigma product SRE0045 by dissolving a 1 mg vial of reagent in 1 mL of 1 *M* Tris buffer, pH 7.6 with brief gentle mixing. In triplicate, aliquots of a 1:10 dilution of the stock solution (2 μL, 2×10^{-4} mg enzyme) were pipetted into wells of a 96-well white plate (Greiner Bio-One North America, Inc., Monroe, NC). The luminometer injector was loaded with a solution comprising Promega LH_2 potassium salt (150 μ*M*) and Sigma Mg-ATP (0.4 μ*M*) in 50 m*M* Tricine buffer, pH 7.6, and 0.1 mL of the solution was injected into the wells containing enzyme. Light emission from the wells was monitored until the signals were steady and not decaying and then the RLUs were recorded, for example, at 10 s. The Luminoskan mean value reading of 1340 (after blank subtraction) was set to 4×10^6 RLU [$(2 \times 10^{-4}$ mg$) \times (2 \times 10^{10}$ RLU/mg$)$]; therefore, 1 Luminoskan

unit=2985 RLU. Using our standard methods (Branchini et al., 2014) for measuring specific activity, we obtained values for our recombinant *P. pyralis* luciferase essentially equivalent to the Promega Luc2 enzyme.

2.3 Preparation of PLG2

1. The PLG2 luciferase required for PLG2 Assay Mix preparation can be obtained from the procedures below by transforming *E. coli* with the pGEX-6P-2 PLG2 plasmid and inducing expression of the GST-luciferase fusion protein. Please contact the authors to obtain the plasmid or a sample of the PLG2 Assay Mix (see Section 2.4 later).
2. Preparation of protein expression reagents. LB/AMP media is prepared by dissolving 25 g Luria Bertani media (Fisher Scientific) in 1 L of distilled deionized H_2O (ddH_2O), autoclaving the media, and adding 1 mL of Ampicillin solution (100 mg/mL in ddH_2O, sterile filtered) once the media has cooled. Store at 5°C. IPTG solution (24 mg/mL) is made up by dissolving isopropyl-α-D-thiogalactoside in ddH_2O. Solutions should be sterile filtered, divided into aliquots and stored at −20°C.
3. Preparation of bacteria cell pellet. For a 0.25-L culture whose expression typically yields ~6 mg of purified enzyme, prepare a starter culture by using a sterile inoculating loop to scrape the surface of a glycerol stock of pGEX-6P-2 PLG2 plasmid (previously transformed into *E. coli* strain BL21(DE3)pLysS cells), inoculating 5 mL of LB/AMP, and growing it overnight at 37°C with vigorous shaking (325 rpm). Add 3 mL of starter culture to 250 mL of LB/AMP in a 1-L flask and grow at 37°C with shaking while monitoring cell growth of 1 mL aliquots by absorbance at 600 nm. Within 1.5–2 h following inoculation, the A_{600} nm should reach 0.4–0.6, and at this point, the incubator temperature is adjusted to 22°C. After 15 min, add 0.25 mL of IPTG solution and continue shaking at 22°C for 19 h. Harvest cells by centrifugation at 5000 × *g* for 10 min at 4°C and discard the supernatant. Freeze the cell pellet at −80°C for 15 min, thaw, and continue the preparation or store (stable for months) at this temperature.
4. Preparation of protein purification reagents. Store all solutions at 4°C and keep on ice during protein purification. Prepare 1 L of 10× PBS buffer: 400 m*M* NaCl, 27 m*M* KCl, 100 m*M* Na_2HPO_4, and 18 m*M* KH_2PO_4 in ddH_2O, pH 7.3, and 1 L of 1× PBS by dilution of 10× PBS in ddH_2O; buffers are sterilized by filtration. Resuspension

buffer is prepared immediately prior to use by adding 50 μL of 25 mg/mL phenylmethylsulfonyl fluoride (PMSF, Sigma) solution (made by dissolving 2.5 mg of PMSF in 0.1 mL of 95% ethanol with vigorous mixing) to 50 mL of 1 × PBS containing 3.8 mg of DTT. Lysozyme (Sigma, from egg white) solution is prepared by dissolving 10 mg/mL lysozyme in 1 × PBS. RNaseA solution (Sigma, from bovine pancreas) in 10 mg/mL, RNaseA in 10 m*M* Tris buffer, pH 7.5, containing 15 m*M* NaCl. DNaseI solution is 5 mg/mL DNaseI (Sigma, from bovine pancreas) in 10 m*M* Tris buffer, pH 7.5 containing glycerol (1:1, v/v), 50 m*M* NaCl, 10 m*M* $MgCl_2$, and 1 m*M* DTT. Protease cleavage buffer (PCB) comprises 50 m*M* Tris, pH 7.0, containing 150 m*M* NaCl, 1 m*M* EDTA, and 1 m*M* DTT.

5. Preparation of PLG2 activity assay solutions. LH_2 solution is 0.52 m*M* LH_2 (Promega, K salt) in 50 m*M* Tricine buffer, pH 7.8. Dissolve ~2 mg LH_2 in 10 mL of buffer and adjust concentration by dilution based on UV using a molar absorptivity coefficient of 7600 *M*/cm at 266 nm. ATP solution is prepared by dissolving 9 m*M* ATP (Sigma, Mg salt) in 50 m*M* Tricine buffer, pH 7.8 as follows: Dissolve ~60 mg Mg-ATP in 10 mL buffer, readjust pH to ~7.8 with 1 N NaOH, and adjust concentration using a molar absorptivity coefficient of 15,400 *M*/cm at 259 nm.
6. Equilibration of Glutathione Sepharose® 4B media. In a cold room (5°C), pipet 1.5 mL of slurry of Glutathione Sepharose® 4B (GE Healthcare Life Sciences) into a 50 mL plastic (polypropylene or polycarbonate plastics, but not polystyrene, are compatible with luciferases) centrifuge tube, add 25 mL of cold 1 × PBS, and centrifuge at 500 × *g* for 5 min. Carefully pipet off ~15 mL of supernatant making sure not to remove resin. Wash resin three times with 25 mL portions of cold 1 × PBS and after the third wash, pipet off supernatant to bring down the volume to 3 mL, creating a 50% resin slurry. Store on ice.
7. Isolation of crude enzyme. Samples should be kept on ice during all procedures. Thaw bacteria pellet (Section 2.3 step 3) for 30 min (on ice) and add 25 mL of resuspension buffer. Gently resuspend the pellet by withdrawing and expelling the mixture with a pipet, making sure to avoid foaming and bubbling. Add 2.5 mL of lysozyme solution, let stand for 20 min, and sonicate. We use a Virsonic Model 475 instrument with a 0.5-in. tip diameter at setting 3 (95 W) and apply 3 bursts at 10 s each. It is important to avoid frothing during the sonication.

Add 25 μL RNaseA solution and 25 μL DNaseI solution and incubate mixture for 15 min. Add 1.25 mL 20% Triton X-100 (aq., v/v), mix, transfer the suspension into a 50 mL centrifuge tube, and centrifuge at 20,000 × *g* for 45 min at 4°C, carefully decant the supernatant (avoid taking any suspended pellet) into a centrifuge tube and record the volume of the crude protein solution (should be ~25 mL).

8. Perform enzyme activity measurement on crude supernatant. It is a good idea to check the total activity of the supernatant prior to continuing to the affinity chromatography step. However, it is necessary to do this quickly to avoid loss of activity. Assay the supernatant by pipetting 3 μL aliquots into wells of a 96-well white plate (Greiner Bio-One North America, Inc., Monroe, NC) containing 0.2 mL of 0.52 m*M* LH_2 solution (prepared as described in Section 2.3 step 5). Load the luminometer injector with the solution of 9 m*M* Mg-ATP (Section 2.3 step 5) and inject 60 μL into the wells containing enzyme and LH_2. Monitor the RLUs for ~30 s and record the maximum RLU value (~1.3×10^{10} RLU).

9. Glutathione Sepharose® 4B affinity chromatography. Additional general information is available in the GST Gene Fusion System Handbook 18-1157-58 (GE Healthcare http://www.gelifesciences.com/file_source/GELS/Service%20and%20Support/Documents%20and%20Downloads/Handbooks/pdfs/GST_gene_fusion_system_handbook.pdf). All procedures are performed in a cold room (5°C). Add 3 mL of resin slurry (Section 2.3 step 6) to the crude supernatant (Section 2.3 step 7) in a centrifuge tube with gentle agitation (we use a VWR rocking platform at setting 3.5) for 2 h. Centrifuge at 500 × *g* at 4°C for 5 min, carefully remove the supernatant (avoid removing resin), and record the volume of the supernatant (typically ~20 mL). Determine the percentage of RLU remaining in the supernatant (Section 2.3 step 8), typically ~25–45%. Wash resin (three times) using 30 mL cold PBS with brief gentle mixing, centrifuge (500 × *g* at 4°C for 2 min), and remove supernatant by pipet. Next wash resin with 30 mL of PCB, centrifuge, remove supernatant, and transfer to a 5 mL polypropylene tube keeping ~3 mL of the ~50% slurry. Add 100 units of PreScission (GE Healthcare Life Sciences) to remove the GST tag and gently mix overnight (~16 h) (we use a VWR Tube Rotator). Transfer the resin slurry to a gravity flow column, collect, and save the flow through (~1.5 mL) and elute with 3 mL PCB

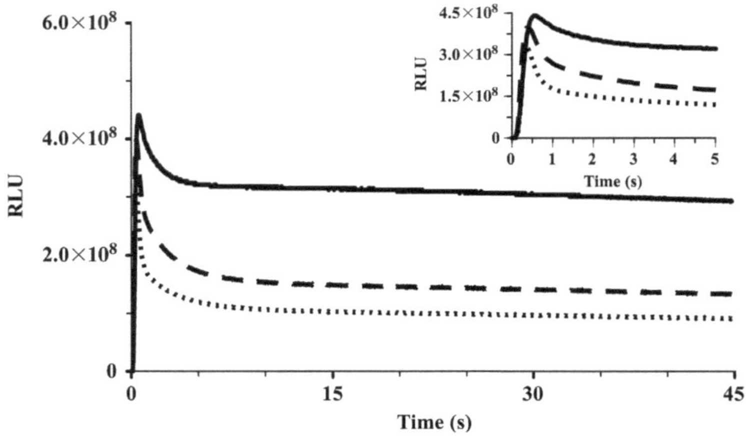

Fig. 1 Bioluminescence time course of reactions of PLG2 (–), recombinant *P. pyralis* luciferase containing the N-terminal peptide GlyProLeuGlySer (- -) and Sigma recombinant *P. pyralis* luciferase (····). Aliquots of purified enzymes (2 μL of 0.01 mg/mL solution) in PCB were added to the wells of a 96-well white plate. The reaction was initiated by the injection of a solution containing 0.40 m*M* LH_2 and 2 m*M* Mg-ATP in 50 m*M* Tricine, pH 8.

(collect 1 mL fractions). Determine the protein concentration of the flow through and column fractions and combine those that are at least ~1 mg/mL. We recommend the Bio-Rad Protein Assay System using bovine serum albumin as the standard.

10. Determine the specific activity of purified PLG2. Measure the activity of the pooled protein and calculate the specific activity, which should be $\sim 2.2 \times 10^{13}$ RLU/mg. By comparison, the specific activity of the Sigma luciferase is 1.6×10^{13} RLU/mg (Fig. 1).
11. Enzyme storage. PLG2 solutions (~2 mg/mL in PCB) can be stored for 1–2 months at 5°C. For long-term storage, flash freeze 0.1–0.3 mL aliquots in thin-walled PCR tubes by submerging in liquid nitrogen for 60 s and store at −80°C. Enzyme solutions can be thawed by gently swirling tubes in a 37°C water bath until only a small amount (size of a grain of rice) remains and immediately place on ice. Thawed aliquots can be used to make PLG2 Assay Mix.

2.4 PLG2 Assay Mix for ATP Determination

Approximately 100 ATP assays can be performed with 5 mL of the PLG2 Assay Mix (3.3 μ*M* PLG2, 0.8 m*M* LH_2 and 10 m*M* $MgSO_4$ in 50 m*M* Tricine buffer, pH 8).

1. Autoclave 0.5 L of 50 m*M* Tricine buffer, pH 8, microcentrifuge tubes, 5–10 mL plastic tubes and caps, and pipet tips.
2. Prepare 100 m*M* $MgSO_4$ stock solution in 50 m*M* Tricine buffer, pH 8.0, and store at 5°C.
3. Prepare a 5 m*M* LH_2 stock solution by dissolving ~2 mg of LH_2 potassium salt in 1 mL of 50 m*M* Tricine buffer, pH 8.0, and adjust the concentration by dilution based on UV using a molar absorptivity coefficient of 7600 at 266 nm.
4. For 5 mL of PLG2 Assay Mix, add 0.8 mL LH_2 stock solution, 0.5 $MgSO_4$ stock solution, 1 mL of 1.0 mg/mL PLG2 in PCB, and 50 m*M* Tricine buffer, pH 8 to volume. For highest sensitivity (ATP levels below $\sim 1 \times 10^{-14}$mol), background emission can be reduced by storing the PLG2 Assay Mix overnight at 5°C. This is necessary to detect ATP levels below $\sim 1 \times 10^{-14}$ mol. PLG2 Assay Mix can be stored at 5°C for 48 h without measurable loss of activity, while after 2 weeks, the solution retained 85% activity. For convenience, the assay solution can be lyophilized, stored at −20°C for months, and reconstituted with ATP-free ddH_2O.

2.5 ATP Standard Curve and Analysis

1. Prepare a standard 2 m*M* solution of ATP (Sigma, disodium salt hydrate) in 50 m*M* Tricine, pH 8. Dissolve ~6 mg of ATP in 5 mL buffer, readjust pH to 8.0 with 1 N NaOH and adjust the concentration using a molar absorptivity coefficient of 15,400 at 259 nm. Serially dilute the 2 m*M* ATP standard solution into the Tricine buffer to make a set of concentrations ranging from 0.2 μ*M* to 0.2 p*M*. The stock ATP solution can be stored at 5°C for at least a week and the dilutions can be made, stored at 5°C, and used the next day.
2. Assay 0.05 mL aliquots of the ATP dilutions (corresponding to 1×10^{-11} to 1×10^{-17} mol of ATP) in triplicate in white 96-well plates (Greiner Bio-One North America, Inc., Monroe, NC) and a blank (Tricine buffer, no ATP) by automatic injection of 0.05 mL of the PLG2 Assay Mix. Record light emission for 1 s after a 1 s delay, record the mean test values after subtraction of the instrument blank. A representative standard curve is shown in Fig. 2. The data were analyzed using the regression function of the Analysis ToolPack of Microsoft Excel version 14.0. To detect statistical correlations, Pearson's correlation coefficient (*r*) was calculated between the light emission and moles of ATP. The associated

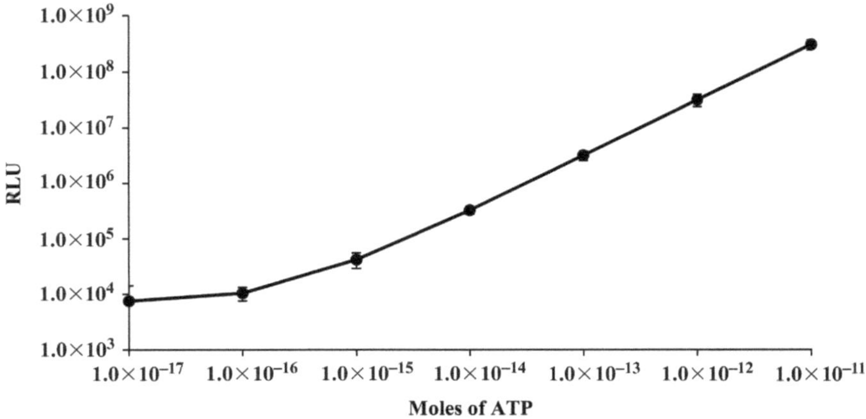

Fig. 2 ATP standard curve generated from the bioluminescent signals with PLG2 Assay Mix (Section 2.5). The mean relative light units (RLUs) for 1 s integrals from three independent experiments are plotted as a function of ATP levels and the standard deviations are indicated by error bars. The correlation coefficient (*r*) for the portion of the curve from 1×10^{-16} to 1×10^{-11} mol of ATP is 0.999 and the associated significance is $P < 0.001$. Source: *Branchini, B. R., Southworth, T. L., Fontaine, D. M., Kohrt, D., Talukder, M., Michelini, E.,…Grossel, M. J. (2015). An enhanced chimeric firefly luciferase-inspired enzyme for ATP detection and bioluminescence reporter and imaging applications.* Analytical Biochemistry, 484, *148–153. doi:10.1016/j.ab.2015.05.020.*

probability (*P*) was determined using the Student *t*-distribution function of Excel.

3. User samples should be assayed on the same day that the standard curve is constructed.

2.6 Viable Cell Quantitation Protocol

1. Generate an ATP standard curve (Section 2.5) by preparing the ATP solutions in 50 m*M* Tricine buffer, pH 8 containing 10% PBS.
2. We validated this application by demonstrating that ATP levels correlated with cell number using HEK293T cells (Fig. 3). Aliquots of cell suspensions were washed twice with PBS and counted using a TC10 automated cell counter (Bio-Rad). Triplicate cell suspensions (200,000 cells in 0.1 mL of PBS) were serially diluted in PBS and lysed by five cycles of freezing at −80°C and thawing at 30°C. The lysed cell suspensions were diluted 10-fold in 50 m*M* Tricine buffer, pH 8, and aliquots (0.05 mL) were assayed in triplicate as described in Section 2.5 step 2.

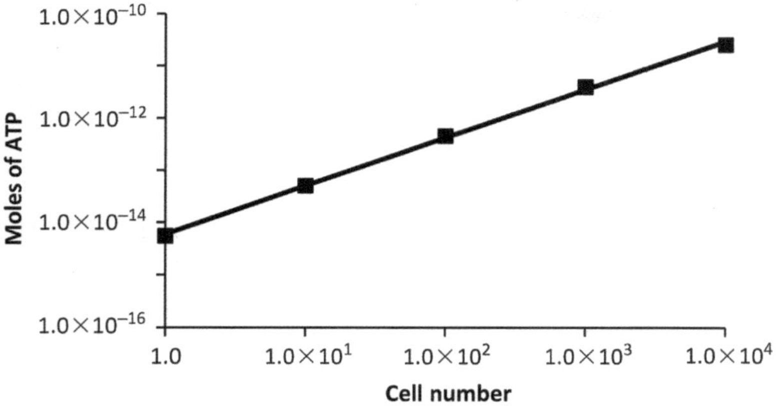

Fig. 3 Correlation between HEK293T cell numbers and ATP levels. Using the protocol described in Section 2.6, a standard ATP curve was made and used to measure actual ATP content in the mammalian cells. Bioluminescence was measured with the PLG2 Assay Mix in experiments that were performed three times with individual assays performed in triplicate. Symbols represent mean values, and the line represents the linear data fit. Error bars indicate the standard deviations among replicates. Pearson's correlation coefficient (r) and the associated significance (P) value were: $r=0.999$, $P<0.001$. Source: *Branchini, B. R., Southworth, T. L., Fontaine, D. M., Kohrt, D., Talukder, M., Michelini, E., . . . Grossel, M. J. (2015). An enhanced chimeric firefly luciferase-inspired enzyme for ATP detection and bioluminescence reporter and imaging applications.* Analytical Biochemistry, 484, *148–153. doi:10.1016/j.ab.2015.05.020.*

3. CONCLUSIONS

3.1 PLG2-Based ATP Analysis

A rapid and simple method for ATP assays using PLG2 has been described that, based on the standard curve provided in Fig. 2, provides linearity over six orders of magnitude with a detection limit of $\sim 1 \times 10^{-16}$mol of ATP. The method provides better sensitivity than the femtomole detection capabilities of the Promega ENLITEN® kit and represents detection of ATP at a level limited only by instrument noise. The sensitivity improvement that is possible with PLG2 enables using less enzyme to reach femtomole detection. Moreover, the enhanced stability of PLG2 should provide performance advantages compared to currently available standard luciferases. While results in individual labs may vary, especially depending on the type of light measuring device that is used, the detection of ATP is straightforward and feasible with a stable, inexpensive single assay reagent, and a basic detection device. Additionally, the methodology can be used to detect viable mammalian and bacterial cells (Branchini et al., 2015). Using mammalian cells, light intensity and cell number were proportional over four orders of magnitude, from 1 to 10,000 mammalian cells (Fig. 3, $r=0.999$, $P<0.001$). Using a standard curve similar to the one in Fig. 2, we determined that there were 5.5 ± 0.4 fmol of ATP in a single HEK293T cell, in very good agreement with the reported value of 2–6 fmol ATP per viable mammalian cell (Sonderhoff, Kilburn, & Piret, 1992), thus validating our protocol as we had previously done with bacterial cells (Branchini et al., 2015).

ACKNOWLEDGMENTS

We acknowledge the financial support of the Air Force Office of Scientific Research under Grant FA9550-14-1-0100, the National Science Foundation under MCB 1410390, and the Hans & Ella McCollum '21 Vahlteich Endowment. We thank Aldo Roda, Martha J. Grossel, Danielle M. Fontaine, Dawn Kohrt, MunyaTalukder, Demetria DeBartolo, Elisa Michelini, Luca Cevenini, Lance Encell, and Mary Hall for their highly valued contributions.

REFERENCES

Afshari, A., Uhde-Stone, C., & Lu, B. A. (2015). A cooled CCD camera-based protocol provides an effective solution for in vitro monitoring of luciferase. *Biochemical and Biophysical Research Communications*, *458*(3), 543–548. http://dx.doi.org/10.1016/j.bbrc.2015.01.150.

Ando, Y., Niwa, K., Yamada, N., Enomot, T., Irie, T., Kubota, H., … Akiyama, H. (2008). Firefly bioluminescence quantum yield and colour change by pH-sensitive green emission. *Nature Photonics*, *2*(1), 44–47.

Binkowski, B. F., Butler, B. L., Stecha, P. F., Eggers, C. T., Otto, P., Zimmerman, K., … Wood, K. V. (2011). A luminescent biosensor with increased dynamic range for intracellular cAMP. *ACS Chemical Biology*, *6*(11), 1193–1197. http://dx.doi.org/10.1021/cb200248h.

Bottcher, R. T., Pollet, N., Delius, H., & Niehrs, C. (2004). The transmembrane protein XFLRT3 forms a complex with FGF receptors and promotes FGF signalling. *Nature Cell Biology*, *6*(1), 38–44. http://dx.doi.org/10.1038/Ncb1082.

Branchini, B. R., Southworth, T. L., DeAngelis, J. P., Roda, A., & Michelini, E. (2006). Luciferase from the Italian firefly Luciola italica: Molecular cloning and expression. *Comparative Biochemistry and Physiology. Part B, Biochemistry & Molecular Biology*, *145*(2), 159–167.

Branchini, B. R., Southworth, T. L., Fontaine, D. M., Davis, A. L., Behney, C. E., & Murtiashaw, M. H. (2014). A Photinus pyralis and Luciola italica chimeric firefly luciferase produces enhanced bioluminescence. *Biochemistry*, *53*(40), 6287–6289. http://dx.doi.org/10.1021/bi501202u.

Branchini, B. R., Southworth, T. L., Fontaine, D. M., Kohrt, D., Talukder, M., Michelini, E., … Grossel, M. J. (2015). An enhanced chimeric firefly luciferase-inspired enzyme for ATP detection and bioluminescence reporter and imaging applications. *Analytical Biochemistry*, *484*, 148–153. http://dx.doi.org/10.1016/j.ab.2015.05.020.

Cevenini, L., Camarda, G., Michelini, E., Siciliano, G., Calabretta, M. M., Bona, R., … Alano, P. (2014). Multicolor bioluminescence boosts malaria research: Quantitative dual-color assay and single-cell imaging in Plasmodium falciparum parasites. *Analytical Chemistry*, *86*(17), 8814–8821. http://dx.doi.org/10.1021/ac502098w.

Che, P. L., Cui, L., Kutsch, O., Cui, L. W., & Li, Q. J. (2012). Validating a firefly luciferase-based high-throughput screening assay for antimalarial drug discovery. *Assay and Drug Development Technologies*, *10*(1), 61–68. http://dx.doi.org/10.1089/adt.2011.0378.

Etzov, E., Cohen, A., & Marks, R. S. (2015). Bioluminescent liquid light guide pad biosensor for indoor air toxicity monitoring. *Analytical Chemistry*, *87*(7), 3655–3661. http://dx.doi.org/10.1021/ac5038208.

Gould, S. J., & Subramani, S. (1988). Firefly luciferase as a tool in molecular and cell biology. *Analytical Biochemistry*, *175*, 5–13.

Hakkila, K., Maksimow, M., Karp, M., & Virta, M. (2002). Reporter genes lucFF, luxCDABE, GFP, and DsRed have different characteristics in whole-cell bacterial sensors. *Analytical Biochemistry*, *301*(2), 235–242.

Hattori, M., & Ozawa, T. (2015). High-throughput live cell imaging and analysis for temporal reaction of G protein-coupled receptor based on split luciferase fragment complementation. *Analytical Sciences*, *31*(4), 327–330.

Kuo, M. S., Auriau, J., Pierre-Eugene, C., & Issad, T. (2014). Development of a human breast-cancer derived cell line stably expressing a bioluminescence resonance energy transfer (BRET)-based phosphatidyl inositol-3 phosphate (PIP3) biosensor. *PloS One*, *9*(3). http://dx.doi.org/10.1371/journal.pone.0092737.

Leclerc, G. M., Boockfor, F. R., Faught, W. J., & Frawley, L. S. (2000). Development of a destabilized firefly luciferase enzyme for measurement of gene expression. *BioTechniques*, *29*(3), 590–601.

Lewis, M. D., Francisco, A. F., Taylor, M. C., Burrell-Saward, H., McLatchie, A. P., Miles, M. A., & Kelly, J. M. (2014). Bioluminescence imaging of chronic Trypanosoma cruzi infections reveals tissue-specific parasite dynamics and heart disease in the absence of locally persistent infection. *Cellular Microbiology*, *16*(9), 1285–1300. http://dx.doi.org/10.1111/cmi.12297.

Lundin, A. (1993). Optimized assay of firefly luciferase with stable light emission. In A. Szalay, L. J. Kricka, & P. E. Stanley (Eds.), *Bioluminescence and chemiluminescence: Status report* (pp. 291–295). Chichester: John Wiley and Sons.

Paddison, P. J., Caudy, A. A., & Hannon, G. J. (2002). Stable suppression of gene expression by RNAi in mammalian cells. *Proceedings of the National Academy of Sciences of the United States of America*, *99*, 1443–1448.

Promega. (2009). ENLITEN ATP assay system bioluminescence detection kit for ATP measurement. Madison, WI, USA.

Roda, A. (2011). Chemiluminescence and bioluminescence: Past, present and future. Cambridge, UK: Royal Society of Chemistry.

Roda, A., Guardigli, M., Michelini, E., & Mirasoli, M. (2009). Bioluminescence in analytical chemistry and in vivo imaging. *TrAC Trends in Analytical Chemistry*, *28*(3), 307–322.

Roda, A., Mirasoli, M., Dolci, L. S., Buragina, A., Bonvicini, F., Simoni, P., & Guardigli, M. (2011). Portable device based on chemiluminescence lensless imaging for personalized diagnostics through multiplex bioanalysis. *Analytical Chemistry*, *83*(8), 3178–3185. http://dx.doi.org/10.1021/ac200360k.

Shi, J. W., Zhang, H., Fang, L. R., Xi, Y. Q., Zhou, Y. R., Luo, R., … Chen, H. C. (2014). A novel firefly luciferase biosensor enhances the detection of apoptosis induced by ESAT-6 family proteins of Mycobacterium tuberculosis. *Biochemical and Biophysical Research Communications*, *452*(4), 1046–1053. http://dx.doi.org/10.1016/j.bbrc.2014.09.047.

Sonderhoff, S. A., Kilburn, D. G., & Piret, J. M. (1992). Analysis of mammalian viable cell biomass based on cellular ATP. *Biotechnology and Bioengineering*, *39*(8), 859–864. http://dx.doi.org/10.1002/Bit.260390807.

Taylor, M. C., & Kelly, J. M. (2014). Optimizing bioluminescence imaging to study protozoan parasite infections. *Trends in Parasitology*, *30*(4), 161–162. http://dx.doi.org/10.1016/j.pt.2014.01.006.

Thorne, N., Shen, M., Lea, W. A., Simeonov, A., Lovell, S., Auld, D. S., & Inglese, J. (2012). Firefly luciferase in chemical biology: A compendium of inhibitors, mechanistic evaluation of chemotypes, and suggested use as a reporter. *Chemistry & Biology*, *19*(8), 1060–1072. http://dx.doi.org/10.1016/j.chembiol.2012.07.015.

Urban, A., Eckermann, S., Fast, B., Metzger, S., Gehling, M., Ziegelbauer, K., … Freiberg, C. (2007). Novel whole-cell antibiotic biosensors for compound discovery. *Applied and Environmental Microbiology*, *73*(20), 6436–6443.

Vannini, A., Agriesti, F., Mosca, F., Roncarati, D., Scarlato, V., & Danielli, A. (2012). A convenient and robust in vivo reporter system to monitor gene expression in the human pathogen Helicobacter pylori. *Applied and Environmental Microbiology*, *78*(18), 6524–6533. http://dx.doi.org/10.1128/aem.01252-12.

Widder, E. A., & Falls, B. (2014). Review of bioluminescence for engineers and scientists in biophotonics. *IEEE Journal of Selected Topics in Quantum Electronics*, *20*(2), 232–241. http://dx.doi.org/10.1109/jstqe.2013.2284434.

Yoon, S. M., Namkung, W., & Lee, J. (2013). A comparison of Ypet and firefly luciferase as reporter proteins for high-throughput screening. *Bioscience Biotechnology and Biochemistry*, 77(11), 2328–2330. http://dx.doi.org/10.1271/bbb.130537.

CHAPTER THIRTEEN

Highly Modular Bioluminescent Sensors for Small Molecules and Proteins

Qiuliyang Yu*,†,‡, Rudolf Griss*,†,‡, Alberto Schena§, Kai Johnsson*,†,‡,1

*École Polytechnique Fédérale de Lausanne (EPFL), Institute of Chemical Sciences and Engineering (ISIC), Lausanne, Switzerland
†École Polytechnique Fédérale de Lausanne (EPFL), Institute of Bioengineering, Lausanne, Switzerland
‡National Centre of Competence in Research (NCCR) Chemical Biology, Lausanne, Switzerland
§Lucentix SA, Lausanne, Switzerland
[1]Corresponding author: e-mail address: kai.johnsson@epfl.ch

Contents

Abstract

Obtaining patient-specific information through the quantification of small molecules and proteins in bodily fluids is essential for personalized therapies. Point-of-care (POC) diagnostic devices hold the promise of delivering such benefit to a wide range of patients. However, there is a lack of enabling technology, as the majority of newly developed POC devices focus on the same underlying core technologies. Here we provide an overview of a new technology based on highly modular bioluminescent sensors that enables the quantification of small molecules and proteins at the POC with low-cost devices.

Methods in Enzymology, Volume 589
ISSN 0076-6879
http://dx.doi.org/10.1016/bs.mie.2017.01.005

1. INTRODUCTION

The sensing of small molecules and proteins plays an important role in healthcare. The patient's metabolic profile constitutes a major part of the person-specific information needed to deliver and improve personalized medicine. For example, many therapeutic drugs require the monitoring of their concentration in the patient's blood in order to assure effectiveness but prevent toxicity (Lennard, 2001). Another example is the detection and quantification of proteins, which is essential in the early diagnosis, treatment, and management of many diseases (Rifai, Gillette, & Carr, 2006).

Today, the detection of small molecules and proteins mainly relies on processes requiring sophisticated instruments and sample treatment such as liquid chromatography coupled with mass spectrometry (Deglon, Thomas, Mangin, & Staub, 2012), fluorescence polarization immunoassay (FPIA) (Roberts, Norris, Paterson, & Martin, 2012), and enzyme-linked immunosorbent assay (Khanna et al., 2013). Though sensitive and accurate, such methods require centralized laboratories and specialized personnel. The associated high cost and low availability are a bottleneck for delivering the benefits of personalized therapy to a wide range of patients.

Point-of-care (POC) diagnostic devices hold the promise of overcoming this barrier (Gubala, Harris, Ricco, Tan, & Williams, 2012). Multiple complex analyses could be integrated on one single device, all while keeping manufacturing costs low. Even though technologies enabling measurements in resource-limited settings have seen considerable development in recent years, for the most part they remain qualitative or semiquantitative. Most importantly, the big majority of newly developed methods still rely on the same handful of core technologies such as immunoassays, electrochemical sensors, enzymatic assays, or DNA-based assays; often in new formats such as microfluidic devices (St John & Price, 2014; Pollock et al., 2012; Wang et al., 2012). To answer the need for new core technologies that enable quantitative POC devices, our laboratory has developed a class of highly modular biosensors called **LUC**iferase-based **I**ndicators of **D**rugs (LUCIDs) (Griss et al., 2014; Schena, Griss, & Johnsson, 2015).

1.1 Principle of LUCIDs

LUCIDs are semisynthetic bioluminescent sensors consisting of a receptor protein, a luciferase, and a synthetic compound containing a fluorophore and a ligand specific to the receptor. The synthetic compound is covalently attached to the recombinant protein via a self-labeling tag such as SNAP-tag (Keppler et al., 2003). The fluorophore and the luciferase must form a good pair for bioluminescence resonance energy transfer (BRET) with well-separated emission peaks. The BRET pair typically used in LUCIDs consists of the fluorophores Cy3 or tetramethylrhodamine and the luciferase NanoLuc (NLuc) (Hall et al., 2012). The LUCIDs for small molecules contain only one tethered ligand (Fig. 1A), while the sensors for proteins contain a secondary ligand specific to the protein analyte (Fig. 1B).

The sensor undergoes a major conformational shift when the analyte concentration changes. In the absence of analyte, the intramolecularly tethered ligand binds to the receptor protein, leading to a closed conformation of the sensor molecules. At high concentrations, the analyte can displace the intramolecular ligand from the receptor protein and the majority of the sensor molecules adopt an open conformation. In the case of sensing proteins, the analyte does not open the sensor by competing directly with the intramolecular ligand for the same binding site, but rather binds to a secondary ligand and sterically hinders the binding between the primary ligand and the receptor.

BRET is used to translate the conformational change of the sensor construct into an observable signal. For a closed sensor, the fluorophore and the luciferase are close to each other, allowing efficient BRET. As a result, the

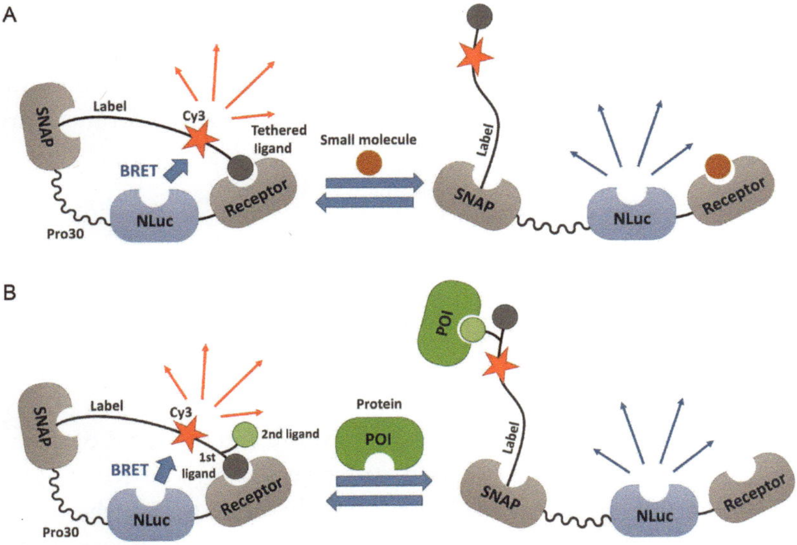

Fig. 1 Cartoon of the LUCIDs for sensing small molecules and proteins. (A) LUCID for small molecules. The protein scaffold SNAP-Pro30-NLuc-Receptor is linked to a synthetic compound through self-labeling protein SNAP tag. The synthetic compound contains a fluorophore (*red star*) and a tethered ligand (*gray ball*). The analyte (*brown ball*) can compete off the tethered ligand and open the sensor. (B) Sensor for proteins. The sensor contains a primary ligand (*gray ball*) specific to the receptor and a secondary ligand (*green ball*) specific to the protein of interest (POI). The POI opens the sensor by binding to the secondary ligand.

emission of the luciferase is weak and that of the fluorophore is dominant. When the sensor is in its open state, the fluorophore and luciferase are far apart, leading to a low BRET efficiency and a strong emission of the luciferase. Thus, the ratio between the emission intensity of the luciferase and the emission intensity of the fluorophore (NLuc/Cy3) can be used as readout of the sensor. Since this readout is ratiometric, the signal is independent of the total emission intensity, making the measurement more robust.

There are two key parameters for sensors based on the LUCID principle: the dynamic range and the analyte response range. The dynamic range is the degree of signal change going from low to saturating analyte concentrations while the analyte response range describes over which analyte concentrations this change occurs. The center of the analyte response range is indicated by the sensor's c_{50}—the concentration of analyte that triggers a half-maximal sensor response.

2. SENSING SMALL MOLECULES

Our laboratory has used the above-described principle to develop sensors for various small molecules. While there are general design principles, the detailed development has to be adapted to each analyte and the available receptor proteins. Importantly, the dynamic range and the analyte response range have to be optimized individually for each new sensor. To illustrate these aspects, the development and optimization of a sensor for methotrexate are explained in detail below.

2.1 LUCID Sensor for Methotrexate

Methotrexate (Fig. 2A) is used in chemotherapy against cancer and treatment of autoimmune diseases. A high dose of methotrexate can induce damage in organs such as the liver and the kidneys. To prevent excessive toxicity, the serum level of methotrexate is routinely monitored during treatment (Lennard, 2001).

Methotrexate targets dihydrofolate reductase (DHFR), an extensively studied protein with known crystal structure (Sawaya & Kraut, 1997). For a first version of the sensor, DHFR was therefore used as the receptor protein and the known DHFR inhibitor trimethoprim (TMP) was used as a tethered ligand. A fusion protein consisting of SNAP-tag, a 30-proline linker (Pro30), NanoLuc luciferase, and DHFR (SNAP-Pro30-NLuc-DHFR) (Fig. 2D) was thus produced and labeled with a synthetic molecule consisting of an O^6-benzylguanine (BG) group, an ethylene glycol linker, the fluorophore Cy3, and trimethoprim (BG-Cy3-TMP) (Fig. 2B).

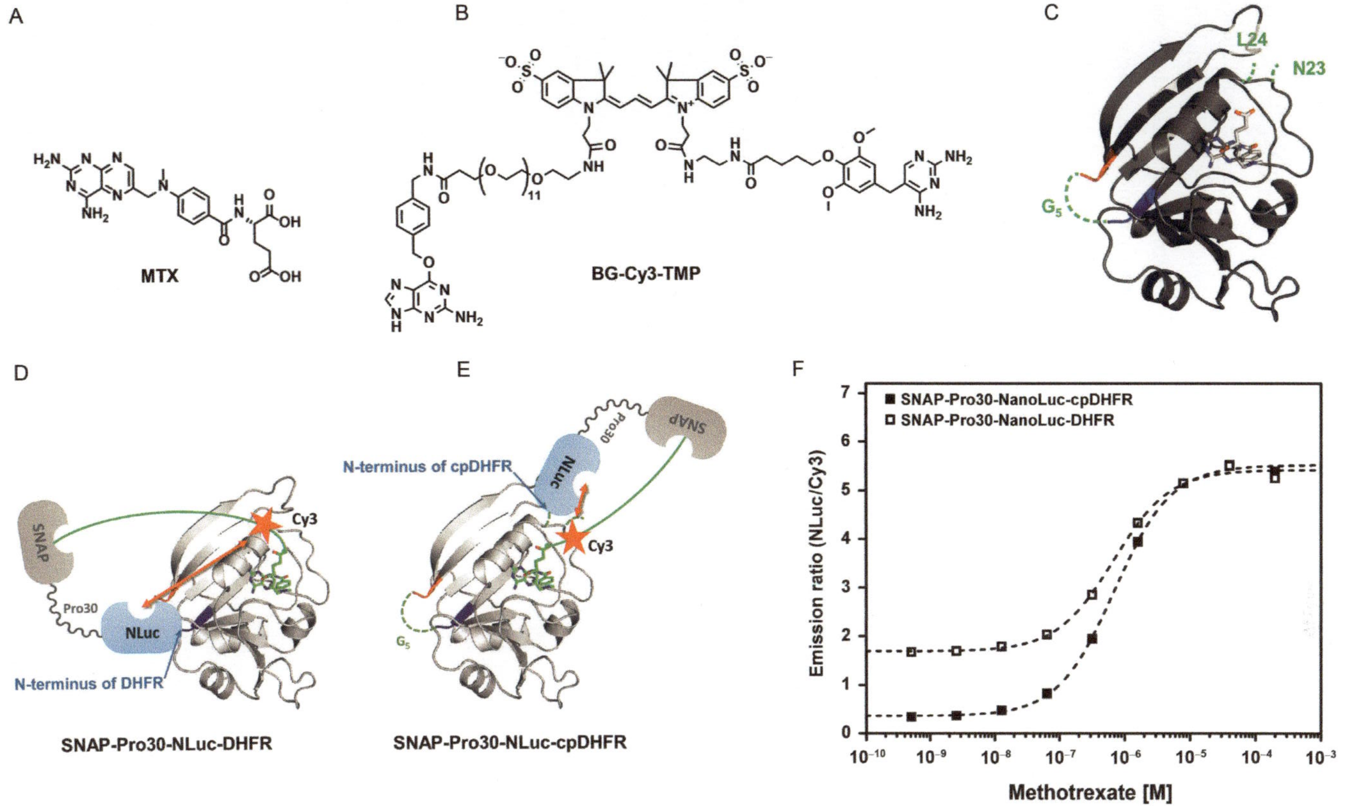

Fig. 2 See legend on next page.

The resulting sensor responded to methotrexate with a dynamic range of 218 ± 6% (Fig. 2F).

In order to achieve a high dynamic range, the BRET efficiency has to be as low as possible in the sensor's open state and as high as possible in its closed state. In the open state, a rigid 30-proline linker between SNAP-tag and the luciferase ensures a large distance between the fluorophore and the luciferase (Brun et al.,. 2011). In the closed state, where the intramolecular ligand is bound to the receptor protein, the two BRET partners have to be brought close together. Since the fluorophore is attached to the intramolecular ligand, the luciferase has to be attached close to the binding site on the receptor protein. Judging from the crystal structure of DHFR, neither the N-terminus nor the C-terminus is close to the active site (Fig. 2C). As a result, when DHFR is used as the receptor, the distance between the two BRET partners is still large when the sensor is closed and BRET efficiency is relatively low (Fig. 2D). To solve this problem, we circularly permuted the protein by linking the native termini with a 5-glycine linker and by splitting the amino acid chain between residues Asn23 and Leu24, in this way creating new protein termini close to the active site (Fig. 2C and E). When the resulting circularly permutated DHFR (cpDHFR) was used as a receptor protein, the sensor's dynamic range increased more than sixfold to 1340 ± 90% (Fig. 2F). The sensors using DHFR and cpDHFR have a similar open state ratio; however, SNAP-Pro30-NLuc-cpDHFR shows a much lower ratio in the absence of methotrexate, indicating a more efficient BRET in the closed state.

As the sensor is based on the competition between the free analyte and the tethered ligand, the c_{50} (the analyte concentration giving a half-maximal sensor response) and thus the analyte response range can be tuned by using

Fig. 2 Design and optimization of the LUCID for methotrexate. (A) Structure of methotrexate. (B) Structure of the synthetic compounds BG-Cy3-TMP. (C) Structure of DHFR bound to methotrexate. The position where circular permutation was performed is shown in *green*. The native termini were linked with a 5-glycine linker and new termini were created between residues Asn23 and Leu24. (D) Cartoon of the LUCID for methotrexate with DHFR as receptor. The distance between the BRET partners *(red arrow)* is large. (E) When cpDHFR is used as receptor, the distance between the BRET partners is decreased. (F) Titration of the two LUCIDs for methotrexate. The sensor with cpDHFR demonstrates a much higher BRET efficiency in the closed state at low methotrexate concentration. *Reproduced from Griss, R., Schena, A., Reymond, L., Patiny, L., Werner, D., Tinberg, C. E., …, Johnsson, K. (2014). Bioluminescent sensor proteins for point-of-care therapeutic drug monitoring.* Nature Chemical Biology, 10*(7), 598–603. http://dx.doi.org/10.1038/Nchembio.1554 with the authors' permission.*

tethered ligands with different binding affinities. A stronger affinity leads to an increased c_{50}, as higher concentrations of free analyte are required to compete off the tethered ligand and open the sensor. By changing TMP to the stronger ligand methotrexate (Fig. 3A), we have shifted the sensor's c_{50} from 0.75 ± 0.04 to 85 ± 6 μM. With these two tethered ligands, the sensor can cover most of the concentrations of methotrexate that occur in serum during therapy (Fig. 3B).

For the sensor to be used in patient samples, its specificity has to be confirmed. The metabolites of methotrexate have been reported to interfere with existing immunoassays based on antibodies (Al-Turkmani, Law, Narla, & Kellogg, 2010). The sensor was titrated with 7-hydroxy methotrexate and 4-amino-4-deoxy-*N*-methylpteroic acid (DAMPA), two metabolites that can reach high concentrations in human serum (Fig. 4A). We found that the c_{50} values of the sensor for the two metabolites are more than 100-fold higher than that for methotrexate (Fig. 4B). As a result, the measurement of methotrexate will not be affected by the metabolites in biologically relevant samples.

2.2 Other LUCID Sensors for Small Molecules

The recognition between receptor protein and analyte governs the specificity of the sensor. By using suitable receptors and ligands, we have developed sensors for the following analytes: topiramate (antiepileptic agent) (Johannessen et al., 2003), tacrolimus (immunosuppressant) (Kahan, 2002), sirolimus (immunosuppressant) (MacDonald, Scarola, Burke, & Zimmerman, 2000), cyclosporin A (immunosuppressant) (Armstrong & Oellerich, 2001), and digoxin (cardiac glycoside) (Schoemakers, Wasylewicz, Scheepers-Hoeks, van Wezel, & Grouls, 2011). The chosen analytes are widely used drugs for which monitoring of their concentrations in blood is necessary to avoid drug toxicity and ensure efficacy.

Since topiramate is an inhibitor of human carbonic anhydrase II (HCA), the LUCID for topiramate was developed by using HCA as the receptor in the protein scaffold SNAP-Pro30-NLuc-HCA. Another inhibitor, benzene sulfonamide, was used as the tethered ligand. The resulting sensor shows a ratio change of $591 \pm 12\%$ and a c_{50} of 7.3 ± 13 μM, suitable for measuring topiramate in the relevant therapeutic range.

The immunosuppressants tacrolimus and sirolimus target the protein FKBP12. This protein was hence used as the receptor in the LUCID for tacrolimus. The active site of the protein is far away from both the

A

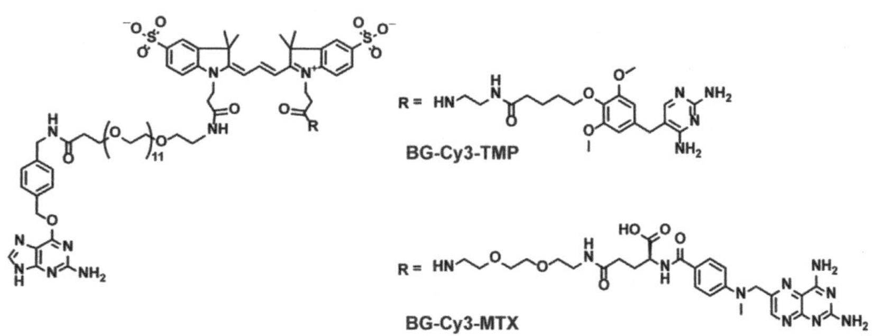

B

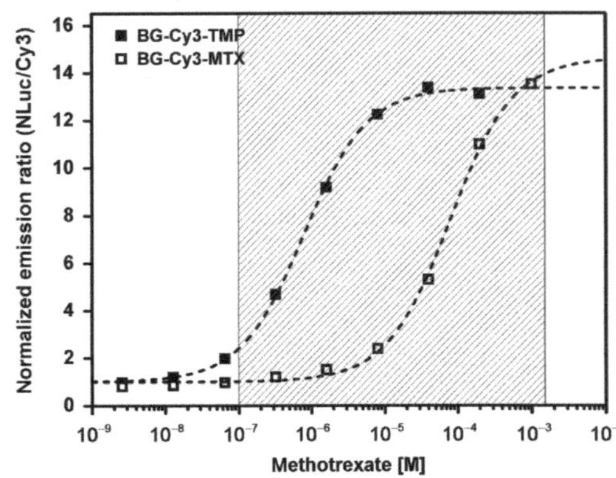

Fig. 3 Tuning the analyte response range of the LUCID for methotrexate. (A) Structure of the synthetic compounds BG-Cy3-TMP and BG-Cy3-MTX. (B) Titrations of the sensors consisting of the protein scaffold SNAP-Pro30-NLuc-cpDHFR labeled with either BG-Cy3-TMP or BG-Cy3-MTX. The sensors can cover most of the methotrexate concentrations that occur in serum during therapy. *Reproduced from Griss, R., Schena, A., Reymond, L., Patiny, L., Werner, D., Tinberg, C. E., …, Johnsson, K. (2014). Bioluminescent sensor proteins for point-of-care therapeutic drug monitoring.* Nature Chemical Biology, 10*(7), 598–603. doi: 10.1038/Nchembio.1554 with the authors' permission.*

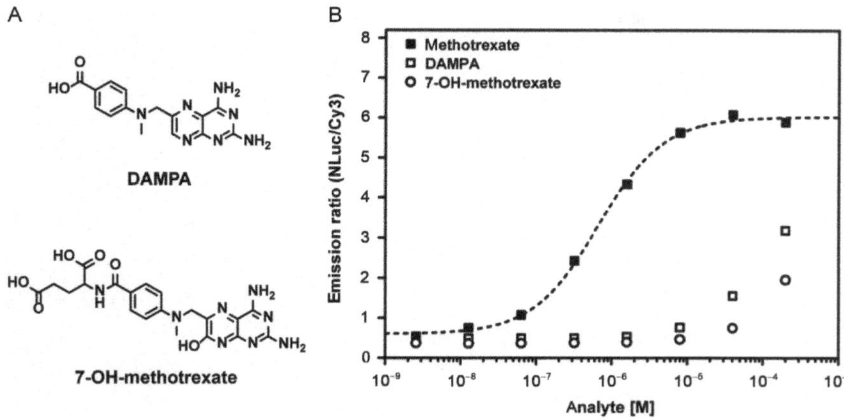

Fig. 4 Sensor performance in the presence of methotrexate metabolites. (A) Chemical structure of the methotrexate metabolites DAMPA and 7-OH-methotrexate. (B) Titration of the sensor with methotrexate and its metabolites. The c_{50} for the metabolites are more than 100-fold higher than for methotrexate. *Reproduced from Griss, R., Schena, A., Reymond, L., Patiny, L., Werner, D., Tinberg, C. E., …, Johnsson, K. (2014). Bioluminescent sensor proteins for point-of-care therapeutic drug monitoring.* Nature Chemical Biology, 10*(7), 598–603. http://dx.doi.org/10.1038/Nchembio.1554 with the authors' permission.*

N- and C-termini. However, the protein has a secondary binding site near the N-terminus (Rohrig, Loch, Guan, Siegal, & Overhand, 2007). To ensure a high BRET efficiency, the tethered ligand was designed to span both binding sites on FKBP12, and the moiety recognizing the secondary site was placed closer to the fluorophore. The resulting LUCID responds to tacrolimus with a ratio change of 460 ± 30% and a c_{50} of 17 ± 3 n*M*; and to sirolimus with a ratio change of 480 ± 30% and a c_{50} of 13.7 ± 1.6 n*M*.

The human protein cyclophilin A was used to develop the LUCID for its ligand cyclosporin A, another important immunosuppressant. Cyclophilin A was circularly permutated to create new termini at Arg148 and Asn149, which are close to the binding site. A tethered ligand with suitable affinity was also developed to allow the sensor to operate in the relevant concentration range. The sensor displays a ratio change of 192 ± 5% with a c_{50} of 500 ± 90 n*M*.

The LUCID for digoxin was developed using a computationally designed binding protein, DIG 10.3 (Tinberg et al., 2013). As its C-terminus is close to the binding site, the receptor was fused to the N-terminus of NLuc in the scaffold DIG10.3-NLuc-Pro30-SNAP. Progesterone was used as the tethered ligand. The resulting LUCID has a ratio change of 458 ± 11% and a c_{50} of 22 ± 5 n*M*, suitable for detecting toxic digoxin levels above 5 n*M* (Terra, Washam, Dunham, & Gattis, 1999).

3. SENSING PROTEINS

The modularity of the design principle allowed our laboratory to develop sensors not only for small molecules but also for proteins. To extend the LUCID principle to proteins, we developed a technology called chemical ligand-associated steric hindrance (CLASH) (Schena et al., 2015). In the synthetic part of the sensor, two ligands, primary and secondary, are attached next to the fluorophore in close proximity to each other. The primary ligand binds to the receptor in the sensor scaffold while the secondary ligand recognizes the protein analyte. The initially closed sensor opens at high analyte concentrations, as the analyte binds to the secondary ligand and sterically hinders the binding between the primary ligand and the receptor (Fig. 1B). Two sensors for protein analytes, streptavidin and DHFR, are described below.

3.1 CLASH Sensor for Streptavidin

The first proof-of-principle CLASH sensor developed in our laboratory was against streptavidin as an analyte. Streptavidin is commonly used in chemical biology because of its strong affinity for biotin (Biot). We used a previously developed LUCID based on human carbonic anhydrase II (HCA) as receptor protein and sensitized it to streptavidin by introducing a secondary ligand (Fig. 5A). The LUCID's intramolecular ligand, benzenesulfonamide (SA), was kept as the primary ligand in the synthetic part of the sensor. Biotin was used as the secondary ligand and attached close to the primary ligand through organic synthesis. The resulting compound BG-Cy3-SA/Biot (Fig. 5B) was used to label the protein scaffold SNAP-Pro30-NLuc-HCA to obtain a functional sensor. The sensor is initially closed due to the binding between the primary ligand SA and HCA. When streptavidin binds to the secondary ligand biotin, SA is sterically prevented from binding to HCA and the sensor opens, showing a ratio change of $640 \pm 50\%$. No response was observed for a control sensor lacking biotin as secondary ligand (Fig. 5C).

The CLASH principle can be used not only to detect the protein analyte using BRET but also to switch an enzyme on and off. We demonstrated this on the esterase activity of HCA, which can be easily monitored using a chromogenic assay. In the closed form of the sensor, SA inhibits HCA by binding to the active site, leading to low enzymatic activity. With sufficient concentrations of streptavidin, SA is displaced from HCA, thereby activating the enzyme (Fig. 5D).

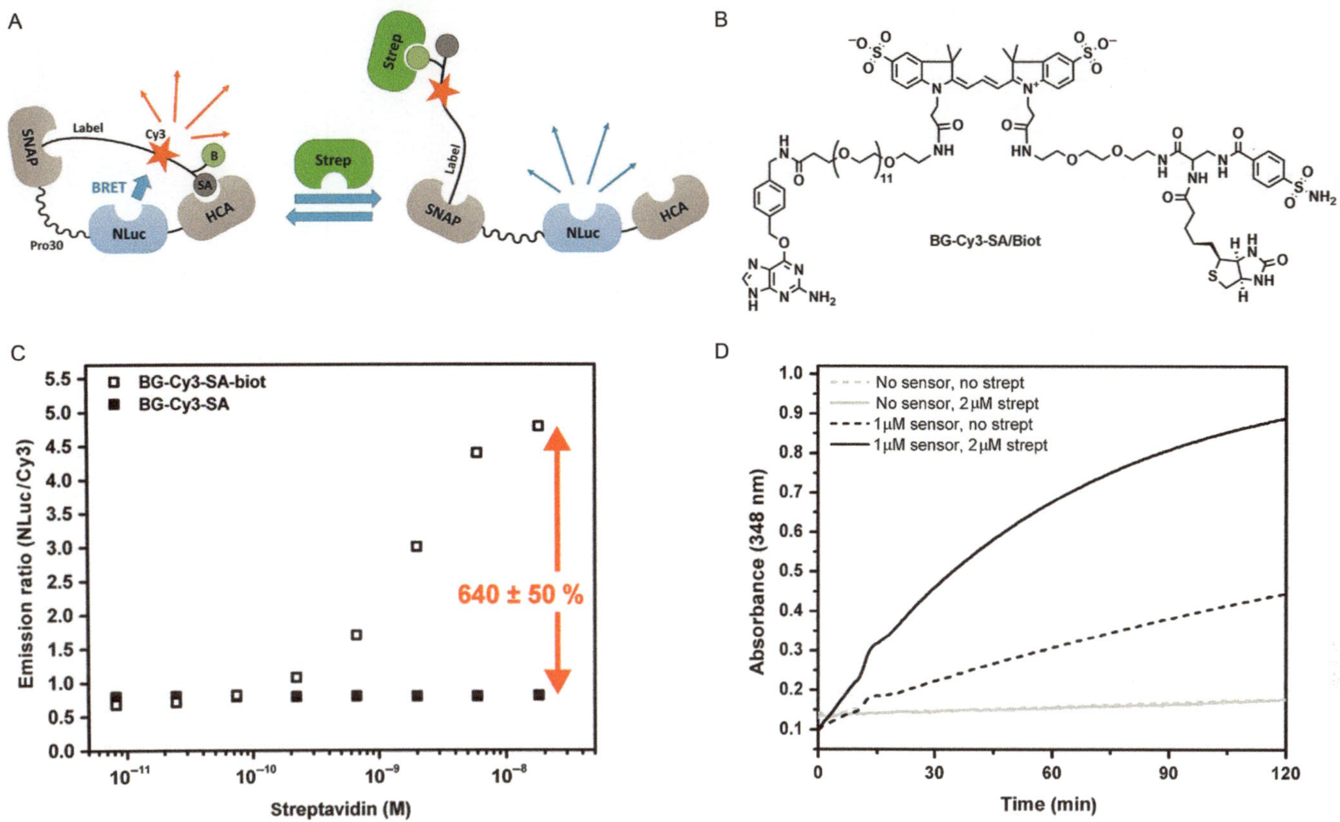

Fig. 5 See legend on next page.

3.2 CLASH Sensor for DHFR

The sensor's specificity is governed by the recognition between the secondary ligand and the protein analyte. To demonstrate the modularity of the principle, we transformed the sensor described earlier into a DHFR sensor by simply changing the original secondary ligand to the DHFR ligand trimethoprim (Fig. 6A and B). As expected, the sensor opens when a high concentration of DHFR is present, as observed by an increased emission ratio (Fig. 6C).

An attractive feature of the CLASH principle is that the geometry of the sensor construct does not have to be optimized anymore, since the specificity can be switched simply by changing the secondary ligand. This versatile modular principle to develop sensors for proteins thus has the potential to generate a multitude of sensors for various proteins with known ligands.

4. TOWARD APPLICATION IN POC SETTINGS

The ratiometric, bioluminescent LUCIDs are attractive starting points for POC applications as they considerably simplify the readout device compared to conventional methods. The ratiometric signal of the sensor is independent of both the sensor concentration and the signal intensity. The use of bioluminescence simplifies optical detection and does not depend on external light sources for excitation. In addition, the bioluminescent readout bypasses the fluorescent background of sample matrices that interferes with conventional fluorescent assays. Our laboratory has demonstrated the feasibility of developing POC devices using the LUCID for methotrexate.

The therapeutic drug monitoring of methotrexate is performed on serum samples. However, light-absorbing components in serum such as bilirubin can interfere with the bioluminescent signal of the sensor. As can be seen from a titration of the sensor in serum with and without 10 μ*M* spiked

Fig. 5 Design and function of the sensor for streptavidin. (A) Cartoon of the sensor for streptavidin based on chemical ligand-associated steric hindrance (CLASH). (B) Synthetic part of the sensor. The compound contains a primary ligand benzenesulfonamide (SA) for the receptor HCA and a secondary ligand biotin (Biot) for the analyte streptavidin. (C) Titration of the sensor with streptavidin. Only the sensor with both primary and secondary ligand (*white squares*) showed a response. The sensor lacking the secondary ligand biotin was used as a control (*black squares*). (D) Using the CLASH principle as a switch of the receptor protein's enzymatic activity. HCA is activated when 2 μ*M* streptavidin are added. *Reproduced from Schena, A., Griss, R., & Johnsson, K. (2015). Modulating protein activity using tethered ligands with mutually exclusive binding sites.* Nature Communications, 6. *AID 7830. http://dx.doi.org/10.1038/Ncomms8830 with the authors' permission.*

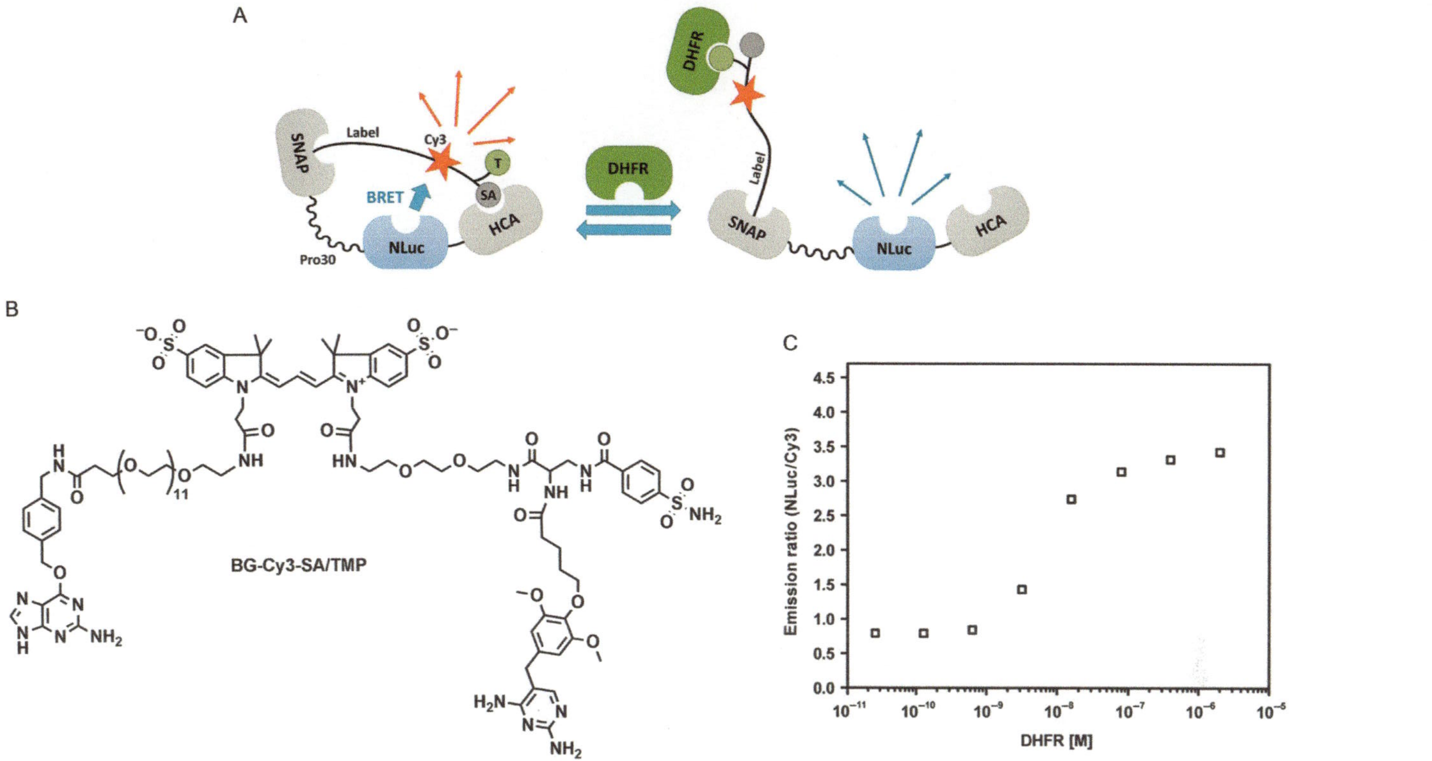

Fig. 6 Design and function of the sensor for DHFR. (A) Cartoon of the sensor for DHFR. Specificity for DHFR is achieved by using trimethoprim (T) as secondary ligand. (B) Chemical structure of the synthetic compound. (C) Response of the sensor to increasing concentrations of DHFR. *Reproduced from Schena, A., Griss, R., & Johnsson, K. (2015). Modulating protein activity using tethered ligands with mutually exclusive binding sites.* Nature Communications, 6. *AID 7830. http://dx.doi.org/10.1038/Ncomms8830 with the authors' permission.*

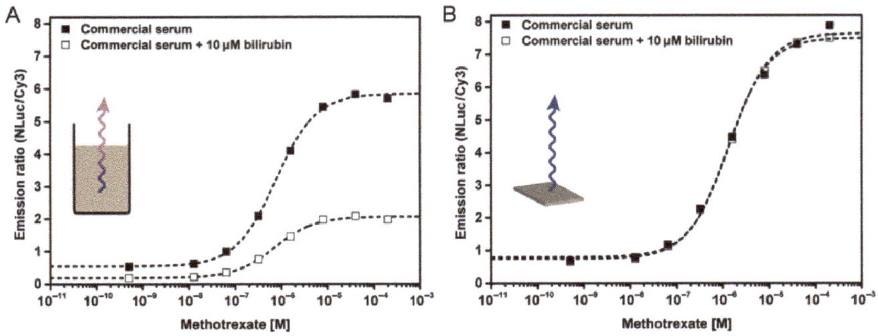

Fig. 7 Removing the interference of bilirubin. (A) Titration of the LUCID for methotrexate in 100 μL human serum with and without 10 *μM* bilirubin. Serum bilirubin affects the emission ratio by absorbing *blue light*. (B) The same titration measured from a thin layer by spotting 5 μL of solution onto cellulose paper. *Reproduced from Griss, R., Schena, A., Reymond, L., Patiny, L., Werner, D., Tinberg, C. E., …, Johnsson, K. (2014). Bioluminescent sensor proteins for point-of-care therapeutic drug monitoring.* Nature Chemical Biology, 10*(7), 598–603. http://dx.doi.org/10.1038/Nchembio.1554 with the authors' permission.*

bilirubin, the absorption of blue light considerably lowers the sensor's emission ratio (Fig. 7A). As a result, the quantification performance of the sensor can be compromised due to the varying concentration of bilirubin in human serum samples.

However, this interference can be eliminated by measuring the sensor from a thin layer. In this format, the titration points with and without bilirubin show similar emission ratios (Fig. 7B). The thin layer reduces the path length and thereby minimizes the light absorption by bilirubin. Meanwhile, the emission of the sensor is not affected. Practically, this can be achieved by spotting 5 μL of the serum samples on a piece of cellulose paper. To facilitate such a measurement, circles can be printed on the paper using wax-based ink. The paper is subsequently heated at 80°C for 2 min to melt the wax and form a hydrophobic barrier to confine the liquid sample (Pollock et al., 2012). In this way, the serum samples form a reproducible thin layer for the measurement.

To demonstrate the potential of LUCIDs in POC diagnostics, we performed measurements using simple point-and-shoot digital cameras for detection of the emitted light. The blue and red color channels of CMOS or CCD RGB image sensors cover the two well-separated emission peaks of the sensor: 460 nm (NLuc) and 570 nm (Cy3). We used the camera to capture pictures of the test strips placed in a homemade black box to exclude surrounding light (Fig. 8A). In the picture of the test strips, the sample glows

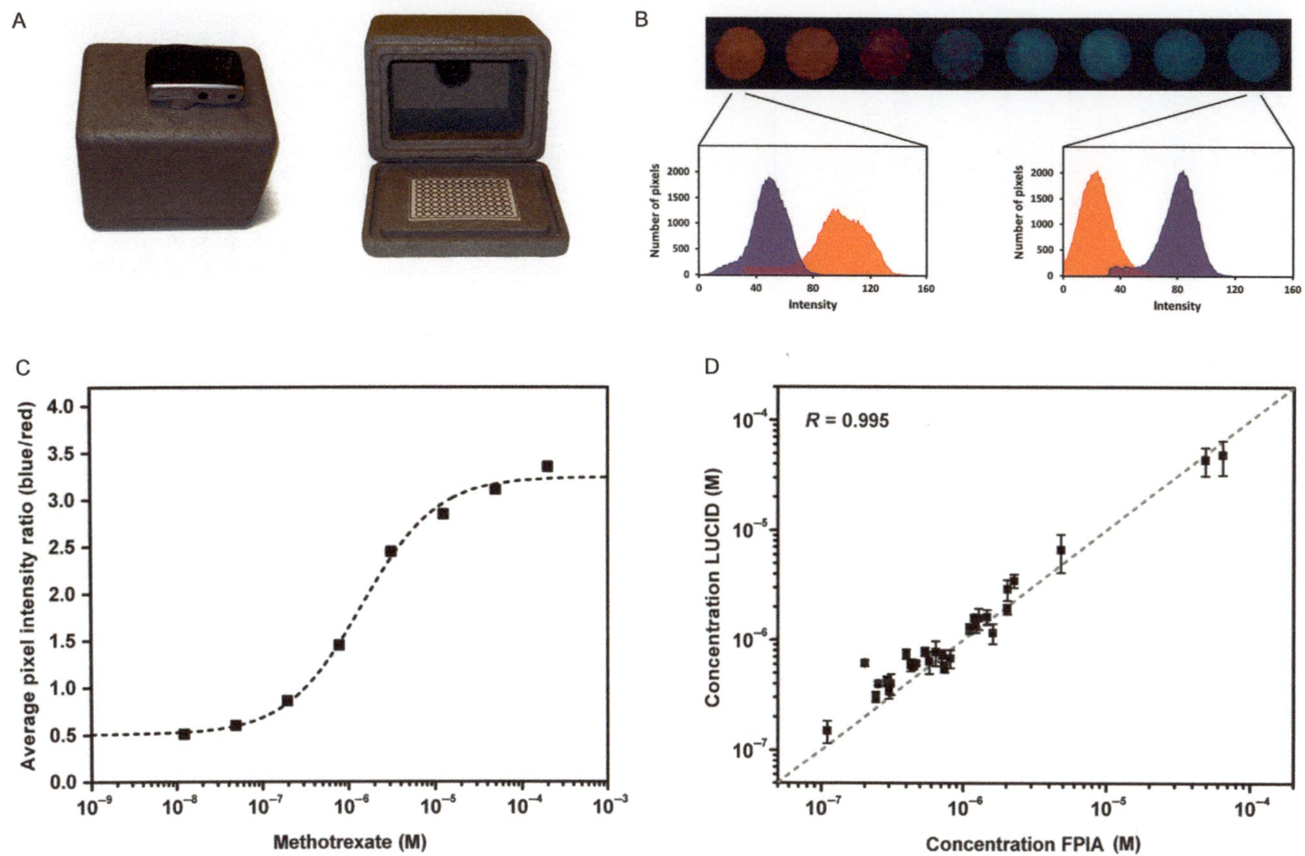

Fig. 8 See legend on next page.

in the wax circle. The color captured from the spots ranges from red to blue with increasing methotrexate concentration. A simple software program was developed to identify the spots and quantify the intensity of red and blue channels (Fig. 8B). The intensity ratios of the two channels were plotted against the respective methotrexate concentrations. The plot was fitted to a single-site binding isotherm and the obtained equation was used to calculate methotrexate level in unknown samples (Fig. 8C).

Thirty serum samples from patients undergoing methotrexate treatment were obtained from the University Hospital of Lausanne (CHUV). To each sample 50 n*M* sensor was added. Five microliters of the mixture was spotted on the paper and measured using the format described earlier. The concentration of methotrexate in patient samples was calculated based on the ratio measured by the camera and the calibration curve. Our result demonstrates a very good correlation between the LUCID and the standard FPIA performed at CHUV (Fig. 8D).

5. CONCLUSIONS

We have demonstrated a general principle to develop semisynthetic bioluminescent sensors for a wide range of small molecules and proteins. The modular design of the sensor makes it possible to easily tune and shift the sensor's specificity and sensitivity by changing the tethered ligand and the receptor protein. LUCIDs are suitable for POC applications thanks to the bioluminescent and ratiometric readout as well as the large dynamic range. Using the sensor for methotrexate as an example, we performed the quantitative measurement of patient samples with a simple, low-cost, and portable device.

Fig. 8 LUCIDs enable the measurement of methotrexate using simple digital cameras. (A) Digital camera mounted on homemade black box containing cellulose paper with wax-based circles. (B) Picture of paper-based test strips containing 5 μL of 50 n*M* LUCID spiked with various methotrexate concentrations. Each spot is 6 mm in diameter. (C) Titration curve obtained by calculating the average pixel intensity ratio of the *red* and *blue color channels* shown in panel (B). (D) Patient samples were measured either using the LUCID or standard fluorescence polarization immunoassay (FPIA). Regression analysis of the two sets of results yielded a Pearson correlation coefficient (R) = 0.995. Error bar represents ± SD of three independent measurements. *Reproduced from Griss, R., Schena, A., Reymond, L., Patiny, L., Werner, D., Tinberg, C. E., …, Johnsson, K. (2014). Bioluminescent sensor proteins for point-of-care therapeutic drug monitoring.* Nature Chemical Biology, 10*(7), 598–603. http://dx.doi.org/10.1038/Nchembio.1554 with the authors' permission.*

REFERENCES

Al-Turkmani, M. R., Law, T., Narla, A., & Kellogg, M. D. (2010). Difficulty measuring methotrexate in a patient with high-dose methotrexate-induced nephrotoxicity. *Clinical Chemistry*, *56*(12), 1792–1794. http://dx.doi.org/10.1373/clinchem.2010.144824.

Armstrong, V. W., & Oellerich, M. (2001). New developments in the immunosuppressive drug monitoring of cyclosporine, tacrolimus, and azathioprine. *Clinical Biochemistry*, *34*(1), 9–16. http://dx.doi.org/10.1016/S0009-9120(00)00175-2.

Brun, M. A., Griss, R., Reymond, L., Tan, K. T., Piguet, J., Peters, R. J., … Johnsson, K. (2011). Semisynthesis of fluorescent metabolite sensors on cell surfaces. *Journal of the American Chemical Society*, *133*(40), 16235–16242. http://dx.doi.org/10.1021/ja206915m.

Deglon, J., Thomas, A., Mangin, P., & Staub, C. (2012). Direct analysis of dried blood spots coupled with mass spectrometry: Concepts and biomedical applications. *Analytical and Bioanalytical Chemistry*, *402*(8), 2485–2498. http://dx.doi.org/10.1007/s00216-011-5161-6.

Griss, R., Schena, A., Reymond, L., Patiny, L., Werner, D., Tinberg, C. E., … Johnsson, K. (2014). Bioluminescent sensor proteins for point-of-care therapeutic drug monitoring. *Nature Chemical Biology*, *10*(7), 598–603. http://dx.doi.org/10.1038/Nchembio.1554.

Gubala, V., Harris, L. F., Ricco, A. J., Tan, M. X., & Williams, D. E. (2012). Point of care diagnostics: Status and future. *Analytical Chemistry*, *84*(2), 487–515. http://dx.doi.org/10.1021/ac2030199.

Hall, M. P., Unch, J., Binkowski, B. F., Valley, M. P., Butler, B. L., Wood, M. G., … Wood, K. V. (2012). Engineered luciferase reporter from a deep sea shrimp utilizing a novel imidazopyrazinone substrate. *ACS Chemical Biology*, 7(11), 1848–1857. http://dx.doi.org/10.1021/cb3002478.

Johannessen, S. I., Battino, D., Berry, D. J., Bialer, M., Kramer, G., Tomson, T., & Patsalos, P. N. (2003). Therapeutic drug monitoring of the newer antiepileptic drugs. *Therapeutic Drug Monitoring*, *25*(3), 347–363. http://dx.doi.org/10.1097/00007691-200306000-00016.

Kahan, B. D. (2002). Therapeutic drug monitoring of immunosuppressant drugs in clinical practice (vol 24, pg 330, 2002). *Clinical Therapeutics*, *24*, 1223.

Keppler, A., Gendreizig, S., Gronemeyer, T., Pick, H., Vogel, H., & Johnsson, K. (2003). A general method for the covalent labeling of fusion proteins with small molecules in vivo. *Nature Biotechnology*, *21*(1), 86–89. http://dx.doi.org/10.1038/nbt765.

Khanna, R., Sattin, B. D., Afif, W., Benchimol, E. I., Bernard, E. J., Bitton, A., … Feagan, B. G. (2013). Review article: A clinician's guide for therapeutic drug monitoring of infliximab in inflammatory bowel disease. *Alimentary Pharmacology & Therapeutics*, *38*(5), 447–459. http://dx.doi.org/10.1111/apt.12407.

Lennard, L. (2001). Therapeutic drug monitoring of cytotoxic drugs. *British Journal of Clinical Pharmacology*, *52*, 75s–87s.

MacDonald, A., Scarola, J., Burke, J. T., & Zimmerman, J. J. (2000). Clinical pharmacokinetics and therapeutic drug monitoring of sirolimus. *Clinical Therapeutics*, *22*, B101–B121. http://dx.doi.org/10.1016/S0149-2918(00)89027-X.

Pollock, N. R., Rolland, J. P., Kumar, S., Beattie, P. D., Jain, S., Noubary, F., … Whitesides, G. M. (2012). A paper-based multiplexed transaminase test for low-cost, point-of-care liver function testing. *Science Translational Medicine*, *4*(152). 152ra129, http://dx.doi.org/10.1126/scitranslmed.3003981.

Rifai, N., Gillette, M. A., & Carr, S. A. (2006). Protein biomarker discovery and validation: The long and uncertain path to clinical utility. *Nature Biotechnology*, *24*(8), 971–983. http://dx.doi.org/10.1038/nbt1235.

Roberts, J. A., Norris, R., Paterson, D. L., & Martin, J. H. (2012). Therapeutic drug monitoring of antimicrobials. *British Journal of Clinical Pharmacology*, *73*(1), 27–36. http://dx.doi.org/10.1111/j.1365-2125.2011.04080.x.

Rohrig, C. H., Loch, C., Guan, J. Y., Siegal, G., & Overhand, M. (2007). Fragment-based synthesis and SAR of modified FKBP ligands: Influence of different linking on binding affinity. *ChemMedChem*, *2*(7), 1054–1070. http://dx.doi.org/10.1002/cmdc.200600296.

Sawaya, M. R., & Kraut, J. (1997). Loop and subdomain movements in the mechanism of Escherichia coli dihydrofolate reductase: Crystallographic evidence. *Biochemistry*, *36*(3), 586–603. http://dx.doi.org/10.1021/Bi962337c.

Schena, A., Griss, R., & Johnsson, K. (2015). Modulating protein activity using tethered ligands with mutually exclusive binding sites. *Nature Communications*, *6*. 7830, http://dx.doi.org/10.1038/Ncomms8830.

Schoemakers, R. J., Wasylewicz, A. T. M., Scheepers-Hoeks, A. M. J. W., van Wezel, R., & Grouls, R. J. E. (2011). Clinical rule guided therapeutic drug monitoring of digoxin: Development, validation and potential value. *British Journal of Clinical Pharmacology*, *72*(5), 858.

St John, A., & Price, C. P. (2014). Existing and emerging technologies for point-of-care testing. *Clinical Biochemist Reviews*, *35*(3), 155–167.

Terra, S. G., Washam, J. B., Dunham, G. D., & Gattis, W. A. (1999). Therapeutic range of digoxin's efficacy in heart failure: What is the evidence? *Pharmacotherapy*, *19*(10), 1123–1126. http://dx.doi.org/10.1592/phco.19.15.1123.30570.

Tinberg, C. E., Khare, S. D., Dou, J. Y., Doyle, L., Nelson, J. W., Schena, A., … Baker, D. (2013). Computational design of ligand-binding proteins with high affinity and selectivity. *Nature*, *501*(7466), 212–216. http://dx.doi.org/10.1038/nature12443.

Wang, S. M., Ge, L., Song, X. R., Yu, J. H., Ge, S. G., Huang, J. D., & Zeng, F. (2012). Paper-based chemiluminescence ELISA: Lab-on-paper based on chitosan modified paper device and wax-screen-printing. *Biosensors & Bioelectronics*, *31*(1), 212–218. http://dx.doi.org/10.1016/j.bios.2011.10.019.

CHAPTER FOURTEEN

Sensitive Protein Detection and Quantification in Paper-Based Microfluidics for the Point of Care

Caitlin E. Anderson[2], Kamal G. Shah[2], Paul Yager[1]
University of Washington, Seattle, WA, United States
[1]Corresponding author: e-mail address: yagerp@uw.edu

Contents

Abstract

The design of appropriate diagnostic assays for the point of care requires development of suitable biosensors, detection methods, and diagnostic platforms for sensitive, quantitative detection of biological analytes. Protein targets in particular are especially challenging to detect quantitatively and sensitively due to the lack of amplification strategies akin to nucleic acid amplification. However, recent advances in transducer and biosensor design, new detection labels, and paper-based microfluidics may realize the goal of sensitive, fast, portable, and low-cost protein detection.

In this review, we discuss the biochemistry, optics, and engineering advances that may be leveraged to design such a sensitive protein diagnostic assay. The binding

[2] Indicates authors who contributed equally.

Methods in Enzymology, Volume 589
ISSN 0076-6879
http://dx.doi.org/10.1016/bs.mie.2017.01.018

kinetics, mechanisms of binding in porous networks, and potential transducers are explained in detail. We discuss the relative merits of various optical detection strategies, potential detection labels, optical readout approaches, and image-processing techniques that are amenable to point-of-care use. To conclude, we present a systematic analysis of potential approaches to enhance the sensitivity of paper-based assays. The assay development framework presented here provides bioassay developers a strategy to methodically enhance the sensitivity and point-of-care suitability of protein diagnostics.

1. INTRODUCTION

Diagnosing disease at the point of care remains challenging, especially in resource-limited settings, due to a dearth of appropriate, sensitive, low-cost, and easy-to-use diagnostic assays. Lateral flow assays are a suitable diagnostic platform for the point of care due to their low cost, ease of use, and portability, but one reason that their widespread use is hindered is their poor analytical sensitivity. This drawback is especially acute for protein diagnostics, as there exists no direct, general protein amplification analogs akin to polymerase chain reaction for nucleic acids. However, recent interdisciplinary advances in protein design, biomolecule production, and semiconductor manufacturing may be leveraged to bridge the gap in sensitivity, thereby enabling the design of rapid, sensitive, low-cost, and point-of-care protein diagnostic assays.

In this review, we explore key concepts associated with point-of-care protein detection and quantification. This review is structured to provide assay development guidelines and is neither meant to be a history of protein detection nor an exhaustive list of references, which are available in several excellent reviews published recently (Fu, 2014; Griffin, Williams, Stratis-Cullum, & Vo-Dinh, 2003; Janssen, Knez, Spasic, & Lammertyn, 2013; Yetisen et al., 2013). Rather, we discuss herein strategies for developing rapid, ultrasensitive protein assays while acknowledging the often-conflicting assay optimization goals of sensitivity, cost, portability, and speed. Toward this end, we explore the suitability of paper-based assays to realize these goals, summarizing the kinetics and biochemistry concepts underlying the selection of protein epitope-binding elements, optical labels, and detectors. Finally, we explicate strategies to enhance the sensitivity of bioassays. Together, the insights contained herein ought to provide a framework for bioassay developers to methodically enhance the sensitivity and point-of-care suitability of protein diagnostics.

2. PAPER-BASED DEVICES AND SANDWICH LATERAL FLOW ASSAYS

Paper-based assays use thin sheets of porous media, such as nitrocellulose, cellulose, or glass fiber, to manipulate and process small volumes of fluids. Such assays are especially conducive to point-of-care testing of biological analytes due to their low cost, portability, and ease of use. Lateral flow assays are the simplest paper-based devices and consist of strips of nitrocellulose that wick fluid based on capillary flow without requiring pumps or additional machinery. A sample pad is usually present at the "input" end, which is often a glass fiber pad onto which a sample is added. The sample then elutes into a conjugate pad, which is typically also a glass fiber material, in which reagents for the assay are stored in dry form and are rehydrated by application of a sample to the sample pad (Fig. 1A). A wicking or collection pad, often made of cellulose, is typically present at the end of the device opposite from the "input" end to draw fluids through the rest of the device. A typical lateral flow assay has a test line and a control line orthogonal to the flow direction, both of which are zones in which capture agents have been deposited on the membrane. Presence of the target analyte and adequate fluid delivery, along with rehydration of assay chemistry, is indicated by a change in color or luminescence of the test and control lines, respectively. Pregnancy tests, which detect human chorionic gonadotropin, are examples of lateral flow assays commonly used for protein detection at the point of care.

Despite the commercial success of the urine-based pregnancy test, many other assays require more complex manipulation of the biological sample to enable clinically useful sensitivity in the detection of an analyte. Platforms such as the two-dimensional paper network (2DPN) format or microfluidic paper-based analytical device (μPAD) format enable the design of these more complex assays through selective or sequential delivery of reagents or fluids (Fig. 1) (Fu, Kauffman, Lutz, & Yager, 2010; Martinez, Phillips, Butte, & Whitesides, 2007; Martinez, Phillips, Whitesides, & Carrilho, 2010). Key concepts associated with manufacturing 2DPN and μPAD devices are explicated in several recent reviews and papers (López-Marzo et al., 2016; Martinez et al., 2010; O'Farrell et al., 2009; Parolo, Medina-Sánchez, de la Escosura-Muñiz, & Merkoçi, 2013; Posthuma-Trumpie, Korf, & Van Amerongen, 2009). Briefly, the paper itself may be cut via laser cutting, mechanical shearing, or slicing; reagents may be deposited via

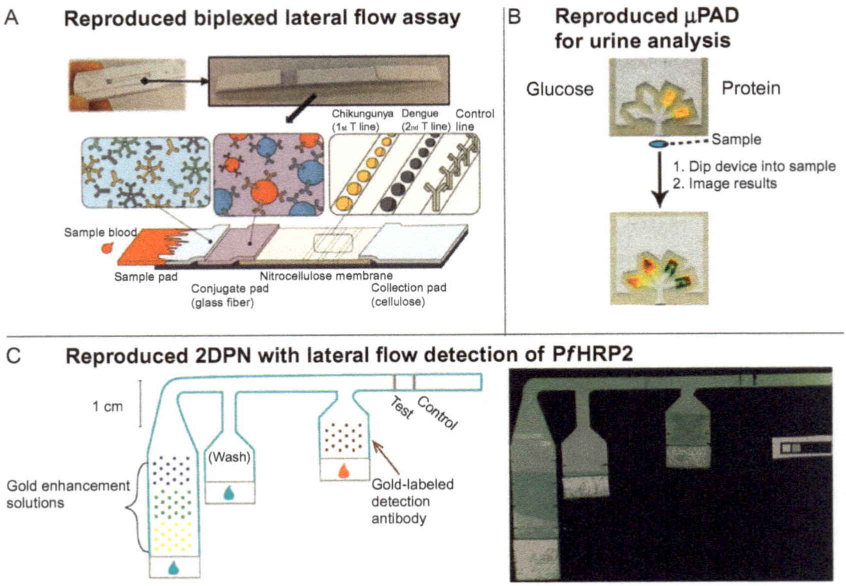

Fig. 1 Examples of paper-based assays. (A) Schematic and photograph of a multiplexed lateral flow assay for differentiating febrile illnesses. (B) Photograph and usage of a microfluidic paper analytical device for the detection of glucose and protein in urine. (C) Schematics and photograph of a two-dimensional paper network coupled with lateral flow readout of *P. falciparum* HRP2. *Images reprinted with permission Fridley, G. E., Le, H., & Yager, P. (2014). Highly sensitive immunoassay based on controlled rehydration of patterned reagents in a 2-dimensional paper network.* Analytical Chemistry, 86*(13), 6447–6453. http://doi.org/10.1021/ac500872j; Lee, S., Mehta, S., & Erickson, D. (2016). Two-color lateral flow assay for multiplex detection of causative agents behind acute febrile illnesses.* Analytical Chemistry. *http://doi.org/10.1021/acs.analchem.6b01828; Martinez, A. W., Phillips, S. T., Carrilho, E., Thomas, S. W., Sindi, H., & Whitesides, G. M. (2008). Simple telemedicine for developing regions: Camera phones and paper-based microfluidic devices for real-time, off-site diagnosis.* Analytical Chemistry, 80*(10), 3699–3707. http://doi.org/10.1021/ac800112r. Copyright 2016, 2008, and 2014 American Chemical Society.*

patterning, printing, or striping; and flow may be controlled through valves, wax deposition, or material layering. These strategies can also be combined with 3D printing, casting, or molding when fabricating a device housing. Such 2DPN and μPAD systems are often then coupled with lateral flow assays for quantitative readout of the analyte concentration.

Traditional lateral flow assays are initiated by the application of a sample onto the device, where the sample passes through a conjugate pad and rehydrates the detection complex. The complex then flows downstream and binds to the capture element deposited at the test line. This type of assay is a sandwich assay, as the macromolecule analyte is sandwiched between two binding elements at different epitopes. The detection element in this assay stack can be either directly or indirectly coupled to an optical label for visual readout. Adequate flow and performance of assay reagents, including the detection element and optical labeling chemistry, is confirmed by the development of signal at the control line. As the major components involved in binding of the analyte, the ligands responsible for the capture and detection of binding events play a significant role in overall assay sensitivity. The success of these ligands is strongly influenced by their binding thermodynamics and kinetics, including physical and chemical properties ranging from steric effects to isoelectric points (Holstein, 2015).

Key benefits derived from the porous nature of the membranes used in paper-based diagnostics, as compared to traditional microfluidic devices, include potentially improved analytical performance, improved ease of use by fewer user steps, reduced cost, and electricity-free usage. This contrasts with traditional microfluidic devices that often require electronic pumps and costly fabrication. Paper-based diagnostics eliminate the need for complex machinery, pumps, and other electrical equipment by manipulating fluids with capillary forces through the pores of the paper membrane (Martinez et al., 2010). The porous substrate also increases the surface area-to-volume ratio of the capture and control regions, thereby enabling localized reagent deposition (Yetisen et al., 2013). The assay's limit of detection and analytical sensitivity are thus improved by the increase in the relative concentration of the capture, detection, and optical elements. Since the surface area-to-volume ratio (typically 50–200 times greater than that of the bulk material) is inversely related to the pore size of the paper substrate, assay performance may be optimized by simply tuning the physical properties of the paper substrate used (EMD Millipore, 2013; Martinez et al., 2010; Yetisen et al., 2013). These advantages may be realized in 2DPNs or microfluidic paper-based devices to implement controlled fluid manipulation, reagent delivery, and rinse steps to further enhance assay performance and reduce user steps (Fu et al., 2010; Grant, Smith, Karvonen, & Richards-Kortum, 2016; Martinez et al., 2010). Paper-based diagnostics are thus an inherently suitable format for where they are needed the most (Fu et al., 2010; Grant et al., 2016; Martinez et al., 2010; Ngom, Guo, Wang, & Bi, 2010; Yetisen et al., 2013), such as for point-of-care protein detection.

3. ASSAY ANALYSIS FRAMEWORK

Assay developers typically have to balance the need to optimize an assay's sensitivity, cost, portability, and speed to suit a particular clinical usage scenario. These needs are discussed in varying levels of detail in publications that document the assay development process (Borysiak et al., 2016; Fu, 2014; Holstein, 2015; Holstein, Griffin, Hong, & Sampson, 2015; Rohrman & Richards-Kortum, 2012). Briefly, some factors to consider include assessing the envisioned usage scenario, such as identifying the end user, environment, and clinical workflow; identifying clinically relevant levels of analyte sensitivity and specificity; determining the appropriate form factor and user experience; determining what analyte(s), sample, and sample volume are needed to attain the intended clinical sensitivity and specificity; and determining the adequate analytical sensitivity and specificity that yields the desired clinical sensitivity and specificity. This distinction between analytical and clinical sensitivity is critical as the two differ considerably in scope: while the analytical sensitivity refers to the slope of the standard curve, clinical sensitivity refers to the ratio of true positives to the total number of diseased individuals in a sample population (Borysiak et al., 2016). The clinical sensitivity of an assay is affected by factors as diverse as the amount of analyte present and assay inhibitors present in real patient samples. Throughout the assay design process, it is important to engage with the intended customers, end users, clinicians, and patients to ensure that the assay meets a practical need. Moreover, it is equally essential to establish a priori design criteria, design constraints, and performance metrics so that the assay may be sufficiently validated.

In order to compare the performance of different assays and assess whether an assay meets performance metrics, appropriate quantitative and statistical methods are necessary to analyze lateral flow assays with statistical validity and reliability (Borysiak et al., 2016; Holstein, Griffin, et al., 2015). Briefly, assay developers assess the extent to which an assay is quantitative by creating a standard curve (binding isotherm) that plots an appropriately defined signal-to-noise ratio against the target analyte concentration, which will typically be fitted to a four-parameter logistic function via nonlinear regression (Fig. 2). The limit of detection (lower limit of detection) should then be defined as the lowest quantified concentration based on the distribution of signal-to-noise ratios of the

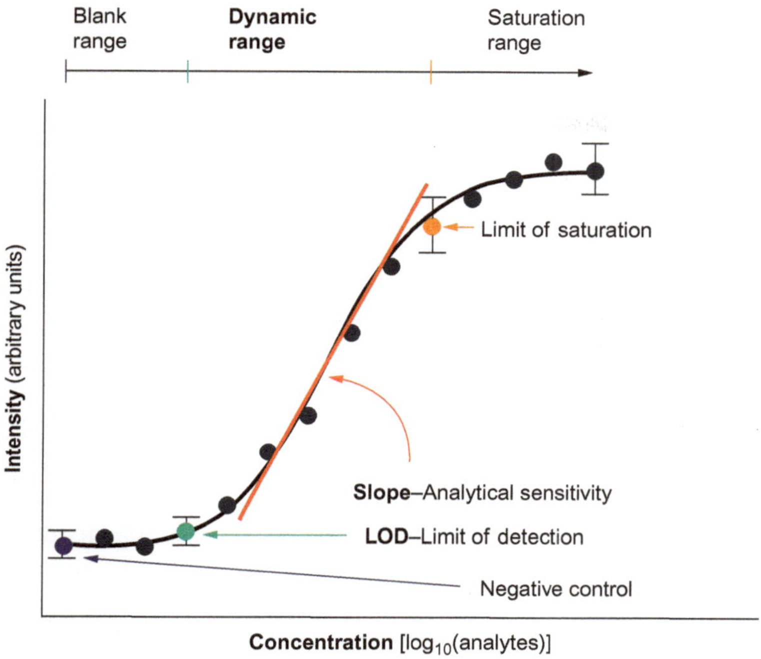

Fig. 2 Visualization of the key assay metrics that is gleaned from a standard curve, which is a regression fitted to the assay output as a function of analyte concentration on a logarithmic scale. Within the dynamic range, the standard curve provides a quantitative relationship between the assay output and analyte concentration.

negative control and of the samples (Borysiak et al., 2016; Holstein, Griffin, et al., 2015). In accordance with international standards, the analytical sensitivity is the slope within the linear portion of the logistic fit and is not synonymous with the limit of detection (Borysiak et al., 2016). Beyond these metrics, the limit of saturation (highest quantifiable concentration or upper limit of detection) and dynamic range (range spanned by the lower and upper limits of detection) are also useful parameters with which assays are characterized (Fig. 2). While the limit of detection is typically reported as the amount of analyte present in the sample volume added to the device, in units such as femtomoles, rather than as an ambiguous concentration, the dynamic range is presented as the unitless number of orders of magnitude spanned, which is expressed as a base-10 logarithm (e.g., "three orders of magnitude").

4. BINDING KINETICS IN POROUS NETWORKS

One of the key parameters that influences the overall analytical sensitivity of a lateral flow assay is the affinity of the binding element to the target analyte (Karlsson, Michaelsson, & Mattsson, 1991). Such affinity, and the corresponding binding kinetics, is dictated by the summation of the interactions between a protein and the ligand to which it binds. These binding interactions consist of a number of molecular forces, including charge–charge interactions, hydrogen bonds, hydrophobic interactions, and van der Waals forces. Note that these forces will usually be modified by the fact that the binding element is immobilized on a matrix that has its own contribution to the forces felt by the ligand. In this section, we will consider the role of binding kinetics in assay sensitivity, with special consideration of the influence of reagent delivery.

The underlying kinetics of a binding reaction can provide insight about the interaction between an analyte and a ligand. Interpretation of a binding event can be directed by an understanding of the underlying kinetics of a reaction between an analyte and a ligand. A simple binding reaction involves an analyte (A) and a ligand (L) which reversibly forms a bound complex (AL). As shown in Eq. (1), the rate of association for the protein and ligand to form a complex is defined as k_{on}, and the rate at which the reaction reverses is defined as k_{off}. Three factors that should be taken into consideration are (1) the concentrations of protein and ligand that are present, (2) the concentration of complex that is formed, and (3) how quickly the complex breaks apart (Williams & Daviter, 2013). These three factors enable quantitative interpretation of a binding event within a Michaelis–Menten kinetics framework (Michaelis & Menten, 1913).

$$A + L \underset{k_{off}}{\overset{k_{on}}{\rightleftharpoons}} AL \tag{1}$$

The dissociation constant (K_D) of a reaction is defined as the rate at which the binding reverses (k_{off}) divided by the rate at which the complex associates (k_{on}). With these rate constants, concentration does play a role in the amount of complex that is formed since the observed net rate of formation (or dissociation), $k_{obs} = k_{on}$ [analyte] $- k_{off}$. As the local concentration of a protein or analyte is increased, the probability of a collision between the two molecules, and subsequent binding event, also increases. When considering binding at a test line, the local concentration in the microenvironment of test line influences the observed on rate for this interaction. Therefore, the transport of a protein to the test line through either convection or diffusion influences the rate at which a binding event can take place.

While traditional methods to measure reaction kinetics are able to provide an estimate for the kinetics of a binding interaction, the rate constants determined by kinetic measurement systems might not necessarily translate directly into those relevant in paper microfluidic systems. This is primarily due to two properties of lateral flow assays: the constant movement of fluid by capillary action and porous nature of the membrane. There are additional effects from the matrix itself, effects that can be seen whether or not a surface has been passivated by a blocking protein such as BSA. Porosity is a measure of the empty space in a material and serves as a key parameter in the design of the lateral flow assays (Ahmad, Low, Shukor, & Ismail, 2008). The larger the void space, or mean pore size, the smaller the surface area-to-volume ratio is for a given segment of membrane. When a ligand is immobilized onto the exposed surface area of a test region, larger amounts of protein can be deposited on membranes with smaller pore sizes. However, as pore size decreases, overall assay speed can suffer as the flow rate is reduced. While this may increase the probability of a collision event as the distance to diffuse to the surface decreases, the primary tradeoff is an increase in time from sample to result.

Experimentally determined rate constants provide insight into the affinity between a protein and a ligand that can translate to improved performance of an assay integrated into a paper microfluidic system. Specifically, empirically determined association and dissociation rate constants may shed light on the kinetics of assay binding for varying reagent delivery formats in porous media (Liang et al., 2016). Reagent delivery can be either sequential, in which each reagent is delivered independently across the test line, or premixed, in which the analyte and detection stack are mixed before they are delivered to the test region. The Fu group has demonstrated a 4- to 10-fold improvement in the limit of detection for the malaria antigen *Pf*HRP2 using the sequential delivery format (Liang et al., 2016). In contrast, recent work has also shown an improved limit of detection for the influenza nucleoprotein using the premixed delivery format (Abe & Yager, 2016). While the underlying mechanisms are not yet fully understood, the difference in performance of these two assays is likely due to differences in antigen binding; this could include multivalency of the antigen, kinetics of binding of the antibodies, or the size of the bound analyte–ligand complex. This body of work highlights that overall assay performance is influenced not just by the affinity of the binding elements, but also by the manner in which they are delivered across the test region. As we discuss the available mechanisms for protein binding in the next section, these concepts regarding assay kinetics must be considered in the context of each assay.

5. MECHANISMS FOR PROTEIN BINDING

The analytical sensitivity of a lateral flow assay is first and foremost limited by the affinity of the binding elements for the target analyte (Karlsson et al., 1991). As mentioned in the previous section, the sensitivity and specificity of protein–ligand binding is driven by the binding kinetics of the interaction. A sandwich assay requires a minimum of two specific binding events: capture of the analyte onto a solid surface by a first ligand immobilized thereon and detection of the analyte with a label conjugated to a second ligand. Capture and detection of an analyte can be through an antibody, a designed affinity protein, or an aptamer, among others. While traditional lateral flow assays utilize antibodies as the capture and detection elements of a sandwich immunoassay, antibodies cannot be applied to all diagnostic targets. In porous media, additional considerations must be taken into account during assay design regarding the strength of immobilization and ability to flow through a porous network for capture and detection. In this section, we will present extant mechanisms for sensitive, specific analyte binding, including recent work demonstrating the use of designed proteins and nucleic acids as binding elements for protein detection.

5.1 Antibodies

Antibodies are the most commonly utilized binding element for lateral flow immunoassays, building from their use in immunoblotting. As the workhorse of the adaptive immune system, antibodies are developed by a host to bind specifically to an antigen (Lipman, Jackson, Trudel, & Weis-Garcia, 2005). Polyclonal and monoclonal antibodies can be raised in a live animal, or bacterial or eukaryotic cells, respectively, and then purified using affinity chromatography (Jayasena, 1999). Polyclonal antibodies are generated by the total antibody response for a given analyte, while monoclonal antibodies are specific to one epitope of interest. Monoclonal and polyclonal antibodies can have different diagnostic applications due to their differences in specificity and production, therefore, their use depends on the assay and the analyte of interest. In both cases, each binding arm (F_{ab}) of an antibody serves as an epitope for binding, while the F_c region of the antibody has biological effector functions that can also be utilized. Integration of antibodies into lateral flow assays is facilitated by the fact that most antibodies adsorb robustly to nitrocellulose, such as in Western blotting applications (Holstein, Chevalier, et al., 2015; Renart, Reiser, & Stark, 1979).

However, antibodies are limited in their diagnostic applications by a variety of factors. The most prominent of these factors is that production of antibodies is typically both labor intensive and expensive (Jayasena, 1999). Challenges include a lack of stability under extreme pH and temperature, batch-to-batch variability, limited control over functional sites, and the inability to target nonimmunogenic or toxic analytes (Mairal et al., 2008; Marx, 2013; Vermeer & Norde, 2000). A number of these limitations have been addressed through the production of antibody fragments that contain just the single chain or multivalent fragments (Binz, Amstutz, & Plückthun, 2005). These antibody fragments can be grown in bacteria, and therefore are much less expensive to manufacture and have greater specificity control. However, this only applies to targets for which antibodies can be produced in the first place, and not all antibodies can be broken down into smaller pieces, and remains stable.

5.2 Proteins

Affinity proteins are a class of proteins that have been utilized for the development of assays for the detection of target analytes. These proteins are expressed recombinantly in bacterial cells, enabling production that is both much simpler and lower cost than for monoclonal antibodies. One approach to develop proteins that bind to a target of interest is computational (de novo) design, in which the tertiary structure of a protein is predicted based on primary amino acid sequence (Whitehead et al., 2012). This design concept, also known as rational molecular design, utilizes molecular approximations that generally require further affinity maturation that enables improvement in binding affinity. Another mechanism for the development of binding proteins is phage display, which is instead based on evolutionary molecular design (Alarie, Sepaniak, & Vo-Dinh, 1990; Binz & Plückthun, 2005). Phage display typically utilizes a randomized library that is made through any of a variety of mutation techniques to create protein variants that bind to the target of interest and is expressed on the exterior of the phage; tight binders are identified by a simple filtration, and the binding protein(s) are then amplified from the nucleic acid encoding it in the phage itself. Affinity maturation is also used with phage display to develop proteins with successively higher affinities.

One of the advantages of affinity proteins is that they can be designed to bind a specific epitope on a target protein, providing more control in the development of a sandwich assay. As these proteins can be expressed in bacteria, they can be produced rapidly and inexpensively. Designed proteins can also be expressed with additional functionalities, such as a binding site for conjugation to a fluorescent label or a paper-binding protein (Liu et al., 2014). Despite these benefits, proteins generated using de novo design require the existence of a high-resolution structure for the analyte of interest. As these crystal structures do not account for movement, and may represent a minority structure present in solution, predictions of proteins that will bind to a specific epitope can often be incorrect. Additionally, there are often some conformations of an analyte for which protein prediction is additionally difficult. One example of this is the hemagglutinin (HA) protein on the influenza virus. In the California/09 strain of H1N1, there is a mutation in the sialic acid binding pocket which makes it difficult to design a protein that can effectively bind in that pocket. While evolutionary design more closely replicates the mutations that take place in antibody production in vivo, it is limited in the degree of mutation that can be made to a single protein without significantly altering the overall structure, function, and stability of the protein (Paschke, 2006). In the case of both rational and evolutionary design, the developed proteins can be evolved to have specific desired properties, such as stability under specific or varying pH or temperature conditions (Fleishman et al., 2011; Heyd et al., 2003).

5.3 Aptamers

Another class of binding moiety that has been utilized in place of antibodies is aptamers. Aptamers are oligonucleotides (DNA or RNA) that can fold into well-defined secondary, tertiary, and quaternary structures (Chen & Yang, 2015). They are easily amplified by polymerase chain reaction, which can be done in a variety of conditions that are not restricted physiologically (Chen & Yang, 2015; Famulok, Mayer, & Blind, 2000). Manufacturing of aptamers that bind to a particular target is accomplished through a process known as systematic evolution of ligands by exponential enrichment (SELEX). SELEX screens large combinatorial libraries of oligonucleotides; they are then further refined using iterations of in vitro selection and amplification (Ellington & Szostak, 1990; O'Sullivan, 2002). Higher affinity can be selected through additional rounds of enrichment. This process is similar to the production of proteins through the evolutionary design process.

Aptamers retain many of the benefits of designed affinity proteins, which include ease of production, low cost, and improved stability. They can be produced to bind any target, even ones that are nonimmunogenic or toxic. Because aptamers are nucleic acids, they have the ability to return to their tertiary structure after being unfolded by heat or pH (Lönne et al., 2015). This property is beneficial in the context of point-of-care diagnostics that need to have a long shelf life under variable temperature and humidity. Aptamers, like affinity proteins, can be designed with a single reactive group for site-specific conjugation to facilitate assay development (Chen & Yang, 2015). To this day, aptamers have not been successfully demonstrated for as wide a range of targets as antibodies. This is in part because polynucleotides do not possess the full range of chemical moieties as are presented by amino acid side chains.

Each of the molecule families listed earlier for binding an analyte of interest has its respective benefits and drawbacks. It is important to note that, in terms of assay performance, all three have been shown in the literature to have comparable analytical sensitivities in the low nanomolar to picomolar range (Chen & Yang, 2015; Gold, 1995; Holstein, 2015). For example, they have been used for the detection of the recombinant HA with varying degrees of success (Holstein, 2015; Holstein, Chevalier, et al., 2015; Shiratori et al., 2014; Woo, Lee, Yim, & Jeong, 2015). Fig. 3 is a graphical representation of all of these binding strategies for simultaneous capture and detection of the analyte of interest.

A critical requirement for the capture molecule in a lateral flow assay is that it effectively immobilized onto the porous membrane. Direct adsorption from a buffer is the technique most commonly used for paper-based assay development, though more specific techniques including covalent attachment and anchor-based immobilization can also be used (Holstein, Chevalier, et al., 2015). Careful consideration of the strength of the interaction and the overall membrane capacity must be taken when selecting a capture molecule for a paper-based assay. It has been proposed that protein adsorption to nitrocellulose is influenced primarily by electrostatic interactions, hydrogen bonding, and the hydrophobic effect (Přistoupil, Kramlová, & Štěrbíková, 1969). Accordingly, effective adsorption of a capture molecule is a critical parameter that must be

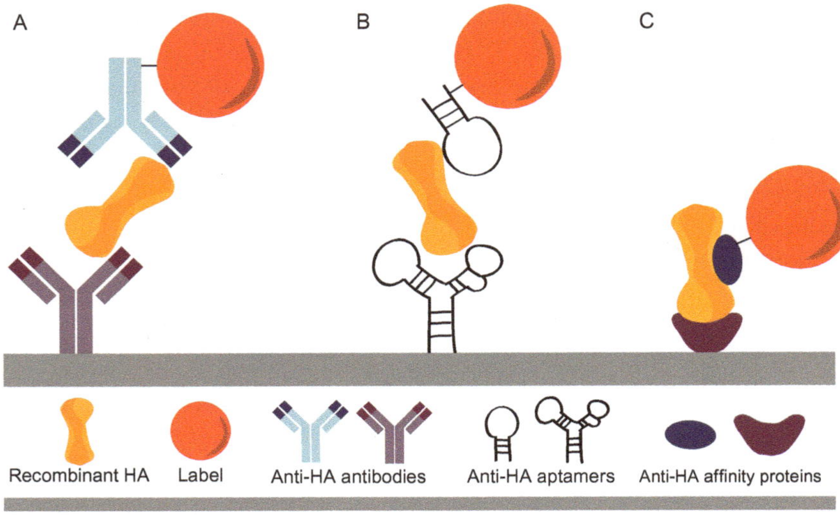

Fig. 3 Graphical representation of three mechanisms that can be used for the capture and detection of recombinant hemagglutinin. In all three cases, the capture and detection elements were raised or designed to target different epitopes on the HA protein. (A) Antibody detection of recombinant hemagglutinin (Holstein, 2015). (B) Aptamer capture and detection (Shiratori et al., 2014; Wang, Zhao, et al., 2013; Woo et al., 2015). (C) Affinity protein detection (Holstein, 2015; Holstein, Chevalier, et al., 2015).

considered when selecting a capture element, whether antibody, aptamer, or affinity protein.

6. OPTICAL METHODS IN PAPER

Beyond the biochemistry underlying an assay, the choice of a detection label and the physics of detecting it will strongly impact the analytical sensitivity and limit of detection of a lateral flow assay. While conductimetric or amperometric readout of assays has been demonstrated, optical labeling is more commonly used, especially in commercial assays such as in pregnancy or HIV tests. Potential optical labels may be classified broadly as color-based, chromophoric labels, or as luminescent, photon-emitting labels. This section will elucidate the relative merits of potential optical labels, for which representative demonstrations to detect protein targets in paper-based assays are documented in Table 1. We will then discuss potential optical readers and image-processing strategies that are conducive to point-of-care protein detection.

Table 1 Representative Performance of Optical Detection Methods to Detect Analytes of Interest

Detection Principle	Potential Labels	Potential Readout Methods	Representative Limit of Detection	Dynamic Range (log_{10})	References
Optical (absorbance)	Gold nanoparticles (AuNP), colored latex beads	Visual, scanner, smartphone, photodiode	6.9 fmol	4	Holstein (2015) and Holstein, Griffin, et al. (2015)
Optical (chromogenic)	HRP-DAB; HRP-TMB	Visual, scanner, smartphone, photodiode	0.06 fmol; 8.3 fmol	1; 4	Grant et al. (2016) and Stevens (2010)
Fluorescence	Organic fluorophores (e.g., Texas Red), quantum dots, lanthanide chelates	Visual, smartphone, photodiode	0.86 fmol	3	Shah, Abe, Kauffman, and Yager (2016)
Phosphorescence	Lanthanide chelates	Visual, smartphone, photodiode	0.35 fmol	3	Paterson et al. (2014)
Chemiluminescence	HRP-Femtoglow	Smartphone, photodiode	0.51 fmol	2	Chen et al. (2016)
Thermal	Gold nanoparticles	Infrared (thermal) camera	Est. 0.2 fmol (32× w.r.t. AuNP)	3.5	Qin et al. (2012)

6.1 Absorbance-Based Labels

Chromophoric optical labels are commonly utilized in lateral flow assays. Typical chromophores include low-cost, highly stable gold nanoparticles, which selectively absorb light at particular wavelengths (Hu et al., 2013), thereby enabling ready discernment of red lines against the uniform white background of the nitrocellulose membrane typically used in lateral flow assays. As a result, most commercial assays use chromophoric labels, including those for human chorionic gonadotropin and HIV antibodies. However, these labels have the drawback of large particle size (20 nm or larger), which contributes to a potential for aggregation under extreme salt or pH concentrations and to nonspecific binding to the nitrocellulose membrane. Both of these limitations may contribute to reduced signal-to-noise ratio, lower analytical sensitivity, and poorer limits of detection.

Chromogenic optical labels were adapted to lateral flow assays to address some of the limitations of traditional chromophores. As denoted by the term, chromogenic labels operate on the principle of producing a color change to denote the presence of analyte. Such labels were adapted from histological labels, including horseradish peroxidase (HRP)-catalyzed oxidation of diaminobenzidine (DAB) or 3,3′-5,5-tetramethylbenzidine (TMB) into a dark precipitate in the presence of the cosubstrate hydrogen peroxide. The speed and specificity of this reaction results in a marked reduction in nonspecific signal and improved signal-to-noise ratio compared to gold-based chromophores (Grant et al., 2016). Unfortunately, the reagent concentrations must be controlled fairly precisely; hydrogen peroxide must be produced at the time of use due to its instability (Grant et al., 2016), and the additional user steps or complexity are necessitated due to the additional reagents (Grant et al., 2016; Stevens, 2010); and the HRP enzyme itself is deactivated by hydrogen peroxide (Sennepin et al., 2009), all of which contribute to a relative dearth of literature utilizing such optical labels. Despite these potential drawbacks, we and others have demonstrated the suitability of chromogenic labels using the HRP enzyme for either wide dynamic range or low limit of detection measurements (Grant et al., 2016; Stevens, 2010).

6.2 Luminescent Labels

In contrast to absorbance-based optical labels, luminescent labels have the potential to improve the analytical sensitivity of lateral flow assays without a commensurate increase in assay complexity. Potential options include fluorescent, phosphorescent, or chemiluminescent labels, all of which have been used to improve the signal-to-noise ratio because even when their signals are weak, the very low background at the wavelengths of interest enables the signal to be detected, but at the cost of requiring sensitive light detectors (O'Farrell et al., 2009, p. 88).

Fluorescent labels include organic fluorophores such as Texas Red or fluorescein derivatives, quantum dots, and lanthanide chelates, among others. Each of these absorb photons at particular wavelengths and reemit them at primarily longer ones as visualized by the labels' excitation and emission spectra, respectively. The difference in the excitation and emission spectral maxima is defined as the Stokes' shift, and larger Stokes' shifts are preferred due to ease of wavelength discrimination, simpler optics, and potentially improved cost and sensitivity (Lee, Nordman, Johnson, & Oldham, 2013; Li et al., 2010; Shah et al., 2016; Venkatraman & Steckl, 2015). Assay developers choose the wavelengths of interest of the excitation source and detectors to minimize cross talk, typically by using lasers or light-emitting diodes as the excitation source and by using optical filters on both the excitation source and the detector. This wavelength discrimination inherently improves the analytical sensitivities and limits of detection of fluorescence-based lateral flow assays compared to their absorbance-based counterparts. However, most organic fluorophores suffer from small Stokes' shifts and tend to photobleach, while quantum dots suffer from the same size considerations as chromophoric nanoparticles, and europium chelates must be excited in the ultraviolet spectrum below 350 nm (Wang, Nchimi Nono, et al., 2013). Moreover, many nitrocellulose membranes used in lateral flow assays are themselves fluorescent, which produces background luminescence that may also contribute to reduced signal-to-noise ratios (Walter et al., 2010).

Phosphorescent labels may improve the signal-to-noise ratio of lateral flow assays by enabling temporal separation of the excitation and emission of luminescent labels, and also by having larger Stokes shifts than many fluorescent compounds. In this optical detection scheme, fluorophores are deliberately excited into the nonfluorescent, phosphorescent triplet state, which can increase the time constant of photon emission by as much as a dozen orders of magnitude, thereby providing a means to visualize photon-emitting labels against a dark, nonluminescent background after the excitation source is toggled off. However, use of phosphors as labels in paper-based assays has been limited, few potential labels have been identified, and their use requires a specialized reader (Paterson et al., 2014).

Chemiluminescent (and bioluminescent) optical labels are akin to their chromogenic counterparts in that a photon-emitting substrate is produced by the catalytic activity of an enzyme, e.g., HRP (Ge, Wang, Song, Ge, & Yu, 2012), luciferase (Urata, Iwata, Noda, Murakami, & Kuroda, 2009), or alkaline phosphatase (Yu et al., 2015). The primary benefit of this method is that no illumination source is needed, eliminating the primary source of background noise for fluorescence and phosphorescence detection, and thereby greatly enhancing the sensitivity compared to chromophoric labels. Chemiluminescence has been incorporated into highly sensitive commercial enzyme-linked immunosorbent assays (ELISA) instruments. However, the drawbacks associated with chromogenic labels using the HRP enzyme are likewise applicable to chemiluminescent and bioluminescent labels, along with an added drawback of potentially requiring a specialized reader to image the lateral flow strips in a dark environment (Chen et al., 2016).

6.3 Optical Readout and Image Processing

The choice of optical label guides the selection of an appropriate reader to quantify the assay output. Potential readout methods include visual discernment by human eye, imaging with a scanner, photography via a smartphone or portable camera, or an electrical output with a photodiode. The choice of readout method will have ramifications on the assay reproducibility, analytical sensitivity, usability, and feasibility for point-of-care use.

Visual readout by human eye is by far the simplest readout method as it may enable design of an electricity-free assay and enable point-of-care use in austere conditions. Typically, an end user will discern the presence or absence of color (for a qualitative assay) or ascertain the intensity of a color (in a semiquantitative assay) to determine the relative concentration of analyte present. An alternate approach to render paper-based assays semiquantitative is to have spots that selectively change color if the analyte concentration is above a threshold, thereby enabling an end user to count the number of colored spots to ascertain a specific range of analyte concentration (Lathwal & Sikes, 2016a). However, the assay's resolution of analyte concentration is inherently quantized using this approach, and there is a potential that the results of visual readout may not be reproducible between users due to differences in eyesight. While these limitations are relatively inconsequential for qualitative assays including those commercialized for pregnancy, the limits of visual readout may render it impractical for other protein assays that need to quantify analyte concentration.

Therefore, many assay developers now use electronic readout methods that enable high-resolution quantification of analyte independent of visual readout bias (Martinez et al., 2010), including stationary flatbed scanners and portable smartphones. The reader of choice during early stages of assay development, especially in our lab, tends to be a flatbed scanner connected to a computer, which enables taking images with high spatial resolution (Göröcs et al., 2014), with high bit depth, and most importantly, in a linear color space. A developer usually quantifies the mean or maximum pixel intensity of a region of interest corresponding to the analyte capture region, subtracts that of a local background region on the lateral flow strip to account for nonspecific binding, and corrects for nonuniform lighting by dividing by a measure of the illuminance, often a local background region adjacent to the capture region not on the lateral flow strip (Holstein, Griffin, et al., 2015). This analysis is typically performed in ImageJ, Adobe Photoshop, or MATLAB, among other programs. A standard curve is then fitted to the data, where y is the normalized signal intensity and x is the analyte concentration, usually using a four-parameter logistic curve as in Eq. (2) (Holstein, 2015):

$$y = d + \frac{a-d}{\left(1 + \frac{\log_{10}(x+2)}{c}\right)^{b}} \tag{2}$$

However, the results from this approach may lead to an unintentional bias in the results due to use of the sRGB color space. The sRGB color space is the current standard for computer and smartphone displays and digital photos. Unfortunately, the sRGB color space is nonlinear; gamma varies from 1 to 2.4 depending on luminosity (sRGB, 1999), which fundamentally prevents device-independent, accurate analyte quantification since the statistical assumptions underlying the generation of a standard curve often assume normally distributed residuals. There are three main methods to address this concern: incorporate the nonlinearity of gamma in the standard curve generation process, linearize the color data, or use a file format with a constant gamma. The latter option is usually the simplest, which leads many to use the TIFF file format in a linear RGB colors pace in lieu of more common JPG files in a sRGB color space, or to use the device-independent Lab color space (Abe, Suzuki, & Citterio, 2008; Lathwal & Sikes, 2016b; Stevens, 2010).

These issues are also important when considering an assay with a smartphone or digital camera readout. Smartphones are highly portable and are conducive to the development of custom apps for assay quantification. Smartphones are, therefore, amenable to point-of-care use (Martinez et al., 2008), unlike scanners that require stable electricity, a computer, and highly trained users. However, it is important to consider the ramifications of rapid update cycles (typically annually), differences between smartphone models, eight-bit color channels, variations in ambient lighting, automatic postprocessing performed by the camera hardware, and unwillingness of end users to potentially contaminate their own device in a particular scenario when developing smartphone-based diagnostic assays.

For all electronic readout methods, the choice of detector will dictate the overall analytical sensitivity and limit of detection of the assay. These include the spectral sensitivity, quantum efficiency, field of view, and dark noise of the detector. While devices with broad spectral sensitivity and high quantum efficiency are advantageous, such systems suffer from poor dark noise, thereby reducing the signal-to-background ratio and offering minimal improvement to assay sensitivity. This tradeoff is an inherent limitation of jack-of-all-trades CMOS and CCD detectors that are found in modern cameras, smartphones, and scanners. In contrast, photodiode readout purports to improve assay sensitivity by enabling one to focus on the wavelengths of interest with minimal noise (since the noise is typically distributed over many wavelengths while the fluorophore emits at a more discrete wavelength) and by allowing signal integration over time, but at the expense of reduced field of view and lower spatial resolution. Their low cost, reproducibility, compactness, and ease of use render them highly suitable for point-of-care use (Ellerbee et al., 2009), as demonstrated by commercial devices such as the Qiagen ESEQuant or BD Veritor lateral flow assay readers.

7. SENSITIVITY ENHANCEMENTS (LOD)

Improving the sensitivity of a bioassay requires improving the signal-to-noise ratio of detecting the target analyte. This is accomplished by increasing the intensity of the signal of interest or by reducing the background intensity, both of which require optimizations either to the biochemistry mediating the assay binding or to the physics of the optical labeling.

7.1 Increasing Signal Intensity

Conceptually, the simplest way to improve the signal-to-noise ratio is to increase the magnitude of the signal intensity, the numerator. Potential options here include increasing the multiplicity of optical labels to targets via a signal amplification scheme or using optical labels that are intrinsically brighter. While increasing the intensity of the light source (to an extent) could conceivably increase the signal intensity (e.g., for fluorescence or absorbance-based measurements), the background noise may also increase commensurately (if the incident light intensity is increased beyond a critical level, perhaps due to Rayleigh or Raman scattering), which may not necessarily improve the signal-to-noise ratio.

Many labeling approaches conjugate labels to recognition moieties at a one-to-one or lower basis, wherein a single optical label binds on an average to one or more analyte molecules. One way to increase the signal intensity is to increase the number of optical labels bound to each analyte (Choi et al., 2010; Fu et al., 2010; Hu et al., 2013), either by selecting an appropriate linker that can bind to multiple epitopes of the analyte (Holstein, 2015), by designing the linker so that multiple optical labels may bind to the linker (Grant et al., 2016), or by coupling nucleic acid amplification to protein detection via immuno-polymerase chain reaction strategies (Janssen et al., 2013; Zhang, Li, & Le, 2012), among other approaches. For example, the HRP-DAB system should improve the signal-to-noise ratio compared to gold nanoparticle-based labels because a single HRP ligated to an analyte could produce hundreds or thousands of colored products; the signal could be increased even further by using the streptavidin–biotin linkage system due to the four binding sites that streptavidin (and avidin) has for biotin (Grant et al., 2016). In theory, the signal amplification strategy should increase the signal intensity by a factor up to the multiplicity of the label to analyte. In practice, however, amplification improves the analytical sensitivities and limits of detection of lateral flow assays by a factor less than the theoretical increase implied by the multiplicity, likely due to a commensurate increase in background intensity levels (Grant et al., 2016; Holstein, 2015; Janssen et al., 2013).

Modulating the incident light intensity with a greater magnitude entails both the selection of an appropriate, "bright" optical label and the selection of an appropriate wavelength at which measurements are made. For example, one novel readout method that may markedly enhance the analytical sensitivity and limit of detection of lateral flow assays is thermal readout by measuring infrared emission from binding events at roughly 10 μm wavelengths in lieu of traditional visible light measurements in the submicron wavelength regime, which only measure absorbance of the bulk nanoparticle rather than detection of the binding event itself. The reader may consist of commercial smartphone thermal camera dongles, which could be used to image a standard gold nanoparticle-based lateral flow assay with no modifications to the bioassay itself. Proof-of-concept has been demonstrated by Qin et al, who have demonstrated a factor of 10 improvement in the limit of detection compared to visible wavelength imaging of gold nanoparticle-labeled assays (Qin et al., 2012).

7.2 Reducing Background Intensity

It is difficult to detect low levels of a signal of interest against high levels of a background signal. Among other factors, elevated background signal may be due to the choice of optical label or reader, to nonspecific binding of the detecting molecule or optical label, or due to the intrinsic brightness of the highly scattering paper substrate. Use of optical labels that are visualized against a highly luminescent background typically leads to a reduction in an assay's analytical sensitivity, as is the case for chromophoric and chromogenic labels against the typically white nitrocellulose membrane used in lateral flow assays. Since external lighting is needed for such assays, the reflected background white signal from the membrane overwhelms the optical reader, thereby reducing the signal-to-background ratio of the assay. In contrast, a luminescent optical label that operates in the dark and does not require external lighting may have an improved signal-to-background ratio, which in turn may improve the assay's analytical sensitivity and limit of detection (Venkatraman & Steckl, 2015).

Reducing nonspecific binding tends to be a more delicate process that involves optimizing the assay biochemistry following insights gained in ELISA. Since the nitrocellulose membrane used in lateral flow assays adsorbs proteins well and typically has a negative surface charge similar to an ELISA plate, there is a potential for the optical label or for the intermediary linker (detecting antibody, binder, or aptamer) to bind nonspecifically to the membrane (Holstein, 2015). Therefore, the lateral flow membrane is often blocked with proteins such as BSA (Lee, Kim, & Moon, 2013; Stevens et al., 2008), the isoelectric point of the intermediary linkers are optimized

(Holstein, 2015), and vigorous wash steps with an appropriate buffer (e.g., phosphate-buffered saline with added surfactant and blocking protein) are included in the assay procedure (Rohrman & Richards-Kortum, 2012; Shah et al., 2016; Stevens et al., 2008). In addition, the chemical bonds between the intermediary linker and optical label can be optimized themselves. The linker and label must be either covalently or noncovalently bound to each other, for instance via strong, noncovalent streptavidin–biotin interactions, among others (Hall, Chen, Chun, Du, & Zhao, 2016). Theoretically, noncovalently bound optical labels may be susceptible to dissociation in harsh conditions such as in the flow regimes present in lateral flow assays during wash or rinse steps, which may, in turn, reduce the signal-to-noise ratio. While a covalent bond between the linker and label is preferable due to its immutability ($k_{off} \sim 0$), noncovalent interactions are more commonly utilized since they too are conducive to modularly designed assays and interface well with commercially available reagents such as conjugated antibodies.

8. SUMMARY AND NEXT STEPS

Recent advances toward ultrasensitive protein detection at the point of care have demonstrated the possibility of improving the sample-to-result runtime of paper-based assays beyond that of laboratory-based ELISA while improving limits of detection compared to traditional gold nanoparticle-labeled lateral flow assays (Grant et al., 2016). Implementing strategies ranging from incorporating novel-binding mechanisms, optimizing reagent delivery formats, varying detection labels, and introducing washing steps have contributed to a two order of magnitude improvement in limits of detection as shown in Fig. 4. Unfortunately, these assays still suffer from requiring complex user steps and have poor limits of detection that are at least an order of magnitude worse than the most sensitive ELISA (Fig. 4). Despite these limitations, these assays offer the prospect of enabling rapid, sensitive detection at the point of care. Incorporation into autonomous detection platforms, such as those recently demonstrated for nucleic acids (Lafleur et al., 2016) or synthetic biology (Stein & Alexandrov, 2014), would fill that gap. Such a design may encourage the development of a new generation of rapid, ultrasensitive, and point-of-care protein assays, which could herald a breakthrough in patient care in both developed and developing countries.

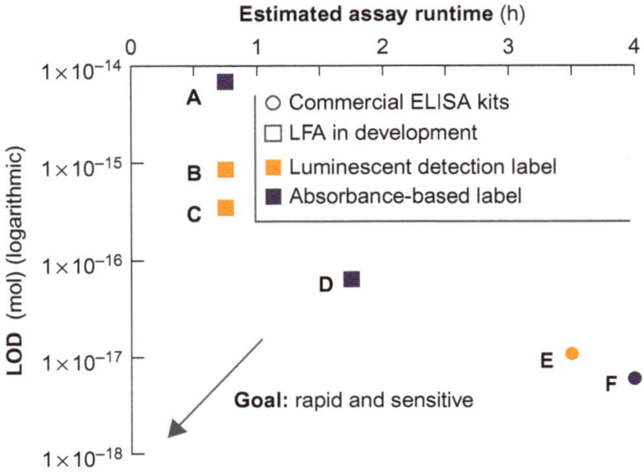

Fig. 4 Visualization of the tradeoffs between the limit of detection (LOD) and estimated assay sample-to-result time for select protein detection mechanisms. Optical labels include (A) gold nanoparticles (Holstein, Griffin, et al., 2015), (B) fluorescent quantum dots (Shah et al., 2016), (C) phosphorescent nanoparticles (Paterson et al., 2014), and (D) horseradish peroxidase-diaminobenzidine (Grant et al., 2016). Extant point-of-care lateral flow assays have limits of detection that are at least one order of magnitude worse than those of the most sensitive commercial ELISA kits as reported in specification sheets (E and F) (ab188389—Free Prostate Specific Antigen (free PSA) Human SimpleStep ELISA® Kit., 2016; Guide to enzyme substrates for ELISA, n.d.).

REFERENCES

ab188389—Free Prostate Specific Antigen (free PSA) Human SimpleStep ELISA® Kit. (2016). Retrieved from http://www.abcam.com/ps/products/188/ab188389/documents/ab188389_Free PSA_28 June 2016_v3 (website).pdf.

Abe, K., Suzuki, K., & Citterio, D. (2008). Inkjet-printed microfluidic multianalyte chemical sensing paper. *Analytical Chemistry*, *80*(18), 6928–6934. http://doi.org/10.1021/ac800604v.

Abe, K., & Yager, P. (2016). Comparing the sensitivity of pre-mixed and sequential reagent delivery for an amplified influenza a nucleoprotein immunoassay. In *MicroTAS, Dublin, Ireland*.

Ahmad, A. L., Low, S. C., Shukor, S. R. A., & Ismail, A. (2008). Synthesis and characterization of polymeric nitrocellulose membranes: Influence of additives and pore formers on the membrane morphology. *Journal of Applied Polymer Science*, *108*(4), 2550–2557. http://doi.org/10.1002/app.27592.

Alarie, J. P., Sepaniak, M. J., & Vo-Dinh, T. (1990). Evaluation of antibody immobilization techniques for fiber optic-based fluoroimmunosensing. *Analytica Chimica Acta*, *229*, 169–176. http://doi.org/10.1016/S0003-2670(00)85125-9.

Binz, H. K., Amstutz, P., & Plückthun, A. (2005). Engineering novel binding proteins from nonimmunoglobulin domains. *Nature Biotechnology*, *23*(10), 1257–1268. http://doi.org/10.1038/nbt1127.

Binz, H. K., & Plückthun, A. (2005). Engineered proteins as specific binding reagents. *Current Opinion in Biotechnology*, *16*(4), 459–469. http://doi.org/10.1016/j.copbio.2005.06.005.

Borysiak, M. D., Thompson, M. J., Posner, J. D., Saah, A. J., Hoover, D. R., Sackett, D. L., … Macmillan, D. (2016). Translating diagnostic assays from the laboratory to the clinic: Analytical and clinical metrics for device development and evaluation. *Lab on a Chip, 16*(8), 1293–1313. http://doi.org/10.1039/C6LC00015K.

Chen, H., Hagström, A. E. V., Kim, J., Garvey, G., Paterson, A., Ruiz-Ruiz, F., … Roda, A. (2016). Flotation immunoassay: Masking the signal from free reporters in sandwich immunoassays. *Scientific Reports, 6*, 24297. http://doi.org/10.1038/srep24297.

Chen, A., & Yang, S. (2015). Replacing antibodies with aptamers in lateral flow immunoassay. *Biosensors and Bioelectronics, 71*, 230–242. http://doi.org/10.1016/j.bios.2015.04.041.

Choi, D. H., Lee, S. K., Oh, Y. K., Bae, B. W., Lee, S. D., Kim, S., … Kim, M. G. (2010). A dual gold nanoparticle conjugate-based lateral flow assay (LFA) method for the analysis of troponin I. *Biosensors and Bioelectronics, 25*(8), 1999–2002. http://doi.org/10.1016/j.bios.2010.01.019.

Ellerbee, A. K., Phillips, S. T., Siegel, A. C., Mirica, K. A., Martinez, A. W., Striehl, P., … Whitesides, G. M. (2009). Quantifying colorimetric assays in paper-based microfluidic devices by measuring the transmission of light through paper. *Analytical Chemistry, 81*(20), 8447–8452. http://doi.org/10.1021/ac901307q.

Ellington, A. D., & Szostak, J. W. (1990). In vitro selection of RNA molecules that bind specific ligands. *Nature, 346*(6287), 818–822. http://doi.org/10.1038/346818a0.

EMD Millipore. (2013). *Rapid lateral flow test strips: Considerations for product development.* Billerica, MA: EMD Millipore.

Famulok, M., Mayer, G., & Blind, M. (2000). Nucleic acid aptamers-from selection in vitro to applications in vivo. *Accounts of Chemical Research, 33*(9), 591–599. http://doi.org/10.1021/ar960167q.

Fleishman, S. J., Whitehead, T. A., Ekiert, D. C., Dreyfus, C., Corn, J. E., Strauch, E.-M., … Baker, D. (2011). Computational design of proteins targeting the conserved stem region of influenza hemagglutinin. *Science, 332*(6031), 816–821. http://doi.org/10.1126/science.1202617.

Fu, E. (2014). Enabling robust quantitative readout in an equipment-free model of device development. *The Analyst, 139*(19), 4750. http://doi.org/10.1039/C4AN01003E.

Fu, E., Kauffman, P., Lutz, B., & Yager, P. (2010). Chemical signal amplification in two-dimensional paper networks. *Sensors and Actuators B: Chemical, 149*(1), 325–328. http://doi.org/10.1016/j.snb.2010.06.024.

Ge, L., Wang, S., Song, X., Ge, S., & Yu, J. (2012). 3D origami-based multifunction-integrated immunodevice: Low-cost and multiplexed sandwich chemiluminescence immunoassay on microfluidic paper-based analytical device. *Lab on a Chip, 12*(17), 3150–3158. http://doi.org/10.1039/c2lc40325k.

Gold, L. (1995). Oligonucleotides as research, diagnostic, and therapeutic agents. *Journal of Biological Chemistry, 270*, 13581–13584. American Society for Biochemistry and Molecular Biology. http://doi.org/10.1074/jbc.270.23.13581.

Göröcs, Z., Ozcan, A., Ozcan, A., Coskun, A. F., Ozcan, A., Wei, Q., … Manolis, S. K. (2014). Biomedical imaging and sensing using flatbed scanners. *Lab on a Chip, 14*(17), 3248. http://doi.org/10.1039/C4LC00530A.

Grant, B. D., Smith, C. A., Karvonen, K., & Richards-Kortum, R. (2016). Highly sensitive two-dimensional paper network incorporating biotin-streptavidin for the detection of malaria. *Analytical Chemistry, 88*, 2553–2557. American Chemical Society. http://doi.org/10.1021/acs.analchem.5b03999.

Griffin, G. D., Williams, M. W., Stratis-Cullum, D. N., & Vo-Dinh, T. (2003). Amplification techniques for optical detection. In Tuan Vo-Dinh (Ed.), *Biomedical photonics handbook*. CRC Press. Retrieved from http://www.crcnetbase.com/isbn/9780203008997.

Guide to enzyme substrates for ELISA. (n.d.). Retrieved August 26, 2016, from https://tools.thermofisher.com/content/sfs/brochures/TR0033-Substrates-ELISA.pdf.

Hall, E. A. H., Chen, S., Chun, J., Du, Y., & Zhao, Z. (2016). A molecular biology approach to protein coupling at a biosensor interface. *TrAC—Trends in Analytical Chemistry*, *79*, 247–256. http://doi.org/10.1016/j.trac.2016.01.024.

Heyd, B., Pecorari, F., Collinet, B., Adjadj, E., Desmadril, M., & Minard, P. (2003). In vitro evolution of the binding specificity of neocarzinostatin, an enediyne-binding chromoprotein. *Biochemistry*, *42*(19), 5674–5683. http://doi.org/10.1021/bi0273664.

Holstein, C. A. (2015). *Development of a novel paper-based flu test for improved diagnosis at the point of care*. Seattle, WA: University of Washington. Retrieved from http://hdl.handle.net/1773/33583.

Holstein, C. A., Chevalier, A., Bennett, S., Anderson, C. E., Keniston, K., Olsen, C., … Yager, P. (2015). Immobilizing affinity proteins to nitrocellulose: A toolbox for paper-based assay developers. *Analytical and Bioanalytical Chemistry*, *408*(5), 1335–1346. http://doi.org/10.1007/s00216-015-9052-0.

Holstein, C. A., Griffin, M., Hong, J., & Sampson, P. D. (2015). Statistical method for determining and comparing limits of detection of bioassays. *Analytical Chemistry*, *87*(19), 9795–9801. http://doi.org/10.1021/acs.analchem.5b02082.

Hu, J., Wang, L., Li, F., Han, Y. L., Lin, M., Lu, T. J., … Luong, J. H. T. (2013). Oligonucleotide-linked gold nanoparticle aggregates for enhanced sensitivity in lateral flow assays. *Lab on a Chip*, *13*(22), 4352. http://doi.org/10.1039/c3lc50672j.

Janssen, K. P. F., Knez, K., Spasic, D., & Lammertyn, J. (2013). Nucleic acids for ultrasensitive protein detection. *Sensors (Switzerland)*, *13*, 1353–1384. Multidisciplinary Digital Publishing Institute http://doi.org/10.3390/s130101353.

Jayasena, S. D. (1999). Aptamers: An emerging class of molecules that rival antibodies in diagnostics. *Clinical Chemistry*, *45*(9), 1628–1650. http://doi.org/10.1126/science.2200121.

Karlsson, R., Michaelsson, A., & Mattsson, L. (1991). Kinetic analysis of monoclonal antibody-antigen interactions with a new biosensor based analytical system. *Journal of Immunological Methods*, *145*(1–2), 229–240. http://doi.org/10.1016/0022-1759(91)90331-9.

Lafleur, L., Bishop, J. D., Heiniger, E. K., Gallagher, R. P., Wheeler, M. D., Kauffman, P. C., … Yager, P. (2016). A rapid, instrument-free, sample-to-result nucleic acid amplification test. *Lab on a Chip*, *52*, 3377–3383. http://doi.org/10.1039/C6LC00677A.

Lathwal, S., & Sikes, H. D. (2016a). A method for designing instrument-free quantitative immunoassays. *Analytical Chemistry*, *88*(6), 3194–3202. http://doi.org/10.1021/acs.analchem.5b04538.

Lathwal, S., & Sikes, H. D. (2016b). Assessment of colorimetric amplification methods in a paper-based immunoassay for diagnosis of malaria. *Lab on a Chip*, *16*(8), 1374–1382. http://doi.org/10.1039/C6LC00058D.

Lee, S., Kim, G., & Moon, J. (2013). Performance improvement of the one-dot lateral flow immunoassay for aflatoxin B1 by using a smartphone-based reading system. *Sensors (Switzerland)*, *13*(4), 5109–5116. http://doi.org/10.3390/s130405109.

Lee, L. G., Nordman, E. S., Johnson, M. D., & Oldham, M. F. (2013). A low-cost, high-performance system for fluorescence lateral flow assays. *Biosensors*, *3*(4), 360–373. http://doi.org/10.3390/bios3040360.

Li, Z., Wang, Y., Wang, J., Tang, Z., Pounds, J. G., & Lin, Y. (2010). Rapid and sensitive detection of protein biomarker using a portable fluorescence biosensor based on quantum dots and a lateral flow test strip. *Analytical Chemistry*, *82*(16), 7008–7014. http://doi.org/10.1021/ac101405a.

Liang, T., Robinson, R., Houghtaling, J., Fridley, G., Ramsey, S. A., & Fu, E. (2016). Investigation of reagent delivery formats in a multivalent malaria sandwich immunoassay and implications for assay performance. *Analytical Chemistry*, *88*(4), 2311–2320. http://doi.org/10.1021/acs.analchem.5b04222.

Lipman, N. S., Jackson, L. R., Trudel, L. J., & Weis-Garcia, F. (2005). Monoclonal versus polyclonal antibodies: Distinguishing characteristics, applications, and information resources. *ILAR Journal/National Research Council, Institute of Laboratory Animal Resources*, *46*(3), 258–268. http://doi.org/10.1093/ilar.46.3.258.

Liu, Y., Zhang, X., Tan, Y. L., Bhabha, G., Ekiert, D. C., Kipnis, Y., … Kelly, J. W. (2014). De novo-designed enzymes as small-molecule-regulated fluorescence imaging tags and fluorescent reporters. *Journal of the American Chemical Society*, *136*(38), 13102–13105. http://doi.org/10.1021/ja5056356.

Lönne, M., Bolten, S., Lavrentieva, A., Stahl, F., Scheper, T., & Walter, J.-G. (2015). Development of an aptamer-based affinity purification method for vascular endothelial growth factor. *Biotechnology Reports*, *8*, 16–23. http://doi.org/10.1016/j.btre.2015.08.006.

López-Marzo, A. M., Merkoçi, A., Petryayeva, E., Algar, W. R., Sun, J., Xianyu, Y., … Sia, S. K. (2016). Paper-based sensors and assays: A success of the engineering design and the convergence of knowledge areas. *Lab on a Chip*, *16*(17), 3150–3176. http://doi.org/10.1039/C6LC00737F.

Mairal, T., Cengiz Özalp, V., Lozano Sánchez, P., Mir, M., Katakis, I., & O'Sullivan, C. K. (2008). Aptamers: Molecular tools for analytical applications. *Analytical and Bioanalytical Chemistry*, *390*(4), 989–1007. http://doi.org/10.1007/s00216-007-1346-4.

Martinez, A. W., Phillips, S. T., Butte, M. J., & Whitesides, G. M. (2007). Patterned paper as a platform for inexpensive, low-volume, portable bioassays. *Angewandte Chemie International Edition*, *46*(8), 1318–1320. http://doi.org/10.1002/anie.200603817.

Martinez, A. W., Phillips, S. T., Carrilho, E., Thomas, S. W., Sindi, H., & Whitesides, G. M. (2008). Simple telemedicine for developing regions: Camera phones and paper-based microfluidic devices for real-time, off-site diagnosis. *Analytical Chemistry*, *80*(10), 3699–3707. http://doi.org/10.1021/ac800112r.

Martinez, A. W., Phillips, S. T., Whitesides, G. M., & Carrilho, E. (2010). Diagnostics for the developing world: Microfluidic paper-based analytical devices. *Analytical Chemistry*, *82*(1), 3–10. http://doi.org/10.1021/ac9013989.

Marx, V. (2013). Finding the right antibody for the job. *Nature Methods*, *10*(8), 703–707. http://doi.org/10.1038/nmeth.2570.

Michaelis, L., & Menten, M. (1913). *The kinetics of the inversion effect.* http://doi.org/0366-0753.

Ngom, B., Guo, Y., Wang, X., & Bi, D. (2010). Development and application of lateral flow test strip technology for detection of infectious agents and chemical contaminants: A review. *Analytical and Bioanalytical Chemistry*, *397*, 1113–1135. Springer-Verlag. http://doi.org/10.1007/s00216-010-3661-4.

O'Farrell, B., Rosen, S., Ponti, J., Brown, M., Chun, P., Mansfield, M., … Tse, H. (2009). In R. Wong & H. Tse (Eds.), *Lateral flow immunoassay.* Totowa, NJ: Humana Press. http://doi.org/10.1007/978-1-59745-240-3.

O'Sullivan, C. K. (2002). Aptasensors—The future of biosensing? *Analytical and Bioanalytical Chemistry*, *372*(1), 44–48. http://doi.org/10.1007/s00216-001-1189-3.

Parolo, C., Medina-Sánchez, M., de la Escosura-Muñiz, A., & Merkoçi, A. (2013). Simple paper architecture modifications lead to enhanced sensitivity in nanoparticle based lateral flow immunoassays. *Lab on a Chip*, *13*(3), 386–390. http://doi.org/10.1039/C2LC41144J.

Paschke, M. (2006). Phage display systems and their applications. *Applied Microbiology and Biotechnology*, *70*(1), 2–11. http://doi.org/10.1007/s00253-005-0270-9.

Paterson, A. S., Raja, B., Garvey, G., Kolhatkar, A., Hagstrom, A. E. V., Kourentzi, K., … Willson, R. C. (2014). Persistent luminescence strontium aluminate nanoparticles as reporters in lateral flow assays. *Analytical Chemistry*, *86*(19), 9481–9488. http://doi.org/10.1021/ac5012624.

Posthuma-Trumpie, G. A., Korf, J., & Van Amerongen, A. (2009). Lateral flow (immuno) assay: Its strengths, weaknesses, opportunities and threats. A literature survey. *Analytical and Bioanalytical Chemistry*, *393*(2), 569–582. http://doi.org/10.1007/s00216-008-2287-2.

Přistoupil, T. I., Kramlová, M., & Štěrbíková, J. (1969). On the mechanism of adsorption of proteins to nitrocellulose in membrane chromatography. *Journal of Chromatography A*, *42*, 367–375. http://doi.org/10.1016/S0021-9673(01)80636-1.

Qin, Z., Chan, W. C. W., Boulware, D. R., Akkin, T., Butler, E. K., & Bischof, J. C. (2012). Significantly improved analytical sensitivity of lateral flow immunoassays by using thermal contrast. *Angewandte Chemie (International Ed. in English)*, *51*(18), 4358–4361. http://doi.org/10.1002/anie.201200997.

Renart, J., Reiser, J., & Stark, G. R. (1979). Transfer of proteins from gels to diazobenzyloxymethyl-paper and detection with antisera: A method for studying antibody specificity and antigen structure. *Proceedings of the National Academy of Sciences*, *76*(7), 3116–3120. http://doi.org/10.1073/pnas.76.7.3116.

Rohrman, B. A., & Richards-Kortum, R. R. (2012). A paper and plastic device for performing recombinase polymerase amplification of HIV DNA. *Lab on a Chip*, *12*(17), 3082. http://doi.org/10.1039/c2lc40423k.

Sennepin, A. D., Charpentier, S., Normand, T., Sarré, C., Legrand, A., & Mollet, L. M. (2009). Multiple reprobing of Western blots after inactivation of peroxidase activity by its substrate, hydrogen peroxide. *Analytical Biochemistry*, *393*, 129–131. http://doi.org/10.1016/j.ab.2009.06.004.

Shah, K. G., Abe, K., Kauffman, P. C., & Yager, P. (2016). Filter-free fluorescence measurements at the point of care: A simple device to improve the limit of detection of lateral flow assays. In *MicroTAS, Dublin, Ireland*.

Shiratori, I., Akitomi, J., Boltz, D. A., Horii, K., Furuichi, M., & Waga, I. (2014). Selection of DNA aptamers that bind to influenza A viruses with high affinity and broad subtype specificity. *Biochemical and Biophysical Research Communications*, *443*, 37–41. http://doi.org/10.1016/j.bbrc.2013.11.041.

sRGB. (1999). Retrieved January 22, 2017, from http://www.color.org/chardata/rgb/srgb.xalter.

Stein, V., & Alexandrov, K. (2014). Protease-based synthetic sensing and signal amplification. *Proceedings of the National Academy of Sciences*, *111*(45), 15934–15939. http://doi.org/10.1073/pnas.1405220111.

Stevens, D. Y. (2010). *Development and optical analysis of a microfluidic point-of-care diagnostic device*. Seattle, WA: University of Washington. Retrieved from http://gradworks.umi.com/34/31/3431744.html.

Stevens, D. Y., Petri, C. R., Osborn, J. L., Spicar-Mihalic, P., McKenzie, K. G., & Yager, P. (2008). Enabling a microfluidic immunoassay for the developing world by integration of on-card dry reagent storage. *Lab on a Chip*, *8*(12), 2038–2045. http://doi.org/10.1039/b811158h.

Urata, M., Iwata, R., Noda, K., Murakami, Y., & Kuroda, A. (2009). Detection of living Salmonella cells using bioluminescence. *Biotechnology Letters*, *31*(5), 737–741. http://doi.org/10.1007/s10529-009-9924-5.

Venkatraman, V., & Steckl, A. J. (2015). Integrated OLED as excitation light source in fluorescent lateral flow immunoassays. *Biosensors and Bioelectronics*, *74*, 150–155. http://doi.org/10.1016/j.bios.2015.06.049.

Vermeer, A. W., & Norde, W. (2000). The thermal stability of immunoglobulin: Unfolding and aggregation of a multi-domain protein. *Biophysical Journal*, *78*(1), 394–404. http://doi.org/10.1016/S0006-3495(00)76602-1.

Walter, J. G., Stahl, F., Reck, M., Praulich, I., Nataf, Y., Hollas, M., ... Scheper, T. (2010). Protein microarrays: Reduced autofluorescence and improved LOD. *Engineering in Life Sciences*, *10*(2), 103–108. http://doi.org/10.1002/elsc.200900078.

Wang, Q., Nchimi Nono, K., Syrjänpää, M., Charbonnieère, L. J., Hovinen, J., & Härmä, H. (2013). Stable and highly fluorescent europium(III) chelates for time-resolved immunoassays. *Inorganic Chemistry*, *52*(15), 8461–8466. http://doi.org/10.1021/ic400384f.

Wang, R., Zhao, J., Jiang, T., Kwon, Y. M., Lu, H., Jiao, P., ... Li, Y. (2013). Selection and characterization of DNA aptamers for use in detection of avian influenza virus H5N1. *Journal of Virological Methods*, *189*(2), 362–369. http://doi.org/10.1016/j.jviromet.2013.03.006.

Whitehead, T. A., Chevalier, A., Song, Y., Dreyfus, C., Fleishman, S. J., De Mattos, C., ... Baker, D. (2012). Optimization of affinity, specificity and function of designed influenza inhibitors using deep sequencing. *Nature Biotechnology*, *30*(6), 543–548. http://doi.org/10.1038/nbt.2214.

Williams, M. A., & Daviter, T. (Eds.). (2013). *Protein-Ligand Interactions* (Vol. 1008). Totowa, NJ: Humana Press. http://doi.org/10.1007/978-1-62703-398-5.

Woo, H.-M., Lee, J.-M., Yim, S., & Jeong, Y.-J. (2015). Isolation of single-stranded DNA aptamers that distinguish influenza virus hemagglutinin subtype H1 from H5. *PloS One*, *10*(4), e0125060. http://doi.org/10.1371/journal.pone.0125060

Yetisen, A. K., Akram, M. S., Lowe, C. R., Whitesides, G. M., Berthier, E., Young, E. W., ... Mars, M. (2013). Paper-based microfluidic point-of-care diagnostic devices. *Lab on a Chip*, *13*(12), 2210. http://doi.org/10.1039/c3lc50169h.

Yu, L., Shi, Z., Fang, C., Zhang, Y., Liu, Y., & Li, C. (2015). Disposable lateral flow-through strip for smartphone-camera to quantitatively detect alkaline phosphatase activity in milk. *Biosensors and Bioelectronics*, *69*, 307–315. http://doi.org/10.1016/j.bios.2015.02.035.

Zhang, H., Li, X. F., & Le, X. C. (2012). Binding-induced DNA assembly and its application to yoctomole detection of proteins. *Analytical Chemistry*, *84*(2), 877–884. http://doi.org/10.1021/ac203207g.

CHAPTER FIFTEEN

Microneedle Enzyme Sensor Arrays for Continuous In Vivo Monitoring

Anthony E.G. Cass[1], Sanjiv Sharma
Department of Chemistry, Imperial College London, London, United Kingdom
[1]Corresponding author: e-mail address: t.cass@imperial.ac.uk

Contents

Abstract

Microneedle enzyme sensors by virtue of their minimally invasive and hence pain-free penetration of skin allow for the measurement of metabolites, biomarkers, and drugs in the interstitial fluid that bathes the dermal tissue. Such devices if they are to be adopted widely into clinical practice need to be capable of delivering reliable measurements over extended periods of time (days) and to be fabricated by low-cost, scalable methods. Using injection molding of the base structures in polycarbonate, metal film deposition by sputtering and enzyme immobilization by electrodeposition can meet these requirements. The workflow to produce devices for clinical evaluation is then completed by sterilization and packaging. In vitro evaluation of the sensors' response to varying analyte concentrations and their mechanical testing establish performance and safety characteristics. While most of the work is focused on glucose sensing, reflecting the significance of the global diabetes "epidemic," the microneedles can also be used to measure lactate (another metabolite) and theophylline (a therapeutic drug).

Methods in Enzymology, Volume 589
ISSN 0076-6879
http://dx.doi.org/10.1016/bs.mie.2017.02.002

1. INTRODUCTION

Continuous in vivo monitoring of important biological markers (proteins, nucleic acids, metabolites) as well as exogenous molecules (drugs, toxins) can yield important insights into human physiology and pathology. In clinical practice, such measurements have the potential to improve the diagnosis of disease and enable a more personalized and responsive approach to therapy.

Many technological approaches have been explored and implemented to try and achieve reliable, continuous monitoring, using both invasive and noninvasive devices. In many instances, they are used in hospitals, especially in intensive therapy units.

The terms "invasive," "minimally invasive," and "noninvasive" are commonly used to refer the types of devices employed in continuous monitoring; the first and last are fairly straightforward to understand. However, it is not always clear what constitutes a minimally invasive measurement. Here we shall take the position that while minimally invasive devices do penetrate the outer layer of the body (the skin or internal epithelial tissues such as the nasal mucosa or the gut), they do so by less than 1 mm and do not require trained medical personnel for their application. While noninvasive methods are in many ways the most desirable for continuous measurements and have been extensively used in both professional and consumer products, they are primarily aimed at physiological rather than molecular parameters. Common examples include pulse rate, skin temperature, blood oxygenation, and electrical activity. Where molecular parameters are measured continuously, they tend to be in secreted fluids (sweat, lachrymal fluid, saliva) whose composition may be far from that of blood or serum, and which may be further perturbed by local microbial flora on the surface of the skin or in the oral cavity.

One body fluid that is accessible in a minimally invasive manner and whose composition does follow that of blood quite closely in many instances is the interstitial fluid (ISF). This is the fluid that bathes the epidermal tissues (<1 mm below the surface of the skin) and which maybe either extracted or analyzed in situ. It is this fluid that can be accessed by microneedles and offers a window into clinical diagnostics that is attracting much current interest (Miller, Narayan, & Polsky, 2016).

Much of the research on and technology development of microneedles (μNs) over the past two decades has focused on their applications for drug delivery, either by virtue of their greater efficacy (for example, with vaccines), or where patients suffer from needle phobia (at least 10% of the general population), or to enable patients to take drugs that are not suitable for oral or diffusive transdermal delivery. It is only more recently that interest in using μNs for continuous sensing has attracted significant interest as shown by the number of recent reviews that have appeared on this topic. To understand the factors that have to be considered in designing μNs for either monitoring or vaccine/drug delivery, we need to look more closely at the anatomy, physiology, and biochemistry of skin.

As the largest tissue in the human body (around 15%) the skin serves multiple functions including water retention, temperature regulation, and nociception as well as protection against visible radiation, toxins, and pathogens. Reflecting these multiple roles the anatomy of the skin comprises several well-defined layers (El-Laboudi, Oliver, Cass, & Johnston, 2013). The stratum corneum comprises the outermost layer; it is the mechanically most robust and is the main physical barrier. The thickness of the stratum corneum can vary widely with location and age. On the forearm, a typical measurement site, it is 20–30 μm thick. Below this is the viable epidermis comprising several sublayers and devoid of either blood vessels or nerve endings. It is this region that is ideally sampled by the μNs (typically some 400 μm below the skin surface but ranging from 800 to 1500 μm). Beneath the viable epidermis is the dermis and then the subcutaneous tissue. The fluid bathing the cells in the viable epidermis, as mentioned earlier, is referred to as interstitial fluid and has a composition that is quite similar to blood, with the main exception being that it has almost no blood cells and only about 25%–30% of the protein content (Fogh-Andersen, Altura, Altura, & Siggaard-Andersen, 1995). Small molecules often equilibrate between blood and ISF, although the concentrations and time profiles can differ. The different time profiles are often manifest as lags in the ISF levels compared with plasma levels and often have both physiological and measurement components. The physiological component may be associated with, for example, intercompartment transport kinetics that in turn depend on other parameters, while the measurement component can arise from the design compromises in the sensor.

μNs, either as single needles or arrays, have been most commonly used to sense glucose in ISF, not only because of the importance of such measurements in the management of diabetes, but also because using the sensor output to control insulin delivery "closes the loop," creating an artificial pancreas (Herrero, Georgiou, Oliver, Johnston, & Toumazou, 2012). Such closed-loop control would be extremely valuable in other areas of medicine, for instance, in delivering antibiotics or drugs with narrow therapeutic windows.

There have been many published designs for microneedles with a variety of fabrication methods and materials employed. Broadly we can think of either hollow microneedles, i.e., those with a lumen through which fluid is extracted, or solid microneedles wherein the μN is also the sensor. The former is usually associated with a subsequent offline analysis of the fluid extracted and therefore is not so well suited to continuous measurements. Furthermore, hollow μNs are likely to be more mechanically fragile and break during insertion or use.

2. APPROACHES TO THE FABRICATION OF MICRONEEDLES FOR SENSING

In this section the emphasis will be on methods to fabricate microneedle arrays for sensing in ISF. Single-needle sensors (in a wide range of sizes) have formed the bedrock of both academic and commercial efforts toward reliable continuous in vivo glucose sensing for several decades. They have been extensively reviewed (Oliver, Toumazou, Cass, & Johnston, 2009) and still form the basis of most commercial continuous glucose monitors (CGMs), both as standalone units and as components of closed-loop control systems, when combined with insulin delivery devices such as pumps.

μNs fabricated from silicon have been widely explored, in part because its material properties are very well understood and the tools for creating microstructures very well established. Starting from a silicon wafer (400 μm thick) and using photolithography to define feature size, subsequent wet or dry etching can be used to produce both solid and hollow mn arrays. The earliest microneedle arrays based on silicon were extractive and the ISF analyzed in a subsequent step (Mukerjee, Collins, Isseroff, & Smith, 2004). Subsequently, work to assess the role of factors such as size and geometry on extraction was carried out by several groups (Chua, Desai, Tierney, & Tamada, 2013). Moreover, while capillary forces can drive fluid into the lumen, this can often be insufficient to maintain a consistent flow and there is the ever-present issue of blockage of the needle by protein deposition. This has meant that additional hydrostatic pressure may be needed, adding to the complexity of the device. To overcome this problem Strambini et al. (2015) have fabricated very high density (10^6 cm^{-1}) hollow mn arrays with much smaller lumens (down to 4 μm in diameter) having an electrochemical glucose sensor on the backside. This configuration results in uptake rates of 1 $\mu L\ s^{-1}$ with good analytical figures of merit for synthetic ISF, although they have not yet been evaluated in skin. Solid silicon microneedle arrays (microprojection arrays) functionalized with antibodies to capture protein antigens in ISF have been described recently by Kendall and coworkers (Coffey, Meliga, Corrie, & Kendall, 2016).

As an alternative to machining silicon, various micromolding procedures have been demonstrated, either with photopolymerizable precursors (Barrett, Dawson, & O'Mahony, 2015) or hydrogel-forming monomers (Caffarel-Salvador, Brady, Eltayib, & Meng, 2015; Donnelly et al., 2011; Hardy et al., 2016). An early example used "backside illumination" of SU8 (a common epoxy-based negative photoresist) to form high aspect ratio structures that were subsequently metallized to form hollow metal mn arrays (Kim et al., 2004). Following this Moniz, Michelakis, Trzebinski, and Sharma (2012) functionalized the SU8 μNs for glucose sensing, and as described below SU8 can also be photopolymerized in polydimethylsiloxane (PDMS) molds (Sharma, Huang, Rogers, Boutelle, & Cass, 2016). Micromolding of hydrogel μNs has been used by several groups primarily to extract ISF via swelling. The swollen mn gels can then be extracted intact and their components analyzed.

An alternative to micromolding involves stereolithographic 3D printing. Miller, Moorman, et al. (2016) and Miller et al. (2014) describe a two-photon polymerization procedure to create hollow pyramidal μNs for transdermal potassium ion sensing and also for protein detection.

One particular fabrication method that is very well suited to low-cost, large-scale manufacture is (micro)injection molding. A wide range of medical grade plastics can be processed by this technique; Sharma et al. produced solid polycarbonate μNs (Sharma, Saeed, Johnson, Gadegaard, & Cass, 2016), while Mohan, Windmiller, Mishra, and Wang (2017) used a liquid crystal polymer to also microinjection mold μNs for alcohol sensing.

3. FUNCTIONALIZATION OF MICRONEEDLES FOR SENSING

The functionalization of μNs is necessary where they are being used for in situ sensing and the choice of functionalization conditions depends upon the target analyte being sensed. In virtually every case this entails an electrochemical signal transduction scheme, either potentiometric or amperometric. Solid μNs can be metallized by sputter or e-beam evaporative coating with a metal such as gold or platinum (Sharma, Huang, et al., 2016; Sharma, Saeed, et al., 2016); hollow microneedles have either been packed with a carbon paste/metallic nanoparticle catalyst, with various types of nanocarbon (Miller et al., 2014; Valdés-Ramírez, Li, Kim, & Jia, 2014; Windmiller et al., 2011; Yoon, Lee, Yoo, & Lee, 2013), or had a metal wire inserted into the lumen (Mohan et al., 2017).

Metallized solid μNs have been modified with oxidase enzymes using a variety of procedures including electropolymerization, hydrogel membrane entrapment, and coadsorption with electron transfer mediators such as TTF (Moniz et al., 2012; Sharma, Huang, et al., 2016; Trzebinski et al., 2012) and PEDOT (Invernale et al., 2013; Lee et al., 2016; Ventrelli, Marsilio Strambini, & Barillaro, 2015).

Although enzyme-modified μNs have been most extensively investigated, driven in large part by glucose sensing, the recent paper from the Kendall group (Coffey et al., 2016; Jenkins, Corrie, Flaim, & Kendall, 2012) where they coated solid microneedles with antibodies to affinity extract protein antigens from ISF suggests that it may be feasible to add affinity-based sensing to enzyme based and greatly broaden the range of analytes that could be continuously monitored in this way.

4. METHODS

4.1 Preparation of Microneedle Structures

The microneedle 4 × 4 arrays can be designed using SolidWorks design software (Dassault Systemes). Each 4 × 4 microneedle array comprises pyramidal microneedles that are 1.2 mm in height and 0.6 mm base. The pitch (tip–tip) between the microneedles is 1.2 mm and distance between the base of the microneedles is 0.6 mm. The tip of the microneedle is designed to be 20 μm sharp.

The designs can then be transferred to an electric discharge milling equipment to obtain a metal master milled out of an aluminum or a copper–tungsten block (25 mm × 25 mm × 10 mm).

We describe here two methods for the fabrication of commercial epoxy negative photoresist (SU8 50) and polycarbonate microneedle structures. While in the first method (PDMS) molds are obtained from the aluminum master, in the other method aluminum molds are obtained from the copper–tungsten master.

4.1.1 SU8 50 Microneedle Structures

Equipment

1. Electric discharge milling equipment (JOEMARS EDM AZ50DR)
2. Desktop centrifuge (Fisher)
3. UV chamber with light (365 nm) transilluminator

Materials

1. Pure aluminum block, 25 mm × 25 mm × 10 mm (Erodex, UK)
2. Sylgard 184 (PDMS) kit (VWR, UK)
3. SU8 50 epoxy (Chestech Ltd., UK)

Preparation method

1. The electric discharge milling equipment is set to make two milling passes over the aluminum block followed by a washing step to obtain the aluminum master.
2. The master is then placed in a Petri dish and a degassed PDMS mix (Sylgard 184 mixed in the ratio of 10:1 elastomer:curing agent w/w) poured on the aluminum master. The PDMS mix can be either left at ambient temperature for 48 h or at 60°C to cure in 2 h. Once cured the PDMS mold is transferred to a refrigerator for an hour. *Tip*: Cooling the PDMS mold ensures a smooth release of the mold from the aluminum master.
3. The peeled out mold is then used to create the final microneedle structures. SU8 50 is added to the mold using a plastic dropper. The mold is then transferred to a desiccator and vacuum applied to degas the PDMS mold to ensure that the SU8 50 epoxy reaches the finer parts of the molds especially the inverse of tip structure.
4. The mold with the SU8 50 is then transferred to a UV chamber with a transilluminator to ensure double-side exposure to 365 nm UV light for 30 min.
5. The final SU8 50 microneedle structure can be peeled off the PDMS mold.

4.1.2 Polycarbonate Microneedle Structures

Equipment

1. FANUC ROBOCUT α-OiC (Series 180is-WB) for wire erosion
2. Electric discharge milling equipment (JOEMARS EDM AZ50DR) for spark erosion (add information on the machine)
3. Hydraulic injection molding machine (Victory 28, Engel)

Fig. 1 The fabrication of aluminum molds for polycarbonate microneedle structures. (A) Cu-W masters obtained by wire erosion, (B) aluminum molds obtained after spark erosion, and (C) an electric discharge milling machine used for spark erosion of aluminum block.

Materials

1. Copper–tungsten (Cu-W) standard plate, 200 mm × 62.5 mm × 25 mm (Erodex, UK)
2. Pure aluminum block, 25 mm × 25 mm × 10 mm (Erodex, UK)
3. Polycarbonate (Makrolon OD2015)

Preparation method

1. The FANUC ROBOCUT α-OiC (Series 180is-WB) for wire erosion is set to make three milling passes over the copper–tungsten (Cu-W) block to create master electrodes for spark erosion.
2. The Cu-W master is then used for spark erosion of the aluminum block in JOEMARS EDM AZ50DR to obtain the metal inlay (Fig. 1).
3. Polycarbonate is dried at 110°C for 24 h under vacuum prior to use.
4. The following conditions are applied to the injection molding process: $T_m = 270°C$ (PC melt temperature), $T_w = 80°C$ (tool temperature), $v_i = 20\ cm^3\ s^{-1}$ (injection speed), shot volume of 4.4 cm^3, and a cooling time (t_c) of 5 s.
5. Each polycarbonate microneedle structure produced measures $\approx 25 \times 25 \times 2$ mm.

4.2 Characterization: Electron Microscopy Studies

The microneedle structures fabricated by either of the methods described above can be characterized by scanning electron microscopy (SEM).

Equipment

1. SEM equipment (JEOL 5610)
2. SEM sputtering system

Materials

1. Cu grids, carbon tapes (Ted Pella Inc., CA)

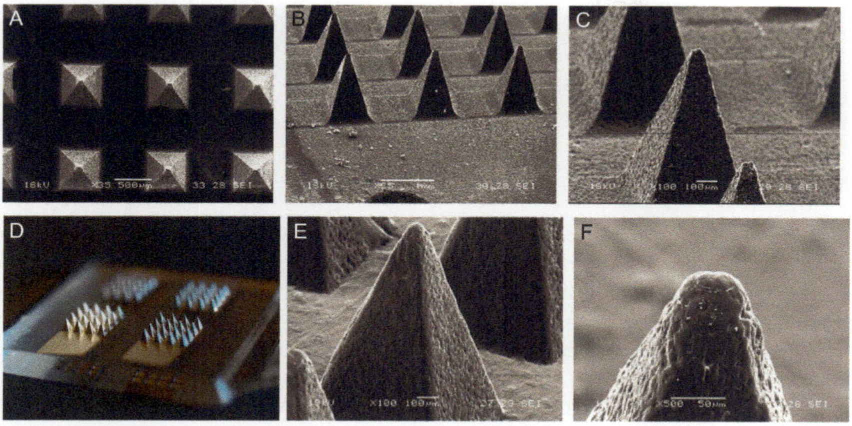

Fig. 2 SEM images of the microneedle devices. *Row above*: (A) top view of the SU8 50 microneedle array, (B) 35 × magnification, and (C) 100 × magnification. *Row below*: (D) a selectively metallized polycarbonate microneedle structure, (E) SEM image of polycarbonate structures (100 ×), and (F) SEM image of polycarbonate structures (500 ×). *Panels (A–C): Reproduced from Sharma, S., Huang, Z., Rogers, M., Boutelle, M., & Cass, A. E. G. (2016). Evaluation of a minimally invasive glucose biosensor for continuous tissue monitoring.* Analytical and Bioanalytical Chemistry, 408*(29), 8427–8435. http://doi.org/10.1007/s00216-016-9961-6. Panels (D–F): Reproduced from Sharma, S., Saeed, A., Johnson, C., Gadegaard, N., & Cass, A. E. (2016). Rapid, low cost prototyping of transdermal devices for personal healthcare monitoring.* Sensing and Bio-Sensing Research, *1–5. http://doi.org/10.1016/j.sbsr.2016.10.004.*

Procedure

1. The microneedle structures can be sputtered with 20 nm of gold or 50 nm of chromium before SEM examination.
2. The microneedle structures can then be fixed to the Cu grid using the carbon tape and imaged at accelerating voltage of 20 kV.
3. Images can be acquired by choosing a higher working distance (20 cm) and tilting the samples at angles up to 60°.
4. After acquiring the images, they are analyzed either manually or using ImageJ software (from NIH, free software) to estimate the dimensions of the microneedle (Fig. 2).

4.3 In Vivo Characterization of the Microneedle Structures

Note: This work involves inserting devices into healthy volunteers, and therefore requires approvals by a research ethics committee and receiving informed consent from the volunteer before the in vivo studies.

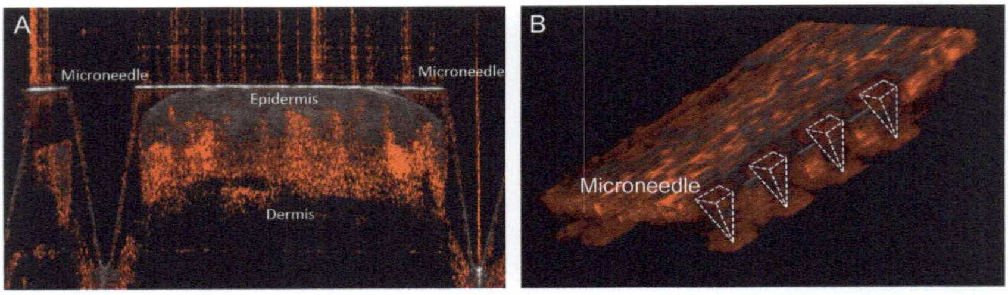

Fig. 3 Optical coherence tomography images from microneedles inserted in a healthy volunteer. (A) Microneedles shown penetrating the epidermis sitting 700 μm in the subdermal space. (B) A 3D image constructed using the ImageJ package showing a row of the pyramidal microneedles inserted in the skin. *Reproduced from Sharma, S., Saeed, A., Johnson, C., Gadegaard, N., & Cass, A. E. (2016). Rapid, low cost prototyping of transdermal devices for personal healthcare monitoring. Sensing and Bio-Sensing Research, 1–5. http://doi.org/10.1016/j.sbsr.2016.10.004.*

Equipment

1. Optical coherence tomography (OCT) imaging instrument (Vivosight, Michelson Diagnostics)

Materials

1. Polycarbonate microneedle structures (unmetallized)
2. 3M Tegaderm film (3M, UK)

Procedure

1. Rest the volunteer's arm on a flat surface, and after wiping the ventral forearm surface with alcohol insert the device, then cover it using a 3M Tegaderm film dressing tape. Apply moderate thumb pressure on the base of the microneedles through the dressing tape for 60 s.
2. The probe of the OCT can be placed on the base plate of the microneedle array structure to scan through the microneedles to image them in situ.
3. The images can be saved in tiff format and analyzed using ImageJ software (Fig. 3).

4.4 Masking, Metallization, and Wire Bonding of Microneedle Electrodes

Equipment

1. DC sputter system, JLS MPS 500
2. Desktop drill machine with 1-mm drill burr

Materials

1. Conformal sputtering metal target (titanium, platinum, and silver)

2. RS Pro Black (Kynar Wire Wrap Equipment Wire, 0.05 mm^2 CSA, 300 V, 0.4 A, 50 m)
3. RS Silver Bottle Paint Conductive Adhesive
4. Araldite resin RX771C/NC and Aradur Hardener HY1300 (CGM072/912) (Robnor Resins Ltd., Swindon, UK)
5. Ferric chloride (Sigma, UK)
6. Ferrocene carboxylic acid (FCA) (Sigma, UK)

Procedure

1. The microneedle structure can be masked with Parafilm for isolation of the microneedle electrodes and selective metallization. In the first step the 3 mm × 0.5 mm-wide Parafilm strips are wrapped to obtain a cross structure; this cross structures will isolate the four (4 × 4) microneedle arrays.
2. In the next step one microneedle array can be covered with Parafilm and coated with an adhesion layer of titanium (15 nm) followed by platinum (50 nm) by conformal sputtering.
3. The platinum microneedle arrays can then be covered with Parafilm and the exposed microneedle array coated with an adhesion layer of titanium (15 nm) followed by silver (150 nm).
4. The silver microneedle array can be modified to Ag/AgCl reference electrode by drop coating 25 μL of a saturated ferric chloride solution. The excess solution of ferric chloride can be washed off after 10 min.
5. The metallized base of the microneedles can be drilled with a pair of holes at a distance from the microneedles, using a 1-mm drill burr, for bonding wires. Wires (15 cm long) stripped off their insulating parts can be passed through the drilled holes and bonded using the RS silver paint and araldite resin.
6. The potential of the integrated reference electrode can be measured against an external reference electrode and should be in the 30 ± 5 mV region.
7. The electrode surface of the microneedle array can be assessed by cyclic voltammetry using 5 m*M* FCA as a redox probe.

4.5 Functionalization and Evaluation of the Microneedle Biosensor

Equipment

1. Multichannel potentiostat: CHI 1030b potentiostat (CHI Instruments) running general-purpose electrochemical software (GPES v4)

Materials

1. Phosphate-buffered saline (PBS) tablets (Sigma, UK)
2. Phenol (Sigma, UK)
3. Glucose oxidase (GOx) (Glucose Oxidase HPS300, Activity 239 U/mg powder at 25°C, source Ex *Aspergillus niger*) (Sekisui Diagnostics (UK), Ltd.)
4. Lactate oxidase (Lactate Oxidase II, Sekisui Diagnostics (UK), Ltd., Units, Activity 41 U/mg powder at 37°C, source Ex *Aerococcus viridans*)
5. Xanthine oxidase (Xanthine Oxidase microbial, Sigma-Aldrich, UK, Activity >7 U/mg powder)
6. Polyurethane (Sigma, UK)
7. Epoxy ATACS5104 epoxy adhesive (Sigma, UK)

Procedure

Two of the three platinum microneedle electrodes can be functionalized to glucose biosensors with an active enzyme film using an electropolymerization method. The third Pt microneedle can be functionalized to a counter electrode using inactivated enzyme film entrapped in the electropolymerized film. Glucose oxidase can be inactivated either using an acidic solution of ammonium sulfate (pH 1–2) or heat treatment (at 70°C for 8 h).

1. Cover one of the Pt microneedle array with Parafilm and place the microneedle structure in a 50 m*M* solution of phenol and 10 mg/mL of GOx enzyme dissolved in 100 m*M* PBS. The pH of the solution has to be buffered to pH 7.2.
2. Each electropolymerization cycle will comprise holding the working electrode at 0 V for 20 s followed by polarizing it to a potential of 0.9 V against the integrated reference electrode and an external counter electrode for 15 min for electropolymerization of the film.
3. Each cycle can be repeated six times to obtain the desired film thickness and glucose oxidase loading.
4. In the next step, cover the Pt microneedle arrays functionalized with GOx films, with Parafilm exposing the Pt microneedle to be functionalized as counter electrode and repeat step 2 with a solution comprising inactivated glucose oxidase.
5. Following electropolymerization, the microneedles should be gently rinsed with deionized water and stored dry overnight at 4°C before use.
6. Electrochemical measurements can be performed in a three-electrode configuration with a potentiostat.

7. Add 200 μL aliquots of varying glucose concentrations prepared in 100 m*M* PBS solution (pH 7.2) stepwise to the devices with the microneedle electrode arrays polarized at 0.7 V against the monolithically integrated Ag/AgCl reference electrode and Pt counter electrode.
8. The steady-state current can be measured at 60 s.
9. Determine the steady-state current as a function of the glucose concentration over the range 0–30 m*M*.
10. Plot the apparent I_{max} and K_m from the current at 60 s against concentration and fit the data with the Michaelis–Menten equation using software such as OriginPro 9.0.

The devices can be sterilized with 25 kGy of ^{60}Co gamma ray irradiation and then used for in vivo measurements.

4.5.1 Lactate Biosensor

1. Follow the procedure mentioned above for glucose biosensor using the enzyme lactate oxidase.
2. For the epoxy-PU membrane, dissolve 0.5 *M* TTF, 17.8 mg each of part A and part B of ATACS5104 epoxy adhesive, and 26.7 mg of PU in 4 mL of THF and 1 mL of Brij 30.
3. Use 10 μL of such membrane for drop coating onto each microneedle array.
4. Leave for drying at room temperature for 20 min.

4.5.2 Theophylline Sensor

1. Follow the procedure mentioned above for glucose biosensor using the enzyme xanthine oxidase.
2. For the epoxy-PU membrane, dissolve 0.5 *M* TTF, 17.8 mg each of part A and part B of ATACS5104 epoxy adhesive, and 26.7 mg of PU in 4 mL of THF and 1 mL of Brij 30.
3. Use 10 μL of such membrane for drop coating onto each microneedle array.
4. Leave for drying in room temperature for 20 min.

ACKNOWLEDGMENTS

The authors wish to acknowledge the MRC funding under the Confidence in Concept program and the NIHR funding under the Invention for Innovation scheme (i4i). This is a summary of independent research partly funded by the National Institute for Health Research (NIHR)'s Invention for Innovation (i4i) Programme (Grant Reference

Number II-LA-0313-20004). The views expressed are those of the author(s) and not necessarily those of the NHS, the NIHR, or the Department of Health.

We would also like to thank Professor Nikolaj Gadegaard and Dr. Anwer Saeed from Glasgow University, Glasgow (UK). We also acknowledge Dr. Jon Holmes from Michelson Diagnostics (Vivosight) for the OCT measurements.

REFERENCES

Barrett, C., Dawson, K., & O'Mahony, C. (2015). Development of low cost rapid fabrication of sharp polymer microneedles for in vivo glucose biosensing applications. *ECS Journal of Solid State Science and Technology*, *4*(10), S3053–S3058. http://doi.org/10.1149/2.0141510jss.

Caffarel-Salvador, E., Brady, A. J., Eltayib, E., & Meng, T. (2015). Hydrogel-forming microneedle arrays allow detection of drugs and glucose in vivo: Potential for use in diagnosis and therapeutic drug monitoring. *PLoS One*, *10*. e0145644. http://doi.org/10.1371/journal.pone.0145644

Chua, B., Desai, S. P., Tierney, M. J., & Tamada, J. A. (2013). Effect of microneedles shape on skin penetration and minimally invasive continuous glucose monitoring in vivo. *Sensors and Actuators, A: Physical*, *203*, 373–381. http://doi.org/10.1016/j.sna.2013.09.026.

Coffey, J. W., Meliga, S. C., Corrie, S. R., & Kendall, M. A. F. (2016). Dynamic application of microprojection arrays to skin induces circulating protein extravasation for enhanced biomarker capture and detection. *Biomaterials*, *84*, 130–143. http://doi.org/10.1016/j.biomaterials.2016.01.015.

Donnelly, R. F., Majithiya, R., Singh, T. R. R., Morrow, D. I. J., Garland, M. J., Demir, Y. K., et al. (2011). Design, optimization and characterisation of polymeric microneedle arrays prepared by a novel laser-based micromoulding technique. *Pharmaceutical Research*, *28*(1), 41–57. http://doi.org/10.1007/s11095-010-0169-8.

El-Laboudi, A., Oliver, N. S., Cass, A., & Johnston, D. (2013). Use of microneedle array devices for continuous glucose monitoring: A review. *Diabetes Technology & Therapeutics*, *15*(1), 101–115. http://doi.org/10.1089/dia.2012.0188.

Fogh-Andersen, N., Altura, B. M., Altura, B. T., & Siggaard-Andersen, O. (1995). Composition of interstitial fluid. *Clinical Chemistry*, *41*(10), 1522–1525.

Hardy, J. G., Larrañeta, E., Donnelly, R. F., McGoldrick, N., Migalska, K., McCrudden, M. T. C., et al. (2016). Hydrogel-forming microneedle arrays made from light-responsive materials for on-demand transdermal drug delivery. *Molecular Pharmaceutics*, *13*(3), 907–914. http://doi.org/10.1021/acs.molpharmaceut.5b00807.

Herrero, P., Georgiou, P., Oliver, N., Johnston, D. G., & Toumazou, C. (2012). A bio-inspired glucose controller based on pancreatic beta-cell physiology. *Journal of Diabetes Science and Technology*, *6*(3), 606–616.

Invernale, M. A., Tang, B. C., York, R. L., Le, L., Hou, D. Y., & Anderson, D. G. (2013). Microneedle electrodes toward an amperometric glucose-sensing smart patch. *Advanced Healthcare Materials*, *3*(3), 338–342. http://doi.org/10.1002/adhm.201300142.

Jenkins, D., Corrie, S., Flaim, C., & Kendall, M. (2012). High density and high aspect ratio solid micro-nanoprojection arrays for targeted skin vaccine delivery and specific antibody extraction. *RSC Advances*, *2*(8), 3490. http://doi.org/10.1039/C2RA20153D.

Kim, K., Park, D. S., Lu, H. M., Che, W., Lee, J. B., & Ahn, C. H. (2004). A tapered hollow metallic microneedle array using backside exposure of SU-8. *Journal of Micromechanics and Microengineering*, *14*, 597. http://doi.org/10.1088/0960-1317/14/4/021).

Lee, H., Choi, T. K., Lee, Y. B., Cho, H. R., Ghaffari, R., Wang, L., et al. (2016). A graphene-based electrochemical device with thermoresponsive microneedles for diabetes monitoring and therapy. *Nature Nanotechnology*, *11*(6), 566–572. http://doi.org/10.1038/nnano.2016.38.

Miller, P., Moorman, M., Manginell, R., Ashlee, C., Brener, I., Wheeler, D., et al. (2016). Towards an integrated microneedle total analysis chip for protein detection. *Electroanalysis*, *28*, 1305–1310. http://doi.org/10.1002/elan.201600063.

Miller, P. R., Narayan, R. J., & Polsky, R. (2016). Microneedle-based sensors for medical diagnosis. *Journal of Materials Chemistry B*, *4*(8), 1379–1383. http://doi.org/10.1039/C5TB02421H.

Miller, P. R., Xiao, X., Brener, I., Burckel, D. B., Narayan, R., & Polsky, R. (2014). Microneedle-based transdermal sensor for on-chip potentiometric determination of K(+). *Advanced Healthcare Materials*, *3*(6), 876–881. http://doi.org/10.1002/adhm.201300541.

Mohan, A. M. V., Windmiller, J. R., Mishra, R. K., & Wang, J. (2017). Continuous minimally-invasive alcohol monitoring using microneedle sensor arrays. *Biosensors and Bioelectronics*, *91*, 574–579. http://doi.org/10.1016/j.bios.2017.01.016.

Moniz, A., Michelakis, K., Trzebinski, J., & Sharma, S. (2012). Minimally invasive enzyme microprobes: An alternative approach for continuous glucose monitoring. *Journal of Diabetes Science and Technology*, *6*(2), 479–480. http://doi.org/10.1177/193229681200600239.

Mukerjee, E., Collins, S. D., Isseroff, R. R., & Smith, R. L. (2004). Microneedle array for transdermal biological fluid extraction and in situ analysis. *Sensors and Actuators A: Physical*, *114*(2-3), 267–275. http://doi.org/10.1016/j.sna.2003.11.008.

Oliver, N. S., Toumazou, C., Cass, A. E. G., & Johnston, D. G. (2009). Glucose sensors: A review of current and emerging technology. *Diabetic Medicine: A Journal of the British Diabetic Association*, *26*(3), 197–210. http://doi.org/10.1111/j.1464-5491.2008.02642.x.

Sharma, S., Huang, Z., Rogers, M., Boutelle, M., & Cass, A. E. G. (2016). Evaluation of a minimally invasive glucose biosensor for continuous tissue monitoring. *Analytical and Bioanalytical Chemistry*, *408*(29), 8427–8435. http://doi.org/10.1007/s00216-016-9961-6.

Sharma S., Saeed A., Johnson C., Gadegaard N. and Cass A.E., (2016). Rapid, low cost prototyping of transdermal devices for personal healthcare monitoring, *Sensing and Bio-Sensing Research*, 1–5. http://doi.org/10.1016/j.sbsr.2016.10.004. (in press).

Strambini, L. M., Longo, A., Scarano, S., Prescimone, T., Palchetti, I., Minunni, M., et al. (2015). Self-powered microneedle-based biosensors for pain-free high-accuracy measurement of glycaemia in interstitial fluid. *Biosensors and Bioelectronics*, *66*, 162–168. http://doi.org/10.1016/j.bios.2014.11.010.

Trzebinski, J., Sharma, S., Radomska-Botelho Moniz, A., Michelakis, K., Zhang, Y., & Cass, A. E. G. (2012). Microfluidic device to investigate factors affecting performance in biosensors designed for transdermal applications. *Lab on a Chip*, *12*, 348–352. http://doi.org/10.1039/C1LC20885C.

Valdés-Ramírez, G., Li, Y. C., Kim, J., & Jia, W. (2014). Microneedle-based self-powered glucose sensor. *Electrochemistry Communications*, *47*, 58–62. http://doi.org/10.1016/j.elecom.2014.07.014.

Ventrelli, L., Marsilio Strambini, L., & Barillaro, G. (2015). Microneedles for transdermal biosensing: Current picture and future direction. *Advanced Healthcare Materials*, *4*(17), 2606–2640. http://doi.org/10.1002/adhm.201500450.

Windmiller, J. R., Zhou, N., Chuang, M.-C., Valdés-Ramírez, G., Santhosh, P., Miller, P. R., et al. (2011). Microneedle array-based carbon paste amperometric sensors and biosensors. *The Analyst*, *136*(9), 1846–1851. http://doi.org/10.1039/c1an00012h.

Yoon, Y., Lee, G. S., Yoo, K., & Lee, J.-B. (2013). Fabrication of a microneedle/CNT hierarchical micro/nano surface electrochemical sensor and its in-vitro glucose sensing characterization. *Sensors*, *13*(12), 16672–16681. http://doi.org/10.3390/s131216672.

CHAPTER SIXTEEN

Visualization of the Genomic Loci That Are Bound by Specific Multiprotein Complexes by Bimolecular Fluorescence Complementation Analysis on *Drosophila* Polytene Chromosomes

Huai Deng[1], Tom K. Kerppola[2]
University of Michigan, Ann Arbor, MI, United States
[2]Corresponding author: e-mail address: kerppola@umich.edu

Contents

Abstract

We have developed a procedure that enables visualization of the genomic loci that are bound by complexes formed by a specific combination of chromatin-binding proteins. This procedure is based on imaging bimolecular fluorescence complementation (BiFC)

[1] Current address: Department of Biology, University of Minnesota Duluth, 253A SSB, 1035 Kirby Drive, Duluth, MN 55812, United States.

Methods in Enzymology, Volume 589
ISSN 0076-6879
http://dx.doi.org/10.1016/bs.mie.2017.02.003

complexes on *Drosophila* polytene chromosomes. BiFC complexes are formed by the facilitated association of two fluorescent protein fragments that are fused to proteins that interact with, or are in close proximity to, each other. The intensity of BiFC complex fluorescence at individual genomic loci is greatly enhanced by the parallel alignment of hundreds of chromatids within the polytene chromosomes. The loci that are bound by the complexes are mapped by comparing the locations of BiFC complex fluorescence with the stereotypical banding patterns of the chromosomes. We describe strategies for the design, expression, and validation of fusion proteins for the analysis of BiFC complex binding on polytene chromosomes. We detail protocols for the preparation of polytene chromosome spreads that have been optimized for the purpose of BiFC complex visualization. Finally, we provide guidance for the interpretation of results from studies of BiFC complex binding on polytene chromosomes and for comparison of the genomic loci that are bound by BiFC complexes with those that are bound by subunits of the corresponding endogenous complexes. The visualization of BiFC complex binding on polytene chromosomes provides a unique method to visualize multiprotein complex binding at specific loci, throughout the genome, in individual cells.

1. INTRODUCTION

Protein interactions on chromatin are thought to regulate transcription in response to extracellular and intracellular signals. The mechanisms whereby such protein interactions produce the unique expression pattern of each individual gene remain incompletely characterized. In particular, the roles of protein interactions in determination of the genes that are bound by specific transcription factors are not known for the majority of proteins. Studies of genomewide binding by individual transcription factors have shown that many transcription factors can bind to the same regions (ENCODE Consortium, 2012). In most cases, it is not known if these proteins bind to the regions as complexes in the same cell or if they bind to the regions independently or in different cells. These issues are particularly difficult to address for tissues that contain many different cell types. Methods that enable the direct visualization of protein complexes that are bound to chromatin at specific genomic loci will provide valuable insight into the roles of protein interactions in the determination of the chromatin-binding specificities of transcription regulatory proteins in individual cells.

Many methods have been developed to detect protein binding to specific genomic loci in cells (e.g., chromatin immunoprecipitation (ChIP), Dam-ID, DNase footprinting) (Galas & Schmitz, 1978; Gilmour & Lis, 1984; van Steensel & Henikoff, 2000). These methods have been adapted to determine the locations of protein binding throughout the genome

(ChIP-seq, Dam-ID, DNase-Seq) (Boyle et al., 2008; Johnson, Mortazavi, Myers, & Wold, 2007; van Steensel, Delrow, & Henikoff, 2001). Most of these methods require a relatively large number of cells to determine the locations of protein binding (Table 1). These methods generally cannot distinguish if a protein binds to different genomic locations in the same cell or in different cells in a population. It is generally not possible to determine if two different proteins bind together as a complex or if they bind to the same location independently or in different cells. Sequential ChIP (re-ChIP) analysis can be used to determine if different proteins bind to the same chromatin segment (Hatzis & Talianidis, 2002). However, this does not distinguish if the proteins bind independently or as a complex.

Some methods enable direct visualization of protein binding to chromatin in individual cells by imaging proteins on chromatin (e.g., immunostaining or fluorescent fusion protein binding to polytene chromosomes; BiFC complexes with histones, immunofluorescence-FISH) (Lis, 2007; Masumoto, Sugimoto, & Okazaki, 1989; Schwartz, Werner, & Lis, 2004; Silver & Elgin, 1976; Vincenz & Kerppola, 2008; Zhimulev et al., 2004). These methods can be used to detect chromatin binding by several proteins in the same cells (Schwartz et al., 2004; Shaffer et al., 2002). The limited spatial resolution of imaging prevents determination whether the proteins bind to the same loci as a complex or independently (Table 1).

Many important questions cannot be answered using the standard assays for the detection of protein binding to specific genomic loci. In particular, it is not possible to determine the chromatin loci that are bound by protein complexes that contain a specific combination of interaction partners. For this purpose, methods that enable simultaneous detection of protein interaction and of chromatin binding are required.

Bimolecular fluorescence complementation (BiFC) analysis enables visualization of protein interactions in living cells (Hu, Chinenov, & Kerppola, 2002). In this assay, the interaction between two proteins is detected based on the fluorescence that is produced by the association of two fragments of a fluorescent protein that are fused to the two interaction partners (Fig. 1A). BiFC analysis has been adapted to enable the visualization of protein interactions in primary cells and in tissues (Bai & Kerppola, 2011; Diaz, Martinez, Isabel-LaMoneda, Rubio-Somoza, & Carbonero, 2005; Gohl, Banovic, Grevelhorster, & Bogdan, 2010; Hudry, Viala, Graba, & Merabet, 2011; Plaza et al., 2008). BiFC analysis has also been used to visualize epigenetic regulatory protein binding to large chromosomal regions on mitotic chromosomes in cultured cells (Vincenz & Kerppola, 2008). Because

Table 1 Comparison of the Characteristics of Different Methods for Genomewide Identification of the Locations of Protein and Protein Complex Binding

Method	BiFC on Polytene Chromosome	BiFC with Histones	IF on Polytene Chromosome	ChIP-seq	Dam-ID	DNase-Seq
Characteristic						
Target	Specific protein complex	Specific fusion protein	Specific endogenous protein(s)	Single endogenous protein	Single fusion protein	Any chromatin-binding protein
Resolution	Hundreds kb	Chromosome regions	Hundreds kb	~100–1000 bp	~200–500 bp	~10–100 bp
Sample size	Single cells	Single cells	Single cells	Cell population	Cell population	Cell population
Potential caveats	Exogenous BiFC fusion protein expression	Exogenous BiFC fusion protein expression	Antibody specificity	Antibody specificity, signal amplification	Exogenous Dam fusion protein expression, signal amplification	DNase I specificity, signal amplification
Applicability	*Drosophila* salivary glands	Cells with exogenous protein expression	*Drosophila* salivary glands	Sequenced genomes	Sequenced genomes	Sequenced genomes

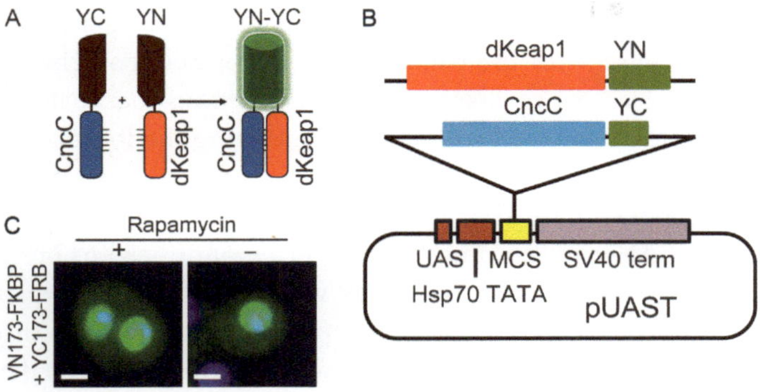

Fig. 1 (A) Diagram of BiFC complex formation by proteins fused to fluorescent protein fragments (YN and YC). (B) Plasmids encoding dKeap1 and CncC fused to fluorescent protein fragments (YN and YC) were constructed using the pUAST vector (Brand & Perrimon, 1993). The pUAST vector contains five tandem GAL4-binding sites (UAS) that mediate transgene activation by GAL4 and sequences that facilitate transcription and P-element sequences that facilitate transgene insertion into the *Drosophila* genome. (C) BiFC analysis of the specificity of association of FKBP and FRB fusions to fragments of the Venus (VN173) and YFP (YC173) fluorescent proteins in S2 cells. The cells were cultured in the absence (−) and in the presence (+) or rapamycin, which induces FKBP–FRB interaction (Rivera et al., 1996). The fluorescence that is observed in the absence of rapamycin indicates that the fluorescent protein fragments associate spontaneously. Scale bars: 2 μm.

of the limited resolution of fluorescence microscopy, it has not been possible to map the specific genomic loci that are bound by protein complexes by using BiFC analysis in diploid cells. To overcome this limitation, we adapted BiFC analysis to visualize protein complex binding at specific endogenous genomic loci on *Drosophila* polytene chromosomes (Fig. 3A) (Deng & Kerppola, 2014).

The visualization of BiFC complex binding on polytene chromosomes requires the selective expression of two fusion proteins in salivary glands at appropriate levels. Fusion proteins can be expressed in the salivary glands of transgenic *Drosophila* larvae by using the GAL4-UAS strategy (Brand & Perrimon, 1993). Using this method, BiFC complexes representing homodimers of the Scr homeodomain protein were detected in salivary gland cells and were found to be enriched in a cluster in salivary gland cells that contain multiple tandem copies of a synthetic Scr binding site, consistent with chromatin binding (Papadopoulos, Skouloudaki, Adachi, Samakovlis, & Gehring, 2012). The endogenous genomic loci that are bound by fluorescent proteins can be mapped by comparing the locations of protein fluorescence with the banding patterns of polytene chromosomes (Deng & Kerppola, 2013). Likewise, the endogenous loci that are bound by BiFC complexes can be identified by comparing the locations of BiFC complex fluorescence with the polytene chromosome banding patterns (Deng & Kerppola, 2014). For this purpose, the conditions for spreading and imaging of polytene chromosomes must be optimized to preserve the fluorescence of BiFC complexes and to visualize the banding patterns of polytene chromosomes. Conditions for spreading and imaging of fluorescent fusion protein binding on polytene chromosomes have been developed (DiMario, Rosby, & Cui, 2006; Johansen et al., 2009). These conditions eliminate the acids that are used in conventional protocols that can quench the fluorescence of many fluorescent proteins. Using this strategy, we identified specific chromatin loci that are bound by BiFC complexes that are formed by dKeap1 (*Drosophila* Keap1) and CncC (*Drosophila* Nrf2 homolog) fusion proteins (Deng & Kerppola, 2014).

The use of BiFC analysis to visualize protein complex binding to polytene chromosomes enables direct detection of the genomic loci that are bound by protein complexes that contain a specific combination of interaction partners in individual cells. The approach can be used to compare the binding specificities of protein complexes and of individual proteins in parallel. This can identify mechanisms that affect the chromatin-binding specificity of transcription regulatory protein complexes. The current approach has several limitations, including the requirement for polytene chromosomes that are found mainly in the salivary glands of *Drosophila*. The stable association of fluorescent protein fragments can also alter the dynamics of fusion protein interactions and could influence their binding to chromatin.

2. DESIGN OF FUSION PROTEINS FOR BiFC ANALYSIS

2.1 Fusion proteins that contain fragments of fluorescent proteins fused to the protein of interest are designed using principles that have been described previously (Hu et al., 2002; Hu & Kerppola, 2003; Kerppola, 2006). The fluorescent protein fragments can be fused to the N- or C-termini of the proteins of interest. The efficiency of BiFC complex formation often differs depending on the locations of the fluorescent protein fragments in the complex. It is therefore important to test many different combinations of fusion proteins to identify the combination with the highest efficiency and specificity of BiFC complex formation. To reproduce the functions of the native proteins, it is best to avoid fusing the fragments near protein regions that have known functions. The functions of the fusion proteins should be tested using several independent assays that enable comparison of their functions with those of the native proteins.

2.2 It is generally best to design fusion proteins that encompass the complete proteins of interest. Fusion proteins can also be constructed that contain selected regions of the proteins of interest. The characteristics of such proteins may or may not correspond to a subset of the characteristics of the intact proteins. In all studies of protein interactions, it is important to include negative control proteins that contain substitutions that are predicted to disrupt the interaction. Such control proteins should be analyzed in parallel with, and under conditions identical to those that are used to analyze the intact proteins. Unrelated proteins are not useful controls since many of their characteristics are likely to differ from those of the proteins of interest.

2.3 Many different fluorescent protein fragments have been used to investigate protein interactions in cultured cells (Chu et al., 2009; Hu et al., 2002; Hu & Kerppola, 2003; Robida & Kerppola, 2009; Saka, Hagemann, Piepenburg, & Smith, 2007; Shyu, Liu, Deng, & Hu, 2006). Fragments of different proteins have different characteristics that confer distinct advantages and disadvantages. We have used fragments of YFP that end at residues 155 and 173, and have obtained results that are consistent with facilitated complex formation by these fragments in *Drosophila* tissues when they are fused to interaction partners, but not when they are fused to noninteracting proteins (Deng & Kerppola, 2014). A variety of different fluorescent protein fragments have been used in studies with cultured mammalian cells, including fragments of the Venus fluorescent protein. BiFC complex formation by fusion proteins that contain fragments of Venus often results in a higher fluorescence intensity than BiFC complex formation by fusion proteins that contain YFP fragments. At least some combinations of fusion proteins that contain fragments of Venus produce spontaneous BiFC complexes in the absence of a specific interaction between the proteins that are fused to the Venus fragments (Fig. 1C) (Robida & Kerppola, 2009; Saka et al., 2007). The advantages and disadvantages of different fluorescent protein fragments in the investigation of protein interactions in *Drosophila* have not been established.

3. CONSTRUCTION AND TESTING OF PLASMID VECTORS FOR BiFC ANALYSIS

3.1 The principal consideration in the choice of the strategy to be used for expression of the fusion proteins is the purpose of the experiments. This choice often involves a trade-off between a high level of expression to ensure that interaction partners form a level of BiFC complexes that can be detected and a lower level of expression that is more similar to, and therefore more likely to reflect the properties of, the endogenous proteins. The GAL4-UAS system enables fusion protein expression in specific tissues, including salivary glands. The original version of this approach does not enable control of the level of protein expression, which is often much higher than that of the corresponding endogenous proteins. We have also used this strategy to express BiFC fusion proteins in the developing heart, muscle, and imaginal disc cells.

3.2 To express BiFC fusions using the GAL4-UAS system, plasmid vectors based on the pUAST plasmid need to be constructed (Brand & Perrimon, 1993). pUAST plasmids that encode the proteins of interested fused to fluorescent protein fragments are constructed by using standard molecular biology methods. Many different linker sequences have been used to connect different proteins to fluorescent protein fragments. The effects of different linker sequences on the efficiency and specificity of BiFC complex formation have not been tested directly. In this chapter, we describe results from experiments using dKeap1 and CncC proteins that are fused to the N-termini of YN173 and YC173 (Fig. 1B).

3.3 Because of the substantial effort that is required to generate *Drosophila* lines that express fusion proteins in specific tissues, it is best to test the efficiency and specificity of BiFC complex formation first in cultured cells. BiFC complex formation can be visualized in cultured *Drosophila* cell lines such as Schneider 2 (S2) cells (Fig. 1C). This can expedite the experiments by enabling use of the same plasmids for transient expression in cultured cells for the purpose of testing BiFC complex formation, and for the generation of transgenic *Drosophila* lines. The constructs can be transfected into S2 cells using Fugene 6 or other transfection reagents.

Important controls: Plasmids that encode the intact fusion proteins as well as fusion proteins in which residues that are essential for the interaction are substituted should be compared in parallel. Plasmids that encode fusion proteins that are known to produce BiFC complexes in the cells should be included as positive controls. The efficiency of plasmid introduction and expression in the cells should be determined by the expression of an intact fluorescent protein. The proportion of cells that express interaction partners that have detectable BiFC fluorescence is often less than 10% of the number of cells that express an intact fluorescent protein. The proportion of cells that express noninteracting proteins that have detectable fluorescence is expected to be at least an order of magnitude lower.

Troubleshooting 1. Result: Low frequency of fluorescent cells in transient expression assays. It can be difficult to transfect S2 cells with high efficiency. In general, low transfection efficiency is not a problem as long as the fluorescence intensity produced by a specific interaction is sufficiently high to be distinguished from that which is produced by noninteracting control fusion proteins. If the efficiency is not sufficient, several alternative strategies can be tried: (1) Construct additional plasmid vectors with different configurations of fluorescent protein fragments fusions; (2) Modify the method used for cell transfection; (3) Use a different cell line for transfection such as human embryonic kidney HEK293T cells.

Troubleshooting 2. Result: The intensity of BiFC fluorescence is very weak. Many BiFC complexes form inefficiently and produce much lower fluorescence intensity than intact fluorescent proteins. The low fluorescence intensity is not a problem as long as it is sufficiently high to be distinguished from that which is produced by noninteracting control fusion proteins. If the fluorescence intensity is not sufficient, several alternative strategies can be tried: (1) Use a high-sensitivity CCD camera and a long exposure time; (2) Modify the method used for cell transfection; (3) Use a different cell line for transfection such as HEK293T cells; (4) Construct additional plasmid vectors encoding proteins in which the fluorescent protein fragments are fused to different positions of the putative interaction partners; (5) Use fluorescent protein fragments that produce BiFC complexes with higher efficiencies—note that some fragments, including fragments of Venus, can form BiFC complexes independently of a specific interaction (Fig. 1C) (Robida & Kerppola, 2009; Saka et al., 2007).

Troubleshooting 3. Result: Amino acid substitutions that are predicted to eliminate protein interactions do not reduce the BiFC complex fluorescence. This could be due to either a low efficiency of BiFC complex formation by the intact fusion proteins or a high efficiency of BiFC complex formation by the mutant fusion proteins. To address the former issue, see troubleshooting 1 and 2 earlier. To address the latter issue, it is recommended to use fluorescent protein fragments that have a low propensity to associate with each other in the absence of a specific interaction between proteins fused to the fragments (Robida & Kerppola, 2009). It is also possible that a high level of fusion protein expression favors complex formation by proteins that do not form complexes efficiently at lower concentrations. Finally, it is possible that the amino acid substitutions do not eliminate complex formation by the proteins of interest.

4. PRODUCTION OF *DROSOPHILA* LINES THAT EXPRESS BiFC FUSION PROTEINS

4.1 Plasmids that express proteins of interest fused to fluorescent protein fragments as well as plasmids that express control proteins containing substitutions at residues that are predicted to be essential for the interaction are microinjected into *Drosophila* embryos. Transgenic lines are established and selected as described (Bachmann & Knust, 2008; Rubin & Spradling, 1982). Multiple independent lines should be produced in order to evaluate potential effects of the integration site on the levels of fusion protein expression and on BiFC complex formation. BiFC complex formation should be compared between lines that express the intact proteins and lines that express the mutant proteins at comparable levels.

4.2 The chromosomal locations of the transgenes are mapped as described (Bachmann & Knust, 2008). Some alleles can be deleterious, especially when homozygous. Such alleles should be maintained using appropriate balancer chromosomes (Robert, 1998). Complementary transgenes that are integrated on chromosomes 2 and 3 make the derivation of double transgenic lines most convenient. Lines that contain two transgenes on the same chromosome can be derived by homologous recombination (Robert, 1998). The transgenes that express fusion proteins should be maintained as a stock separate from GAL4 transgenes to avoid potential selection for mutants that alter fusion protein expression.

4.3 The GAL4 driver transgene should be selected based on two criteria: (1) production of the lowest level of fusion proteins that are necessary for the detection of BiFC complex formation by interaction partners and (2) specific expression in salivary glands or the tissue(s) of interest.

Expression of the fusion proteins in many different tissues increases the risk that they have deleterious effects. Such expression could reduce the viability or the fertility of the flies, complicating the experiments.

4.4 To visualize protein complex binding to specific chromosomal loci, we expressed BiFC fusion proteins in the salivary glands of third instar larvae. We used the Sgs3-GAL4 driver, which promotes UAS-controlled transcription selectively in salivary gland cells during the late third instar stage (Cherbas, Hu, Zhimulev, Belyaeva, & Cherbas, 2003). Other GAL4 drivers that promote transcription in salivary gland cells, including fkh-GAL4, 71B-GAL4, and 34B-GAL4, could potentially be used (Gerlitz, Nellen, Ottiger, & Basler, 2002; Maybeck & Roper, 2009). These drivers potentially promote higher levels of fusion protein expression in salivary glands or in other tissues. This can reduce the selectivity of BiFC complex formation and binding to specific genomic loci, and could have deleterious effects for the flies.

Troubleshooting 1. Result: The combination of two transgenes encoding BiFC fusion proteins has deleterious effects even in the absence of GAL4 driver expression. In the event that long-term maintenance of the two BiFC transgenes in the same line produces a deleterious phenotype, they can be maintained separately and lines that produce both fusion proteins can be produced by crossing one line first with the GAL4 driver line and subsequently with the other BiFC fusion protein line.

Troubleshooting 2. Result: The expression two BiFC fusion proteins have deleterious effects. In the event that expression of the two BiFC transgenes under the control of a GAL4 driver has deleterious effects, alternate expression strategies can be used: (1) Test multiple transgenic lines—lines with lower levels of expression can have fewer deleterious effects; (2) Test GAL4 drivers that promote different levels of expression in different tissues; (3) Test inducible expression strategies, including GAL80ts, GeneSwitchGAL4, and GAL4-UAS combined with tet-on or tet-off systems (Han, Stein, & Stevens, 2000; Osterwalder, Yoon, White, & Keshishian, 2001; Stebbins et al., 2001; Zeidler et al., 2004).

5. COMPARISON OF THE LEVELS OF BiFC FUSION PROTEIN EXPRESSION

5.1 The levels of fusion protein expression can affect the specificity of their interaction. It can also alter their subcellular localization, functions, and chromatin binding. The GAL4-UAS system can produce levels of fusion proteins that exceed those of endogenous proteins by several orders of magnitude. To identify potential artifacts that are caused by fusion protein overexpression, it is essential to compare the characteristics of the ectopically expressed fusion proteins with those of their endogenous counterparts. Usually different transgenic lines express different levels of the fusion proteins. Comparison of the results that are obtained using different transgenic lines with different levels of fusion protein expression can facilitate determination if differences in the levels of fusion protein expression affect the results.

5.2 To compare the levels of fusion protein expression with their endogenous counterparts, extracts of the tissues where they are expressed can be analyzed by immunoblotting. In many cases, fusion proteins are expressed at levels that exceed their endogenous counterparts by more than an order of magnitude (Fig. 2A). Ideally, the tissues that are

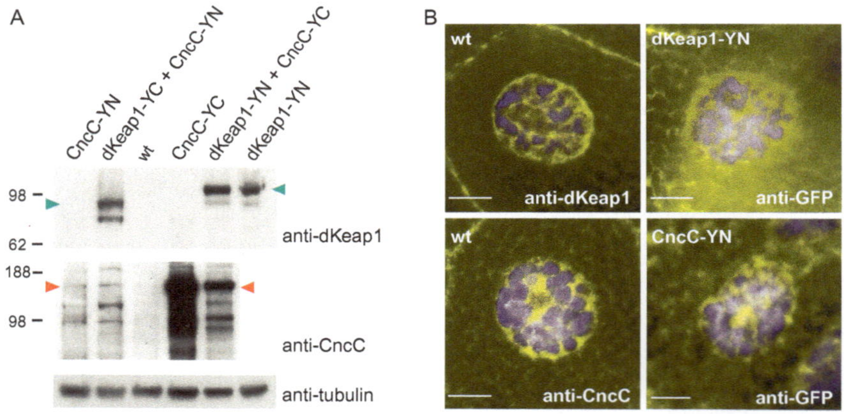

Fig. 2 Comparison of the levels of protein expression and subcellular localization of dKeap1 and CncC fusion proteins that are expressed at different levels with their endogenous counterparts. (A) The salivary glands of larvae that express the fusion proteins indicated above each lane were excised and the levels of protein expression were analyzed by immunoblotting using the antibodies indicated to the right of each image. The endogenous dKeap1 and CncC proteins were not detected in these exposures. The bands representing dKeap1 (*cyan*) and CncC (*red*) fusion proteins are labeled by *arrowheads*. (B) The salivary glands of wild-type larvae (*left column*) and of larvae that expressed the fusion proteins indicated at the *top* of the images (*right column*) were analyzed by immunofluorescence (*yellow*) using the antibodies indicated at the *bottom* of the images. The nuclei were labeled by Hoechst staining (*blue*). Scale bars: 10 μm. *Panel (A) Image modified from Deng, H., & Kerppola, T. K. (2014). Visualization of the Drosophila dKeap1-CncC interaction on chromatin illumines cooperative, xenobiotic-specific gene activation.* Development, 141, *3277–3288.*

analyzed should be matched with those where the GAL4 driver is known to be expressed.

5.2.1 To analyze the levels of fusion proteins in salivary glands, *Drosophila* larvae are prepared as described in part 6.

5.2.2 Carefully remove wandering third instar larvae from the vial using tweezers. Briefly rinse the larva with water and transfer the larva into PBS. Under a stereoscope, dissect salivary glands quickly as described in part 7.1.3.

5.2.3 Immediately transfer the salivary gland into 30 μL IP buffer (with freshly added protease inhibitors) in a 0.2-mL glass tissue grinder (KONTES) on ice. Dissect additional larvae to harvest 3–6 pairs of salivary glands in 30 μL IP buffer per protein whose expression is to be analyzed.

5.2.4 Apply prechilled pestle to grind tissue 10 times on ice. Transfer the sample into a 0.5-mL tube. Add 7 μL of 5 × SDS-loading buffer and mix well. Boil for 10 min. Clear the lysate by centrifuging for 1 min at >12,000 × *g* at room temperature. Load the supernatant onto an SDS-PAGE gel or store at −80°C.

5.2.5 Analyze the cleared lysate using SDS-PAGE followed by immunoblotting, as described (Wang, Zhang, Jin, Johansen, & Johansen, 2001). The fusion proteins and their endogenous counterparts are detected using antibodies directed against the endogenous proteins. Anti-tubulin antibody should be used to evaluate the amounts of protein loaded in different lanes.

Troubleshooting. Result: Weak or undetectable signals on immunoblotting. The levels of many endogenous proteins are difficult to detect in salivary glands. The following approaches can be used to improve their detection: (1) Ensure optimal immunoblotting protocol and time of exposure. Use proteins from other sources (recombinant, cultured cells) to optimize the assay; (2) Use larger numbers (10–20) of salivary glands for extraction; (3) Ensure efficient extraction of the proteins from the tissue by evaluating the extraction of other endogenous protein such as tubulin or lamin Dm0.

5.3 The subcellular localization and the chromatin-binding specificities of the fusion proteins should be compared with their endogenous counterparts (Fig. 2B). This should be done for each of the fusion proteins expressed separately and in combination. Differences in the subcellular localization or chromatin binding by the fusion proteins compared to their endogenous counterparts should be considered carefully when interpreting results of BiFC analysis. Differences in the subcellular localization or chromatin binding of fusion proteins when they are expressed in combination vs separately could reflect effects of their interactions on their localization or chromatin binding.

5.3.1 To determine the subcellular localization of fusion proteins in salivary gland cells, *Drosophila* larvae are prepared and their salivary glands are excised from the larvae as described in parts 6 and 7.1.3. In general, it is not necessary to completely remove mouth hooks and other attached tissues (brain, imaginal discs). The salivary glands are fixed and immunostained using antibodies directed against the fluorescent protein fragments and the endogenous proteins as described (Phalle Bde, 2004).

5.3.2 The immunostained salivary glands are imaged by fluorescence microscopy. The distributions of the fusion proteins in salivary glands from the transgenic larvae are compared with the distributions of the endogenous proteins in salivary glands from wild-type larvae.

Troubleshooting. Result: Weak or undetectable immunofluorescence signals. The levels of many endogenous proteins are difficult to detect in tissues. Use a sensitive CCD camera and a long exposure time. Optimize the immunostaining protocol by varying the concentrations and incubation time of primary and secondary antibodies. Analyze proteins that are known to be present at high levels in salivary glands, or other tissues known to contain a high level of the protein of interest in parallel as controls.

6. PREPARATION OF *DROSOPHILA* LARVAE FOR SALIVARY GLAND ISOLATION

6.1 To produce larvae that express BiFC fusions in the salivary glands, cross the Sgs3-GAL4 driver line with a double transgenic BiFC fusion protein line (e.g., Sgs3-GAL4/Sgs3-GAL4 × UAS-dKeap1-YC/UAS-dKeap1-YC; UAS-CncC-YN/UAS-CncC-YN). Either male or female flies carrying the BiFC transgenes can be crossed with female or male flies carrying the GAL4 driver. Ideally, equal numbers of male and female larvae should be used for optimal mating efficiency. In cases where limited numbers of flies of one sex are available a ratio of up to 1:5 males:females can be used. We prefer crossing male flies carrying the BiFC transgenes to female flies carrying the GAL4 driver since: (1) The GAL4 transgenic females generally have high fertility, which can increase the number of progeny; (2) These females can be readily amplified to perform multiple crosses with males that carry different combinations of BiFC transgenes to compare BiFC complex formation by different combinations of fusion proteins.

6.2 To enhance replication of the polytene chromosomes to high copy number and to improve polytene chromosome spreading, culture the F1 progeny larvae at 21°C under conditions of minimal crowding (place four females and four males in a vial and change the vial every day). Larvae normally climb up the sides of the vial within 6–7 days after mating. Select the fattest third instar wandering larvae among the group that are the first to emerge after the mating. The larvae should be harvested immediately before pupation. Pupation is preceded by reduced mobility of the larvae on the wall of the vial. Larvae can also be raised at 18°C, which can increase the thickness of the polytene chromosomes (Johansen et al., 2009). An additional benefit of the lower temperature is that it reduces transcription of the transgenes, which can reduce the levels of fusion protein expression.

Troubleshooting 1. Result: Very few or no progeny are produced. If no progeny are produced in multiple matings, it is likely that either one of the lines is not fertile. Test each line with wild-type flies in parallel to verify the conditions and to identify the infertile line. If the double transgenic BiFC fusion line is infertile, mate the two transgenic lines encoding different BiFC fusions sequentially to the GAL4 line. Also, try to switch the sexes of the flies that carry the various transgenes.

Troubleshooting 2. Result: Third instar larvae that express BiFC fusions have small salivary glands. Overexpression of some proteins can inhibit the growth or cause the death of salivary gland cells prior to full amplification of the polytene chromosomes. To overcome this problem, test lines with lower levels or a shorter time of transgene expression. If necessary, protein expression can be regulated using GAL80ts, GeneSwitchGAL4, or tet-on/off systems (Han et al., 2000; Osterwalder et al., 2001; Stebbins et al., 2001; Zeidler et al., 2004).

7. PREPARATION OF POLYTENE CHROMOSOME SPREADS

7.1 Isolation and fixation of salivary glands

7.1.1 In the conventional squash protocol, acetic and lactic acids are added to facilitate chromosome spreading. This increases the resolution of the banding patterns. These acids quench the intrinsic fluorescence of YFP and BiFC complexes. To avoid quenching of the BiFC fluorescence, polytene chromosome spreads should be prepared using an acid-free squash protocol (DiMario et al., 2006; Johansen et al., 2009). In this procedure, salivary glands are fixed in dilute formaldehyde and squashed after incubation in 50% glycerol. The high viscosity of the squash medium helps preserve the chromosome morphology during spreading (Johansen et al., 2009).

7.1.2 Carefully remove a wandering larva from the vial using tweezers. Briefly rinse the larva with water in a dish. Transfer the larva into ≥2 mL of freshly made Brower's Fixation Buffer in a Petri dish at room temperature.

7.1.3 To isolate and fix salivary glands, under a dissecting microscope, dissect the larva as follows (Johansen et al., 2009). Using two pairs of tweezers, grasp the mouth hooks and hold the body at a position about 2/3 of the way down. Pull on the mouth hooks to open the body and expose the salivary glands. Remove the brain, imaginal discs, and fat tissue surrounding the salivary glands. The dissection should be completed during fixation, which generally takes 2–3 min. The time of fixation may need to be optimized for different proteins. Transfer the salivary gland into 2 mL PBST and incubate 2–3 min. Then transfer into 2 mL of 50% glycerol and incubate 5 min at room temperature.

7.2 Salivary gland squash to spread polytene chromosomes

7.2.1 Use Kimwipes to clean siliconized coverslips and poly-L-lysine-coated microscope slides. Transfer the salivary gland into 15 μL of 50% glycerol on the coverslip. Use tweezers to fragment the salivary glands into 10–20 pieces. Carefully lower a slide onto the coverslip, and flip it after the coverslip attaches to the slide. Using the eraser of a pencil, gently move the coverslip back and forth about 2 mm three to five times to separate and lyse the cells and to spread the chromosomes. Apply a light pressure to the eraser that is just enough to move the coverslip. Clouding of the sample is an indication that the cells have separated and lysed. Tap the coverslip with the eraser of a pencil or the back of forceps to encourage chromosome spreading. One technique is to drop a pencil from a height of about 10 cm on the eraser end onto the coverslip. Repeat 10–20 times. Hold one corner of the coverslip to prevent large horizontal movements as this can damage the polytene chromosomes.

7.2.2 Examine the preparation immediately using a phase contrast microscope with a 20 × objective. If many of the chromosomes appear to be well-spread, proceed to the next step. If not, tap the coverslip again and continue tapping until an appropriate extent of spreading is obtained. Under the acid-free conditions, spreading of the polytene chromosomes is hindered by cross-linking, making the chromosomes more difficult to spread. Therefore, the acid-free protocol requires more and harder tapping compared with conventional squash protocols. This often results in broken chromosome arms. The broken chromosome arms can be used to map individual genomic loci as long as they have clear banding patterns (Fig. 3B). A larger number of squash attempts are needed to obtain intact polytene chromosome preparations using the acid-free squash protocol than when the conventional protocol is used.

7.2.3 Place the slide between two layers of tissue paper. Flatten the chromosomes by placing your thumb on the coverslip and press firmly. Avoid horizontal movement of the coverslip since this can shear the chromosomes. Fill a thermal flask with liquid nitrogen. Hold the slide with a long forceps, put the slide in the liquid nitrogen until the boiling stops. Remove the slide and immediately (before the disappearance of frost) use a new razor blade to flip off the coverslip from the slide. Place the slide in a Coplin jar with cold PBS. If more slides are prepared, store the slides in PBS at 4°C for up to 2 h.

Troubleshooting 1. Result: Inadequate spreading of polytene chromosomes. The polytene chromosomes can be hard to spread. Make sure that the salivary glands are well developed with lengths corresponding to 1/3–1/2 of the body length. Use freshly prepared fix

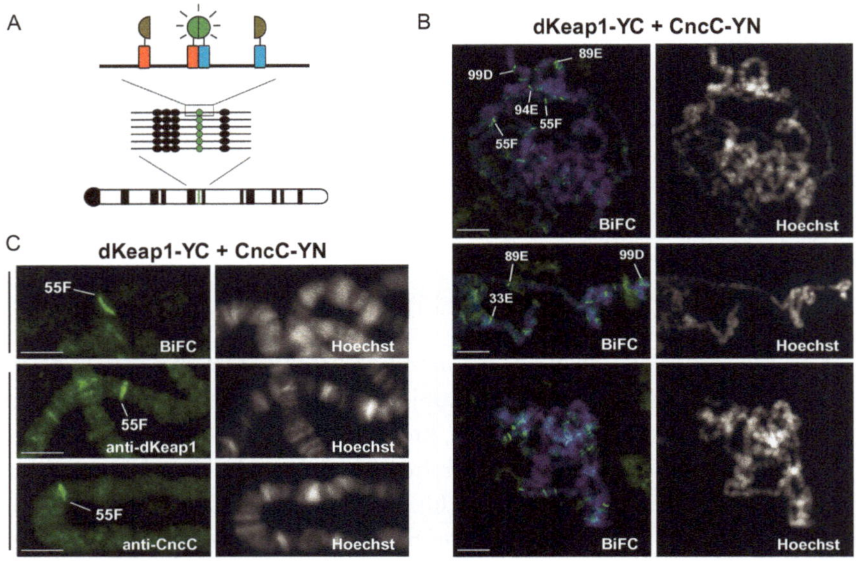

Fig. 3 Visualization of BiFC complexes on polytene chromosomes. (A) Diagram of BiFC complex formation on polytene chromosomes. The intensity of BiFC complex fluorescence at individual genomic loci is enhanced by the parallel alignment of hundreds of chromatids within the polytene chromosomes. (B) Examples of BiFC complex fluorescence (*green*) on polytene chromosomes stained using Hoechst (*blue*). The chromosomes were spread under acid-free conditions. The banding patterns of the polytene chromosomes stained using Hoechst are shown in the images to the *right*. Note the differences in chromosome spreading and extension that are necessary to map the localization of the BiFC complexes to individual genomic loci. *Upper panel*: an intact spread allowing mapping of a number of loci, including the two copies of the 55F locus, corresponding to the two homologs that were presumably separated during the squash procedure; *Middle panel*: a partial spread with broken arms, which are well extended and suitable for mapping a subset of the loci; *Lower panel*: a poorly extended spread on which the loci that are bound by BiFC complex cannot be easily identified. Scale bars: 10 μm. (C) Comparison of the localization of BiFC complexes and of the individual BiFC fusion proteins using acid-free (*upper images*) and conventional (*middle and bottom images*) squash protocols. Polytene chromosomes that prepared using the acid-free protocol have less sharp banding patterns than those that are prepared using conventional squash protocols. Both dKeap1–CncC BiFC complexes and each of the fusion proteins bound the 55F locus. Both dKeap1 and CncC fusions bound many other loci, whereas the dKeap1–CncC BiFC complex bound the 55F locus with higher specificity. Scale bars: 5 μm.

solution within 10 min after preparation. Optimize the fixation time: A shorter fixation time (2 min or less) can improve spreading as long as it is sufficient to cross-link the BiFC complexes to the chromatin.

Troubleshooting 2. Result: A small number of polytene chromosome spreads is found on the slides. Check the coverslip that was removed from the frozen slide. If polytene chromosomes are attached to the coverslip (indicated by a nonreflective surface), prepare new siliconlized coverslips and poly-L-lysine coated slides.

8. STAINING AND IMAGING OF POLYTENE CHROMOSOMES

8.1 Prior to staining the sample, briefly tap the edge of the slide to remove excess PBS. Place the slide tissue side up in a humidity chamber. Apply 100 μL of Hoechst solution on the tissue. Cover the tissue with a 20 × 35-mm piece of Parafilm to fully cover the sample. Cover the chamber and incubate 10 min in the dark at room temperature.

8.2 Remove the Parafilm and tap the slide to remove the Hoechst solution. Return the slide into the Coplin jar and incubate 2 × 5 min in PBS in the dark. Drain the slide by touching the edge to a tissue paper and dry the back. Remove as much liquid as possible without touching the tissue. Air dry the slide in the dark for 1–2 min.

8.3 Place 1–2 drops (10–20 μL) of mounting solution on the tissue. Touch one edge of a 22 × 40-mm coverslip to the solution and lay down slowly to avoid bubbles. Leave the slide in dark for 5 min to allow the mounting solution to permeate the sample. Wick off excess mounting solution using tissue paper. Seal the edges of the coverslip using nail polish.

8.4 The polytene chromosomes with BiFC complexes can be costained using antibodies directed against endogenous or ectopic proteins (Silver & Elgin, 1976). Secondary antibodies conjugated to fluorophores whose spectra do not overlap with that of the BiFC complex should be used.

8.5 Images of the polytene chromosomes are acquired using a fluorescence microscope. The locations of BiFC complexes and the banding pattern of Hoechst staining are imaged using different excitation and emission wavelengths. BiFC complexes formed by YFP fragments can be imaged using 504 nm excitation and 542 nm emission wavelengths. Hoechst signal can be imaged using 358 nm excitation and 461 nm emission wavelengths. The best results are obtained by imaging the samples immediately after preparation. If the samples must be stored, store them at −20°C in the dark. To distinguish BiFC complex fluorescence from cross talk with other fluorophores and other potential sources of fluorescence, samples that do not contain both fusion proteins should be processed and analyzed in parallel.

8.6 Polytene chromosomes that are visualized using the acid-free protocol generally have less sharp banding patterns than those that are produced by conventional squash protocols (Fig. 3C). It is often necessary to perform a larger number of squashes to produce satisfactory data. Persistent effort to optimize the squash procedure is generally rewarded with spreads that enable mapping of BiFC complex binding throughout the genome.

Troubleshooting. Result: Weak or no BiFC fluorescence is detected. In general, the fluorescence intensity of BiFC complexes is much weaker than those of intact fluorescent proteins that are expressed at similar levels. This is likely to be due to the fact that only a small portion of fusion proteins form BiFC complexes (Hu et al., 2002). To avoid bleaching weak BiFC complex fluorescence, samples should be initially viewed by excitation of Hoechst fluorescence to locate chromosome spreads and to adjust focus. A sensitive CCD camera and long-exposure times are often necessary. The use of fluorescent protein fragments with brighter BiFC fluorescence can help. However, these can result in complex formation by proteins that do not normally interact with each other (Fig. 1C) (Robida & Kerppola, 2009). Likewise, higher levels of protein expression can increase the fluorescence signal. However, higher levels of fusion protein expression can result in complex formation by proteins that do not normally interact with each other as well as potential deleterious effects.

9. INTERPRETATION OF BiFC COMPLEX BINDING ON POLYTENE CHROMOSOMES

9.1 The genomic loci that are bound by BiFC complexes can be determined by comparing the locations of BiFC complex fluorescence with the banding pattern that is produced by Hoechst staining (Fig. 3B and C) (Deng & Kerppola, 2014). The reproducible pattern of Hoechst staining allows determination of the chromatin bands, and interbands that are occupied by BiFC complexes in individual cells. This approach enables mapping of the genomic locations that are bound by BiFC complexes at a resolution of 100–700 kb, which typically encompasses 10–40 genes depending on the genomic region.

9.2 To identify the specific genes that are bound and potentially regulated by the protein of interest, additional approaches can be used. Chromatin binding can be analyzed by chromatin immunoprecipitation if the complex binds the same region in a large proportion of the cells and if the location of the region is known or can be predicted with a resolution of 1 kb. In addition, the effects of depletion or overexpression of the proteins of interest on transcription of the genes within the chromatin region implicated by mapping of the location of BiFC complex binding can identify genes whose transcription is affected by the proteins of interest.

Troubleshooting. Result: Poorly resolved bands prevent mapping of loci bound by BiFC complexes. The polytene chromosome arms obtained by the acid-free squash protocol are generally more diffuse and less extended compared with the sharp and well-separated bands produced by conventional squash protocols (Fig. 3C) (Johansen et al., 2009). Optimization of the conditions of larval culture to produce maximally amplified polytene chromosomes and of the squash procedure will improve the quality of data that can be obtained. Larvae that are raised at lower temperatures (16–18°C) can produce more highly amplified polytene chromosomes. A shorter fixation time can improve polytene chromosome spreading and thus increase the quality of bands. The reproducibility of the squash procedure can be improved by the use of mechanical aids (Johansen et al., 2009; Novikov, Kireev, & Belmont, 2007).

10. STRATEGIES TO LIMIT THE ARTIFACTS CAUSED BY THE DIFFERENCES BETWEEN FUSION PROTEINS AND ENDOGENOUS COUNTERPARTS

10.1 Ectopically expressed fusion proteins can have subcellular distributions, chromatin-binding specificities, or protein-interaction specificities that differ from those of their endogenous counterparts. These differences complicate the interpretation of potential effects of their interactions. There are several potential causes for differences between the characteristics of fusion proteins and their endogenous counterparts.

10.2 Expression of fusion proteins at levels that are much higher than the levels of their endogenous counterparts can alter their functions. The

GAL4-UAS system provides little control over the level of fusion protein expression. This strategy often results in much higher levels of transgene expression than the levels of endogenous gene expression (Fig. 2A). The levels of transgene expression can be reduced by: (1) Screening for lines with a lower level of transgene expression. The eye color that is produced by the mini-white reporter that is integrated with the transgene can be used to prescreen the lines, since it reflects the chromatin environment of the integration site; (2) Raising the larvae at a lower temperature (16°C) to reduce the activity of GAL4 (Duffy, 2002); (3) The use of GAL4 drivers that are known to promote lower levels of transgene expression; (4) The use of inducible expression strategies, such as GAL80ts, GeneSwitchGAL4, or tet-on/off systems; (5) The use of native promoters to express the fusion proteins (Ding et al., 2009; Rugjee et al., 2013).

10.3 The fusion of fluorescent protein fragments to the proteins of interest can alter their characteristics. To reduce the risk that the fluorescent protein fragments affect the functions of the complex, the fragments can be fused to locations that are far from known functional domains. The use of longer or different linker sequences can also reduce the effects of the fluorescent protein fragments on complex functions.

11. EFFECTS OF BiFC COMPLEX FORMATION ON THE SPECIFICITY OF CHROMATIN BINDING

11.1 The association of the fluorescent protein fragments generally stabilizes the interaction of the fusion proteins to produce a complex that is more stable than complexes formed by the endogenous proteins. This stabilization could affect the specificity or the avidity of protein complex binding to chromatin.

11.2 To distinguish between the potential effects of BiFC complex formation and of fusion protein coexpression on the specificity of chromatin binding, the loci that are bound by BiFC complexes can be compared with those that are bound by the proteins of interest fused to intact fluorescent proteins when they are coexpressed (Deng & Kerppola, 2014). If the proteins that are fused to intact fluorescent proteins bind together at the same loci that are bound by the BiFC complexes, it is likely that BiFC complex binding reflects that binding specificity of the protein complex and is not affected by the association of the fluorescent protein fragments (Deng & Kerppola, 2014).

Troubleshooting. Result: BiFC complexes bind to loci that are not bound by endogenous proteins. Many factors can alter the specificity of chromatin binding by BiFC complexes. Artifactual effects include potential effects of the protein fusions on chromatin binding by the proteins of interest, and potential effects of ectopic overexpression of the fusion proteins on their expression levels. These can be evaluated as described above. The possibility that fluorescent protein fragment association affects chromatin binding directly can be evaluated by parallel analysis of chromatin-binding specificity using other experimental approaches (Cai et al., 2014; Deng & Kerppola, 2013, 2014; Rugjee et al., 2013; Wang et al., 2001).

12. MATERIALS AND SOLUTIONS

12.1 Equipment and materials

- Dissecting microscope with a 5–10 × objective
- Illuminator
- 35 × 10-mm Petri dish
- Drummond dissection tweezers (2) (72695-D)
- 0.2-mL Glass tissue grinder (Kimble 885470-0000)
- Frosted microscope slides (Fisher 12-544-7) coated with poly-L-lysine (Sigma P8920). Incubate slides in the poly-lysine solution for 5 min and dry the slides at room temperature. Coated slides can be stored at room temperature for one year. Poly-lysine coating increases the attachment of chromosome preparations to the glass surface.
- Siliconized 22 × 22 mm 1.5 cover slips (Fisher No. 12–520B) coated with Sigmacote (Sigma SL2). To coat, immerse clean cover slips in Sigmacote in a Petri dish, then air dry in a fume hood. The siliconized cover slips can be stored at room temperature for at least 1 year. Clean with water and dry with Kimwipe before use. Siliconization prevents the attachment of chromosome preparations to the cover slip.
- 22 × 40 mm 1.5 cover slips (Fisher 12-544-B)
- Tissue paper and Kimwipes
- Pencil with eraser (or similar tools that can be used to tap the slides)
- Phase contrast microscope with ~20 × objective
- Thermoflask with 300–500 mL of liquid nitrogen
- Long forceps
- Razor blades

- Coplin Jar
- Humidity chamber (Rubber-Maid or similar sealable tray. Cover bottom with moistened tissue papers. Use Pasteur pipettes to support slides.)
- Parafilm (cut into 22 × 35 mm pieces)
- Nail polish (to seal cover slips)

12.2 Solutions

- IP buffer (20 m*M* Tris–HCl (pH 8.0), 150 m*M* NaCl, 10 m*M* EDTA, 1 m*M* EGTA, 0.2% Triton X-100, 0.2% NP-40, 2 m*M* Na_3VO_4, 1 m*M* PMSF, 1.5 μg/mL Aprotinin)
- 5 × SDS-loading buffer (0.25 *M* Tris–HCl (pH 6.8), 0.5 *M* DTT, 50% Glycerol, 10% SDS)
- 37% formaldehyde (Fisher BP531)
- Brower's fixation buffer (0.15 *M* PIPES, 3 m*M* $MgSO_4$, 1.5 m*M* EGTA, 1.5% NP40, adjust pH to 6.9, add formaldehyde to a final concentration of 2% just before use)
- PBS (137 m*M* NaCl, 2.7 m*M* KCl, 10 m*M* Na_2HPO_4, 1.8 m*M* KH_2PO_4)
- PBST (PBS with 0.1% Triton X-100)
- Hoechst 33258 (Molecular Probes) solution (0.2 μg/mL in PBS, store in the dark at 4°C)
- Mounting solution (80% glycerol, 10 m*M* Tris pH 9.0, 5% *n*-propyl gallate) (avoid using antifade mounting solutions as they can quench fluorescent proteins).

REFERENCES

Bachmann, A., & Knust, E. (2008). The use of P-element transposons to generate transgenic flies. *Methods in Molecular Biology, 420*, 61–77.

Bai, S., & Kerppola, T. K. (2011). Opposing roles of FoxP1 and Nfat3 in transcriptional control of cardiomyocyte hypertrophy. *Molecular and Cellular Biology, 31*, 3068–3080.

Boyle, A. P., Davis, S., Shulha, H. P., Meltzer, P., Margulies, E. H., Weng, Z., et al. (2008). High-resolution mapping and characterization of open chromatin across the genome. *Cell, 132*, 311–322.

Brand, A. H., & Perrimon, N. (1993). Targeted gene expression as a means of altering cell fates and generating dominant phenotypes. *Development, 118*, 401–415.

Cai, W., Wang, C., Li, Y., Yao, C., Shen, L., Liu, S., et al. (2014). Genome-wide analysis of regulation of gene expression and H3K9me2 distribution by JIL-1 kinase mediated histone H3S10 phosphorylation in Drosophila. *Nucleic Acids Research, 42*, 5456–5467.

Cherbas, L., Hu, X., Zhimulev, I., Belyaeva, E., & Cherbas, P. (2003). EcR isoforms in Drosophila: Testing tissue-specific requirements by targeted blockade and rescue. *Development, 130*, 271–284.

Chu, J., Zhang, Z., Zheng, Y., Yang, J., Qin, L., Lu, J., et al. (2009). A novel far-red bimolecular fluorescence complementation system that allows for efficient visualization of

protein interactions under physiological conditions. *Biosensors & Bioelectronics*, *25*, 234–239.
Deng, H., & Kerppola, T. K. (2013). Regulation of Drosophila metamorphosis by xenobiotic response regulators. *PLoS Genetics*, *9*, e1003263.
Deng, H., & Kerppola, T. K. (2014). Visualization of the Drosophila dKeap1-CncC interaction on chromatin illumines cooperative, xenobiotic-specific gene activation. *Development*, *141*, 3277–3288.
Diaz, I., Martinez, M., Isabel-LaMoneda, I., Rubio-Somoza, I., & Carbonero, P. (2005). The DOF protein, SAD, interacts with GAMYB in plant nuclei and activates transcription of endosperm-specific genes during barley seed development. *The Plant Journal*, *42*, 652–662.
DiMario, P., Rosby, R., & Cui, Z. (2006). Direct visualization of GFP-fusion proteins on polytene chromosomes. *Drosophila Information Service*, *89*, 115–118.
Ding, Y., Yao, C., Lince-Faria, M., Rath, U., Cai, W., Maiato, H., et al. (2009). Chromator is required for proper microtubule spindle formation and mitosis in Drosophila. *Developmental Biology*, *334*, 253–263.
Duffy, J. B. (2002). GAL4 system in drosophila: A fly geneticist's Swiss army knife. *Genesis*, *34*, 1–15.
ENCODE Consortium. (2012). An integrated encyclopedia of DNA elements in the human genome. *Nature*, *489*, 57–74.
Galas, D. J., & Schmitz, A. (1978). DNAse footprinting: A simple method for the detection of protein-DNA binding specificity. *Nucleic Acids Research*, *5*, 3157–3170.
Gerlitz, O., Nellen, D., Ottiger, M., & Basler, K. (2002). A screen for genes expressed in Drosophila imaginal discs. *The International Journal of Developmental Biology*, *46*, 173–176.
Gilmour, D. S., & Lis, J. T. (1984). Detecting protein-DNA interactions in vivo: Distribution of RNA polymerase on specific bacterial genes. *Proceedings of the National Academy of Sciences of the United States of America*, *81*, 4275–4279.
Gohl, C., Banovic, D., Grevelhorster, A., & Bogdan, S. (2010). WAVE forms hetero- and homo-oligomeric complexes at integrin junctions in Drosophila visualized by bimolecular fluorescence complementation. *The Journal of Biological Chemistry*, *285*, 40171–40179.
Han, D. D., Stein, D., & Stevens, L. M. (2000). Investigating the function of follicular subpopulations during Drosophila oogenesis through hormone-dependent enhancer-targeted cell ablation. *Development*, *127*, 573–583.
Hatzis, P., & Talianidis, I. (2002). Dynamics of enhancer-promoter communication during differentiation-induced gene activation. *Molecular Cell*, *10*, 1467–1477.
Hu, C. D., Chinenov, Y., & Kerppola, T. K. (2002). Visualization of interactions among bZIP and Rel family proteins in living cells using bimolecular fluorescence complementation. *Molecular Cell*, *9*, 789–798.
Hu, C. D., & Kerppola, T. K. (2003). Simultaneous visualization of multiple protein interactions in living cells using multicolor fluorescence complementation analysis. *Nature Biotechnology*, *21*, 539–545.
Hudry, B., Viala, S., Graba, Y., & Merabet, S. (2011). Visualization of protein interactions in living Drosophila embryos by the bimolecular fluorescence complementation assay. *BMC Biology*, *9*, 5.
Johansen, K. M., Cai, W., Deng, H., Bao, X., Zhang, W., Girton, J., et al. (2009). Polytene chromosome squash methods for studying transcription and epigenetic chromatin modification in Drosophila using antibodies. *Methods*, *48*, 387–397.
Johnson, D. S., Mortazavi, A., Myers, R. M., & Wold, B. (2007). Genome-wide mapping of in vivo protein-DNA interactions. *Science*, *316*, 1497–1502.
Kerppola, T. K. (2006). Visualization of molecular interactions by fluorescence complementation. *Nature Reviews. Molecular Cell Biology*, 7, 449–456.

Lis, J. T. (2007). Imaging Drosophila gene activation and polymerase pausing in vivo. *Nature, 450*, 198–202.

Masumoto, H., Sugimoto, K., & Okazaki, T. (1989). Alphoid satellite DNA is tightly associated with centromere antigens in human chromosomes throughout the cell cycle. *Experimental Cell Research, 181*, 181–196.

Maybeck, V., & Roper, K. (2009). A targeted gain-of-function screen identifies genes affecting salivary gland morphogenesis/tubulogenesis in Drosophila. *Genetics, 181*, 543–565.

Novikov, D. V., Kireev, I., & Belmont, A. S. (2007). High-pressure treatment of polytene chromosomes improves structural resolution. *Nature Methods, 4*, 483–485.

Osterwalder, T., Yoon, K. S., White, B. H., & Keshishian, H. (2001). A conditional tissue-specific transgene expression system using inducible GAL4. *Proceedings of the National Academy of Sciences of the United States of America, 98*, 12596–12601.

Papadopoulos, D. K., Skouloudaki, K., Adachi, Y., Samakovlis, C., & Gehring, W. J. (2012). Dimer formation via the homeodomain is required for function and specificity of sex combs reduced in Drosophila. *Developmental Biology, 367*, 78–89.

Phalle Bde, S. (2004). Immunostaining of whole-mount imaginal discs. *Methods in Molecular Biology, 247*, 373–387.

Plaza, S., Prince, F., Adachi, Y., Punzo, C., Cribbs, D. L., & Gehring, W. J. (2008). Cross-regulatory protein-protein interactions between Hox and Pax transcription factors. *Proceedings of the National Academy of Sciences of the United States of America, 105*, 13439–13444.

Rivera, V. M., Clackson, T., Natesan, S., Pollock, R., Amara, J. F., Keenan, T., et al. (1996). A humanized system for pharmacologic control of gene expression. *Nature Medicine, 2*, 1028–1032.

Robert, D. G. (1998). *Drosophila: A practical approach*. Oxford, UK: IRL Press.

Robida, A. M., & Kerppola, T. K. (2009). Bimolecular fluorescence complementation analysis of inducible protein interactions: Effects of factors affecting protein folding on fluorescent protein fragment association. *Journal of Molecular Biology, 394*, 391–409.

Rubin, G. M., & Spradling, A. C. (1982). Genetic transformation of Drosophila with transposable element vectors. *Science, 218*, 348–353.

Rugjee, K. N., Roy Chaudhury, S., Al-Jubran, K., Ramanathan, P., Matina, T., Wen, J., et al. (2013). Fluorescent protein tagging confirms the presence of ribosomal proteins at Drosophila polytene chromosomes. *PeerJ, 1*, e15.

Saka, Y., Hagemann, A. I., Piepenburg, O., & Smith, J. C. (2007). Nuclear accumulation of Smad complexes occurs only after the midblastula transition in Xenopus. *Development, 134*, 4209–4218.

Schwartz, B. E., Werner, J. K., & Lis, J. T. (2004). Indirect immunofluorescent labeling of Drosophila polytene chromosomes: Visualizing protein interactions with chromatin in vivo. *Methods in Enzymology, 376*, 393–404.

Shaffer, C. D., Stephens, G. E., Thompson, B. A., Funches, L., Bernat, J. A., Craig, C. A., et al. (2002). Heterochromatin protein 2 (HP2), a partner of HP1 in Drosophila heterochromatin. *Proceedings of the National Academy of Sciences of the United States of America, 99*, 14332–14337.

Shyu, Y. J., Liu, H., Deng, X., & Hu, C. D. (2006). Identification of new fluorescent protein fragments for bimolecular fluorescence complementation analysis under physiological conditions. *Biotechniques, 40*, 61–66.

Silver, L. M., & Elgin, S. C. (1976). A method for determination of the in situ distribution of chromosomal proteins. *Proceedings of the National Academy of Sciences of the United States of America, 73*, 423–427.

Stebbins, M. J., Urlinger, S., Byrne, G., Bello, B., Hillen, W., & Yin, J. C. (2001). Tetracycline-inducible systems for Drosophila. *Proceedings of the National Academy of Sciences of the United States of America, 98*, 10775–10780.

van Steensel, B., Delrow, J., & Henikoff, S. (2001). Chromatin profiling using targeted DNA adenine methyltransferase. *Nature Genetics*, *27*, 304–308.
van Steensel, B., & Henikoff, S. (2000). Identification of in vivo DNA targets of chromatin proteins using tethered dam methyltransferase. *Nature Biotechnology*, *18*, 424–428.
Vincenz, C., & Kerppola, T. K. (2008). Different polycomb group CBX family proteins associate with distinct regions of chromatin using nonhomologous protein sequences. *Proceedings of the National Academy of Sciences of the United States of America*, *105*, 16572–16577.
Wang, Y., Zhang, W., Jin, Y., Johansen, J., & Johansen, K. M. (2001). The JIL-1 tandem kinase mediates histone H3 phosphorylation and is required for maintenance of chromatin structure in Drosophila. *Cell*, *105*, 433–443.
Zeidler, M. P., Tan, C., Bellaiche, Y., Cherry, S., Hader, S., Gayko, U., et al. (2004). Temperature-sensitive control of protein activity by conditionally splicing inteins. *Nature Biotechnology*, *22*, 871–876.
Zhimulev, I. F., Belyaeva, E. S., Semeshin, V. F., Koryakov, D. E., Demakov, S. A., Demakova, O. V., et al. (2004). Polytene chromosomes: 70 years of genetic research. *International Review of Cytology*, *241*, 203–275.

AUTHOR INDEX

Note: Page numbers followed by "*f*" indicate figures, and "*t*" indicate tables.

B

C

D

G

J

K

L

O

P

T

W

X

Y

Z

SUBJECT INDEX

Note: Page numbers followed by "*f*" indicate figures, and "*t*" indicate tables.

D

E

G

M

N

Q

R

S

T

U

V

W

Y

Z

FSC
www.fsc.org
MIX
Papier | Fördert gute Waldnutzung
FSC® C083411

Kolibri 360 GmbH
Ferdinand-Jühlke-Straße 7
99095 Erfurt, Deutschland
produktsicherheit@kolibri360.de